INSECT SOUNDS AND COMMUNICATION

Physiology, Behaviour, Ecology and Evolution

CONTEMPORARY TOPICS in ENTOMOLOGY Series
THOMAS A. MILLER EDITOR

INSECT SOUNDS AND COMMUNICATION

Physiology, Behaviour, Ecology and Evolution

Edited by
Sakis Drosopoulos
Michael F. Claridge

CRC Taylor & Francis
Taylor & Francis Group
Boca Raton London New York

A CRC title, part of the Taylor & Francis imprint, a member of the
Taylor & Francis Group, the academic division of T&F Informa plc.

Published in 2006 by
CRC Press
Taylor & Francis Group
6000 Broken Sound Parkway NW, Suite 300
Boca Raton, FL 33487-2742

© 2006 by Taylor & Francis Group, LLC
CRC Press is an imprint of Taylor & Francis Group

No claim to original U.S. Government works
Printed in the United States of America on acid-free paper
10 9 8 7 6 5 4 3 2 1

International Standard Book Number-10: 0-8493-2060-7 (Hardcover)
International Standard Book Number-13: 978-0-8493-2060-6 (Hardcover)
Library of Congress Card Number 2005048600

Library of Congress Cataloging-in-Publication Data

Insect sounds and communication : physiology, behaviour, ecology and evolution / edited by Sakis Drosopoulos and Michael F. Claridge.
    p. cm. -- (Contemporary topics in entomology series)
    ISBN 0-8493-2060-7 (alk. paper)
    1. Insect sounds. 2. Sound production by insects. 3. Animal communication. I. Drosopoulos, Sakis. II. Claridge, Michael F. III. Series.

QL496.5.I57 2005
595.7159--dc22
                                                                                    2005048600

Taylor & Francis Group
is the Academic Division of Informa plc.

Visit the Taylor & Francis Web site at
http://www.taylorandfrancis.com

and the CRC Press Web site at
http://www.crcpress.com

Dedication

Frej Ossiannilsson

René Cobben

Tom Wood

We dedicate this volume, without permission, to the memory of three outstanding, but sadly now deceased, entomologists who have influenced us both during our careers — Frej Ossiannilsson (Uppsala, Sweden), René Cobben (Wageningen, The Netherlands) and Tom Wood (Delaware, USA). They were all three, in their own way, open minded and truth seeking scientists, uncluttered by preconceived ideas.

Sakis Drosopolous
Mike Claridge

Preface

There is no life without various sorts of sounds. Astronauts must be the only persons who can have such an experience, being in space where silence is absolute. Therefore a special sense has been developed for receiving sounds in all animal species, including insects. In human beings sounds expressed by languages are specific, but differentiated in time and space. As a European visiting Southeast Asia for the first time I felt really isolated not being able to communicate with the local people who spoke only their native language. I cannot forget how nice I felt when a friendly cat approached me obviously calling in a way familiar to me. At that moment I asked myself why humans developed so many languages that can be used only between them, except when some of them have learned a common spoken language, as for example in our days, English. Philosophising this experience without having thought or been informed previously, I came to the question: could this development be a spontaneous linguistic differentiation that for the human species could be analogous to what we call for other animal species an "isolating mechanism?"

Such questions I had raised during my childhood living in my village where nature was, and still is, wonderful for naturalists. Several questions occurred to me there with no reasonable answers available from local people. I was therefore lucky as a postgraduate student in Holland to meet Prof. René Cobben, who visited with me this area of Greece, which he appreciated very much. In later visits he could not stay in Athens more than one night. Of course in that area the whole day was spent with me observing and collecting insects, especially those that he could find only in strictly protected areas at home in Holland. It is not surprising that in this area we found for the first time the planthopper, *Muellerianella fairmairei*, consisting only of diploid individuals in contrast to what we had found previously in Holland where a mixture of both diploid and triploid specimens occurred. With this scientist I felt that I could learn, since learning was always and still is my main principle. Therefore at this point and in respect for his personality I must refer to him briefly. He was a great teacher who introduced his students to the field of biology, which is in danger of extinction in our days, classic systematics. He taught us the basic morphology of insects in the laboratory of his own collection, and he explored with us in the field, where we were introduced to collecting species both qualitatively and quantitatively. In this way we were stimulated to make studies on the ecology of the species under investigation, especially food plants, biotopes, *etc.* However, as an expert in such studies he could see that the systematics and ecology of species were not enough to solve taxonomic problems of closely related species that every good naturalist could detect, but that are impossible to separate only by such studies. Having excellent facilities in his laboratory for rearing insects under controlled temperature, humidity and photoperiod, he suggested that we start biosystematic studies (genetics, ethology, physiology, reproduction and other aspects of the biology of each species). This way of studying species stimulated his first students to be involved in fields with which previously nobody was familiar, but the need to solve questions sooner or later made them experts in particular fields, such as, for example, cytology and acoustics.

During the last 10 years I developed, together with Portuguese colleagues and Mike Claridge, a common project on investigating for the first time in Greece the acoustic signals of two genera of Cicadas, since previous faunistic investigations I had made on some Greek islands appeared to show some acoustic differentiation. Data presented at the Thessaloniki European Entomological Congress in October 2002, at a Symposium on Insect Acoustics organised by the two current

Editors of this book, impressed Prof. Tom Miller so much that he invited us to write a book on the subject. The answer to Tom was that we will try, but we will tell you after 2002. Mike and I were aware that such a book could not be written by one person, although I myself prefer books written by only one or two persons. Four months later and after having contacted most of the eventual contributors, whom we already knew, we answered to the insistence of our friend Tom that we were willing to take on this ambitious, but difficult job, as it has now proved two years later. At this point I must mention our sincere thanks to four particular colleagues and friends who really encouraged us as editors — Prof. Hannelore Hoch (Berlin), Prof. Matija Gogala and Dr. Meta Virant-Doberlet (Slovenia) and Dr. Jérôme Sueur (France). These colleagues, world experts in this subject, not only accepted invitations to contribute as authors to the book, but they suggested to us some other contributors not previously known to us. In addition I invited some contributors well known to us, but whose work was not well known in international journals. To all contributors we express our sincere thanks, not only for their chapters finally all presented in this book, but also for accepting the "hard" revising and commenting, especially by Mike, on their manuscripts. Worth particular mention here is the contribution of Prof. Winston Bailey (Australia), who accepted to contribute to this book despite the fact that he had already published an excellent book of his own on the same subject in 1991.

The idea of supplementing this book with a DVD was made by Prof. H. Hoch and later implemented by my dearest colleague, Elias Eliopoulos, also an expert in electronics, like my friend Matija Gogala, who also revised several chapters. On the accompanying DVD the reader will find color photographs of many of the insects, acoustic analyses, and also the sound emitted by them. In addition is the full text in French of the contribution of the famous cicadologist, Prof. Michel Boulard, Paris. To all of these colleagues and the other contributors we express our real and honest thanks. However, there are no words to thank a very special virologist, my wife, Hanneke Drosopoulou-van Albada, who spent many months of hard work in the background bringing this book to its published form.

Finally, and following the subtitle of the book, there are not only data on acoustics presented here, but these data are related also to many other topics, such as morphology, systematics, ethology, ecology, physiology, genetics, cytogenetics, polymorphism and of course bisexual and unisexual reproduction. All these provide serious knowledge for new discussions on evolution.

Sakis Drosopoulos

The Editors

Sakis Drosopoulos Ph.D. is Professor of Systematics and Biosystematics in the Department of Agricultural Biotechnology of the Agricultural University of Athens. Since 1972 his research has focused on two topics: first faunistic and floristic investigations on phytophageous Hemiptera and second biosystematics (ecology, ethology, genetics) of species complexes, which are difficult to separate only by pure morphological characters. All these still ongoing studies led him to develop his own ideas on the evolution of bisexual, pseudogamous and parthenogenetic organisms world wide. Some of these ideas are presented for the first time in this book. He enriched his knowledge as an active member of the Hellenic Zoological Society, having contacts with various zoologists from all over the world, organising or participating in several international congresses and collaborating with many distinguished scientists, amongst whom is Professor Michael Claridge, with whom he has shared a warm and firm friendship since 1975.

Sakis Drosopoulos (left) and Michael Claridge (right). Photo by Hanneke Drosopoulou

Mike Claridge is Emeritus Professor of Entomology at the University of Wales, Cardiff, U.K., where he served as Head of the School of Pure and Applied Biology from 1989 to 1994. He graduated in zoology at the University of Oxford where he continued for a D. Phil. in the Hope Department of Entomology. His career subsequently was based continuously in Cardiff, from 1959 to 1999, though many field projects on a wide variety of insects have taken him throughout Europe, Australia and tropical South and Southeast Asia. His research interests have always centred on species problems and speciation in insects. In 1997 he coedited, with Hassan Dawah and Mike Wilson, and authored two chapters in a multi-author volume *Species: The Units of Biodiversity* (Chapman & Hall). He came relatively late to studies on acoustic behaviour in order to identify specific mate recognition systems and species isolating barriers in leafhoppers and planthoppers. He has had the privilege to serve as President of the Linnean Society of London (1988–1991), the Systematics Association (1991–1994) and the Royal Entomological Society of London (2000–2002). He was awarded the Linnean Medal of the Linnean Society in 2000 for services to zoology. His long friendship with Sakis Drosopoulos and love affair with Greece have resulted in the present book.

Contributors

Winston J. Bailey
School of Animal Biology
University of Western Australia
Nedlands, Western Australia
wbailey@cyllene.uwa.edu.au

Friedrich G. Barth
Department of Neurobiology
  and Behavioral Sciences
Faculty of Life Sciences
University of Vienna
Vienna, Austria
Friedrich.G.Barth@univie.ac.at

Michel Boulard
EPHE et MNHN
Paris, France
mbkcicada01@yahoo.fr

Maria Bukhvalova
Department of Entomology
Faculty of Biology
M. V. Lomonosov Moscow State University
Moscow, Russia
dt@3.entomol.bio.msu.ru

Jérôme Casas
Institut de Recherches en Biologie
  de l'Insecte
Université de Tours
Tours, France
casas@univ-tours.fr

Michael Claridge
Cardiff University
Wales, United Kingdom
claridge@cardiff.ac.uk

Reginald B. Cocroft
Division of Biological Sciences
University of Missouri
Columbia, Missouri
cocroftr@missouri.edu

Andrej Čokl
Department of Entomology
National Institute of Biology
Ljubljana, Slovenia
andrej.cokl@nib.si

Paul De Luca
Division of Biological Sciences
University of Missouri
Columbia, Missouri
pad6b7@mizzou.edu

Sakis Drosopoulos
Department of Agricultural Biotechnology
Agricultural University of Athens
Athens, Greece
drosop@aua.gr

Elias Eliopoulos
Department of Agricultural Biotechnology
Agricultural University of Athens
Athens, Greece
eliop@aua.gr

Matija Gogala
Slovenian Academy of Sciences and Arts
Ljubljana, Slovenia
matija.gogala@guest.ames.si

K.-G. Heller
Grillenstieg
Magdeburg, Germany
Heller.Volleth@t-online.de

Charles S. Henry
Department of Ecology and
  Evolutionary Biology
University of Connecticut
Storrs, Connecticut
chenry@uconnvm.uconn.edu

Petra Hirschberger
Museum für Naturkunde
Humboldt Universität zu Berlin
Berlin, Germany
petra.hirschberger@siemens.com

Hannelore Hoch
Museum für Naturkunde
Humboldt Universität zu Berlin
Berlin, Germany
hannelore.hoch@museum.hu-berlin.de

Anneli Hoikkala
Department of Biological and
  Environmental Science
University of Jyväskylä
Jyväskylä, Finland
anhoikka@bytl.jyu.fi

Michael Hrncir
Department of Neurobiology and
  Behavioral Sciences
Faculty of Life Sciences
University of Vienna
Vienna, Austria
Michael.hrncir@gmx.at

Martin Jatho
AG Neurobiologie, Biologie-Zoologie
Philipps Universität Marburg
Marburg, Germany
mail@mjatho.de

Klaus Kalmring
AG Neurobiologie, Biologie-Zoologie
Philipps Universität Marburg
Marburg, Germany

K. Kanmiya
School of Medicine
Kurume University
Fukuoka, Japan
kanmiya@med.kurume-u.ac.jp

Julia Kasper
Museum für Naturkunde,
Humboldt Universität zu Berlin
Berlin, Germany
julia.kasper@museum.hu-berlin.de

Claudio R. Lazzari
Institut de Recherche sur la Biologie de
  l'Insecte
Université François Rabelais
Tours, France
claudio.lazzari@univ-tours.fr

Christelle Magal
Institut de Recherches en Biologie
  de l'Insecte
Université de Tours
Tours, France
magal@univ-tours.fr

Gabriel Manrique
Laboratorio de Fisiología de lnsectos DBBE
Fac. Cs. Exactas y Naturales
Universidad de Buenos Aires
Buenos Aires, Argentina
gabo@bg.fcen.uba.ar

Gabriel D. McNett
Division of Biological Sciences
University of Missouri-Columbia
Columbia, Missouri

José A. Quartau
Centro de Biologia Ambiental
Departamento de Biologia Animal
Faculdade de Ciências
Campo Grande Lisboa, Portugal
jaquartau@fc.ul.pt

Wolfgang Rössler
Biozentrum, Zoologie II
Universität Würzburg, Am Hubland
Würzburg, Germany
roessler@biozentrum.uni-wuerzburg.de

Allen F. Sanborn
School of Natural & Health Sciences
Barry University
Miami Shores, Florida
asanborn@mail.barry.edu

John B. Sandberg
Department of Biological Sciences
University of North Texas
Denton, Texas
jbs001@unt.edu

Pablo E. Schilman
Ecology, Behavior, and Evolution Section
Division of Biological Sciences
University of California at San Diego
La Jolla, California
pschilma@biomail.acsd.adu

Paula C. Simões
Centro de Biologia Ambiental
Departamento de Biologia Animal
Faculdade de Ciências
Campo Grande Lisboa, Portugal
pcsimoes@fc.ul.pt

Kenneth W. Stewart
Department of Biological Sciences
University of North Texas
Denton, Texas
stewart@unt.edu

Hildegard Strübing
Kruseweg
Berlin, Germany

Jérôme Sueur
NAMC-CNRS Université Paris
Orsay, France
Jerome.Sueur@ibiac.u-psud.fr

Jürgen Tautz
Beegroup Würzburg, Biozentrum
University of Würzburg Am Hubland
Würzburg, Germany
tautz@biozentrum.uni-wuerzburg.de

D. Yu. Tishechkin
Department of Entomology
Faculty of Biology
M. V. Lomonosov Moscow State University
Moscow, Russia
dt@3.entomol.bio.msu.ru

Penelope Tsakalou
Department of Agricultural Biotechnology
Agricultural University of Athens
Athens, Greece

Meta Virant-Doberlet
Department of Entomology
National Institute of Biology
Ljubljana, Slovenia
meta.virant@nib.si

Andreas Wessel
Museum für Naturkunde
Humboldt Universität zu Berlin
Berlin, Germany
andreas.wessel@museum.hu-berlin.de

Maja Zorović
Department of Entomology
National Institute of Biology
Ljubljana, Slovenia
maja.zorovic@nib.si

Table of Contents

Part I

General Aspects of Insect Sounds

1 Insect Sounds and Communication — An Introduction

Michael Claridge

CONTENTS

SOUND AND VIBRATION

Any elementary textbook of physics will give the basic facts of sound as a longitudinal wave motion transmitted through a medium and emanating from a source of mechanical vibration. The problem is that, when it comes to describing and analysing insect sounds and their use in communication, it is almost impossible to avoid problems of anthropomorphism. This is, of course, a difficulty in all studies on animal behaviour. Sound is defined in dictionaries as a "sensation caused in the ear by vibrations of the surrounding air" (The Concise Oxford Dictionary of Current English, 6th edition, 1976). Following this definition, the term sound can only be used of vibrations which are heard by the human ear and transmitted through the surrounding air. If this were followed, sound would be a useless concept for describing animal behaviour. The human ear is sensitive to a range of vibrations with frequencies from about 30 Hz to 15 kHz. The terms "ultrasound" and "ultrasonic", though useful, are anthropomorphic and refer to vibrations having frequencies above those detectable by the unaided human ear. A wide variety of animals, including many insects (*e.g.* many bushcrickets, cicadas, *etc.*), have ranges of acoustic sensitivity which extend well above that of the human ear and ultrasound is critical in the lives of many animals (Sales and Pye, 1974). Clearly such a limited definition of sound is not useful in the scientific study of animal behaviour.

    Insect sounds have been known and documented since the writings of the classical Greek philosophers, most notably Aristotle. It is only in the past 50 to 60 years that the subject of insect acoustic signals and associated behaviour became available as a subject of intense scientific investigation. The invention of ever-newer techniques for the recording and analysis of sounds has rapidly expanded during this period and continues to do so today in the digital age (Eliopoulos,

3

Chapter 2). The dictionary-based definition, with its anthropomorphic slant, is inadequate and no longer useful to us. One of the earliest English-speaking workers to confront this problem was Pumphrey (1940) in a classical review on insect hearing. He defined sound as "any mechanical disturbance whatever which is potentially referable by the insect to an external and localised source". This then involves vibrations transmitted through any medium, fluid or solid, and is certainly broader than some workers, including many authors in this book, might wish. Good discussions on these problems are given in the excellent volumes of Haskell (1961) and Ewing (1989). A strict definition of sound, as vibration transmitted through fluid media, such as air and water, clearly excludes vibrations through solid substrates which are now known to be very widely used in insect communication and have considerably contributed to this book.

All sounds derive from vibrations which in turn set particles into motion in the surrounding media, whether these be fluid (air, water, *etc.*) or solid. Even the same animal by means of appropriate receptors may detect external vibrations as airborne and substrate transmitted emissions. We know that vibrations passing through different media may have very different physical properties (Michelsen *et al.*, 1982; Ewing, 1989). Animals may perceive such vibrations in similar ways and, more important, may respond in similar ways. It is trivial to argue about the strict uses of the words sound and vibration. In this book we, as editors, have taken a liberal view and allowed the individual authors their own usage, so long as it is clear and consistent. Some authors attempt clearly to separate sound and vibration as two different phenomena, others take a wider view. Many problems in biology are not real ones, but result from the human desire to make strict and exclusive definitions for parts of effectively continuous phenomena.

MECHANISMS OF SOUND PRODUCTION

As Henry (Chapter 10) emphasises here, insects are preadapted to be noisy animals. With a hard and sclerotised exoskeleton, the segmented form of the body and jointed limbs will inevitably cause vibrations in the surrounding environment when an insect moves. It will be very difficult for insects to move silently without making a noise. It is not surprising that many groups have developed specialised systems of sound production and associated receptors which are used in communication within and between species.

Many authors have attempted to classify sound producing mechanisms in insects. The most useful is probably the entirely mechanistic one of Ewing (1989), who recognised five categories of sound producing mechanisms:

1. Vibration
2. Percussion
3. Stridulation
4. Click mechanisms
5. Air expulsion

These categories are not completely exclusive and some insects may use combinations of them. In addition to Ewing, many others have also reviewed sound producing mechanisms, including particularly Haskell (1961) and Dumortier (1963a). It is therefore unnecessary here to review them in detail, but some comments may be useful to the general readers.

VIBRATION — INCLUDING TREMULATION

All animal sounds result from the vibration of some structures. However, in this category are included sound emissions which result from vibrations of relatively unspecialized parts of the insect body, most usually oscillations of the abdomen, either dorso-ventrally or laterally. The term tremulation is useful for this type of sound production and differentiates it from the very general term "vibration". Such sounds are usually transmitted through the legs to the substrate on which the insect is walking or standing. These will therefore usually be detected as substrate transmitted vibrations. Such signals have been documented in various insects, but are well known in lacewing flies and their allies, due particularly to the work of Henry (Chapter 10).

Vibrations of other body parts may be important in insect signalling, most obviously the wings. Sounds are inevitably produced as byproducts of flapping flight, but many insects have developed the use of wing vibrations in communication. The flight sounds in swarming mosquitoes are known often to be species-specific and to function in part, for species recognition, (*e.g.* Roth, 1948). The use of low frequency wing vibration in the courtship dances of *Drosophila* species is better known. When in close proximity, the pulsed songs of these flies stimulate antennal receptors of other individuals by air particle vibration in the vicinity (Bennet-Clark, 1971). In this book Hoikkala (Chapter 11) describes the diversity and genetics of these signals in Drosophilidae.

PERCUSSION

Tremulation does not involve percussion either of the substrate or of other body parts. This is regarded as a separate mechanism. Percussion of one body part against another may develop as a communication system, as documented for example, in some Australian moths (Bailey, 1978) and in some cicadas (Boulard, Chapter 25).

Signalling by percussion of the substrate with the tip of the abdomen is well known in various insect groups, for example termites (Isoptera) (Howse, 1964) and particularly stoneflies (Plecoptera) (Stewart and Sandberg, Chapter 12). An unusual example among the bushcrickets (Tettigoniidae) are the species of *Meconema*. It is a group otherwise well known for the production of loud stridulatory signals (Heller, Chapter 9). Males of *Meconema* lack the distinctive stridulatory mechanism typical of the family, but actively stamp the substrate with one of their hind legs and produce patterned signals in that way (Ragge, 1965).

STRIDULATION

The term stridulation has sometimes been used as a general term for any mechanism of sound production in insects (*e.g.* Haskell, 1961), but that negates the utility of the term. It is more usually confined to sounds produced by frictional mechanisms, involving the movements of two specialized body parts against each other in a regular patterned manner. This is an extremely widespread and relatively well-studied mechanism. Such systems have been described in at least seven different insect orders (Ewing, 1989), in most of which it has evolved separately on numerous occasions, as for example in the Coleoptera (Wessel, Chapter 30). Almost all body parts which it is possible to bring into juxtaposition have been modified as stridulatory mechanisms in one group or another (Dumortier, 1963a). The mechanisms in the groups of Orthoptera *sensu lato* are particularly well-known and documented (for example the Tettigoniidae by Heller, Chapter 9).

CLICK MECHANISMS

These rely on the deformation of a modified area of cuticle, usually by contraction and relaxation of special musculature within the body. This results in a series of clicks which may be repeated rapidly in distinctive patterns. Such signals may be amplified in a variety of ways in different insects. The specialized area of cuticle, as exemplified most obviously in the loud singing cicadas (Hemiptera, Cicadidae), is known as a tymbal. The basics of this mechanism were known to the ancient Greek philosophers, but were first described precisely by Réaumur (1740) (Boulard, Chapter 25). Such mechanisms are now well known, though not necessarily well understood, in many other groups, including most, if not all other Auchenorrhyncha (Ossiannilsson, 1949; Strübing, Chapters 19 and 26; Cocroft and McNett, Chapter 23; Tishechkin, Chapter 24), many Heteroptera (Gogala, Chapter 21) and various families of Lepidoptera (Sales and Pye, 1974).

AIR EXPULSION

This is an unusual and rare mechanism within the Insecta. Various authors have described in a number of insects exhalatory sounds, often expelled via the tracheal spiracles, but little is known about any function (Ewing, 1989). The best-known example is the large and spectacular European hawkmoth, the Death's Head Hawk, *Acherontia atropos*, which expels air forcibly through the mouthparts to make a distinctive piping sound (Busnel and Dumortier, 1959). Sales and Pye (1974) reported similar sounds produced by several African Sphingid moths.

Thus many sound producing mechanisms have been described for a wide variety of insects, but many exist only as possible mechanisms based simply on surmise from morphological evidence.

SONGS, CALLS AND TERMINOLOGY

Insect sounds are often called "songs", but to many human ears are simply noises. It has long been agreed that insect sounds are amplitude (pulse) modulated and that information is therefore encoded as temporal patterns of pulses and groups of pulses, most clearly visible to the human observer in oscillograms. Early studies on insect acoustic receptors confirmed that these were particularly sensitive to amplitude-modulated patterns over a band of frequency sensitivity (Pumphrey, 1940; Haskell, 1961). This clear differentiation between birds and mammals, whose receptors are more sensitive to frequency modulated signals, and insects has been obscured to some degree over the past 20 or 30 years with the investigation of more species and the availability of more refined techniques. Many insects are now known to be sensitive to frequency modulation and many examples are discussed in this book.

The demonstration of a function for a particular sound emission requires behavioural experiments which must also be based on an understanding of the acoustic receptor system. The use of playback of recorded signals is a powerful tool in the investigation of acoustic behaviour. Authors in this book demonstrate a wide array of signal functions in different insects. Perhaps the most universal context in which acoustic signals have been shown to function is that of calling. Calling songs or signals are used, usually at a distance, during mate finding and courtship to identify a conspecific and receptive mate. Other signals may be used during complex courtship and mating behaviour sequences and of course may also include nonacoustic signals such as visual and chemical ones. Different acoustic signals may be used by the same insects in inter-male and aggressive interactions. The complete acoustic repertoire may be very complex, nowhere more so than in social species, best known in bees (Hrncir, Barth and Tautz, Chapter 32).

Much attention has been devoted to the description of insect song patterns. The earliest of these were hampered by the lack of technology and often attempted to use musical notation to imitate the sounds, as for cicadas by Myers (1929) and smaller leafhoppers and planthoppers by Ossinanilsson (1949). With the widespread use of oscilloscopes and sound spectrographs in the 1950s and 1960s, more objective descriptions of song patterns became possible (Haskell, 1961). The many oscillograms and spectrograms which were subsequently published, such as those in this book, demonstrate the great complexity of many signals and raise problems of description and terminology. Many authors have attempted to produce universally applicable systems of nomenclature and terminology for insect call structures (*e.g.* Broughton, 1963; Ragge and Reynolds, 1998). Some have pressed very hard for such universality (*e.g.* Broughton, 1963), but, in my view, this leads only to unnecessary controversy. In a very valuable intensive review of acoustic communication in insects and Anura (frogs and toads), Gerhardt and Huber (2002) recently concluded "Whatever the rationale for identifying acoustic elements, attempts to establish a universal terminology are almost certainly doomed by traditional usage and the sheer diversity of signal structures".

Insect sounds are usually structured in some obvious way, so that oscillograms show characteristic temporal patterns (see most chapters of this book). Occasionally, in very well worked examples, it is possible to determine an exact correspondence between the detailed structure of the mechanism and the sound pattern. For example, many crickets and grasshoppers produce obvious units of sound, each of which corresponds to a complete cycle of movement of the stridulatory surfaces. Different authors have used different terms for these basic units, such as chirp and echeme. In most insects an exact equivalence between movements of the mechanism and the sound produced is not obvious. Often, the precise mechanism is not even known. It is essential that descriptions of songs use a clear and objective terminology for units of sound which have no pretence at illuminating underlying physiology or claiming exact homology between species in the absence of detailed evidence. The units of sound and the terms used should then be clearly identified on accompanying oscillograms and sonograms (see also Eliopoulos, Chapter 2). Unfortunately terminological argument has often in the past confused biologically interesting problems in insect acoustic behaviour.

RECORDING DEVICES, HEARING AND BEHAVIOUR

Pierce (1948) published one of the first comprehensive attempts, known to me, to record and analyse insect songs in a systematic manner. None of the equipment which we now take for granted, with the exception of the microphones, was then available to him. He manufactured a number of ingenious devices for recording and analysing calls, but his results are now of little more than historical interest. All of the insects that Pierce described produced loud, airborne, acoustic signals and were amenable to recording with standard microphones. The airborne calls of many insects extend well into the ultrasonic range. These may be conveniently detected and recorded by the use of a "bat detector", as reported for cicadas by Popov *et al.* (1997). Eliopoulos (Chapter 2) gives a full discussion on the diversity of microphone types and their characteristics, as well as the diversity of recording devices from tape recorders to direct recording with a computer.

The realisation that many insects produce calls not easily detected by the unaided human ear, transmitted as substrate vibration, is relatively more recent. The detection of such usually low frequency signals requires some specialised, though not always expensive, equipment. The simplest method is the use of a crystal gramophone-type pick-up as a transducer in contact with the surface through which the insects are communicating (Ichikawa, 1976; Claridge, 1985a, 1985b). Other devices, including commercial accelerometers and the so-called magneto-dynamic system of Strübing and Rollenhagen (1988), have their adherents and are reviewed by Gogala (Chapter 21). All of these techniques share the same disadvantage of necessarily imparting some physical load on the substrate and may therefore affect the structure of any signals detected. The best solution is to use laser vibrometry, as first brilliantly applied to insect acoustics by Michelsen (see Michelsen and Larsen, 1978). Until recently this equipment has been prohibitively expensive and very difficult to take into the field. However, new developments are making the techniques more easily available (see Virant-Doberlet, Chapter 5; Gogala, Chapter 21; Cocroft and McNett, Chapter 23; and others in this book).

Insect vibratory sense organs, including tympanal receptors specialised for detecting airborne sounds, have been much more widely studied and are reviewed here by Rössler *et al.* (Chapter 3), Čokl *et al.* (Chapter 4) and Virant-Doberlet *et al.* (Chapter 5). The word hearing is, of course, yet another with anthropomorphic overtones, but is used widely at least for the reception of airborne signals by insects.

The modern ability to record and manipulate acoustic signals simply has made acoustic behaviour one of the most active and advanced areas of insect ethology. The playback of previously recorded signals, or various modifications of them, makes precise understanding of the functions of particular calls possible. This has been used for many insects, for example, cicadas (Alexander and

Moore, 1958, 1962) and planthoppers (Claridge, 1985a, 1985b; Claridge and de Vrijer, 1994). The complex functions of calls and the often wide song repertoire of insects have also been extensively studied (for example Bailey, 1991 and Chapter 8; Cocroft and McNett, Chapter 23).

The nature of sound production and reception in any particular insect is clearly subject to a wide variety of physical constraints, one of which is body size. Small insects will inevitably be forced to use relatively high frequency sound, and even ultrasound, for aerial communication over distances of more than about one body length (Bennet-Clark, 1971). However, small insects may use low frequency sounds, providing that these are transmitted through the substrate. Little attention has been given previously to any relationship between body size and signal parameters for substrate-transmitting acoustic insects, but here Cocroft and De Luca (Chapter 6) present novel analyses on this. Among other physical constraints, temperature is a very obvious one. Rates of biological processes, including muscle vibration, must ultimately be temperature dependent and the effects of temperature on insect rates of calling have been frequently documented. Sanborn (Chapter 7) here gives an elegant review of effects of temperature on the calling of a variety of aerial insect singers.

DIVERSITY

Vibratory senses are ubiquitous in insects so that it is not surprising that more and more examples of vibratory signalling systems are being discovered regularly in almost all insect orders. It is possible that vibratory communication is a basic characteristic of the Insecta, or at least anyway of the Neoptera. The recent classification of insect orders adopted by Gullan and Cranston (2005) is followed here.

Among the Polyneoptera, one of the basal groups, the stoneflies and their relatives (Plecoptera) produce a variety of substrate transmitted signals by either percussion or tremulation or a combination of both (Stewart and Sandberg, Chapter 12). The Orthoptera *sensu lato* of course include many of the most obvious insect singers, crickets, grasshoppers, bushcrickets, *etc.*, in which loud airborne signals are widely used and well studied (recently by Gerhard and Huber, 2002; and in this book by Heller, Chapter 9, and Bukhvalova, Chapter 14).

The Hemiptera, the biggest order in the Paraneoptera, includes the other well-known group which produces loud airborne acoustic signals. These are restricted to the one family Cicadidae, of which the typical male tymbal mechanism with associated resonant air sac is completely characteristic (Pringle, 1954; Young and Josephson, 1983a, 1983b; and in this book, Boulard, Chapter 25). Other families of Hemiptera, including many Heteroptera (Gogala, Chapter 21) and Homoptera (Cocroft and McNett, Chapter 23; Tishechkin, Chapter 24 and Chapter 27; Kanmiya, Chapter 28) are characterised by the production of often complex, low intensity, substrate transmitted acoustic signals. In most Homoptera Auchenorrhycha there are complicated, but little understood, mechanisms which resemble to varying degrees the tymbal system of cicadas, but always lack the associated air sac. Complex patterned, damped, low intensity calls are characteristic of many of these species (Claridge, 1985a, 1985b). Comparative physiological studies are urgently needed for these families. We still have only the wonderful classic, but now very old, work of Ossiannilsson (1949).

Within the Endopterygota sound production has been noted in some species of most orders. The basal Neuropterids are characterised by tremulation and percussive mechanisms widely used in courtship and species recognition and which have been greatly elucidated by Henry (Chapter 10). The enormous order, Coleoptera, includes many groups in which supposed stridulatory organs were early described. Charles Darwin (1871) documented many of these as apparent examples of sexual selection in the late 19th century. Wessel (Chapter 30) provides a useful brief overview of this diversity within the order, and Kasper and Hirschberger (Chapter 31) describe their detailed work on acoustic communication in one group of dung beetles, species of *Aphodius*.

Among the Diptera many groups are known to use vibratory signals in communication. Particularly well studied are the Drosophilidae which signal by pulsed wing vibrations when in very close proximity (Bennet-Clark, 1971). Hoikkala here (Chapter 11) illustrates the patterns of such calls for numerous *Drosophila* species. Other vibratory mechanisms, resembling tremulation, are known in some groups of Diptera (Kanmiya, Chapter 29).

Few examples of acoustic communication are known for the Lepidoptera, but some families do possess tymbal systems, very similar to the true cicada type (Sales and Pye, 1974). The Hymenoptera, one of the biggest orders, is probably the least studied. Some parasitic chalcid wasps are known to use wing vibratory signals during courtship (*e.g.* Assem and Povel, 1973; Assem and Werren, 1994). Perhaps they may function as near field signals, much like those of *Drosophila*. Certainly vibratory signals are known to be important in eusocial Hymenoptera. Hrncir *et al.* (Chapter 32) give a fascinating account here of such signals and associated behaviour in social bees.

Most groups of insects have well-documented examples of acoustic signals and communication, but there is a wide open field for research with the technology which is now available. Many of the authors of this book outline the diversity as at present known.

ECOLOGY

Interests in acoustic behaviour have centred on sexual behaviour in mate searching, recognition and courtship. The role of sound in more purely ecological contexts has been less widely studied. Buchvalova (Chapter 15) here discusses the possible role of grasshopper songs in the organisation of natural field assemblages and communities of insects. In interspecific interactions, particularly predator–prey and parasitoid–host interactions, sound has only occasionally been studied. The classic work of Blest (1964) on the ultrasonic calls of night flying Arctiid moths and their aposematic function against hunting insectivorous bats has stimulated much work on acoustic antipredator responses much more widely and is reviewed by Bailey (1991; and Chapter 8). The detection of sound signals by adult parasitoids during host finding is also little studied, but it is a rapidly expanding field which has been reviewed by Meyhöfer and Casas (1999) and here by Casas and Magal (Chapter 20).

BIOSYSTEMATICS AND EVOLUTION

The use of acoustic signals as characters in systematic studies is unusual, probably because of a lack of required detailed data, even for relatively well-studied groups. Tishechkin here (Chapter 24) shows how, for the Hemiptera Auchenorrhyncha, when a sufficient database is accumulated, useful conclusions and pointers may be made to systematic relationships.

At the level of biological species, acoustic studies have been widely and effectively used to establish the status of related populations of a wide variety of insects. The term biosystematics, sometimes biotaxonomy, has come in recent years to be used as synonymous with systematics. This is unfortunate as it was originally used in the narrower and more useful sense to include all sorts of studies which might illuminate the genetic, and therefore specific, status of groups of related organisms (Claridge and Morgan, 1987; Claridge, 1988, 1991). A plea is made elsewhere in this book by Drosopoulos (Chapter 18) to return to this usage, a plea which I heartily endorse.

In this narrower sense, biosystematics is a necessary prelude to understanding the process of speciation and therefore of evolutionary divergence in any group. Critical to such studies is the identification of the signalling methods which particular insects use in their specific mate recognition systems (Paterson, 1985; Claridge, 1988; Claridge, Dawah and Wilson, 1997a). These may be of any sensory modality and often mixtures of several. When these are acoustic, recording, analysis and playback are relatively simple. It is therefore not surprising that, unlike general systematics, acoustic studies have been central to much biosystematic work on insects. In this book,

Sueur (Chapter 15) has reviewed this widely in the context of species concepts. Many examples are included elsewhere in this volume — Heller for bushcrickets (Chapter 9); Henry for Neuropterida (Chapter 10); Hoch and Wessel for cave-inhabiting planthoppers (Chapter 13); Drosopoulos *et al.* and Quartau and Simões, for different genera of European cicadas (Chapters 16 and 17); Cocroft and McNett for treehoppers (Chapter 23); Strübing for some leafhoppers (Chapter 19); Kanmiya for whiteflies and some Diptera (Chapters 28 and 29) and Kasper and Hirschberger for some dung beetles (Chapter 31). In addition Drosopoulos (Chapter 18) shows how the use of acoustic analyses, in conjunction with other genetic techniques, may illuminate unexpected complications and even parthenogenesis and pseudogamy.

It is clear then that the past 40 years or so have seen enormous developments in the study and understanding of insect acoustic behaviour. It is worthy of study, not only as a subject in its own right, but also for the light it may shed on many fundamental, and indeed also applied, fields of biology, from cell biology and physiology to ethology and evolution. In this volume we have brought together an unprecedented range of expertise from many different countries on all aspects of our subject. We hope that the resulting book adds to the diversity of information now easily available to the general reader, but also that it will stimulate new interests and exciting lines of research.

2 Sound and Techniques in Sound Analysis

Elias Eliopoulos

CONTENTS

INTRODUCTION

The science dealing with the methods of generation, reception and propagation of sound is called *acoustics*. The branch dealing with sounds produced by animals is called *bioacoustics*. The songs of insects are among the major sources of biological noise in the environment. Studying insect sounds sometimes requires specialised instruments and techniques for the acquisition and analysis of the signals (Burbidge *et al.*, 1997; Bradbury and Vehrenkamp, 1998). The aim of this chapter is to highlight some sound properties, associated with insect signal production, transmission and detection as well as some of the instrumentation and methodology used for acoustic analysis.

Animals can perceive mechanical media perturbations transmitted through air, water or solids such as soil, sand and different plant materials. These vibrations provoke a nervous response, constituting the process known as hearing. The elastic waves in which the disturbance, manifested as a strain, a pressure or a bulk displacement involving many atoms, propagates with a velocity depending on the elastic properties of the medium, are called sound. This disturbance is an orderly collective, usually oscillating motion, in which all atoms in a small volume suffer essentially the same displacement. For liquids and gases this orderly motion is superimposed on the random or disorderly molecular agitation known as *Brownian motion*. The net result is the propagation of the disturbance through the medium with the concurrent attenuation of the wave properties due to the collision of molecules of the medium. The velocity of propagation is dependent therefore on the density of the medium. It is also sensitive to temperature and pressure changes because of the dependence of velocity on density.

The sound source may be an oscillating diaphragm, or movements of the wings and legs of the insects. It might also generate as a result of the insect striking the ground with its head, abdomen or legs (W.J. Bailey, *The Acoustic Behaviour of Insects: An Evolutionary Perspective*, Chapman & Hall, London, 1991). Irrespective of the shape or form of the sound or vibration source, an oscillation is produced in the form of combined sinusoidal waves of fixed frequency, which propagate through the respective medium (air, liquid, plant stem).

SOUND PROPERTIES

FREQUENCY AND WAVELENGTH

The frequency of sound vibration is the number of cycles of pressure change with respect to time. This is measured in units of cycles per second or Hertz (Hz) in the SI system. The human hearing sensation responds to frequencies between 16 Hz and about 20,000 Hz (Figure 2.1) while for other animals these limits are different (Figure 2.2) (Yost and Nielsen, 1985). For different insects the limits range between 15 Hz and 120 kHz (Roeder, 1965; Smith, 1979). The velocity of propagation is practically independent of the frequency for a very large range of frequencies extending up to more than 100 MHz. The wavelength though, being the distance covered by one

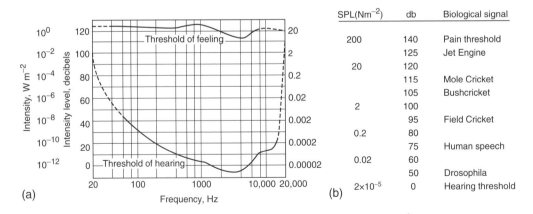

FIGURE 2.1 (a) Plot of human hearing limits as a function of frequency and (b) Measure of SPL for several sources. (Table from Ewing, A. W., *Arthropod Bioacoustics: Neurobiology and Behaviour*, Edinburgh University Press, Edinburgh, p. 260, 1989a. With permission.)

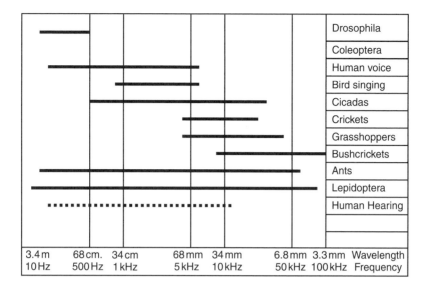

FIGURE 2.2 Frequency/wavelength ranges of various biological signals.

cycle of the oscillation, depends on the velocity of the perturbation of the medium and therefore on the density of the medium. The relationship of this dependence can be explained as frequency = velocity/wavelength:

$$f = \nu/\lambda$$

In air, at mean sea level, the velocity of sound is 340 m/sec, so the wavelength of a 340 Hz oscillation will be 1 m, while if the oscillator emits at the frequency of 3.4 KHz, the wavelength will be 10 cm. In fresh water at 25°C, the velocity of sound is 1493.2 m/sec, while in sea water it is 1,532.8 m/sec, giving wavelengths of 43 and 45 cm, respectively, it is for the same frequency of 3.4 kHz. In solids, the sound propagates with velocities varying from 6000 m/sec for the very dense granite to 1230 m/sec to the softer lead.

SOUND PROPAGATION PROPERTIES: ATTENUATION, NEAR FIELD EFFECTS AND DISTORTION

A measure of amplitude of sound wave is given by the relative measure of sound pressure levels (SPL) with respect to a reference level:

$$SPL = 20 \log_{10}(p/p_r) \text{ in dB (decibel) units}$$

where p is the measured pressure level and p_r the reference pressure level. As $p_r = 2 \times 10^{-5}$ N/m^2 which is the human hearing threshold at 4 kHz.

Sound level is dependent on the amplitude of the oscillation produced by the sound source, *e.g.* the vibrating diaphragm, and decreases with distance from the sound source. Since the perturbation of the sound wave is spreading radially in uniform space, sound intensity, which is the rate of energy transfer or energy flow at a given point, is inversely proportional to the distance from the source (Figure 2.3).

Sound intensity is also proportional to the product of pressure and particle velocity. As the wave propagates, apart from changes in pressure, the media particles also accelerate and decelerate. In fact at the source point, maximum pressure is exerted when the oscillating particles are at a

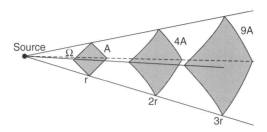

FIGURE 2.3 Spherical spreading of a sound wave from a point source demonstrating the inverse square law attenuation of sound intensity.

maximum distance from the equilibrium point. At the maximum distance of the particle from the equilibrium point, the velocity is zero and therefore the particle movement and pressure are 90° out of phase, while at a distance from the sound source this difference is diminished.

Since animals communicate in nonuniform media, modified by a large number of environmental factors, energy loss may be greater (Michelsen, 1985). In addition heterogeneities within the medium may cause scattering and interference effects, leading not only to sound damping, but also to directionality changes. If the medium through which the signal travels is not uniform, the signal will suffer some distortion as well as attenuation. The geometrical spreading which causes reduction of the pressure levels by half at doubling of the distance is further impeded by temperature and humidity (mainly in the air) changes (affecting the density) and heterogeneities of the medium.

Although particle movement is associated with pressure changes anywhere in the sound field, their directional components are most clearly experienced close to the sound source (near field). Recording sound with a conventional microphone close to a large radiating source at distances less than a third of the emitted wavelength will be affected by complex interferences. Since most insects use rather low frequencies for near field communication with wavelengths in air many times longer than that of the insect's largest dimension, a pressure sensitive microphone is very inefficient for the detection of such signals. Detection would be better performed by particle movement sensors, such as ribbon microphones or systems with a light piezoelectric foil. For the detection of particle movement in solids, accelerometers and more recently laser vibrometers are used (see relevant Chapters 4, 5 and 22 in this book). The amplitude of particle displacement falls with the third power of the distance and because of this any near field effect will become almost insignificant at a distance of one or two wavelengths. This has important consequences for insects, many of which are small and produce low frequency sounds. Because acoustic efficiency is low at a distance of at least one wavelength (which in air for a frequency of 2 kHz is about 1.5 to 2.0 cm), for near field communication small insects use receptors to detect the particle velocity component rather than the pressure component (see Bennet-Clark, 1971). To be acoustically efficient insects must either be large or produce high frequency sounds.

Because the density and the elasticity of insect cuticles are very different from that of air, energy transfer to the surrounding medium is very inefficient. Thus in air, insects are rather inefficient at converting muscular energy to airborne sound. It is interesting that some insects produce signals with frequency components close to the resonant frequency of the sound propagating medium, diminishing the mechanical impedance and maximizing the vibration thereof. Efficiency increases to 23% in some beetles using substrate vibrations for communication (Leighton, 1987). High frequencies are attenuated more rapidly than low frequencies. So, the lower the frequency, the greater is the damping of the acoustic signal with distance. This also explains the nonuniformity of frequency components in signal detection with respect to distance (Michelsen *et al.*, 1982).

SUBSTRATE VIBRATIONS

In order to offset the inefficiency of air transmission, many insects use resonators to amplify sound. Others take the alternative route of producing substrate borne vibrations alone, at a much lower energy cost due to lower attenuation of mechanical energy in solids compared with air. The efficiency of an acoustic signal of a cricket is only 1% with respect to its muscular energy and for the cicada it is only 0.5% (Kavanach, 1987).

Most animals live on the interface of some solid with air or water and the waves which affect this interface will be significant for communication. Because of the nature of linear structures such as leaves and stems, vibratory information is transmitted as longitudinal, transverse and bending waves. Longitudinal and transverse waves will change with the length and width of the structure (*e.g.* stem), but bending waves will create changes along its surface.

Vibrations in solids are mainly transmitted as longitudinal waves (waves where the particle movement component is along the length of transmission), but their surface effects in the form of compression and extensions is less that 1% of total length. Transverse waves (waves where the particle movement component is perpendicular to the direction of wave transmission) produce an even less significant effect at the surface of the solid and are unlikely to be detectable. Bending waves are perturbations which are created in long, thin structures such as plant stems, where constructive interference of transverse waves can create amplified perturbation along the surface of the substrate. Because propagation velocity of bending waves depends on the physical properties and dimensions of the medium and the wavelength of the vibration, vibrations of different frequencies travel away from the sound source at different speeds. If the vibrations are in the form of pulses, which is something usual in insect communication, their group propagation velocity is twice the phase velocity.

Since the information collected by insect receptors concerns the relative movement at the surface of the substrate, measurements taken from accelerometers (instruments used to measure surface vibration), given as a rate of change (m/sec^2), relate to the amplitude with which a portion of the substrate rises and falls. For low intensity bending waves, amplitude will be low and vertical vibration velocity will be small. For more intense vibrations, surface vibration velocities will be high.

SIGNAL STRUCTURE AND TERMINOLOGY

In order to compare signals emitted from the same individual, in different conditions or different individuals of the same or different species, certain terminologies have been adopted to describe the hierarchical structure. Calling songs are repeated over long periods of time. The repeat is sometimes called a phrase. Each repeatable phrase usually consists of two subgroups with strong amplitude variations, followed by a silent interphase gap (ICD). The two subgroups of the phrase are referred as low amplitude (LPD) and high amplitude (HPD) parts (Sueur and Aubin, 2004). In the temporal domain, there is big variation in the frequency of phrases (phrases per second), but for most acoustic signals the phrases last for periods of seconds. Each phrase subgroup consists of chirps or echemes with a temporal length of the order of hundredths of seconds or some tenths of milliseconds (msec). In more detail each echeme has an internal structure of syllables which are of the order of milliseconds. Further, these consist of pure pulses or impulses of the order of tenths of milliseconds. Impulses normally represent the principal unit of sound or vibration production, such as a wing flick, a single tooth strike or a leg movement (Figure 2.4).

SOUND ACQUISITION

The study of insect signals requires specialised instruments and methodologies for detection, recording and analysis (Kettle and Vieillard, 1991; Kroodsma *et al.*, 1996). Apart from unwanted noises, such as vehicles or wind due to outdoor recording, most insects cannot be approached

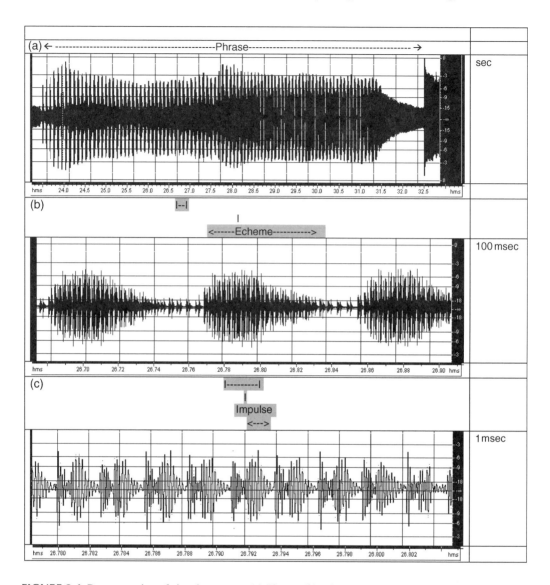

FIGURE 2.4 Representation of signal structure. (a) Phrase, (b) echeme or chirp and (c) impulses.

closely, resulting in recordings with low signal to noise (S/N) ratio. Nowadays, most sound recordings are made with specialised sensitive microphones and either analogue or digital tape recorders. Vibration detection is performed with accelerometers and laser vibrometers (for detailed description of vibration detectors see Chapters 5, 13, 22 and 24). One important aspect in the process is the recording and transfer of the signal, be it sound or substrate vibration, in as undistorted form as possible with minimal background noise. A description of the different available sound acquisition devices, analogue and digital, for recording insect signals are as follows.

MICROPHONES, PICK-UP PATTERNS AND PROPERTIES

The microphone performs the crucial task of converting acoustic signals as pressure variations to its electrical analogue, a modulated electric signal. The type of transducer, its efficiency (or sensitivity), the frequency response and the polar pattern (or directivity) are all equally important features which characterise a microphone.

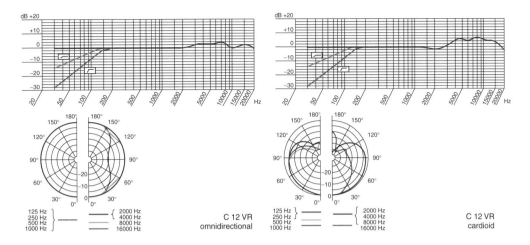

FIGURE 2.5 Radial and logarithmic plots of microphone sensitivity as a function of frequency for two representative types (omnidirectional and cardioid) of commercially available microphones. (From AKG Acoustics, The ABC's of AKG: Microphone Basics and Fundamentals of Usage, white paper found on www.AKG.com. With permission.)

The most fundamental characteristic of a microphone is its three dimensional pattern of pick-up. The polar pattern is a graphic representation of the sensitivity of the microphone with respect to the frequency and the angle of incidence of sound. Typically, directionality increases with increasing frequency (Figure 2.5).

There are three basic directional patterns: omnidirectional, bidirectional and unidirectional (also called simply directional). Depending on the various degrees of directionality, the microphones are further categorised as super, ultra or hyper directional. The basic directional and simplest pattern is the cardioid. Perhaps 90% of all microphones fall into the omnidirectional (also called omni) category and the cardioid pick-up categories.

Omnidirectional Microphones

For omnidirectional microphones, there is very little distinction based on the direction of the impinging sound. The microphone will respond equally to sound from all directions (Figure 2.6). At very high frequencies there will be some departure from this and the microphone will show a preference for sounds arriving from the front. An example of an excellent cardioid microphone measured at 0°, 90° and 180° is also shown. The back rejection at 180° of the order of 20 to 25 dB in the midfrequency range can be seen, while the cardioid action diminishes at very high and very low frequencies.

The Cardioid Family

Cardioid microphones are basically unidirectional and there are three variants, the primary cardioid, the hypercardioid and the supercardioid. A few designs, often referred to as rifle, line or shotgun microphones, have a long interference tube which makes them highly directional at mid and high frequencies and these are very useful when sound pick-up must take place at some distance from the source.

A cardioid microphone has greater "reach" than an omni. Because of its forward oriented pick-up pattern, it has a high ratio of on-axis response to random directional response (Figure 2.5). Figure 2.7 shows a comparison of omni and the cardioid microphones in terms of equivalent working distances. The cardioid microphone may be used at 1.7 times the working distance of an

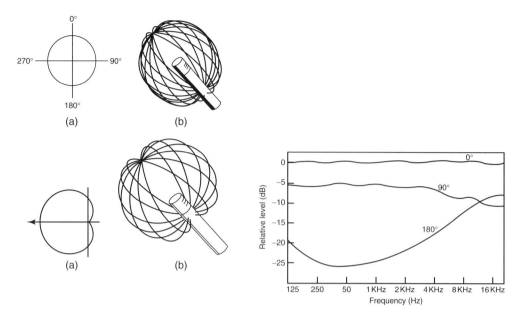

FIGURE 2.6 Representation of the basic omnidirectional pattern in a two-dimensional representation known as a polar plot (a), while a three-dimensional representation is shown at (b). The diagram shows the sensitivity of the microphone as a function of frequency and direction. (From AKG Acoustics, The ABC's of AKG: Microphone Basics and Fundamentals of Usage Catalog, white paper found on www.AKG.com. With permission.)

omni, while still giving the same overall suppression of random noise. The hypercardioid pattern can be used at two times the distance of an omni for the same overall effect, while the supercardioid pattern can be used at 1.9 times the distance for the same effect.

In terms of dB, when used at the same working distance, the cardioid will randomly reject arriving sounds 4.8 dB more effectively than the omni (Figure 2.8). By comparison, the supercardioid would provide 5.8 dB more rejection and the hypercardioid would provide 6 dB more

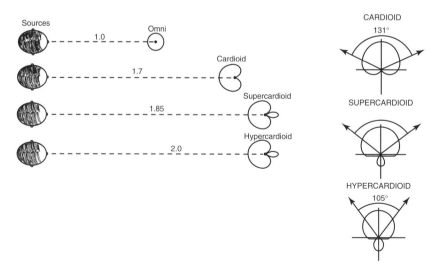

FIGURE 2.7 The nominal acceptance angle (±3 dB) that microphones in the cardioid family provide. (From AKG Acoustics, The ABC's of AKG: Microphone Basics and Fundamentals of Usage Catalog, white paper found on www.AKG.com. With permission.)

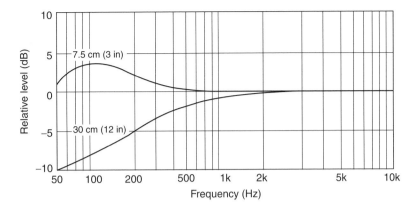

FIGURE 2.8 Noise rejection of a cardiod microphone with respect to frequency.

rejection. As we have seen, the hypercardioid and supercardioid patterns offer slight improvement over the cardioid in terms of immunity to random sounds. While in a studio the standard cardioid may be preferred for its 180° null-in output, the outdoor sound researcher will often prefer the super and hypercardioid for their added reach. By opening up the rear (180°) lobe in the directional response (see views of these patterns in Figure 2.7), the frontal pattern is actually "tighter" than that of the standard cardioid.

Hypercardioid and Supercardioid

These are variations on the basic cardioid pattern which may be very useful for certain applications. If the back path length is varied slightly, the off-axis angle, at which the output is minimum, can be varied. There are two additional patterns resulting from these changes, and these are known as hypercardioid and supercardioid. These patterns have the effect of changing the "reach" of the microphone, and these may be very useful in certain sound reinforcement applications by providing more gain before feedback than the standard cardioid pattern.

Ultradirectional Microphones

Ultradirectional microphones, also called shotgun microphones, are cardioid microphones fitted with an interference tube on their frontal face. The shotgun microphone, characterised by a flat frequency response, is less sensitive to wind and handling noise but offers a lower sensitivity than a microphone mounted in a parabola; the interference tube cancels off-axis signals while the in-axis signals reach the microphone's diaphragm without attenuation or gain. Normally, these microphones are *condenser microphones*. Directional microphones and parabolas help sound collection in nature, by giving emphasis to the sounds coming frontally and attenuating unwanted ambient noise.

Parabolas

A parabola focuses incoming sound waves which are parallel to its axis, on to a single point, called the focus of the parabola, where a microphone is placed (Figure 2.9). Its effectiveness is determined by the diameter of the reflector in relation to the wavelength of the sound; its gain and directivity increase proportionally with increasing the diameter/wavelength ratio. For wavelengths larger than the diameter of the parabola, the response is predominantly that of the microphone itself. Increasing the frequency, as the wavelengths become shorter than the parabola's diameter, increases the gain and directivity. For a parabola to become effective at frequencies as low as 100 Hz, its diameter must be larger than 3 m. Common diameters are 45, 60 and 90 cm, with directionality starting,

FIGURE 2.9 An omnidirectional microphone mounted on a parabolic disk.

respectively, at about 750, 550 and 375 Hz. Specific design, well matching microphones, proper positioning and proper filtering allow the linearization of in-axis frequency response of the whole system. Suitable microphones to use in conjunction with parabolas are dynamic microphones which do not require a power supply with a cardioid (directional) response. Condenser microphones are generally better quality but require a power supply. Since microphones are susceptible to wind, windshields are essential for field recordings (Vieillard, 1993).

Other Microphone Properties, Impedance, Sensitivity and Self Noise Level

Condenser microphones have internal impedances in the range of 200 ohms, while dynamic microphones vary from 200 to 800 ohms. The advantage of low impedance microphones is that they can be used over fairly long distances from the console with negligible losses. Distances up to 600 ft, while rarely encountered in normal applications, can be handled with no problem.

Microphone sensitivity is equally important for the choice of microphone and the recording device. This may vary from 12.5 mV/Pa for a condenser microphone to 0.72 mV/Pa for a dynamic (noise-canceling) microphone. Sensitivity is also stated in decibels relative to one volt, a designation known as dBV. In measuring microphone sensitivity, the microphone is placed in a reference sound field in which a SPL of 94 dB at 1000 Hz is maintained at the microphone. The SPL equivalent to a pressure of one Pascal (Pa), the metric unit of pressure is 94 dB. The unloaded output voltage is measured and stated as the nominal sensitivity.

The self noise of a condenser microphone is the audible noise level the microphone produces when it is placed in isolation from outside sound sources. As an example, a microphone which has a self noise level of 15 dBA produces roughly the same output as a "perfect" microphone placed in a room with a measured noise floor of 15 dBA.

SOUND RECORDERS

Sound recorders operational in remote locations must satisfy several practical considerations: weight, durability of power and power supplies, ruggedness, signal distortion and frequency dynamic range and response. Sound recorders can be analogue or digital, depending on the way the signal is processed and stored. The electrodynamic signal of the microphone, vibrometer or accelerometer is transferred to the sound recorder, where it is processed and stored in an analogue or digital form. A recorder should record a signal without significant distortions by at least preserving all its features and matching the signal's dynamic and frequency range. Traditional analogue tape recorders, compact cassette and open reel recorders, degrade the signals these record by adding hiss, distortion, frequency response alterations, speed variations (wow and flutter), print through effects and dropouts. Digital recorders avoid these problems. Within the dynamic range, the frequency limits dependent on the number of bits and the sampling frequency they incorporate, they record and reproduce signals with great accuracy, low noise, flat frequency response and no speed variations.

Digital Audio Tape

Digital audio tape (DAT) recorders deal mainly with audible signals; their frequency response allows very good recordings from low frequency signals, as low as 10 Hz, up to 22 kHz. Like an analogue tape recorder, a DAT recorder records audio on magnetic tape, but in a different way (Pohlmann, 1989). The electric signal from the input device (microphone, accelerometer) is passed through a low-pass filter (antialiasing filter) which removes all frequencies above 20 kHz. Next, the filtered signal passes through an A/D converter. This converter measures the voltage of the audio waveform several thousand times a second (sampling frequency). Each time the waveform is measured, a binary number (made of ones and zeros) is generated which represents the voltage of the waveform at the instant it is measured. Each one and zero is called a bit, which stands for binary digit. These binary numbers are stored magnetically on tape in the form of a modulated square wave recorded at maximum level. The DAT standard is based on 16 bits of resolution and a sampling frequency of 48,000 samples/sec to allow, respectively, about 90 dB of dynamic range and a frequency response of 10 Hz to 22 kHz. The higher the sampling rate is, the wider the frequency response of the recording. The upper frequency limit is slightly less than half of the sampling rate. The sampling rate of 44.1 kHz is adequate for high-fidelity reproduction up to 20 kHz. Some DAT recorders allow two additional modalities, 16 bit at 44,100 s/sec to allow a direct mastering of CDs and 12 bit at 32,000 s/sec to double the recording duration of the tape. Recently, DAT recorders with doubled speed and 96k s/sec were developed to record and play two channels with 40 kHz bandwidth; other DAT models offer four channels with 20 kHz bandwidth or eight channels with 10 kHz bandwidth.

At playback, because the digital head reads only binary numbers, it is insensitive to tape hiss and tape distortion. Numbers are read into a buffer memory first and read out at a constant rate, eliminating speed variations. The resulting absence of noise and distortion makes digital recordings accurate and clear. Unlike analogue recordings, digital ones can be copied with little or no degradation in quality. Lost data are restored by error correction circuitry.

The DAT recorder delivers a sound quality slightly better than ordinary CD players. Digital transfer from DATs to computers is allowed by specific boards with digital I/O capabilities. Since consumer DAT recording devices are being replaced by minidisc recorders, high quality sound recording is carried out by new emerging technologies such us optical disk, solid state and direct laptop recording.

MiniDisc and Solid State Recorders

MiniDisc MD recorders are popular with field recordists because most portable models are reliable, compact and have a good sound quality. However, the storage format uses a data reduction process to reduce digital storage space, which is designed to be transparent to human hearing. This negatively affects insect signal recordings if they include abrupt variations and very low or high frequencies.

Solid state (flash card/disk drive) recorders have been developed in the last few years. These recorders use compact flash memory cards or computer disk drives in order to store the recordings.

FIGURE 2.10 Latest technology portable PC with digital and analogue to digital/D input.

Some models offer a choice of recording in different file formats with sampling rates reaching 96k s/sec and amplitude resolution up to 24 bit. The uncompressed format (*e.g.* PCM "wav" file) is preferable to the compressed (*e.g.* MP3) formats since no data reduction and compression are performed. Recording times at the uncompressed quality are limited since the memory cards or drives have a limited capacity (currently up to 5 Gb). These types of recorders are likely to become an attractive alternative to other types. The major problem is their built-in microphones which do not perform well for wildlife recordings.

Laptop Recording

Though computers have been used for 25 years for recording, analysing and editing sounds, only in recent years have PC capabilities and the availability of good and cheap sound devices made computers powerful and affordable enough for direct recordings in the field (Pavan, 1992). Recording on a PC, either desktop or laptop, is performed with the use of 16 bit A/D converter sound cards (Figure 2.10). Ordinary PC soundcards typically have a much worse S/N ratio (around 80 dB) than CD or DAT recorders ($>$ 90 dB), therefore adding more noise to the signal recorded. In addition low and high frequencies are attenuated noticeably, not having a good straight frequency response from 20 Hz to 20 kHz. Specialised sound cards for desktops are now available with 192k samples/sec to provide more than 80 kHz of useful bandwidth, while dedicated instrumentation acquisition boards can sample up to 500k samples/sec. For laptop use, USB and FireWire sound devices allow up to eight channels at 96k samples/sec. S/N ratio (measured in dBu), which measures the ratio of noise generated by the recording device, and added to the outgoing signal plays also an important role. The microphone amplifiers in commercial soundcards have usually low S/N ratios and because these are noisy, pick up interference from the computer, add distortion and have typically bad frequency response. Also the low input voltage of some sound cards (the Sound Blaster family of soundcards from Proto 32 has an input voltage of -20 dBV (100 mV or 0.1 V) which is too low) make them unusable with any professional dynamic microphone without an external microphone amplifier connected to the soundcard line input. Therefore special attention must be given to the quality of the individual components as well as their compatibility.

TECHNIQUES FOR SOUND ANALYSIS

DIGITAL REPRESENTATION OF SOUND

The development of digital signal processing techniques and low cost recording and processing hardware has made the visualisation of sound and vibratory signals an invaluable tool for bioacoustic research (Pye and Langbauer, 1998). Signal analysis allows their display graphically,

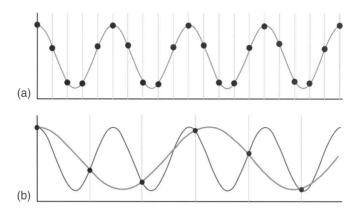

FIGURE 2.11 Sampling to create digital representation of a pure tone signal and aliasing as a result of inadequate sample rate. The same waveform is represented in both diagrams. Vertical lines indicate sampling times. (a) Sampling frequency approximately five times the wave frequency. (b) Sampling frequency approximately 1.5 times the wave frequency. The resulting digitised waveform (in grey), because of inadequate sampling, appears with lower frequency than the original wave.

and thus to comprehend and measure their structural details to correlate with observed species, behaviour patterns and environmental factors. The simplest graphical display is the oscillogram, which shows the waveform of the signal, and the envelope, which shows the amplitude of the signal against time. The transformation of signals to digital form allows a new approach in the management of the data, easing signal analysis (Oppenheim and Schafer, 1975).

Before a continuous, time-varying signal such as sound can be processed on a computer, the signal must be *acquired* or *digitised* by an A/D converter in a similar way described for the DAT recorder (Rabiner and Gold, 1975). During the process called digital sampling, the A/D converter at regular intervals samples the instantaneous voltage amplitude of an input signal at a particular sampling rate, typically thousands or tens of thousands of times per second (Figure 2.11). The digital representation of a signal created by the converter consists of a sequence of numeric values representing the amplitude of the original waveform at discrete, evenly spaced points in time. The precision with which the digitised signal represents the continuous signal depends on two parameters of the digitising process, namely the rate at which amplitude measurements are made (the sampling rate or sampling frequency), and the number of bits used to represent each amplitude measurement (the sample size).

Sampling Rate

The sampling rate at which a signal is to be digitised is chosen according to the frequencies accommodated in it, the storage capacity of the computer, the length of the signal and the precision required. The choices available are determined by the A/D converter hardware and the program (called a *device driver*) which controls the converter. Most converters have two or more sampling rates available. Commercial digital audio applications use higher sampling rates (44.1 kHz for audio compact discs, 48 kHz for digital audio tape). Once a signal is digitised, its sampling rate is fixed. In order to interpret a sequence of numbers as representing a time varying signal, one needs to know the sampling rate. When a digitised signal is saved in a file format that is designed for saving sound information, information about the sampling rate is usually saved along with the actual data points comprising the signal.

The precision of the digitised signal is a function of the sampling frequency. The sampling rate which is required to make an acceptable representation of a waveform depends on how rapidly

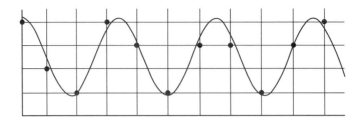

FIGURE 2.12 Error in digitising a sample with a 2 bit A/D converter. 2 bit digitisation can represent only four different amplitude levels. At each sampling timepoint (vertical lines), the amplitude levels are rounded to the nearest value that can be represented in a 2 bit converter (horizontal lines). The amplitude values stored for each sample (black dots) are slightly different from the true amplitude level of the signal at the timepoint the sampling was made.

the signal amplitude changes (*i.e.* on the signal's frequencies). For an accurate representation, the sampling rate must be more than twice the highest frequency contained in the signal. Otherwise, the digitised signal will contain frequencies which were not contained in the original signal. This appearance of phantom frequencies as an artefact of inadequate sampling rate is called aliasing (Figure 2.12). The highest frequency which can be represented in a digitised signal without aliasing is called the Nyquist frequency, which is half the frequency at which the signal was digitised. The highest frequency calculated in a spectrogram or spectrum must not exceed the Nyquist frequency of the digitised signal. If broadband noise contributes in frequencies above the Nyquist frequency in the analogue signal, the effect of aliasing is to increase the noise in the digitised signal. However, if the spectrum of the analogue signal contains any genuine peaks above the Nyquist frequency, the spectrum of the digitised signal will contain spurious peaks below the Nyquist frequency as a result of aliasing. The usual way to guard against aliasing is to pass the analogue signal through a low-pass filter (called an antialiasing filter) before digitising it to remove any contributions at frequencies greater than the Nyquist frequency. (If the original signal contains no energy at frequencies above the Nyquist frequency or if it contains only low level broadband noise, this step is unnecessary.) Due to the characteristic of the built-in antialiasing filter of a sound card, the maximum sampling frequency is usually slightly lower that one half of the sampling frequency. That maximum frequency is also the maximum frequency visible on the spectrogram (Strong and Palmer, 1975).

Sample Size (Amplitude Resolution)

The accuracy with which a sample represents the real amplitude of the signal at the instant the sample is measured, depends on the sample magnitude or the number of bits used in the binary representation of the amplitude value. Some computers provide only 8 bit sampling capability; others allow a choice between 8 bit and 16 bit. An 8 bit sample can resolve 256 (2^8) different amplitude values and a 16 bit converter 65,536 ($= 2^{16}$) values. Sound recorded on audio CDs is stored as 16 bit samples. When a sample is taken, the actual value is rounded to the nearest value that can be represented by the number of bits in a sample. Since the actual analogue value of signal amplitude at the time of a sample is usually not exactly equal to one of the discrete values that can be represented exactly by a sample, there is some inherent error in the process of digitising (Figure 2.12), which results in quantization noise in the digitised signal. The more bits used for each sample, the less noise is contained. The dynamic range of a signal in decibels is equal to $20 \log(V_{max}/V_{min})$, where V_{max} and V_{min} are the maximum and minimum voltages, respectively, in the signal. For a digitised signal, $V_{max}/V_{min} = 2n$, where n is the number of bits per sample. Since $\log(2n) = n(0.3)$, the dynamic range of a digitised signal is 6 dB per bit.

The sample size determines the maximum dynamic range of a digitised sound. Dynamic range is the ratio between the highest amplitude and the lowest nonzero amplitude in a signal, usually expressed in decibels. The dynamic range of a digitised sound is 6 dB/bit. During D/A conversion, care should be taken that the volume setting in the Recording Level panel is adjusted correctly because too high levels will cause clipping and distorted recordings. Too low levels will waste the available dynamic range of the sound card which will affect further signal analysis.

Storage Requirements

Higher sampling rates and the increased dynamic range increase the accuracy of the digitised signal representation. Larger samples come at the expense of the amount of memory or storage capacity required to store a digitised signal. The minimum amount of storage (in bytes) required for a digitised signal is the product of the sample rate (in samples/sec), the sample size (in bytes; one byte equals 8 bits), and the signal duration (seconds). Thus, a 5-sec signal sampled at 22.3k s/sec with 8 bit precision requires about 110 Kb of storage. The actual amount of storage required for a signal may exceed this minimum, depending on the format in which the samples are stored although with the application of compression algorithms (*e.g.* MP3) sound files may become smaller. No matter which file format is used, digitised sound files take up a lot of storage space. On a computer, a 30 Gb hard disk can store 45 hours of digitised signals with DAT quality (16 bit, stereo, 48 kHz sampling) sound.

SPECTRUM ANALYSIS

Spectrographic representation of biological signals has been widely used since the first analogue instruments were developed for military acoustic research (Koenig *et al.*, 1946). It shows the signal's frequency composition in various forms. The instantaneous spectrum (frequency–amplitude plane) shows frequency components of a short segment of a signal, while the representation of more spectra, computed on consecutive or overlapping segments of the signal, shows the evolution of its frequency structure in time. This is achieved by showing the consecutive spectra on the frequency–time plane, with the frequency component intensity coded through a scale of greys or a suitable colour scale. This analysis is called a spectrogram and it is based in principle on the windowed Fast Fourier Transform (FFT) technique. Using graphic representations, signals can be easily compared to detect similarities or differences and to characterise signals according to morphology.

Several approaches are available for digital spectrum analysis. The approach mostly used for making and interpreting spectrograms is based on STFT.

The aim of this section is to explain some of the limitations and tradeoffs intrinsic to spectrum analysis of time varying signals. More extensive mathematical treatments of continuous and discrete spectral analysis, at several levels of sophistication are available (Oppenheim and Schafer, 1975; Rabiner and Gold, 1975).

Time-Domain and Frequency-Domain Representations of Sound

Any acoustic signal can be graphically or mathematically depicted in either of two forms, called the *time-domain* and *frequency-domain* representations (Cohen, 1995). In the time domain, the amplitude of a signal is represented as a function of time (Figure 2.13a).

In the frequency domain, the amplitude of a signal is represented as a function of frequency. The frequency-domain representation of a pure tone is a vertical line (Figure 2.13b). Any sound, no matter how complex, can be represented as a sum of pure tones (sinusoidal components) (Figure 2.14a). Each tone in the series has a particular amplitude, relative to the others, and a particular phase relationship (*i.e.* it may be shifted in time relative to the other components). The frequency composition of complex signals is usually not apparent from inspection of the time-domain representation. Spectrum analysis is the process of converting the time-domain

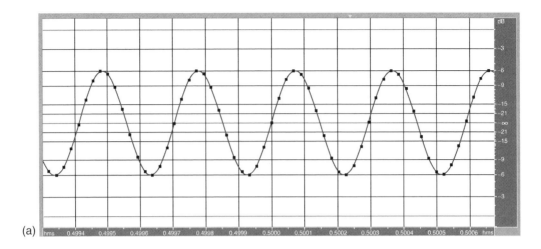

(a)

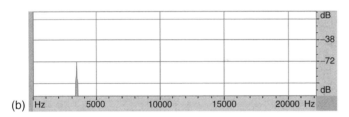

(b)

FIGURE 2.13 Time-domain and frequency-domain representations of an infinitely long pure sinusoidal signal with a frequency of 3400 Hz. (a) Time domain. (b) Frequency domain. (Diagrams made using the sound analysis program Adobe Audition 1.5.)

representation of a signal (which is the representation directly produced by most measuring and recording devices) to a frequency-domain representation that shows how different frequency components contribute to the sound. The magnitude spectrum (Figure 2.14b) contains information about the magnitude of each frequency component in the entire signal.

Fourier transform is a mathematical function which converts the time-domain form of a signal to a frequency-domain representation or spectrum (Figure 2.15). When the signal is represented as a sequence of discrete digital samples, the Fourier transform used is called the discrete Fourier transform (DFT). The input to the DFT is a finite sequence of values, the amplitude values of the signal, which are sampled (digitised) at regular intervals (Jaffe, 1987). The output is a sequence of values specifying the amplitudes of a sequence of discrete frequency components, evenly spaced from zero Hz to half of the sampling frequency (Figure 2.16). The output can be plotted as a magnitude spectrum. In practice, a spectrum is always made over some finite time interval. This interval may encompass the full length of a signal, or it may consist of some shorter part of a signal. The DFT can be implemented on a computer using an algorithm known as the FFT.

Spectral Analysis of Time-Varying Signals: Spectrograms and STFT Analysis

An individual spectrum provides no information about temporal changes in frequency composition during the interval over which the spectrum is made (*e.g.* the sound phrase). To see how the frequency composition of a signal changes over time, we can examine a sound *spectrogram*. A spectrogram is a two-dimensional grid of discrete data points on a plane in which the axes are time (duration of the signal) and frequency (frequency range used in signal recording) (Hopp *et al.*, 1998) (Figure 2.17).

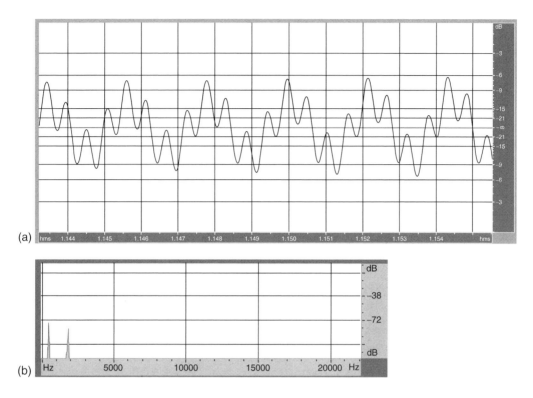

FIGURE 2.14 Time-domain and frequency-domain representations of an infinitely long sound consisting of two tones, with frequencies of 340 and 1800 Hz. (a) Time domain. (b) Frequency domain. (Diagrams drawn using the sound analysis program Adobe Audition 1.5.)

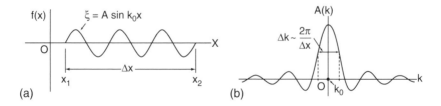

FIGURE 2.15 (a) Representation of a (pure tone) sine wave with frequency k_0. (b) Its Fourier transform around the frequency k_0.

Spectrograms are produced by a procedure known as the STFT (Hlawatsch and Boudreaux-Bartels, 1992). During an STFT operation the entire signal is divided into successive short time intervals or *frames* (which may overlap each other in time). Each frame is then used as the input to a DFT, generating a series of spectra (one for each frame) which approximate the "instantaneous" spectrum of the signal at successive moments in time. To display a spectrogram, the spectra of successive frames are plotted side by side with the frequency range on the abscissa and the magnitude of the frequency component (frequency contribution amplitude) represented by greyscale or colour values.

A given STFT can be characterised by its frame length, usually expressed as the number of digitised amplitude samples that are processed to create each individual spectrum. The frame length of a STFT determines the time analysis resolution (t) of the spectrogram (Figure 2.18). Changes in the signal that occur within one frame length (*e.g.* the end of one sound and the beginning of

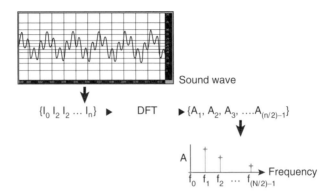

FIGURE 2.16 Schematic representation of the DFT as a black box. The input to the DFT is a sequence of digitised amplitude values $(I_0, I_1, I_2, \ldots, I_{N-1})$ at N discrete points in time. The output is a sequence of amplitude values $(A_0, A_1, A_2, \ldots, A_{(N/2)-1})$ at $N/2$ discrete frequencies. The highest frequency, $f_{(N/2)-1}$, is equal to half the sampling rate $(= 1/(2T)$ where T is the sampling period, as shown in the figure).

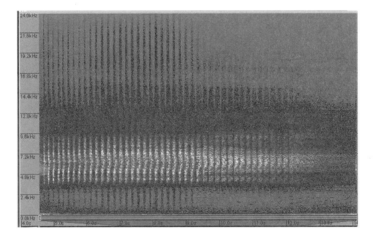

FIGURE 2.17 Sound spectrogram of one phrase from song of a *Lyristes cicada* species, digitised at 48 ks/sec. The greyscale value in each box represents an estimate of the energy amplitude at the time-frequency gridpoint that is at the upper left corner of the box. (Sonograms calculated and drawn using the sound analysis program SONOGRAM.)

another, or changes in frequency) cannot be resolved as separate events. Thus, shorter frame lengths allow better time analysis resolution.

Alternatively, the entire signal can be divided into short frequency ranges called bandwidths, and processed numerically through filters that contain a range of input frequencies around a central analysis frequency. The time varying output amplitudes of each of the bandpass filters, that are centred at a slightly different analysis frequency, are plotted above each other at successive analysis frequencies. All of the filters of a single STFT have the same bandwidth, irrespective of the analysis frequency. The bandwidth determines the frequency analysis resolution (f) of the spectrogram. Frequency components, which differ by less than one filter bandwidth, cannot be distinguished from each other in the output of the filterbank. Thus a STFT with a relatively wide filter bandwidth will have poorer frequency analysis resolution than one with a narrower width.

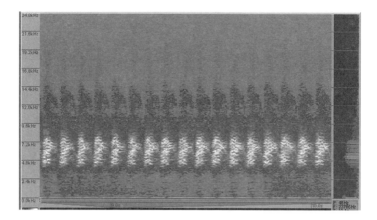

FIGURE 2.18 High resolution spectrogram of part of a phrase of the cicada *Lyristes*, digitised at 48 ks/sec. Frequency resolution 93 Hz, frame length = 512 points (= 79 mS). Grid resolution = 79 mS × 93 Hz. (Sonograms calculated and drawn using the sound analysis program SONOGRAM.)

Filter bandwidths are often measured as the width of the band between the frequencies where the amplitude of the filter's output is 3 dB below the peak output frequency.

Analysis resolution for time and frequency are determined respectively by the frame length and filter bandwidth of a STFT. The frame length and filter bandwidth of a STFT are inversely proportional to each other, and cannot be varied independently. Although a short frame length yields a spectrogram with finer time analysis resolution, it also results in wide bandwidth filters and correspondingly poor frequency analysis resolution. A trade-off exists between how precisely a spectrogram can specify the spectral (frequency) composition of a signal and how precisely it can specify the time at which the signal exhibited that particular spectrum (Beecher, 1988). The choice of analysis resolution depends on how rapidly the signal's frequency spectrum changes and, given the particular application, on what type of information is most important to show in the spectrogram. For many applications an intermediate frame length and filter bandwidth between 256 and 512 points is adequate. In order to observe very short events or rapid changes in the signal, a shorter frame may be better. If precise frequency representation is important, a longer frame may be better.

Since spectrograms are calculated from STFTs using frame length and filter bandwidth boxes, an uncertainty is introduced by the calculation (Beecher, 1988). If the features investigated are distinguishable in the waveform, the best precision and accuracy is achieved by making time measurements on the waveform rather than the spectrogram (Figure 2.19).

The spacing between gridpoints in the horizontal and vertical directions of a spectrogram, and thus the width and height of the boxes representing each grid point in the spectrogram are called, respectively, the time grid resolution and frequency grid resolution of the spectrogram. Grid resolution should not be confused with analysis resolution.

Time grid resolution is the time interval between the beginnings of successive frames. In a spectrogram, this interval is visible as the width of individual consecutive gridpoints. Successive frames which are analysed may be overlapping (positive overlap), be contiguous (zero overlap) or discontinuous (negative overlap). Overlap between frames is usually expressed as a percentage of the frame length. In a low resolution spectrogram each box is as wide as a frame, which in turn is about the same size as each pulse in the signal. The result is a spectrogram that gives an extremely misleading picture of the signal. A spectrogram with a greater frame overlap is much "smoother" than the one with less overlap, and it reveals the frequency modulation of each pulse in the signal. It still provides poor time analysis resolution, however, because of its large frame length.

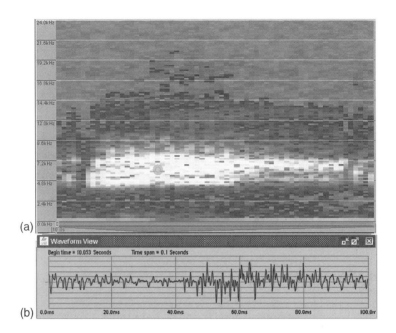

FIGURE 2.19 (a) A STFT spectrogram of signal over a timespan of 1 sec. (Frequency resolution 187 Hz, time resolution 160 msec, frame length = 256 points, FFT size = 1024 points.) (b) The waveform of a time span of 0.1 sec. (Sonograms calculated and drawn using the sound analysis program SONOGRAM.)

Frequency grid resolution is the difference (in Hz) between the central analysis frequencies of adjacent filters in the filterbank modelled by a STFT, and thus the size of the frequency bins. In a boxy spectrogram, this spacing is visible as the height of the individual boxes. Frequency grid resolution depends on the sampling rate (which is fixed for a given digitised signal) and a parameter of the FFT algorithm called FFT size. The relationship can be defined as frequency grid resolution = (sampling frequency)/FFT size where frequency grid resolution and sampling frequency are measured in Hz, and FFT size is measured in points. (A point is a single digital sample.) A larger FFT size draws the spectrogram on a grid with finer frequency. For this reason the frequency resolution of a spectrogram with low frequency sounds is poor when the sampling frequency is too high. The number of frequency bins in a spectrogram or spectrum is half of the FFT size.

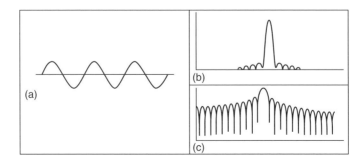

FIGURE 2.20 A single frame of a sinusoidal signal (a) has a spectrum of frequency amplitudes around the central frequency, flanked by frequency ripples (b). The ripples appear as sidelobes in the logarithmic spectrum (c).

Spectral Imperfections and Filtering Functions

The spectrogram produced by an STFT has several imperfections due to the approximations applied by the FFT method (Figure 2.15) as well as the inability of the different mathematical filters to discriminate different frequencies in the filterband. Apart from the smearing produced due to the overlapping frames there is an infinite series of diminishing ripples at the edges of the frame (Figure 2.20) that in a logarithmic spectrum are shown as sidelobes. The magnitude of the sidelobes (relative to the magnitude of the central lobe) in a spectrogram or spectrum of a pure tone is related to how abruptly the signal's amplitude changes at the beginning and end of a frame (Figure 2.21). These can be reduced by multiplying the frame by a window function that selectively suppresses the waveform.

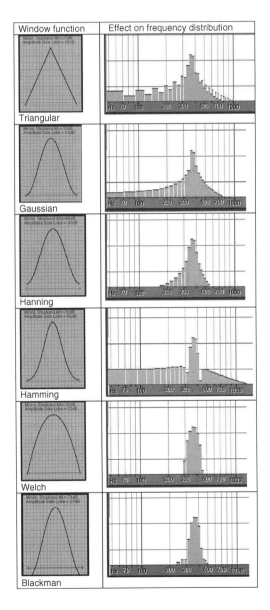

FIGURE 2.21 Single frame spectra of a 400 Hz tone made with six different window functions.

Each window function reduces the height of the highest sidelobe to some particular proportion of the height of the central peak; this reduction in sidelobe magnitude is termed the sidelobe rejection, and is expressed in decibels. The width of the centre lobe in the spectrum of a pure tone is the filter bandwidth. The application of several common window functions used on a sinusoidal pure tone of frequency 400 Hz to reduce sidelobe rejection is shown on Figure 2.21.

The window type also influences the bandwidth. In general, narrow bandwidths should be chosen if the signal to be analysed does not have quick frequency modulations and if there is no important information in the time domain. In contrast, wide bandwidths should be chosen if there is any remarkable frequency modulation or if there are important temporal patterns. Although the smallest bandwidth is realised with the rectangular window, due to spurious edge effects, it is not used for standard spectral analysis.

CONCLUSION: PRESENT AND FUTURE

Technical aspects of the study of insect acoustic and vibratory signals has progressed considerably together with the advances in microelectronics (Stoddard, 1990). Signals nowadays can be recorded, processed, stored, analysed and displayed in real time with considerable accuracy, minimal distortion and with higher limits in recording time. The reduction of electronic circuitry and minimal power consumption have made an impact on portability as well as durability in the field. More sensitive input devices offer the possibility of recording weak signals, very high pitch signals or signals at a distance with improved S/N ratio.

Signal analysis methodology has also advanced. From the days of the analogue oscilloscope we have moved to the digital computer analysis systems where even spectrograms can be real time with instantaneous FFTs. Sonograms, spectrograms or power spectrums provide complementary information on animal communication and behaviour. Direct comparisons of signals is no longer cumbersome since they are performed on the computer display with the use of signal editors. (In the Appendix a list of some suitable sound analysis programs is given.)

The bioacoustic researcher today has a very wide choice of equipment and methodology for either acoustic or vibratory signal detection. As digital equipment is being reduced in size and price, performance is enhanced beyond expectation. The future might well include the integration of moving images in order to provide a complete overview of behaviour. (In this book several pioneering sound movies of insect behaviour are included in digital form indicating future developments.)

APPENDIX: ANIMAL-SPECIFIC SOFTWARE FOR SOUND ANALYSIS

A number of companies and individuals offer software for signal analysis specifically designed for animal research. Most of these are developed in an academic context. Some are commercially available, some are research-developed and available for free. The main consideration for these systems is that the designers were operating within a framework of animal-related research.

The software usually includes analysis, recording and display tools, with real-time spectrograms and cepstrograms, spectral averaging, frequency tracking, event counting, scheduled recording, *etc.* Other features include wide control on all analysis parameters, frequency–time cursor while in real-time mode, FFT size adjustment, frequency zoom capabilities with real-time spanning and frequency tracking. Depending on the installed sound acquisition devices, analogue I/O is allowed in the audio frequency range and on some in the ultrasonic range up to 48 and 96 kHz (96 and 192 kHz sampling, respectively).

A nonexhaustive list of programs is given alphabetically followed by their URLs:

Avisoft-SASLAb. A sophisticated software package designed for bird and other animal research (www.avisoft-saslab.com).

Dadisp. A "signal management" program with a lot of options: filtering, editing, analyses. There is a "student" version available for download (www.dadisp.com).

FFT Properties. A PC based Scope/Spectrum spectral analysis and FIR filter design for real time processing. A program for studying and understanding Fourier transforms (www.teemach.com/FFTProp).

Gram 6.0 by R.S. Horne. A shareware ($25) program that does simple but effective spectrograms, available as download. New version does real-time acquisition (www.visualizationsoftware.com/gram.html).

Ishmael. Sound acquisition program with automatic call (signal) recognition, file annotation, acoustic localisation (cetus.pmen.noaa.gov/cgi-bin/MobySoft.pl).

Micro LAB. A bundled software package with real time stereo spectrum analyser, oscilloscope and signal generator for the PC (www.psdigital.com.br.MicroLABUS.htm).

Raven. The new software package for sound analysis from the Cornell ornithology group (www.birds.cornell.edu/Raven).

Scope DSP and Scope FIR: two programs for F/T analysis and filtering (FIR) (www.iowegian.com).

SEA. Sound emission analyser for real time sound analysis capabilities on a Windows PC. Mainly developed for bioacoustic studies, this software can be used for a wide range of applications requiring real time display of sounds and vibrations (www.nauta-rcs.it).

Signal and RTS from engineering design: a thorough animal oriented system (www.engdes.com)

Sigview. Real time spectral analysis software with various analysis and imaging tools (www.sigview.com)

Sonogram. A Java based excellent program by C. Lauer to create LPC's Cepstrums Wavelets and Sonograms for every media file supported on Java Media Framework. It uses short time Fourier transform (STFT) to generate sonograms/spectrograms. It is freely available for download (www.dfki.de/~clauer).

Sound Ruler. A free analysis and graphics package designed for animal sound analysis. Looks promising (www.soundruler.sourceforge.net).

Spectra-Plus. A flexible system that does a lot of signal analysis (www.telebyte.com/pioneer).

Syrinx. Program designed for real time and interactive playbacks, as well as signal imaging and editing (www.syrinxpc.com).

wSpecGram. Runs in win95/98/NT/2000 and uses any installed Windows compliant sound device, including digital I/O boards including 96 and 192 kHz sound boards (www.unipv.it/cibra/softw.html).

3 The Auditory–Vibratory Sensory System in Bushcrickets

Wolfgang Rössler, Martin Jatho and Klaus Kalmring

CONTENTS

COMPARISON OF MORPHOLOGY, PHYSIOLOGY AND DEVELOPMENT OF THE RECEPTOR ORGANS

INTRODUCTION

In bushcrickets, auditory–vibratory communication plays an important role in reproductive behaviour, agonistic interactions, detection of predators and for general acoustic orientation in the environment. The most important and physiologically dominant receptor organs of the bimodal auditory–vibratory sensory system are the complex tibial organs of all six legs. Complex tibial organs are present in the pro metathoracic, mesometathoracic, and metathoracic legs. In each leg, the complex tibial organs consist of three scolopale organs, *e.g.* the subgenual organ (SO), the

35

intermediate organ (IO), and the crista acustica (CA). Only in the forelegs the tibial organs are specialised as tympanal organs, where the crista acustica and the distal parts of the intermediate organ serve as auditory receptors (Schumacher, 1979). In tettigoniids, the prothoracic spiracles are the main input for airborne sound. The acoustic trachea transmits sound to the tympanal organs in the proximal tibiae of the forelegs; the vibrations of the tympana are caused by sound acting on the inner surface of the tympanum (Lewis, 1974; Michelsen and Larsen, 1978; Heinrich et al., 1993).

The presence of the sound transmitting system and the anatomical differentiation within the auditory receptor organs in the forelegs allow sensitivity to airborne sound. However, the absolute sensitivity and the frequency tuning of the auditory threshold can vary significantly between species.

In spite of the absence of tympana and spiracle, the receptor complexes of the subgenual organ, the intermediate organ and the CA are also fully developed in the mid- and hind legs. However, the CA in the mid- and hind legs consists of a smaller number of receptor cells than in the forelegs. Some structural differences at the receptor-cell level are present (Lin et al., 1994). Schumacher (1979) referred to the organs in the mid- and hind legs as atympanate organs. Some success has been achieved in elucidating the function of the tibial organs of the forelegs (Autrum, 1941; Rheinländer, 1975; Kalmring et al., 1978; Zhantiev and Korsunovskaya, 1978; Oldfield, 1982; Lin et al., 1993). During embryonic development, the tibial organs differentiate in all legs from an invagination of ectodermal cells of the tibia (Meier and Reichert, 1990). The authors postulate that the auditory organs in Ensifera evolved from a serially reiterated group of leg-associated mechano-receptors, which in the prothoracic leg became specialised for the perception of airborne sound.

Receptor cells of the complex tibial organs in the meso- and metathoracic legs also respond to low frequency airborne sound at high intensities (Autrum, 1941). However, in contrast to the prothoracic organs, the sensitivity in the mesothoracic and metathoracic tibiae is not sufficient for the perception of the natural song. The real function of the CA in the mid- and hind legs is still unknown.

Sound can reach the receptor cells of the forelegs in two different ways, via the tympana and via the acoustic tracheal system which consists of the large spiracle, the vesicle and the long tapering trachea. There are many published controversial opinions concerning the function and importance of this stimulus transducing system. It is not that the spiracle is the main site of sound entry (Lewis, 1974; Nocke, 1975; Seymour et al., 1978; Hill and Oldfield, 1981; Larsen, 1981; Oldfield, 1984, 1985; Heinrich et al., 1993; Kalmring et al., 1993; Lin et al., 1993; Shen, 1993), but there is considerable disagreement upon the transmission properties of the acoustic trachea and its contribution to the frequency selectivity of the auditory receptors.

Some detailed physiological investigations exist on the function of the auditory receptor cells of the CA and the IO of the foreleg (Rheinländer, 1975; Kalmring et al., 1978; Zhantiev and Korsunovskaya, 1978; Römer, 1983; Oldfield, 1985). Individual receptor cells differ in their tuning curves, but the whole receptor population of one tympanal organ covers a frequency range from at least 2 kHz up to 70 kHz. There is much overlapping of the frequency–intensity response characteristics of these cells. Attempts to correlate function and structure of the receptor cells of the CA by recording directly from the cells within the organs (Zhantiev and Korsunovskaya, 1978; Oldfield, 1982, 1984) have failed to give a satisfactory explanation of the mechanical stimulus transformation process within the tympanal organ. It is difficult to investigate the structures of the receptor organs which are involved in stimulus transduction, biophysically and physiologically. These structures are located inside the leg and can only be exposed for experimental means by dissecting the cuticle. Such procedures are expected to alter the biophysical and physiological properties of the organ.

The critical structures for mechanical stimulus-transforming processes inside the organs are probably:

(a) The dorsal wall of the acoustic trachea on which the receptor and satellite cells are located
(b) The tectorial membrane covering the receptor and satellite cells situated in the haemolymph channel
(c) Possibly the dividing wall separating the acoustic trachea inside the receptor organ into two chambers

Mechanical displacement of these structures induced by sound stimuli via the motion of the tympana (Bangert *et al.*, 1998) could directly or indirectly evoke the specific activity of the different auditory receptor cells.

In tettigoniids the neuropiles of the anterior Ring Tract (aRT) (Tyrer and Gregory, 1982) are the projection areas of the auditory receptor cells of the CA and the IO in the forelegs (Oldfield, 1985; Römer *et al.*, 1988; Ahi *et al.*, 1993). The projection of the auditory and auditory–vibratory receptors is restricted to the ipsilateral hemiganglion. Receptor cells of the subgenual organ as well as those of other vibro-receptors of the forelegs (*e.g.* campaniform sensilla, femoral and tibial chordotonal organs) project to lateral parts of the related hemiganglion (Mücke, 1991). The same receptor organs as in the forelegs are found also in the mid- and hind legs. The receptor outline of the CA and the IO is similarly constructed, with the decisive difference of lacking a sound transmitting system (spiracle, acoustic trachea and tympana are not developed in the mid- and hind legs) (Schumacher, 1973, 1975). The function and projection of the receptors of the tibial organs there are not yet known. The same is true for the synaptic connection of their receptor input onto auditory–vibratory central neurons.

The projection of some auditory receptors of the forelegs in tettigoniids has been already described (Oldfield, 1983; Römer, 1983, 1985; Römer *et al.*, 1988); this includes the projection pattern and targets of identified receptor cells within a small part of the aRT (up to 50 μm parasagittal to the midline) and a counter clockwise tonotopic organisation across the aRT of the prothoracic ganglion. A first comparative quantitative analysis of the central projections of auditory receptor cells was carried out in three closely related species by Ahi *et al.* (1993).

Since the investigations by Čokl *et al.* (1977) and Silver *et al.* (1980) it is well known that the auditory system of the ventral nerve cord in locusts and bushcrickets is a bimodal sensory system. Most of the auditory neurons in the ventral nerve cord, which ascend to the supraesophageal ganglion, receive inputs from both the auditory receptors and the different vibration receptors of all six legs (intermediate organs, subgenual organs, chordotonal organs of the femur and tibia and campaniform sensilla). Therefore, it is better referred to as a combined auditory–vibratory system.

The function, and in part the morphology, of the bimodal auditory–vibratory ventral cord neurons ascending to the head ganglia have been described (Čokl *et al.*, 1977; Silver *et al.*, 1980; Kalmring and Kühne, 1980; Boyan, 1984). The same is true for the auditory interneurons of the thoracic ventral nerve cord intercalating the auditory receptor cells with the bimodal central neurons (Römer, 1983; Römer *et al.*, 1988; Lakes *et al.*, 1990).

For bushcrickets, so far nothing is known about the neuronal connections between the vibratory receptor cells of all six legs and the auditory–vibratory central neurons ascending to the brain. There is the possibility that information from the receptors of one leg could be collected by one or several interneurons, which transmit this information to ascending neurons of the same segment or of different thoracic segments. It is still unclear if the ascending bimodal ventral cord neurons possess only one or more than one dendritic region for the vibratory and auditory input. In the latter case, the dendritic regions may be distributed on several segments, modulating the response at different positions along the neurons.

THE RECEPTOR ORGANS

Location and Morphology

The tibial organs are the main receptor organs for the detection of acoustic and vibratory signals used in the intraspecific communication of bushcrickets. In each leg the tibial organs are composed of three scolopale organs: the subgenual organ (SO), the IO and the CA, which all lie inside the proximal tibia in the haemolymph channel in close connection with the dorsal wall of the large leg trachea or one of its branches (Figure 3.1).

Similar systems are found in several other tettigoniid species (Figure 3.2) (Rössler, 1992a, 1992b; Lin *et al.*, 1993, 1994). The overall arrangement of the scolopidia in the SO, the IO and the CA show similarities in the three legs, but the differences in the tracheal morphology and size of the structures are evident.

The number of scolopidia in the SO varies between 20 and 25 in each leg. In the intermediate organ, and especially in the CA, the number of scolopidia decreases from the prothoracic to the meso- and metathoracic legs.

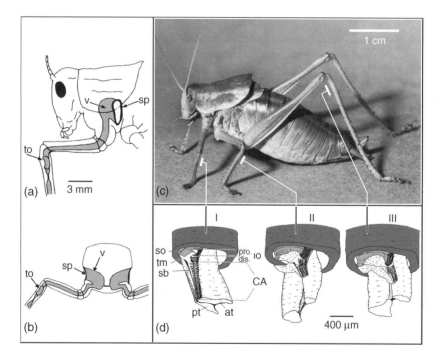

FIGURE 3.1 *Polysarcus denticauda.* (a) Semischematic lateral view of the acoustic trachea of the foreleg. It starts with a large oval opening in the pleurite of the prothoracic segment, the spiracle (sp), which is connected to a funnel-shaped air sac, the vesicle (v), inside the prothorax. From there the trachea continues through the proximal parts of the leg as a gradually narrowing conical tube until it reaches the tibial organ (to). (b) Semischematic view of the acoustic trachea in the prothorax and the forelegs in a frontal view. (c) An adult male. (d) Three-dimensional reconstruction of the complex tibial organs derived from a series of transverse sections. Anterior-dorsal view of these organs in the fore-, mid- and hind legs (I, II, III), respectively, after removing the cuticle. The position of the complex tibial organs in all three legs is indicated (*c.f.* C). at/pt, anterior/posterior branch of the acoustic trachea; pro/dis, proximal/distal part of the IO; SO, subgenual organ; IO, intermediate organ; CA, crista acustica; tm, tectorial membrane; sb, supporting band. (After Sickmann, T., Kalmring, K., and Müller A., *Hear. Res.*, 104, 155–166, 1997. With permission.)

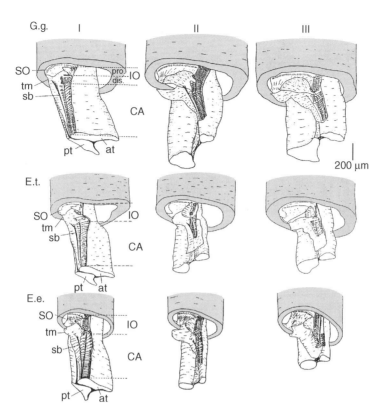

FIGURE 3.2 Adult structure of the complex tibial organs in the pro metathoracic (I), meso metathoracic (II), and metathoracic legs (III) of the bushcricket *Gampsocleis gratiosa* (G.g.), *Ephippigerida taeniata* (E.t.), and *Ephippiger ephippiger* (E.e.). Three-dimensional reconstructions from a series of transverse sections showing the organs in an anterior-dorsal view; the integument and the content of the nerve–muscle channel is removed. Abbreviations as in Figure 3.1. (After Rössler *et al.*, 1992b; Lin *et al.*, 1993. With permission.)

Another striking difference is the morphology of the trachea. The two large branches of the acoustic trachea in the foreleg are only separated by a thin wall (named the central membrane after Lewis (1974), the dividing wall after Nocke (1975) or the partition after Schumacher (1975)). The smaller tracheal branches in the mid- and hind legs run with greater separation and are connected to each other by the tracheal epithelium. In the foreleg, the volume of the anterior branch is somewhat larger than the posterior one; in the mid- and hind legs, the conditions are reversed. The tectorial membrane covering the CA and the IO is developed in each leg. Supporting bands on both sides of the cap cells are very marked in the foreleg CA only poorly developed in the midleg tibia and completely missing in the hind leg CA.

The detailed morphology of the complex tibial organ in the forelegs of bushcrickets is shown (Figure 3.3) Additionally, a composed transverse section from the middle part of the CA in bushcrickets with (as typical for Decticinae, Tettigoniinae, Ephippigerinae and Conocephalinae) and without tympanal covers (as typical for Phaneropterinae and Meconematinae) is inserted.

Figure 3.4 shows transverse sections of the proximal tibia of the three legs at the plane of the SO, the IO and the CA in *Ephippiger ephippiger*.

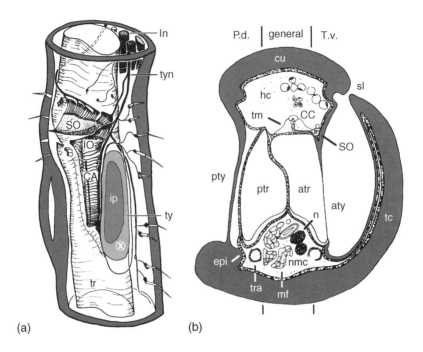

FIGURE 3.3 Morphology of the tibial organ in the forelegs of bushcrickets. (a) Semischematic drawing of the complex tibial organ. For better insight the cuticle is removed at the anterior-dorsal side of the proximal tibia. tr, trachea; ip, inner plate of the tympanum; SO, subgenual organ; IO, intermediate organ; CA, crista acustica; tyn, tympanal nerve; ln, leg nerve; ty, tympanum. (b) A composite transverse section of the tibia at the plane of the middle CA; (left) *P. denticauda* with exposed tympana and (right) the tympanum protected by covers with slits, as *T. viridissima*. The morphology of the middle part ("general") is almost the same in all bushcrickets. Note, the tympana extend from the middle of the haemolymph channel down to the ventral end of the tracheal chambers. atr, anterior tracheal branch; aty, anterior tympanum; cc, cap cells; cu, cuticle; epi, epidermis; hc, haemolymph channel; mf, muscle fibre; n, nerve; nmc, nerve–muscle channel; ptr, posterior tracheal branch; pty, posterior tympanum; sl, slits; so, soma of a receptor cell; tc, tympanal cover; tm, tectorial membrane, tra, trachea in the nerve muscle channel. (After Bangert, M., Kalmring, K., Sickmann, T., Stephen, R., Jatho, M., and Lakes, R., *Hear. Res.*, 115, 27–38, 1998. With permission.)

The size and shape of the SO is nearly the same in the fore- and midlegs; the SO of the hind leg is slightly larger especially in the anterioposterior direction. The arrangement of dendrites in the proximal group of scolopidia in the IO is comparable in all legs. At the plane of the CA, the absence of tympana, tympanal cavities and covers, and slits in the mid- and hind legs are obvious. In addition, the arrangement and location of scolopidia within the CA is different in the three legs because of the different size and course of the tracheal branches. In each case, the scolopidia are located on the anterior branch of the trachea. The acoustic trachea in the foreleg divides into two branches at the plane of the proximal part (in other species not earlier than at the middle) of the CA. The two branches are separated by the thin dividing wall. In the midleg, the trachea is divided at the beginning of the proximal part of the IO. In the hind leg, the trachea is divided into a small anterior and a larger posterior branch before the SO. At the plane of the IO and CA, the size of the haemolymph channel differs in the three legs. The volume of the haemolymph channel is significantly smaller in the foreleg than in the mid- and hind leg because of the large volume of the acoustic trachea.

Despite of these similarities in structure, it is only in the foreleg that the CA and the IO serve as sensitive detectors of airborne sound (Kalmring *et al.*, 1994). This is because these are associated

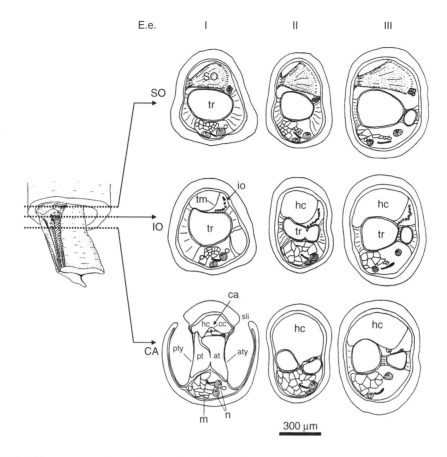

FIGURE 3.4 Transverse sections of the proximal tibia in the pro metathoracic (I), meso metathoracic (II) and metathoracic legs (III) of *Ephippiger ephippiger* at the plane of the SO, proximal part of the IO and CA. at, anterior tracheal branch; cc, cap cell of the scolopidium; hc, haemolymph channel; m, muscles; n, tibial and tarsal nerve; pt, posterior tracheal branch; pty, posterior tympanum; sli, slit; tm, tectorial membrane. Anterior to the right. (Modified from Rössler, W., *Zoomorphology*, 112, 181–188, 1992b. With permission.)

with structures suitable for conducting, filtering and amplifying airborne sound, acoustic spiracles, an acoustic trachea and tympana have developed exclusively in the foreleg (Figure 3.1a, b and Figure 3.3).

The tuning of the individual receptor cells to particular frequencies is probably due to the acousto-mechanical properties of the receptors in combination with their accessory cells, the dorsal wall of the trachea and the tectorial membrane. For instance, the bushcricket species *Decticus albifrons*, *D. verrucivorus* and *Pholidoptera griseoaptera* belong to the same subfamily Decticinae but differ significantly in size (Figure 3.5). In spite of the great differences in the dimensions of the forelegs, the most sensitive range of hearing lies from 6 to 25 kHz in each species. Only in the frequency range from 2 to 5 kHz and >25 kHz are significant differences present. Figure 3.5 shows that, despite substantial differences in the overall dimensions of the leg, the size of the dorsal wall of the anterior trachea is very similar.

The anatomy of the auditory receptor organs was quantitatively investigated using techniques of semithin sectioning and computer guided morphometry (Rössler and Kalmring, 1994). The overall number of scolopidia and the length of the CA differ in the three species, but the relative

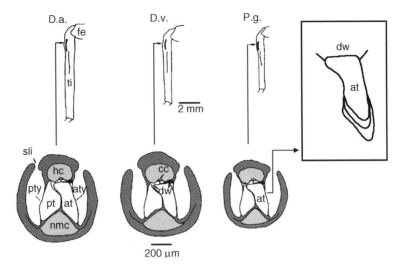

FIGURE 3.5 Morphology of the proximal tibiae of the forelegs of three bushcricket species, *Decticus albifrons* (D.a.), *Decticus verrucivorus* (D.v.) and *Pholidoptera griseoaptera* (P.g.). Upper, left: scale-drawings of the tibiae. The arrows indicate the position of the transverse sections shown below. Lower: scale-drawings from transverse sections at the plane of the scolopidium in the middle of the cristae acusticae. Upper, right: the anterior tracheal branches are superimposed. Anterior is to the right. at, anterior tracheal branch; aty, anterior tympanum; cc, cap cell of scolopidium; dw, dorsal wall of anterior trachea; fe, femur; hc, haemolymph channel; nmc, nerve muscle channel; pt, posterior tracheal branch; pty, posterior tympanum; sli, slit; ti, tibia. (After Rössler, W. and Kalmring, K., *Hear. Res.*, 80, 191–196, 1994. With permission.)

distribution of scolopidia along the CA is very similar. Additionally, the scolopidia and their attachment structures (tectorial membrane, dorsal tracheal wall and cap cells) are of equal sizes at equivalent relative positions along the CA. The results indicate that the constant relations and dimensions of corresponding structures within the CA of the three species are most likely responsible for the similarities in the tuning of the auditory thresholds (Figure 3.6).

Postembryonic Development

The postembryonic (larval) development of the anatomy, morphology and physiology of the complex tibial receptor organs was investigated in all three pairs of legs in *E. ephippiger* (Rössler, 1992a). All the receptor cells in the three parts of the complex tibial organ (the SO, IO, and the crista acustica) are present from the first larval instar (Figure 3.7). The sound transmitting structures of the foreleg tympanal organ, the acoustic trachea, the tympana, the tympanal covers and the acoustic spiracle develop step by step in subsequent instars. The acoustic trachea inside the tibial organ of the foreleg resembles the adult structure for the first time in the fourth instar, although its volume is still small.

In *E. ephippiger*, the auditory threshold curves recorded from the tympanal nerve of the foreleg in instars four, five and six exhibit the same frequency maxima as those in the adult. Similarly, in *Tettigonia cantans* the overall sensitivity increases step by step (by about 10 to 20 dB) with each moult (Figure 3.8b) (Unrast, 1996). In contrast to the responses to airborne sound, sensitivity to vibratory stimuli was already high at larval instar three and did not increase in subsequent instars, including the adult. Both neurophysiological findings were confirmed in behaviour experiments.

The larval development of structures within the CA which are probably involved in stimulus transduction and in frequency tuning have been analysed. The dorsal wall of the anterior tracheal

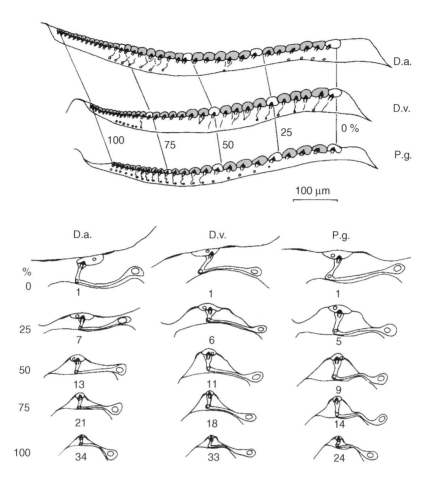

FIGURE 3.6 Dimensions of the scolopidia and their attachment structures at similar relative positions within the CA of bushcricket species: *Decticus albifrons* (D.a.), *D. verrucivorus* (D.v.) and *Pholidoptera griseoaptera* (P.g.). Upper: scale drawings of vertical longitudinal sections of the CA of each species (proximal is to the right). Equivalent positions are connected by lines. Lower: transverse sections of the scolopidia together with the tectorial membrane and the dorsal tracheal wall at positions of 0, 25, 50, and 100% of the total length of the CA. The numbers of scolopidia (counted from proximally) at similar relative positions are indicated below each drawing. Anterior is to the right. (Scale bar same for all.) (After Rössler, W. and Kalmring, K., *Hear. Res.*, 80, 191–196, 1994. With permission.)

branch, the tectorial membrane and the cap cells have similar dimensions, especially in the last three instars and in adults. During development, the main changes in the region of the CA concern the tracheal morphology, the tympana with tympanal covers and the position of the scolopidia. In Figure 3.8a (left), the differentiation of the third scolopidium (counted from proximal) within the CA is clear. In the first instar, the dendrites and scolopales of the receptor cells within the CA are still oriented horizontally like those in the proximal IO of adults or in the proximal tympanal organ of adult gryllids. The tectorial membrane, which covers the CA and the IO, is already differentiated in the first instar. In the second instar, the dendrites become bent upwards towards the haemolymph channel, and supporting bands on the anterior and posterior side of the cap cells appear for the first time. In subsequent instars, the dendrites, cap cells and scolopale caps and rods enlarge. The final length of the dendrite is attained in the fifth instar and the scolopale caps and rods are significantly enlarged after the second instar.

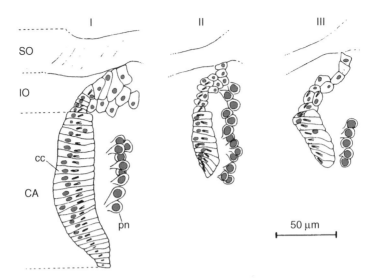

FIGURE 3.7 Development of the tibial receptor organs in the pro metathoracic (I), meso metathoracic (II) and metathoracic leg (III) of *Ephippiger ephippiger*. Arrangement of scolopidia in the first instar larva. In the SO, the IO and the CA of the three legs the total number of scolopidia found in adults is present at the time of hatching. pn, perikaryon of the sensory neurons; cc, cap cell. (After Rössler, W., *Cell Tiss. Res.*, 269, 505–514, 1992a. With permission.)

Stimulus Transduction in the Receptor Organs

The morphology and acoustic characteristics of the acoustic tracheal system were examined in several tettigoniid species (Heinrich *et al.*, 1993; Hoffmann and Jatho, 1995). Measurements and statistical analyses reveal that, in all bushcricket species investigated so far, the shape of the acoustic trachea can be approximated by the equation of an exponential horn (Figure 3.9 and Figure 3.10).

Based on this approximation, the transmission functions of the different tracheae were calculated. Because of its small size, the acoustic trachea must not be treated as an infinite exponential horn, but its transmission function must be calculated by means of the equations for a finite-length horn. The finite horn amplifies sound from a certain frequency (cut-off frequency) in a broad range of frequencies as the infinite horn does; but the broadband transmission is superposed by a few resonances, which are caused by reflections inside the horn. Bioacoustical measurements with a probe microphone at the entrance of the tibial organ (see asterisk in Figure 3.9) proved that the measured transmission corresponds much better with the one calculated for the finite exponential horn than with that calculated for the infinite horn. This is shown in relation to the power spectrum of the conspecific song (Figure 3.11).

Recent morphological investigations (Sickmann *et al.*, 1997) demonstrated that both tympana borders extend to the outer wall of the acoustic trachea and tracheal chambers and also dorsally to border a considerable part of the haemolymph channel (Figure 3.3b). The dorsal wall of the acoustic trachea is attached to the inner surface of the anterior and posterior tympana at the positions of the distal IO and at the region of the CA. Structurally the tympana are partially in contact with air in the trachea and with haemolymph in the channel containing the receptor cells.

These results show that the acoustic trachea is the principal input of acoustic energy into the auditory receptor organs. Inside the trachea, sound signals travel undispersed with a lowered propagation speed. The tympanic membranes play an important role in determining the overall acoustic impedance of the bushcricket ear and in particular the impedance terminating the acoustic

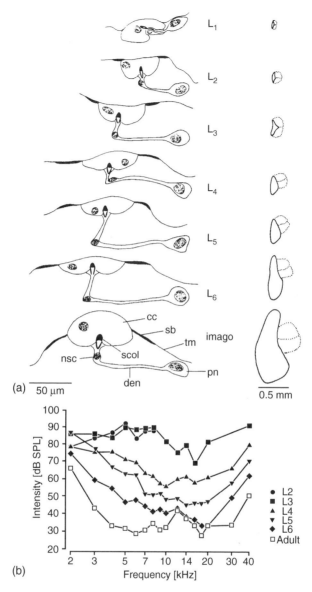

FIGURE 3.8 (a) Left: development of the third scolopidium of the CA within the foreleg of *Ephippiger ephippiger* (counted from proximal). Drawings from transverse sections of the tibia in the first to sixth larval instar and the imago. Note the erection of the dendrite and cap cell in the second instar is correlated with the appearance of the supporting bands (sb). cc, cap cell; den, dendrite; nsc, nucleus of the scolopale cell; pn, perikaryon of the bipolar sensory neuron; sb, supporting band; scol, scolopale cap and rods; tm, tectorial membrane (from Rössler, 1992). Right: development of the prothoracic acoustic spiracle of *E. ephippiger*. The morphology of the spiracle and the associated respiratory spiracle is drawn schematically for each stage. Measurements of the area of the spiracle opening. Note the increase of the area is most marked after the final moult. (b) Auditory threshold curves in larvae and adults of *Tettigonia cantans* measured by hook electrode recordings from the tympanal nerve. Hearing threshold in the fourth (*n* = 4), fifth (*n* = 6), sixth (*n* = 6) larval instar and in the imago (*n* = 10). Ipsilateral stimulation with pure tone bursts of 20 msec duration, repetition rate 2/sec, rise and fall time 1 msec. (Left drawing from Rössler, W., *Zoomorphology*, 112, 181–188, 1992b. With permission. Modified from Rössler, W., *Cell Tiss. Res.*, 269, 505–514, 1992a. With permission.)

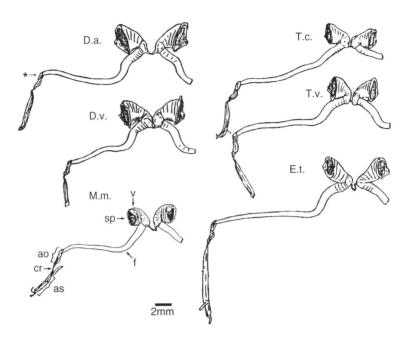

FIGURE 3.9 Acoustic tracheal system of six species. Drawings from dissected adults of *Decticus albifrons* (D.a.), *D. verrucivorus* (D.v.), *Tettigonia cantans* (T.c.), *T. viridissima* (T.v.), *Ephippigerida taeniata* (E.t.), and *Mygalopsis marki* (M.m.). ao, auditory organ; as, air sac; cr, collapsed region, tracheal constriction; f, femoral part of the trachea; sp, spiracle; v, vesicle. (After Heinrich, R., Jatho, M., and Kalmring, K., *J. Acoust. Soc. Am.*, 93, 3481–3489, 1993. With permission.)

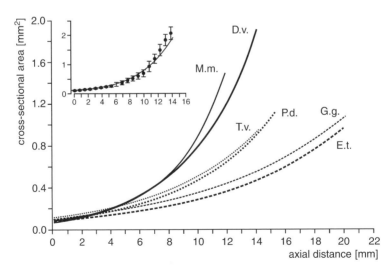

FIGURE 3.10 Cross-sectional area of the acoustic trachea as a function of axial distance from the auditory receptor cells (mean regression function for the different species with $r \geq .957$, $p \leq 0.05$, $n \geq 8$). Inset shows measurements for *Decticus verrucivorus* (average $\pm$ standard deviation) with the mean regression function. M.m., *Mygalopsis marki*; D.v., *Decticus verrucivorus*; T.v., *Tettigonia viridissima*; P.d., *Polysarcus denticauda*; G.g., *Gampsocleis gratiosa*; E.t., *Ephippigerida taeniata*. (After Hoffmann, E. and Jatho, M. *J. Acoust. Soc. Am.*, 98, 1845–1851, 1995. With permission.)

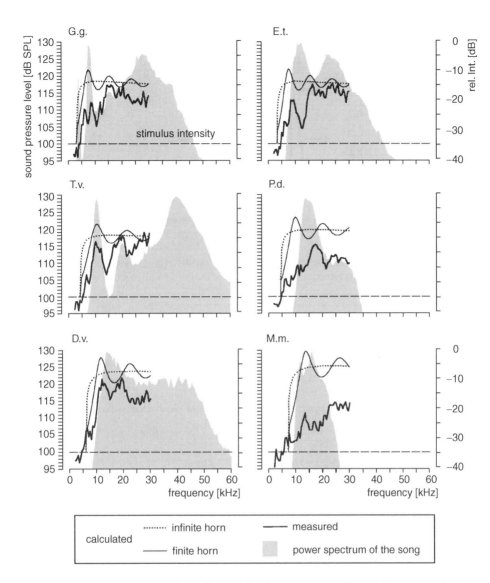

FIGURE 3.11 Theoretical amplification of sound in the acoustic trachea of six species based (as in Figure 3.10) on the equations for an infinite and a finite exponential horn, the corresponding bioacoustical measurements, and the power spectra of the conspecific songs. Numbers of measurements contributing to the averaged measured transmission function: G.g.: 10; E.t.: 15; T.v.: 18; P.d.: 22; D.v.: 10; M.m.: 12. The standard deviation of the amplification in the maxima of the transmission function was about 3 dB. (After Hoffmann, E. and Jatho, M., *J. Acoust. Soc. Am.*, 98, 1845–1851, 1995. With permission.)

trachea (Bangert *et al.*, 1998). Figure 3.12 illustrates schematically the role of the tympana in the transfer of acoustic energy to the receptor cells.

The terminating properties of the trachea will then determine the sound transmission properties of the trachea and the flow of acoustic energy into the organ. The results of a study by Bangert *et al.* (1998) indicate a more significant role for the tympana than that of simple pressure releasing borders of the acoustic trachea. Laser-vibrometry measurements show that the tympana do not behave like vibrating membranes, but rather like hinged flaps. These appear to rotate like a rigid plate about the hinge (at the dorsal edge) and the tympana act phase-coupled (Figure 3.12). This is

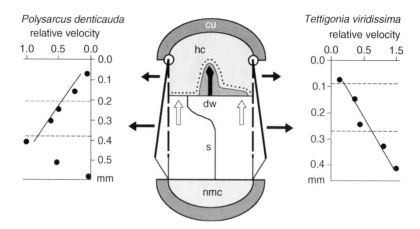

FIGURE 3.12 Semischematic transverse section of the tibia at the distal part of the CA with a diagram of the resulting membrane motion and its effect on the translocation of the dorsal wall together with the dendrite of the receptor cell, the cap cell and the tectorial membrane (compare Figure 3.3b). cu, cuticle; hc, haemolymph channel; dw, dorsal wall; s, septum or dividing wall; nmc, nerve–muscle channel. (after Bangert *et al.*, 1998). The graphs to the left and right show the dorsal–ventral distribution of velocity amplification of tympanal motion measured by laser-vibrometry for an anterior tympanum of *Polysarcus denticauda* (left) and *Tettigonia viridissima* (right) measured at the centre line of the tympanum. The results for the posterior tympana appear the same (not shown here). The horizontal lines mark the dorsal and ventral edge of the tympanal membrane (Figure 3.3), the inner insertion of the dorsal wall of the trachea and the ventral edge of the inner plate (Figure 3.3a), a characteristic outer structural part of the tympanum. (Modified from Bangert, M., Kalmring, K., Sickmann, T., Stephen, R., Jatho, M., and Lakes, R., *Hear. Res.*, 115, 27–38, 1998. With permission.)

also supported by morphological investigations (see Figure 3.3b). It is clear that a positive sound pressure in the tracheal air pushes the tympana outward and, therefore:

1. Stretches the dorsal wall
2. Exerts a negative pressure on the haemolymph by widening the haemolymph fluid tube

The latter effect again (since the haemolymph is incompressible) supports the pressure acting onto the dorsal wall from within the trachea.

Frequency tuning of the receptor cells must be determined by the structures of the CA themselves, *i.e.* the structure and dimensions of the dorsal wall of the anterior tracheal chamber and the size of the dendrites, cap cells and the length and shape of the tectorial membrane. These findings, together with the morphology of the organ and physiological data from the receptor cells, suggest the possibility of an impedance matching function for the tympana in the transmission of acoustic energy to the receptor cells in the tettigoniid ear.

Frequency Tuning of the Receptor Cells

At any instant the sensory organs of an animal receive a large amount of environmental information of which only a minute fraction is relevant in a given behavioural context. The rest is redundant, irrelevant or noise. A major task of the nervous system is to detect the relevant information. It is the more surprising that small nervous systems like those of insects are capable of performing such complex tasks, enabling the animals to detect and react to relevant signals even if these signals are embedded in a background of noise and similar or more intense irrelevant stimuli. It appears as if their nervous system would be able to generate a highly specific matched filter that can be tuned to a particular signal. Insects are an excellent model for the investigation of basic principles of biological pattern recognition, and particularly to study the neural mechanisms on which matched filters might be based (Kalmring *et al.*, 1996).

In many cases there are distinct morphological differences in the tibial organs (especially in the CA) of different bushcricket species. The number of receptor cells in the CA (from 14 up to 50) as well as the length of the acoustic tracheae varies. Nevertheless, similar functional types of receptor cells are present in the different auditory organs. At least 60 to 80% of receptor cells in the different CAs of higher developed tettigoniids (with cell numbers from 30 to 40) have threshold curves of almost identical shape, which is surprising in terms of species separation (Kalmring *et al.*, 1995a, 1995b). This means that stimulus transduction processes in the auditory receptor organs of the investigated species should be very similar and no distinct adaptations to the frequency parameters of the conspecific song seem to be realised. As mentioned above, the number of cells in the CA varies in different species and, therefore, some of these receptor cells might function in a different way. In each species one can usually find some receptor cells that are tuned sensitively to the frequency range of the carrier frequency of the conspecific song (Kalmring *et al.*, 1995a) (Figure 3.13).

There are also differences between the species with respect to the suprathreshold response characteristics. The receptor cells of *D. verrucivorus*, for example, have high discharge rates, whereas those in *E. ephippiger* and *Mygalopsis marki* are lower. Within each species, the difference in the response rate is small. Species-specific differences also exist in the latency–intensity characteristics.

Superficially, there seems to be no functional adaptation of the auditory organs of the different species to the parameters of the conspecific songs. One finds the same hearing range of the ears, the receptor cells belong to the same functional types in the different species and there are no large differences in the suprathreshold response characteristics of the receptor cells either. The question is if the overall characteristics are so similar, what might be the specific processing mechanisms that enable an individual to recognise the conspecific song (Kalmring *et al.*, 1993, 1994).

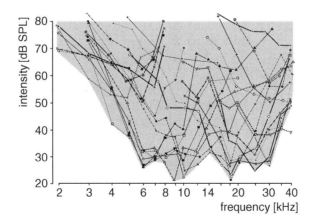

FIGURE 3.13 Threshold curves of the 24 different tympanal receptor cells recorded in *Mygalopsis marki*. Stimulation with pure tone bursts of 20 msec duration, rise and fall time 1 msec, repetition rate 2/sec. The threshold curves show a distribution in the frequency range from about 2 kHz to at least 40 kHz. Note that the CA in the tympanal organ of *Mygalopsis marki* consists in its entirety of only 24 receptor cells. Considerable overlap is evident. (After Kalmring, K., Reitböck, H. J., Rössler, W., Schröder, J., and Bailey, W. J., Synchronous activity in neuronal assemblies as a coding principle for pattern recognition in sensory systems of insects, In *Trends in Biological Cybernetics*, Research Trends, Vol. 1, Menon, J., Ed., Council of Scientific Research Integration, Trivandrum, pp. 45–64, 1991. With permission.)

In spite of this lack of obvious differences between the species, adaptations to the frequency content of the conspecific song can be observed in the auditory neuropile of the prothoracic ganglion, the projection area of auditory receptor cells (Figure 3.14). Here, "frequency weighted" synaptic processes were observed. Neurophysiological investigations using multiunit recordings and Current Source Density (CSD) analyses showed that CSD amplitudes representing ensemble activities within the auditory neuropile to stimulation with different frequencies differed from species to species and were correlated with the frequency content of the conspecific song. Maximal amplitudes in the CSD responses were measured in the range of the fundamental frequency of the songs (Kalmring *et al.*, 1991, 1993; Rössler *et al.*, 1994) (Figure 3.14).

Similarly, neurophysiological comparison of the receptor cells of two species, one with an extreme low fundamental frequency (*Gampsocleis gratiosa*) and one with a narrow-banded song power spectrum (*M. marki*), revealed a clear difference in the sensitivity in the low frequency range (below 5 kHz), whereas similar sensitivities were found at frequencies above 5 kHz. Despite a

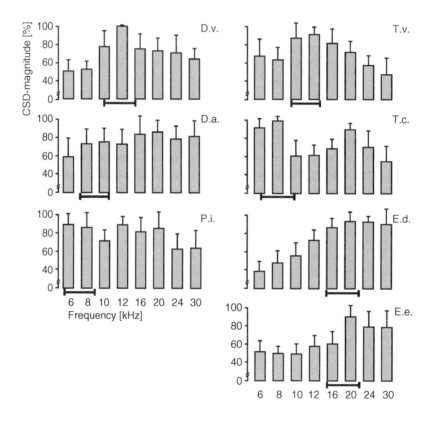

FIGURE 3.14 Dependence of the relative CSD response magnitudes on the stimulus frequency in seven European tettigoniid species: E.e., *Eephippiger ephippiger*; E.d., *Ephippiger discoidales*; D.v., *Decticus verrucivorus*; D.a., *Decticus albifrons*; T.v., *Tettigonia viridissima*; T.c. *Tettigonia cantans*; P.i., *Psorodonotus illyricus*. The response magnitude was evaluated by integration of the averaged CSD courses recorded in the auditory neuropile (*n* = 8 in each case). Stimulus duration 20 msec, rise and fall time 1 msec, mean intensity 71 dB SPL. The fundamental frequency range of the conspecific song of each species is marked by horizontal bars, respectively. (After Kalmring, K., Reitböck, H. J., Rössler, W., Schröder, J., and Bailey, W. J., Synchronous activity in neuronal assemblies as a coding principle for pattern recognition in sensory systems of insects, In *Trends in Biological Cybernetics*, Research Trends, Vol. 1, Menon, J., Ed., Council of Scientific Research Integration, Trivandrum, pp. 45–64, 1991. With permission.)

substantial difference in the overall number of receptor cells in the CA, combined bioacoustical measurements indicated that the difference in sensitivity to low frequencies was most likely caused by the lower cut-off frequency of the acoustic trachea in *G. gratiosa* (Kalmring *et al.*, 1995a, 1995b).

PROJECTION OF THE RECEPTOR CELLS AT THE VENTRAL NERVE CORD LEVEL

The projection pattern and the arborization positions of the different receptor cells from the CA and the IO of the fore- and midlegs within the pro metathoracic and mesothoracic ganglia show differences in terms of their morphology and physiology. The sensitive low-, mid- and high range auditory receptors from the CA of the forelegs project into the neuropile of the aRT with the typical counterclockwise tonotopic organisation already described in other bushcricket species (Römer, 1983; Ahi *et al.*, 1993; Ebendt *et al.*, 1994; Stölting and Stumpner, 1998). Moreover, the arborisation patterns of the projections have similar shape as in many other bushcrickets (Figure 3.15-I A to Figure 3.15-I D; darkly shaded area indicates the ventral view of the auditory neuropile).

The bimodal receptors, probably from the IO of the foreleg, are bifurcated and they project into a neuropile located more laterally within the prothoracic ganglion than the auditory neuropile of the aRT (Figure 3.15-I E and Figure 3.15-I F; lightly shaded area). These receptors are sensitive to vibratory stimulation with frequencies of 800 to 900 Hz (E) and 500 Hz (F), but less sensitive to airborne sound at 3 kHz. Very similar projection patterns and positions within the mesothoracic ganglion were also found for midleg receptors tuned to vibratory frequencies below 1000 Hz (Figure 3.15-II E and Figure 3.15-II F). However, these receptors respond poorly to airborne sound stimulation only within the frequency range of vibratory tuning.

The receptor cells of the distal IO and a few cells of the CA of the midlegs have very different projection patterns when compared with the foreleg receptors from the same part of the organ. The midleg receptors have few branches which end lateral to the midline of the ganglion (Figure 3.15-II A to Figure 3.15-II A D). The midleg receptors are tuned to frequencies of 1000 to 3000 Hz when stimulated with vibrations. As typical for atympanate receptor organs, their auditory responses are restricted to the frequency range of vibration reactions.

The whole neuropile of the aRT of the mesothoracic ganglion seems to be reduced compared to that of the prothoracic ganglion, especially with respect to the midline caudal parts. On account of the great number of receptor cells in the CA of the forelegs in *Polysarcus denticauda*, the auditory neuropile of the prothoracic ganglion is densely packed with presynaptic structures of the end branches, which should provide a powerful synaptic transmission to central auditory neurons. Some examples of the projection of single receptor cells from the fore- and midlegs in the pro metathoracic and mesothoracic ganglia of the Phaneropterine bushcricket, *P. denticauda*, are shown (Figure 3.15).

AUDITORY–VIBRATORY INTERNEURONS IN THE VENTRAL NERVE CORD

Coprocessing of auditory and vibratory information from the tympanal receptor organs and the vibratory receptors of all six legs is a common phenomenon in the central nervous system of Acrididae and Tettigoniidae, and may be important in acoustic behaviour (Cokl *et al.*, 1977; Kalmring and Kühne, 1980). The auditory and vibratory senses converge on the same ventral cord neurons. The majority of the neurons that discharge impulses in response to either airborne sound or vibration stimuli also receive synaptic inputs (directly from the receptor endings or via interneurons) from the other system. The latter elicit either subthreshold excitation or inhibition. Single cell recordings in the ventral nerve cord and the head ganglia of *D. albifrons* and *D. verrucivorus* show that most of the acoustic units respond to sound and vibration (Nebeling, 1994; Sickmann, 1996). However, on the basis of their response characteristics they

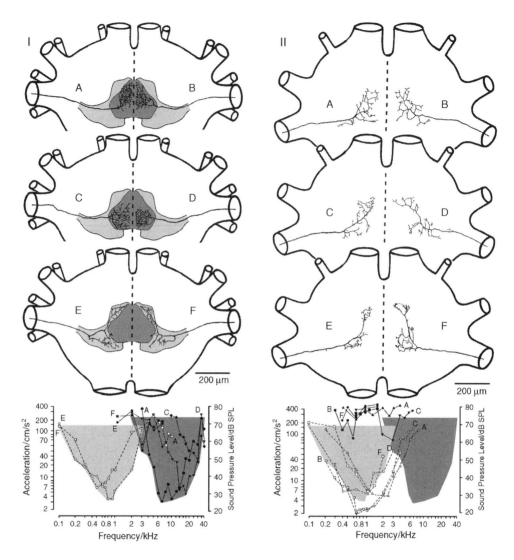

FIGURE 3.15 (I) Ventral view of prothoracic ganglia (whole mounts) showing the central projection of mechanosensitive receptors of the forelegs of *Polysarcus denticauda*. Projections of low-frequency auditory receptor cells (A, B), mid-range receptor cells (D), ultrasonic receptor cells (C), and bimodal vibratory–auditory receptor cells (E, F) are shown. In addition, the threshold curves of these receptors are shown in the graph. In drawings of ganglia, rostral is at top and dark and light areas represent neuropiles in the aRT of auditory and bimodal receptors. (II) Ventral view of mesothoracic ganglia (whole mounts) showing the central projection of mechanosensitive receptors of the midlegs of *P. denticauda*. Projections of bimodal vibratory–auditory receptor cells of CA (A–C) and IO (D–F) are shown as well as the threshold curves of these receptors in the graph below. In drawings of ganglia, rostral is at the top. Shaded areas in the graphs indicate the frequency–intensity region of neuronal reaction of receptors in the forelegs. (After Kalmring, K., Hoffmann, E., Jatho, M., Sickmann, T., and Grossbach, M., *J. Exp. Zool.*, 276, 315–329, 1996. With permission.)

may be classified as sound (S), vibration and sound (VS) and vibration (V) neurons. The response patterns of the different sensory cell types of the ventral cord in both decticine species are basically the same as those previously described for other tettigoniid species (*T. cantans*, *T. viridissima*, *G. gratiosa* and *P. denticauda*) (Silver *et al.*, 1980; Kalmring *et al.*, 1993, 1994, 1997). In each group of these neurons some phasically and tonically reacting neuron types were found on each side of the ventral nerve cord.

Using a combined recording and staining technique, the physiologically characterised S neurons, VS neurons and V neurons can also be described in their morphology: *i.e.* in soma position, axon course and branching pattern. At least two neurons (one phasic S neuron and one phasic VS neuron) possess a T-shaped morphology with a near soma bifurcating axon, one stem ascends to the head ganglia and the other descends to the caudal thoracic ganglia or even passes into the abdominal cord. Another group of neurons are ascending neurons, the axon of which passes to the head ganglia. This group includes tonic and phasic S neurons and especially most of the VS neurons and V neurons. In a third group the neurons send their axons to caudal positions of the ventral nerve cord (descending neurons). Phasic neurons and tonic S neurons belong to this group but descending VS neurons and V neurons could not be identified at all.

In comparison, the morphology of the S neurons is differently organised than the morphology of the VS neurons and V neurons. The somata of all S neurons are located in the prothoracic ganglion. Here they lie without exception in a frontal position, from where the neurites (the only processes of the unipolar neurons) run caudo-medially to the midline of the ganglion where they cross approximately at the centre of the ganglion to the contralateral hemisphere. All the S neurons possess densely packed fronto-medially running dendritic structures with close contact to the projecting end branches of the auditory receptor cells of the foreleg in the neuropile of the aRT.

The axons bifurcate (in the case of the T-shaped neurons), ascend or descend contralaterally to the soma position and run in parallel and close to the midline within the Ventral Intermediate Tract (VIT) in the rostral and caudal direction to their projection areas within the lateral protocerebrum or the caudal ventral cord. Inside the different ganglia the S neurons possess typically medially running short but dense branches, with probably presynaptic endings. They are restricted to one side of the ganglia. The basic morphology of typical S neurons in the pro metathoracic and mesothoracic ganglia is shown (Figure 3.16).

The V Neurons and VS neurons investigated so far correspond in their ground plan and general position within the CNS. In contrast to the S neurons, the somata of these vibrosensitive interneurons (with only one exception) are located in the caudal cortex of the thoracic ganglia. Their axons run within the VIT, DIT (dorsal intermediate tract) or LDT (lateral dorsal tract). The presumed dendritic arborisations originate from commissural bridges near the somata.

The generally ascending axon possesses medially running branches and long laterally projecting collaterals in the thoracic ganglia. The projecting areas within the supraesophageal ganglion are strictly more laterally positioned than those of the S neurons (Nebeling, 1994).

The morphology of standardised VS neurons and V neurons in a dorsal-ventral view of the pro metathoracic (A) and mesothoracic (B) ganglia of tettigoniids are shown in Figure 3.17. Note the similar positions of the somata of the caudal commissural bridges and the laterally directed collaterals. Only the TH2-TC-V1 neurons have an aberrant soma and dendrite position (from Sickmann, 1996).

BIOPHYSICS OF SOUND PRODUCTION AND ACOUSTIC BEHAVIOUR

INTRODUCTION

Bushcricket Songs

In insects, sound or vibration is usually produced by the friction of two body parts moving across one another (Ewing, 1989). This event is termed stridulation. Stridulation using the forewings is widespread in Orthoptera of the suborder Ensifera. Bushcrickets produce sound (and vibration) signals by elytro–elytral stridulation. During evolution they have developed two specialised (asymmetrical) regions on their forewings used for sound production. The right elytron bears a

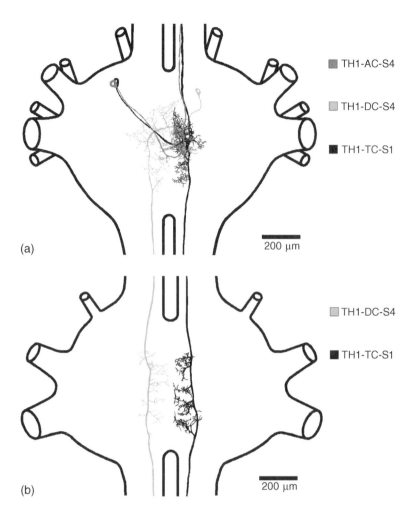

■ TH1-AC-S4

□ TH1-DC-S4

■ TH1-TC-S1

(a)   200 μm

□ TH1-DC-S4

■ TH1-TC-S1

(b)   200 μm

FIGURE 3.16 Dorso-ventral view of standardised S neurons within the pro metathoracic (a) and mesothoracic (b) ganglia of *tettigoniids*. The nomenclature of the neurons is: TH1, TH2: location of the soma in prothoracic, mesothoracic ganglion. AC, TC: ascending or T-shaped axon contralaterally to the soma position. S1, S4, V1, V3, VS5: numbers of S neurons, V neurons, and VS neurons. Note the basic correspondence of the different auditory neurons concerning axon course, branching pattern and general position within these ganglia. The somata of these plurisegmental S neurons are distributed to two clusters within the cortex of the frontal prothoracic ganglia. (After Sickmann, T., Vergleichende funktionelle und anatomische Untersuchungen zum Aufbau der Hör- und Vibrationsbahn im thorakalen Bauchmark von Laubheuschrecken, Dissertation, Marburg, 1996. With permission.)

plectrum on its median edge and a structure called the mirror, a raised area of thin cuticle bounded partly or completely by thickened veins. The latter is possibly responsible for some resonance phenomena. On the ventral side of the left elytron, there is a row of teeth (pars stridens) (Dumortier, 1963a, 1963b). The stridulatory movement of the forewings results in the teeth of the pars stridens serially scraping over the plectrum. Each tooth impact produces a damped oscillation of both wings, thereby generating a very brief sound impulse. The opening and closing movements of the elytra produce a succession of sound impulses causing the opening (small) and closing (main) syllables (Heller, 1988; Ewing, 1989; Jatho *et al.*, 1992). A species-specific number of opening and closing syllables forms a verse. The pressing of both forewings against each other and the velocity of the wing movement results in the typical time–amplitude

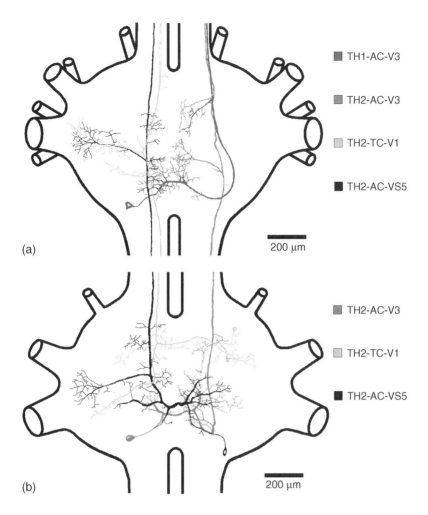

■	TH1-AC-V3
■	TH2-AC-V3
■	TH2-TC-V1
■	TH2-AC-VS5

200 µm

(a)

■	TH2-AC-V3
■	TH2-TC-V1
■	TH2-AC-VS5

200 µm

(b)

FIGURE 3.17 Morphology and position of standardised VS neurons and V neurons within the pro mesothoracic (a) and mesothoracic (b) ganglia of tettigoniids (dorso-ventral view). With the exception of the TH2–TC–V1 all the identified vibro sensitive interneurons correspond concerning the position of their soma and commissural bridges within the caudal half of these ganglia. Note the pronounced lateral arborisations. (After Sickmann, T., Vergleichende funktionelle und anatomische Untersuchungen zum Aufbau der Hör- und Vibrationsbahn im thorakalen Bauchmark von Laubheuschrecken, Dissertation, Marburg, 1996. With permission.)

pattern of the song in tettigoniids. The time–amplitude pattern and the fundamental frequencies of each sound impulse and of the whole song are determined by the structure and vibration properties of the wings (Keuper *et al.*, 1988).

Acoustic signalling in bushcrickets is predominantly a male activity, since in many species the female has no acoustic role other than that of a silent receiver. The songs play an important role in the reproductive behaviour of tettigoniids, *e.g.* in attracting the females and in the spacing and distribution of individuals within the habitat.

The acoustic signals of bushcrickets are often simple in structure when compared with those of some animal groups. Nevertheless significant differences in frequency and temporal structure occur in the songs of different species.

Comparative studies of bushcrickets show that there is a wide variation in the time–amplitude structure of the songs. Syllable durations may vary from in excess of 100 msec to as short as 8 msec

(Keuper *et al.*, 1988). Some of the differences in temporal structure arise from variations in the closing (opening) velocity of the wings during stridulation. In many species the number of teeth and their spacing on the file are similar, as are the number of teeth used to generate the syllable (Heller, 1988; Jatho *et al.*, 1992); impulse repetition rates in the syllables may differ considerably (from 500 Hz up to 8 kHz).

The frequency span of bushcricket songs is broad and ranges from 7 to 10 kHz up to frequencies as high as 60 to 80 kHz. Some species have very low fundamental frequencies (3.8 kHz or 6 kHz), in others the frequency range is more restricted to the higher frequency ranges with a bandwidth of only 10 kHz or less.

In species with a low impulse repetition rate (*e.g. E. ephippiger*), single sound impulses of the main syllable are clearly separated from each other (nonresonant sound production), with the exception of the very last impulse of each syllable (Stiedl, 1991; Jatho *et al.*, 1992). The impulse patterns of the syllables of Ephippigerinae are individually well defined and species specific.

On the other hand, fast singing species (*e.g. T. cantans*) produce songs of short syllable duration with high impulse repetition rates. Within the syllables, the sound impulses are not clearly separated and are frequently superimposed. A tendency to a resonant sound production becomes evident. Consecutive syllables show considerable variability (Jatho *et al.*, 1994). The syllable patterns described for the above two species are examples of the wide diversity of song types found in the Tettigoniidae.

Signal Transmission in the Biotope

Bushcrickets often live in biotopes of dense, high vegetation. Acoustic communication is important in sexual behaviour, as well as in territorial or rivalry behaviour. Females in some species are able to locate phonotactically stridulating males over a distance of more than 10 to 15 m. The presence of dense plant growth which interferes with the transmission of sound signals is a hindrance to localisation of the sound source. Reflections from the ground and leaves cause interference leading to frequency filtering, absorption and scattering by densely packed plants; refraction by wind and temperature gradient occur in such habitats (Michelsen, 1978, 1992; Keuper, 1981; Stephen and Hartley, 1991). For example, direct and reflected sound may, due to interference, cause frequency dependent local maxima and minima of sound pressure at various distances from the singing male. This effect leads to significant distortion of the syllable structure. Under such circumstances, it is difficult for a female to find a sound source solely by walking in the direction of increasing sound pressure level (Michelsen, 1978).

All bushcrickets have sound and vibration receptors. The acoustic neurones in the ventral nerve cord ascending to the head ganglia are without exception bimodal auditory–vibratory in character. During stridulation, the males also produce vibratory signals, which would be detectable over short distances. The frequency of vibratory signals tends to lie in the optimum range of the vibratory receptors. The convergence of the two sensory inputs at the ventral nerve cord level could be a fundamental element in the process of localising a stridulatory partner at short distances (up to 1 to 2 m). Moreover, it could facilitate and improve the recognition of signals from conspecifics.

Broadcasting and Acoustic Behaviour

The song is broadcast by males during a species-specific daily activity period (Dumortier, 1963a, 1963b), which in some habitats can be affected by interspecific acoustic interactions of sympatric species, leading to shifts in the diel periodicity (Greenfield, 1988; Römer *et al.*, 1989). Climatic conditions may also cause a shift of the daily activity period (Stiedl and Bickmeyer, 1991). Among Tettigoniidae species the mobility of singing males varies considerably. Males of the Australian species *M. marki* remain at one calling site for several days (Dadour and Bailey, 1985). Some species keep their perches (Greenfield, 1983) while others change it several times during one activity period (Meixner and Shaw, 1979; Shaw *et al.*, 1981; Schatral and Latimer, 1988). Ground living species, such as *D. verrucivorus* and *Psorodonotus illyricus*, are highly mobile throughout one activity period (Keuper *et al.*, 1985; Schatral and Kalmring, 1985; Schatral and Latimer, 1988). Distances between calling males in aggregations of many species are determined by the intensity and the frequency content of the perceived calls of the neighbours (Thiele and Bailey, 1980; Römer and Bailey, 1986).

Different species of bushcrickets are found in characteristic biotopes. There are species which usually inhabit bushes or tall herb vegetation. The males of these species often sing from high perches, occasionally even trees; emission heights of 1.5 m up to several metres above ground are common. Examples for these species are *Ephippiger bitterensis* (Busnel *et al.*, 1956), *M. marki* (Bailey, 1985), *T. cantans* (Latimer and Schatral, 1983), *T. viridissima* (Keuper *et al.*, 1985). Other species like *Platycleis affinis* (Samways, 1976a), *Ephippiger discoidalis* and *Pholidoptera littoralis* (unpublished observations from Keuper *et al.* (1986)) prefer biotopes with lower herbaceous or graminaceous vegetation, where the maximum height for the emission of songs is around 0.6 to 1.2 m. Most of the species described above show a high degree of territorality with aggressive rivalry behaviour between conspecific males (Morris, 1971; Schatral *et al.*, 1984). The mostly woody plants in the biotopes of these groups of bushcrickets enable the possibility of using additional or even pure vibratory signals for species-specific communication (Busnel *et al.*, 1956; Keuper and Kühne, 1983; Latimer and Schatral, 1983; Keuper *et al.*, 1985). The broadcasting and acoustic behaviour of ground living bushcrickets, such as the two Decticinae species, *P. illyricus* and *D. verrucivorus*, suggest that song might affect more or less continuous locomotion of both sexes in these species. Clearly, with the high mobility in their singing behaviour, the individual males of *D. verrucivorus* and *P. illyricus* hold no fixed territories. Furthermore, there is a lack of aggressive physical contacts, even though the high mobility leads to numerous encounters in the biotope. Singing *P. illyricus* males normally do not approach each other any closer than 0.3 to 0.5 m. Males of *D. verrucivorus* coming close together sing in unison for a few minutes until one or both of the animals walk away. These two species can be found in low grassland where the vegetation height is rarely more than 20 to 30 cm. From biophysical measurements it is known that the propagation of sound close to a plane boundary is strongly influenced not only by the boundary itself but also by the existing microclimatic conditions (Ingard, 1953; Embleton *et al.*, 1976; Michelsen, 1978; Michelsen and Larsen, 1983). The propagation of substrate-borne signals in grass stalks and the soil is very limited compared to the propagation in wooden plant parts (Keuper *et al.*, 1985). The possibility that the distinctive behaviour of these species may also represent an adaptation to the problems of effective sound signalling by continuous mobility is also considered in an effort to increase the understanding of how environmental conditions affect the evolution of acoustic behaviour.

SOUND PRODUCTION

Combined Resonant and Nonresonant Sound Production

Two different mechanisms of sound production have been described for Ensifera with respect to the physical mode of oscillation (Elsner and Popov, 1978). So-called "resonant sound production" which is used by the Gryllidae (Nocke, 1971) and some Tettigoniidae (Suga, 1966) and the "nonresonant" mechanism used by the majority of tettigoniids. The mechanism of resonant sound production in gryllids has been analysed in detail by Nocke (1971), Elliot and Koch (1985) and Koch *et al.* (1988).The songs of gryllids consist of syllables with a narrow banded power spectrum, the syllables having the appearance of a pure sinusoidal sound pulse. Recently, Stephen and Hartley (1995) showed that the frequency of consecutive syllables was not constant, but varied about a mean value. In contrast to previous work (Nocke, 1971; Elliot and Koch, 1985), this showed that the principal frequency component of the syllable was not generated by the vibration of a structure of the wing. If this were the situation, the principal frequency component of consecutive syllables would be constant. Signals of nonresonant sound production are characterised by a series of clearly separated strongly damped sound impulses in the time–amplitude pattern. This type of time signal results in power spectra being broad banded, extending from a couple of kilohertz up to 60 to 80 kHz.

The detailed analysis of frequency content and impulse structure of the sound signals of some tettigoniid species shows that these two mechanisms do not exclude each other and could be used in combination by some species.

Comparison of the power spectra of different tettigoniid species reveals two fundamentally different types of spectra (Figure 3.18). The one consists of a single more or less broadbanded peak with a maximum at a certain frequency (Figure 3.18, left column). The other type of spectrum exhibits different clearly distinguishable maxima. The first peak of these spectra always has a centre frequency in the sonic frequency range and the second peak is almost harmonic to the first (Figure 3.18, right column). The first type of spectrum is, for example, produced by *Platycleis albopunctata* and *Metrioptera roeseli* (Decticinae), different species of Ephippigerinae (*E. ephippiger*, *E. discoidalis*, *E. perforatus*), and *M. marki* (Conocephalinae). The second type of spectra are found in the sound signals of *D. albifrons* and *G. gratiosa* (Decticinae) and the two species of Tettigoniinae, *T. cantans* and *T. viridissima*.

These species may be divided in two groups; each group having the same impulse structure for the closing syllables. The first group produces closing syllables with clearly separated impulses. This is obvious in the songs of *E. ephippiger* and *P. albopunctata*. The song of *M. marki* consists of single sound impulses and the bandwidth of the power spectrum is small in comparison to the other species (Figure 3.19).

In the second group the closing syllables contain groups of fused impulses with a high repetition rate as well as separated impulses. The repetition rate of the fused impulses is remarkably constant and fits the frequency of the first maximum of the power spectra (Figure 3.20). Furthermore the repetition rates within this impulse group are not influenced by ambient temperature at least in the two *Decticinae*, *D. albifrons* and *G. gratiosa*, whereas the other time variables of the song are negatively correlated with ambient temperature, especially impulse intervals within the opening syllables (Figure 3.21) (Jatho, 1995).

In the time–amplitude pattern of *G. gratiosa* a low frequency oscillation with superimposed high frequency impulses becomes visible after the first quarter of the closing syllable (Figure 3.20). After digital low and high pass filtering of the recorded sound signals, it turns out that this takes place also during the production of impulse groups in the other species (Figure 3.22). Obviously, the amplitude of a certain low frequency mode of wing oscillation is increased during this phase of sound production. The amplitude of the high frequency oscillation is similar inside and outside the impulse groups.

In *D. albifrons* the dependency of the amplitude of the low frequency oscillation from the tooth impact rate (*i.e.* impulse repetition rate) is clearly visible in the example. At ambient temperatures around 25°C the second half of the closing syllable consists of one group of fused impulses. At the end of the closing syllable the impulse interval is decreased from 0.11 to 0.055 msec at one step. At this moment the amplitude of the low frequency oscillation drops with its characteristic damping constant.

Consequently, this mode of sound production that occurs during the impulse groups has to be called "resonant sound production". It fulfils at least one basic physical criterion of a resonator which is an increase of the oscillation amplitude when stimulated with the characteristic frequency compared with oscillation amplitude at other frequencies.

The correlation of impulse repetition rate within the impulse groups and the low frequency peak frequency is valid for all the four species investigated (Figure 3.23).

The amplification of low frequency components by means of resonance is useful in several aspects. Because of the small body size of tettigoniids, the sound producing structures are of

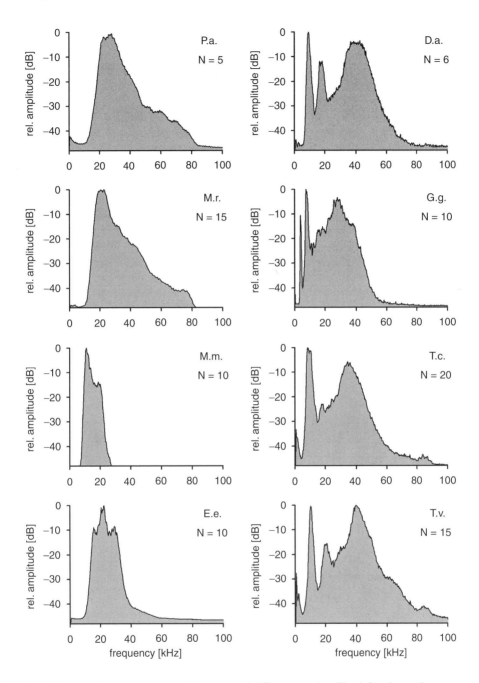

FIGURE 3.18 Averaged power spectra of the songs of different species. The left column shows spectra of species with only one maximum (*Platycleis albopunctata* [P.a.], *Metrioptera roeseli* [M.r.], *Mygalopsis marki* [M.m.], and *Ephippiger ephippiger* [E.e.]). The right column shows spectra of species with more than one peak (*Decticus albifrons* [D.a.], *Gampsocleis gratiosa* [G.g.], *Tettigonia cantans* [T.c.], and *Tettigonia viridissima* [T.v.]). Numbers of individuals from which spectra were derived are given as N. (After Jatho, M., *Untersuchungen zur Schallproduktion und zum phonotaktischen Verhalten von Laubheuschrecken* [Orthoptera: Tettigoniidae], Cuvillier Verlag, Göttingen, 1995. With permission.)

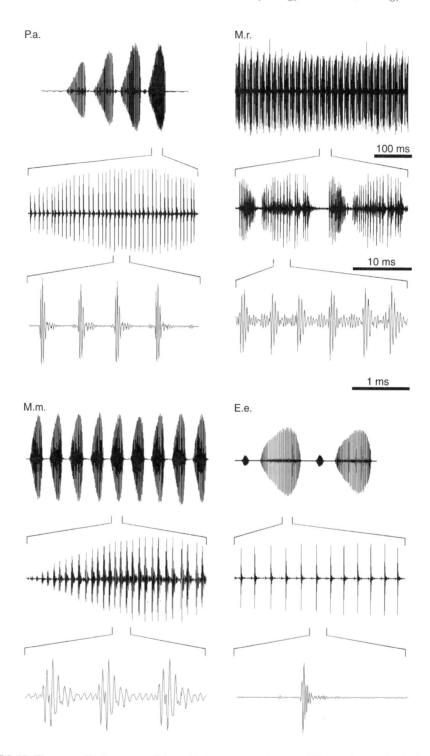

FIGURE 3.19 Time–amplitude pattern of the stridulatory songs shown at different time scales for four species with clearly separated impulses within the closing syllables. The species abbreviations are the same as in Figure 3.18. The time scales given for M.r. are the same for all species. (After Jatho, M., *Untersuchungen zur Schallproduktion und zum phonotaktischen Verhalten von Laubheuschrecken* [Orthoptera: Tettigoniidae], Cuvillier Verlag, Göttingen, 1995. With permission.)

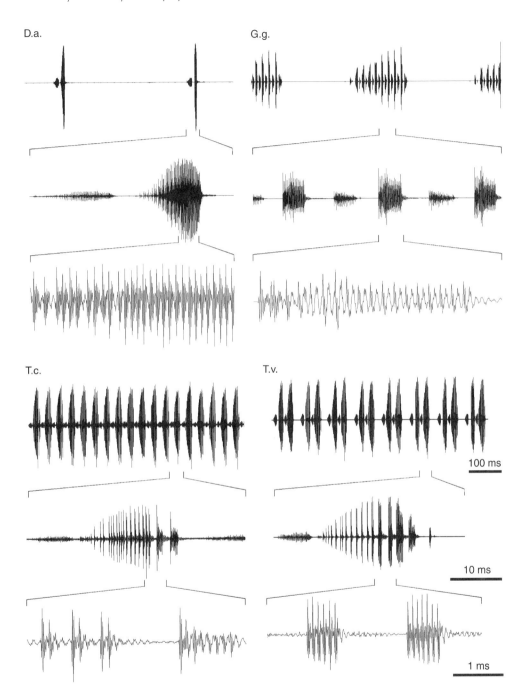

FIGURE 3.20 Oscillograms of songs shown at different time scales for four species with fused impulses within the closing syllables. Species abbreviations are as in Figure 3.18. Time scales same for all species. (After Jatho, M., *Untersuchungen zur Schallproduktion und zum phonotaktischen Verhalten von Laubheuschrecken* [Orthoptera: Tettigoniidae], Cuvillier Verlag, Göttingen, 1995. With permission.)

limited size. As the efficiency of sound radiation drops dramatically if the dimensions of the sound source are less than a quarter of the wavelength of the radiated sound, tettigoniids have serious problems producing sound signals below 10 kHz with sufficient amplitude. The propagation of sound signals with frequencies below 10 kHz does have significant behavioural

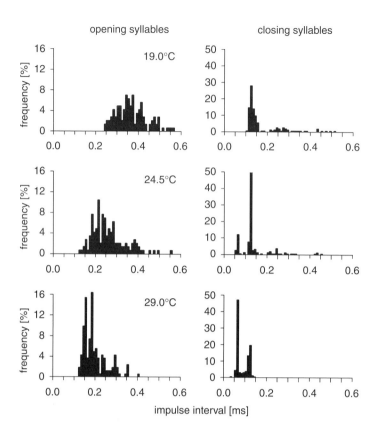

FIGURE 3.21 Frequency distributions of the impulse interval times within the opening (left column) and closing syllables (right column) of *Decticus albifrons* at different ambient temperatures. Note the shift of the distribution to shorter impulse interval times with increasing ambient temperature within the opening syllables. In contrast the distribution of the impulse intervals within the closing syllables is not shifted but, with increasing temperature, proportion of shorter impulse intervals increase. (After Jatho, M., *Untersuchungen zur Schallproduktion und zum phonotaktischen Verhalten von Laubheuschrecken* [Orthoptera: Tettigoniidae], Cuvillier Verlag, Göttingen, 1995. With permission.)

advantages. Below 10 kHz sound waves propagated through the biotope are less subject to attenuation than are frequencies above 10 kHz. Therefore to increase the lowest frequency components would provide a significant increase of communication distance leading to possibly enhanced reproductive success.

The Role of Subtegminal Air Volume during Sound Production

Stephen and Hartley (1995) showed, in a simple model system as well as theoretically, that the vibrating wings and subtegminal space could act as a sharply tuned Helmholtz type resonator. The involvement of the subtegminal resonator in the song generation process in tettigoniids is clearly illustrated by results obtained with six bushcricket species using the same recording conditions as Stephen and Hartley (1995). It turns out that in all species differences exist between the power spectra of the sound signals recorded in air and in a helium–oxygen mixture (heliox).

Two examples of power spectra recorded in air and heliox are given (Figure 3.24). In both the amplitude of low frequency components (at 10 kHz in *M. marki*; at 3.5 and 7 kHz in *G. gratiosa*) is substantially reduced in the heliox recordings. It is assumed that in air the subtegminal air space acts

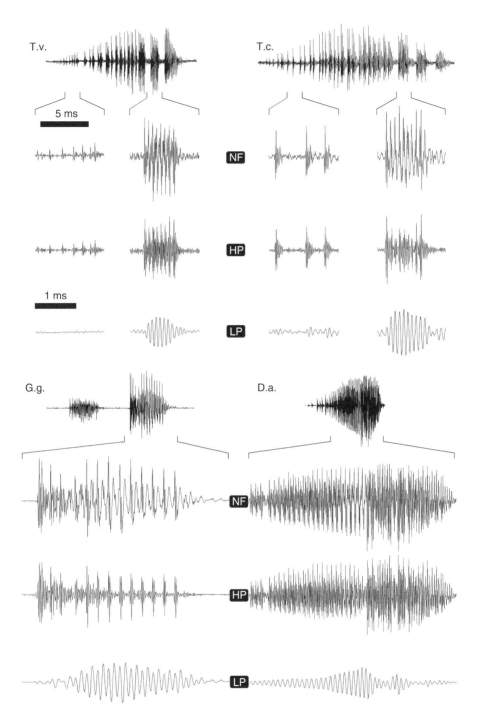

FIGURE 3.22 Time–amplitude patterns of closing syllables after digital high and low pass filtering. The species and abbreviations are as in Figure 3.18. For each an unfiltered closing syllable is shown in the first row. The second row shows a part of the closing syllable with a higher time resolution (NF). The third and fourth rows show the same part after high pass filtering (HP) and low pass filtering (LP). Time scales same for all species. Cut off frequencies are 16 kHz (LP) and 23 kHz (HP) for T.c. and T.v., 10 kHz (LP and HP) for G.g., and 13 kHz (LP and HP) for D.a. (After Jatho, M., *Untersuchungen zur Schallproduktion und zum phonotaktischen Verhalten von Laubheuschrecken* [Orthoptera: Tettigoniidae], Cuvillier Verlag, Göttingen, 1995. With permission.)

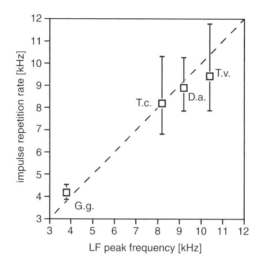

FIGURE 3.23 Correlation of the frequency of the first maximum of the averaged power spectra and the mean impulse rate within the impulse groups of the closing syllables for four species (D.a. *Decticus albifrons*, G.g. *Gampsocleis gratiosa*, T.c. *Tettigonia cantans*, T.v. *Tettigonia viridissima*). Vertical lines represent standard deviations for the mean impulse rate.

as a narrow banded resonator, which amplifies the transmission of certain low frequency components which are produced by mechanical wing oscillations. Due to the change in gas density, when the air is replaced by heliox, the resonance frequency of the subtegminal air volume shifts to higher frequencies and the amplitude of the radiated sound signal decreases at lower frequencies while it increases at higher frequencies. Stephen and Hartley (1995) derived an expression for the resonant frequency of the subtegminal resonator and showed that the resonant frequency was inversely proportional to the square root of the effective mass of the vibrating wings and the effective volume of subtegminal space. The mass of the vibrating wings is the sum of the actual mass of the wings and an additional load associated with the mass of the gas contained in the volume. The effect of this extra load added to the vibrating wings means that the frequencies of the modes of vibration of the wings will depend on the density of the gas in the subtegminal space.

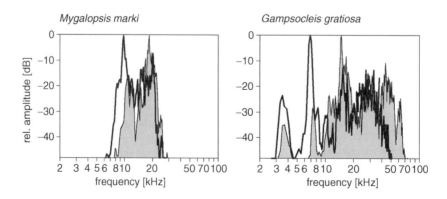

FIGURE 3.24 Representative power spectra of stridulatory songs of one male of *Mygalopsis marki* and one male of *Gampsocleis gratiosa* recorded in air (thick curves) and a helium–oxygen gas mixture (shaded areas).

ACOUSTIC BEHAVIOUR

Species Discrimination

Probably the most important function of sound production and perception in tettigoniids is the attraction of a sexually receptive conspecific partner. The emitted signals have to be detectable, discriminable and locatable by the receiving female. In the majority of species males produce sound signals and females locate the singing male by a behaviour of positive phonotaxis.

The parameters which are decisive for song discrimination are best studied in species with only small differences in the song characteristics and common activity periods and distributions within the biotope.

The two species *T. cantans* and *T. viridissima* live syntopically in some regions and have nearly identical daily activity periods. The females have to discriminate conspecific and heterospecific songs to find a conspecific male for reproduction. The stridulatory songs of *T. cantans* and *T. viridissima* both consist of a long series (several minutes) of brief syllables (15 msec at 20°C) which are repeated at 20 to 30 syllables per second. The most prominent difference between the songs is the syllable pattern. In *T. cantans*, the syllables are repeated continuously with a relatively constant intersyllable interval whereas in *T. viridissima* the syllables are grouped in pairs, leading to an alternation between a short and a long interval between the syllables. There are also differences in the frequency content, particularly in the sonic frequency range up to 20 kHz (Keuper *et al.*, 1988; Jatho, 1995) (see Figure 3.18 and Figure 3.20).

The usage of a computer aided stimulus synthesising system allows the generation of sound signals with an arbitrary combination of frequency content and time–amplitude structure (Schul, 1994; Jatho, 1995). In phonotactic two-choice experiments, females of both species were tested with four stimulus pairs representing different combinations of syllable structure and frequency content (Figure 3.25). The signals were always broadcast simultaneously.

The females of both species were able to discriminate between the conspecific and the heterospecific model song (Figure 3.25, first column). None of them could, however, discriminate between a pair of signals, both having the conspecific syllable pattern with the first having the conspecific and the second with the heterospecific frequency distribution (Figure 3.25, last column). *T. cantans* also failed to discriminate between a pair of signals both with the conspecific frequency content but different syllable patterns, whereas *T. viridissima* significantly preferred the signal with the conspecific syllable pattern (Figure 3.25, second column). In both species a significant preference for the conspecific syllable pattern was found when the conspecific syllable pattern with the heterospecific frequency content and vice versa were presented (Figure 3.25, third column).

In *T. viridissima* the conspecific syllable pattern (double syllables) alone seems to be sufficient to discriminate between conspecific and heterospecific males. The importance of this song parameter was also illustrated by a second set of experiments where the intervals between the syllables were gradually altered. In the natural songs the ratio of the short interval to the long interval in the song of *T. viridissima* is 10/30 msec (TV). This ratio was altered to 13/27 msec (TV + 3), 16/24 msec (TV + 6) and 19/21 msec (TV + 9) at the experimental temperature. In two-choice experiments no preference for either song model could be found when the combination TV vs. TV + 3 was presented but a significant preference for when presented with TV vs. TV + 6 or TV vs. TV + 9. If TV + 3 was presented vs. the song of *T. cantans* (TC) it was significantly preferred but the animals failed to show any phonotaxis when TV + 6 was presented vs. the song of *T. cantans* (Figure 3.26).

These results reveal that the alternation of long and short intervals or the presence of long intervals is necessary for eliciting positive phonotaxis in *T. viridissima*. Furthermore the last result offers the opportunity to perform experiments with only one sound source and monitoring the percentage of phonotactically reacting females.

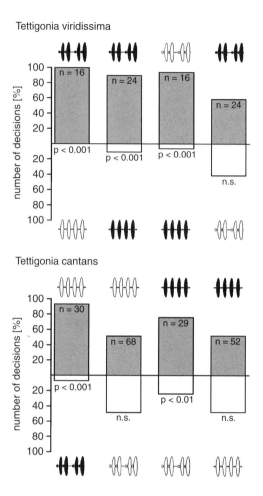

FIGURE 3.25 Results of phonotactic two-choice experiments with females of *Tettigonia viridissima* and *T. cantans*. The symbols above and below the bars indicate the stimulus configuration (double syllables: time pattern of *T. viridissima* males, single syllables: time pattern of *T. cantans* males; black filling: spectrum of *T. viridissima* males, white filling: spectrum of *T. cantans* males. The significant level of the sign test is given below the bars. Number of females that showed positive phonotaxis to either of the stimuli is indicated as n within the bars. (After Jatho, M., *Untersuchungen zur Schallproduktion und zum phonotaktischen Verhalten von Laubheuschrecken* [Orthoptera: Tettigoniidae], Cuvillier Verlag, Göttingen, 1995. With permission.)

Two time parameters were tested in this manner. The long interval between the double syllables (verse interval) with a constant duration of the double syllables (verses) and the verse duration at a constant interval between the verses.

Phonotaxis depends on both parameters in a similar manner (Figure 3.27). There is an optimum range of each parameter where 60 to 80% of the tested females walk towards the broadcasting loudspeaker. That is for the verse interval the range from 30 to 65 msec and for the verse duration the range from 40 to 50 msec. The verse interval appears to be the critical parameter for female phonotaxis as there is a drastic decrease in the response when the interval becomes shorter than 30 msec. The corresponding time parameters of the natural songs of *T. viridissima* lie within the optimum range whereas those of *T. cantans* lie outside the range. Therefore the bandpass characteristic of the neural network underlying this behaviour is sufficient to filter out the conspecific songs, *i.e.* females of *T. viridissima* are able to discriminate between conspecific and heterospecific males solely due to the syllable pattern of their calling songs.

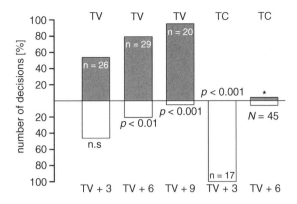

FIGURE 3.26 Results of phonotactic two-choice experiments with females of *Tettigonia viridissima*. The significance levels of sign tests given above or below the bars. Number of females who showed positive phonotaxis to either of the stimuli is indicated as *n* within the bars. Total number of tested females shown for each bar. For stimulus abbreviations see text. (After Jatho, M., *Untersuchungen zur Schallproduktion und zum phonotaktischen Verhalten von Laubheuschrecken* [Orthoptera: Tettigoniidae], Cuvillier Verlag, Göttingen, 1995. With permission.)

SUMMARY AND CONCLUSIONS

COMPARISON OF MORPHOLOGY, PHYSIOLOGY AND DEVELOPMENT

Comparative investigations of the tympanal and atympanal tibial organs in various species of bushcrickets revealed significant differences in the responses of the receptor cells in the fore-, mid- and hind legs to sound stimulation. Whereas, the CA and the distal part of the IO of the foreleg are very sensitive auditory receptor organs; the same organs in the mid- and hind legs appear to respond

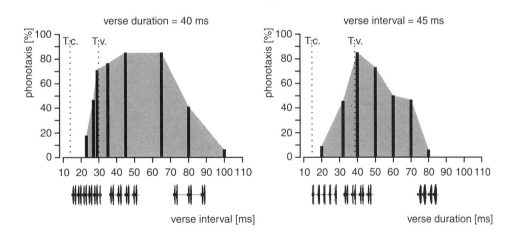

FIGURE 3.27 Results of phonotactic experiments with females of *Tettigonia viridissima* by means of one sound source only. Percentage of females that walked towards loudspeaker shown in dependency of the stimulus parameter. Left: variation of the verse interval; right: variation of the verse duration. The vertical dotted lines represent the values of the corresponding parameter of the natural songs of *T. cantans* (T.c.) and *T. viridissima* (T.v.). Schematic drawings of some stimuli are shown. Number of females tested was 20 for each verse interval and 11 to 15 for each verse duration. (After Jatho, M., *Untersuchungen zur Schallproduktion und zum phonotaktischen Verhalten von Laubheuschrecken* [Orthoptera: Tettigoniidae], Cuvillier Verlag, Göttingen, 1995. With permission.)

not at all or only unspecifically to airborne-sound stimuli. By contrast, the responses to vibratory stimuli are similar if not identical. The functions of the subgenual organs in all six legs are basically identical, but the functions of the middle and distal parts of the CA remain to be characterised.

Summarizing the results of extensive investigations on the morphology, development, bioacoustics properties and physiology of the auditory–vibratory system of bushcrickets at the receptor and ventral cord level, basic similarities become evident when comparing the conditions in different species belonging to different subfamilies. The prothoracic spiracle and acoustic trachea in the foreleg represent the main input for airborne sound. Comparative morphometric measurements and theoretical calculations revealed that the acoustic trachea acts as a finite length exponential horn resulting in broadband transmission with superimposed resonances caused by internal reflections. Laser-vibrometry measurements of the vibrations of the tympana in the foreleg tibiae indicate that these structures play a crucial role in determining the overall acoustic impedance of the bushcricket ear, in particular the terminating properties of the acoustic trachea. This is most important for stimulus transmission to the receptor cells situated on the dorsal wall of the acoustic trachea.

In spite of differences of the dimensions, *e.g.* in leg size and resulting length of the acoustic trachea or of the morphology of the stimulus transmitting structures, like open and covered tympana, decisive structures inside the organs responsible for the acousto-mechanical stimulus transmission and transduction into bioelectrical receptor cell responses are very similar. The important function of these supporting structures for auditory stimulus transmission is also supported by studies on the postembryonic development of the sound conducting acoustic trachea and attachment structures inside the receptor organs. Similarities in the dimensions of these structures obviously cause very similar response types of auditory receptor cells, which can even be found in different species. Nevertheless, species-specific physiological adaptations were found in the different species investigated, especially when the frequency response properties of the entire auditory receptor neuron population was compared.

Basic similarities were also found in the structure and function of auditory–vibratory interneurons at the ventral cord level. Acoustic neurones of the ventral nerve cord ascending to the head ganglia are without exception bimodal auditory–vibratory in character. The convergence of the two sensory inputs at the ventral nerve cord level could be a fundamental element in the process of localising a stridulating partner at short distances (up to 1 to 2 m). Moreover, it could facilitate and improve the recognition of signals from conspecifics.

SOUND PRODUCTION AND ACOUSTIC BEHAVIOUR

The most important function of acoustic communication in bushcrickets is the attraction of sexually receptive conspecifics. In most cases males produce sound and females locate the singing males. Sound (and vibration) signals are produced by elytro–elytral stridulation, and in most cases contain a broad range of frequencies starting in some species at frequencies as low as 3 to 4 kHz and ranging up to 80 kHz. Comparative studies show that there is a wide variation of time–amplitude patterns of song syllables in the song of different species of bushcrickets. Individual sound impulses, damped oscillations within song syllables, are caused by friction of the plectrum over a row of teeth on the pars stridens. These sound impulses are clearly separated in some species (nonresonant sound production) or can be superimposed in others (resonant sound production). In many species, a combination of resonant and nonresonant sound production is evident and was shown to affect the power spectra of the song, especially in amplifying low frequency components. In addition, the subtegminal air volume was shown to affect both the amplitude of the low and high frequency components of the song as shown by sound recordings in a helium–oxygen mixture. Sound is broadcast during species-specific daily activity periods, and sound transmission can be substantially influenced by vegetation, interference with songs produced by other species and, for example, climatic conditions. Therefore, song recognition and discrimination represent crucial features of acoustic behaviour. A series of detailed choice experiments in two closely related species, *T. cantans* and *T. viridissima*, using computer generated song models revealed that differences in the species-specific syllable pattern represent a most important cue for the discrimination of females between conspecific and heterospecific males.

ACKNOWLEDGEMENTS

We are especially grateful to Marc Bangert, Elke Hoffmann and Thomas Sickmann for their contributions to this report, and we thank all former members of the Kalmring laboratory for their help in many experiments that were carried out over the years.

4 Sense Organs Involved in the Vibratory Communication of Bugs

Andrej Čokl, Meta Virant-Doberlet and Maja Zorović

CONTENTS

INTRODUCTION

Vibration is one of the most general constituents of any animal sensory environment and vibratory signals used for communication represent an important part. According to substrate mechanical properties the temporal and spectral characteristics of vibratory signals change during transmission. Sense organs have to be highly sensitive and adapted to extract signals from the general vibratory surroundings on one side, and on the other these need to be opened wide enough to receive the signal and preserve its informational value for neuronal processing in the central nervous system. Because every mechanoreceptor is sensitive to vibration, it is not surprising that with the increasing intensity of vibratory signals, more and more nonspecialised receptors together with specialised sense organs take part in their reception. Communication is a complex process of information exchange between the sender and the receiver, by means of signals of different modalities, running through different media. Observations of short range stink bug behaviour during mating, for example, demonstrated that vibratory communication is supported by visual, chemical, touch and other signals (Fish and Alcock, 1973; Harris and Todd, 1980; Borges *et al.*, 1987; Kon *et al.*, 1988; Miklas *et al.*, 2003). Experimental behavioural and neurobiological data on multimodal communication are lacking in stink bugs and other Heteroptera.

SPECTRAL PROPERTIES OF VIBRATORY SIGNALS

Vibratory signals of arthropods are produced by very different mechanisms like percussion, tymbal mechanisms, vibrations of the body or some of its parts and stridulation (Ewing, 1989). Stridulation is widespread in water- (Hydrocorisae and Amphibiocorisae) and land- (Geocorisae) dwelling bugs. Dominant frequencies of stridulatory signals emitted by a *Palmacorixa nana* (Corixidae) range around 5 kHz (Aiken, 1982) and broad band stridulatory signals of *Enoplops scapha* (Coreidae), for example, can extend to frequencies above 12 kHz (Gogala, 1984). Unlike stridulatory signals, spectra of bug vibratory signals produced by body vibrations are characteristically narrow banded with the dominant frequency below 200 Hz (Čokl and Virant-Doberlet, 2003). Cydnidae and representatives of several other bug families produce vibratory signals by stridulation and vibration of the body. Their spectra contain narrow band, low frequency and broad band components expanding the spectra well above 1 kHz (Gogala, 1984). The hypothesis that broad band signals are better suited for communication through plants than narrow band signals (Michelsen *et al.*, 1982), needs experimental confirmation.

Vibratory signals radiate from the emitter through air, water or solid. Gogala and coworkers (1974) demonstrated that species of Cydnidae communicate exclusively through the substrate, with signals produced by body vibration and stridulation. Because of the small body size of bugs, the intensity of the airborne component of low frequency and stridulatory signals is well below the threshold level of any known insect's auditory organ. Their body size is around 1 cm which is much smaller than one-third of the wavelength emitted and they could radiate only signals of frequencies above 10 kHz (Markl, 1983). Nevertheless we cannot exclude the potential information value of air particle movement in the acoustic nearfield created by mates during communication in close vicinity (Bennet-Clark, 1971; Kon *et al.*, 1988).

VIBRATION RECEPTION IN WATER-DWELLING BUGS

Despite physical constraints, airborne communication is represented in water-dwelling Heteroptera. Several species of the family Corixidae emit stridulatory signals of the carrier frequency well above 1 kHz. In *Corixa punctata*, the dominant frequency of stridulatory signals ranges between 1.5 and 2.8 kHz (Prager and Streng, 1982; Theiss *et al.*, 1983). The air bubbles used for respiration act as a resonator which pulsates during stridulation (Theiss, 1982). The resonant frequency increases with decreasing air bubble volume due to depletion of the air inside. Small species of body size around 2 mm, such as *Micronecta*, emit signals in the 11 to 12 kHz range while 11 to 13 mm sized *Corixa* species produce 1 to 3 kHz songs. The auditory organs (tympanal organs) in *C. punctata* are situated ventrally at the bases of the fore wings (Prager and Larsen, 1981). Each organ has only two sensory cells. When under water, the tympana lie in a branch of the respiratory air bubble, and water transmitted vibrations change air pressure within it causing indirect stimulation of the receptor cells. Different resonant properties of the left and right tympanal membranes (Prager and Larsen, 1981) probably cause bilateral asymmetry between the right and the left sensory cell, respectively, tuned to carrier frequency of conspecific stridulatory signals. It has been demonstrated that the resonant properties of respiratory air bubbles surrounding the submerged animal determine the frequency spectra of the emitted stridulatory signals (Theiss, 1982), and influence the overall tuning of the auditory system (Prager and Streng, 1982). Two-celled thoracic and the first abdominal chordotonal organs, together with one-celled first and second abdominal segment chordotonal organs, are present in many Amphibicorisae and Hydrocorisae. Although these organs are potential sensors also for hydrostatic or general vibration stimuli, the presence of tympana and high sensitivity to airborne sound suggest that at least those on the thorax are simple but true auditory organs (Leston and Pringle, 1963).

Although another water dwelling bug, *Notonecta*, is known to sing (Leston and Pringle, 1963), vibratory receptors have been investigated mainly in the context of localising prey on the water surface (Markl and Wiese, 1969; Wiese, 1972, 1974; Murphey and Mendenhall, 1973). One group of receptors is located in the prothoracic and mesothoracic legs at or near the tibio-femoral joint; the receptors situated at the tibio-tarsal joints are not required for localisation of prey. The mechano-receptive hairs along the margins of the three posterior-most segments and on the genitalia contribute significantly to vibratory directionality in *Notonecta undulata* (Murphey and Mendenhall, 1973). Wiese (1972) demonstrated that two types of water wave receptors are present in the proximal and distal parts of fore- and midlegs of *Notonecta*. The tarsal scolopidial organs as distal receptors (Wiese and Schmidt, 1974) are situated in all legs. Their frequency response ranges extend up to 1000 Hz with highest sensitivity between 100 and 150 Hz (threshold amplitude 0.5 μ). Five phasic and four phasic–tonic sensory units were identified electrophysiologically and histologically within the organs.

VIBRATION RECEPTION IN PLANT-DWELLING BUGS

The song repertoire of terrestrial Heteroptera is described in detail by Gogala (Chapter 21). Although the communication value of contact vibration or rhythmic touch together with near field air particle movement cannot be excluded at short distances, boundary vibrations in the form of bending waves (Michelsen *et al.*, 1982) are the most important carriers of substrate-borne information. Legs are in close contact with the substrate and, as such, are the sites of most sensitive substrate-borne vibratory receptors.

VIBRATION RECEPTION IN PYRRHOCORIDAE AND CYDNIDAE

The morphology of leg vibratory receptors was first described in *Pyrrhocoris apterus* (Pyrrhocoridae) (Debaisieux, 1938) and their function was first investigated for three *Sehirus* species of the family Cydnidae (Devetak *et al.*, 1978). In *Sehirus*, the authors described subgenual receptors broadly tuned between 50 and 2000 Hz, and femoral receptors tuned to frequencies below 200. In the range of highest sensitivity both receptors respond with threshold sensitivity of around 10^{-2} m/sec^2. Organ morphology and single-cell functional properties were not described.

VIBRATION RECEPTION IN *NEZARA VIRIDULA*

Most detailed study of the structure (Michel *et al.*, 1983) and function (Čokl, 1983) of leg vibratory receptor organs has been conducted on the southern green stink bug, *N. viridula*. In and on all six legs, the authors described the subgenual organ (Figure 4.1b), joint chordotonal organs (Figure 4.1a, c and d) and nongrouped campaniform sensilla.

Subgenual Organ

The subgenual organ lies in the blood channel in the proximal part of the tibia and is composed of two scolopidia, each with only one sensory cell (Figure 4.1b). The sensory cells are proximally attached to the epithelium of the tibial wall, whereas the scolopals with cilia and ligament are stretched out in the haemolymph. Axons of both sensory cells run within the sensory nerve and join the one innervating the femoral chordotonal organ. The subgenual organ of the southern green stink bug differs from that described in Orthoptera (Schnorbus, 1971) mainly in the number of receptor cells and in the attachment of its distal part. The ligament formed by two cap cells flattens distally into a thin flag-like structure, being only loosely fixed at the epidermis of the tibial wall and at both main tibial nerves.

Single cell recording from the leg nerve revealed that the middle frequency and high frequency receptor cells (MFR and HFR respectively) are highly sensitive to vibrations (Čokl, 1983) with "V"-shaped threshold curves running in parallel with the line of equal acceleration value

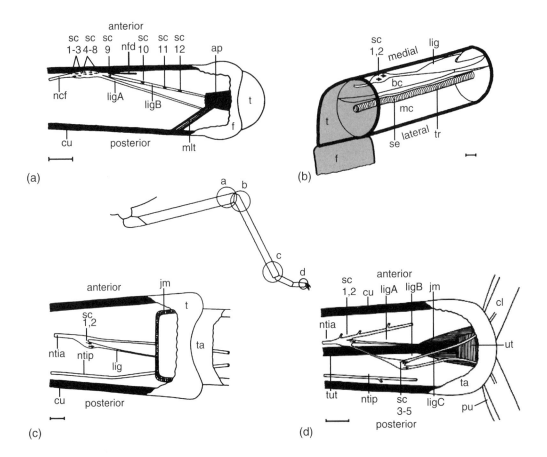

FIGURE 4.1 The morphology of the leg scolopidial organs in *Nezara viridula*. Schematic drawings of femoral chordotonal organ (a), subgenual organ (b), tibial distal chordotonal organ (c) and tarso-pretarsal chordotonal organ (d). ap (apodeme), bc (blood channel), cl (claws), cu (cuticle), f (femur), jm (joint membrane), lig (ligament), mc (muscle channel), mlt (musculus levator tibiae), ncf (femoral chordotonal organ), nfd (dorsal femoral nerve), ntia (tibial anterior nerve), ntip (tibial posterior nerve), pu (pulvillus), sc (scolopidia), se (septum), t (tibia), ta (tarsus), tr (trachea), tut (tendon of unguitractor), ut (unguitractor plate). The scale represents 200 μm (a,b) and 100 μm (c,d). (Partly redrawn from Michel, K., Amon, T., and Čokl, A., *Rev. Can. Biol. Exptl.*, 42, 139–150, 1983. With permission.)

(around 10^{-2} m/sec^2) below the frequency of best velocity sensitivity and in parallel with line of equal displacement value (between 10^{-7} and 10^{-8} for MFR and between 10^{-9} and 10^{-10} m for HFR) above it (Figure 4.2). The MFR is tuned to frequencies around 200 Hz with velocity threshold at or slightly above 10^{-5} m/sec and the HFR between 750 and 1000 Hz with velocity threshold between 1×10^{-6} and 2×10^{-6} m/sec. The frequency response range of MFR lies between 50 and 1000 Hz. Responsiveness of HFR neurons sharply decreases below 200 Hz; in the range between 1000 and 5000 Hz, suprathreshold responses could be obtained by vibratory stimuli of displacements between 10^{-10} and 10^{-9} m. Tonic responses with characteristic prolongation at stimulus frequencies around 200 Hz are characteristic for both subgenual receptor neurons. Phase-locked responses are characteristic for MFR neurons in the frequency range below 200 Hz. In the frequency range above 200 Hz, the spike repetition rate increases with increasing frequency up to 300 spikes/sec in the phasic part of the response and 200 spikes/sec in the tonic part, where every second cycle is followed by a spike up to 500 Hz.

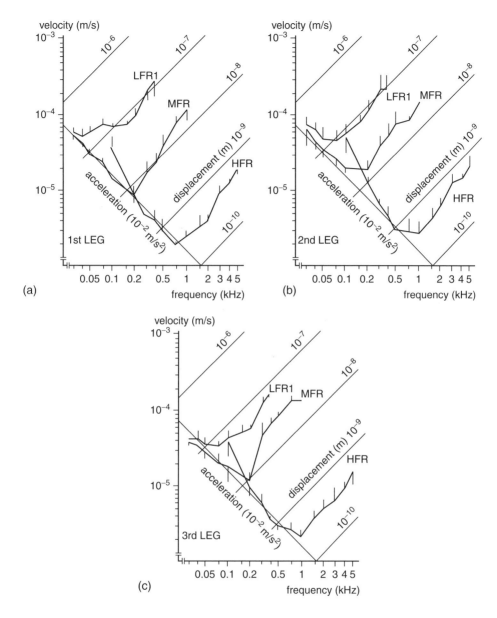

FIGURE 4.2 The threshold curves of LFR 1, MFR, and HFR units of the 1st (a), 2nd (b) and 3rd (c) legs, respectively, of *Nezara viridula*. Bars represent standard deviations. Straight lines in the frequency–velocity plots are lines of equal displacement and equal acceleration values. (From Čokl A., Functional properties of vibroreceptors in the legs of *Nezera viridula* [L.] [Heteroptera: Pentatomidae], *J. Comp. Physiol. A.* 150, 261–269, 1983. With permission.)

Central projections of the mid and hind leg MFR neurons were determined in the central ganglion of *N. viridula* (Figure 4.3) by Zorović *et al.* (2004). Both fibres enter the central ganglion in the posterior third of the leg nerve and have dense terminal arborisations in the medial half of the ipsilateral side of the ganglion. Their similar branching patterns suggest that both fibres originate in the same vibro-receptor organ supporting the assertion that sensory structures repeated in successive segments have similar ganglionic organisation (Eibl and Huber, 1979). Similar arborisation with one projection region has been described in the vibratory–auditory fibres in

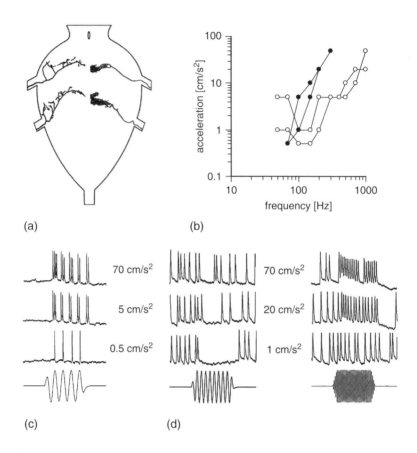

FIGURE 4.3 (a) represents dorsal view of the central ganglion (whole mount) showing central projections of LFR 1 (left) and MFR (right) receptor neurons. (b) represents threshold curves of LFR 1 (filled circles) and MFR (open circles) neurons. (c) represents responses of the middle leg LFR 1 neuron to 50 Hz, 100 msec duration stimulus of varying intensities. (d) represents responses of the middle leg MFR neuron to 100 and 400 Hz stimuli of 100 msec duration (shown as black bars) and varying intensities.

bushcrickets (Kalmring *et al.*, 1966) and in Type 5 auditory fibres in crickets (Eibl and Huber, 1979), whereas the fibres most likely to originate in the subgenual organ of crickets show completely different arborisation (Eibl and Huber, 1979; Esch *et al.*, 1980). Recent investigations (Zorović *et al.*, 2004) indicate that MFR receptor neurons are presynaptically inhibited as described for crickets (Poulet and Hedwig, 2003), locusts (Burrows, 1996) and stick insects (Stein and Sauer, 1999). Wolf and Burrows (1995) proposed that central neurons contribute to presynaptic inhibition of proprio-receptive sensory neurons.

Joint Chordotonal Organs

Chordotonal organs were described in *N. viridula* close to leg joints in the femur (the femoral chordotonal organ) (Figure 4.1a), distal part of the tibia (the tibial distal chordotonal organ) (Figure 4.1c) and in the last tarsal joint (the tarso-pretarsal chordotonal organ) (Figure 4.1d) (Michel *et al.*, 1983). The femoral chordotonal organ, situated in the lateral distal third of the femur at its anterior ridge (Figure 4.1a), is composed of 12 scolopidia, each with two sensory cells. Their axons run in a sensory nerve, branch together with axons of both subgenual organ sensory cells and join the main leg nerve in the posterior part of the femur, close to the joint with the trochanter. The organ is divided into a proximal part with eight scolopidia, and into a distal part with four scolopidia identified by Debaisieux (1938) as the condensed and dispersed scoloparium, respectively. The proximal part of the organ is fixed to the wall of the femur, and its proximal three scolopidia are separated from the rest; their cap cells form a ligament distally attached to the *musculus levator tibia*. The other five scolopidia are positioned distally and their cap cells form a separate ligament, distally fixed directly at the apodeme. In this ligament, the authors described four scolopidia of the dispersed femoral chordotonal organ; the cap cells run within the ligament and attach at the apodeme.

The tibial distal chordotonal organ lies in the blood channel proximally (400 μm) to the tibio-tarsal joint (Figure 4.1c) and has just two scolopidia, one with one and the other with two sensory cells, respectively. The round-shaped ligament of cap cells of both scolopidia extends distally and is attached to the joint membrane between the tibia and the first tarsal joint. The sensory cells lie at the dorsal side of *n. tibialis anterior*, with dendrites separated from the nerve and scolopals with cilia freely stretched in the blood channel. Axons of sensory cells run with other nerve fibres of the *n. tibialis anterior*.

The tarso-pretarsal chordotonal organ consists of two separate scoloparia (Figure 4.1d). The proximal one contains one scolopidium with two and one scolopidium with one sensory cell. The proximal scoloparium is distally fixed by a ligament at the tendon of the unguitractor moving the pretarsus. The distal scoloparium consists of three scolopidia, which are arranged into two separate parts, each with a separate ligament. The cap cell of the one-celled scolopidium is distally attached to the inner wall of the posterior claw. The second part of the tarso-pretarsal organ distal scoloparium contains two scolopidia, one with two and the other with one sensory cell. Their cap cells are attached to the inner wall of the anterior claw. Axons of the distal scoloparium form a thin nerve, which joins with others in the *n. tibialis anterior*.

The joint chordotonal organs and campaniform sensilla detect low frequency substrate vibrations (Figure 4.2). The low frequency receptor neurons (LFR) in *N. viridula* respond in the frequency range below 120 Hz in a phase-locked manner with best velocity sensitivity (threshold between 3 and 6×10^{-5} m/sec) between 50 and 70 Hz. Their threshold curves run in parallel with the line of equal displacement value (around 10^{-7} m) (Čokl, 1983). According to the phase of response, three types of LFR were identified electrophysiologically.

Two LFR1 neurons were marked with Lucifer yellow during intracellular recording from the central ganglion (Figure 4.3) (Zorović *et al.*, 2004). Their central arborisation and projection areas are restricted to the ipsilateral half of the ganglion. The fibres enter the central ganglion in the posterior third of the leg nerve. The main branch arches anteriorly until it reaches the ganglion midline. The side branches diverge from the main axon mostly on the anterior side and decrease in size towards the midline. This branching pattern is somewhat similar to those of the fCO fibres in locusts (Burrows, 1996) although the phase-locked response characteristics suggest that fibres may originate in campaniform sensilla (Kühne, 1982).

Antennal Mechanoreceptors

Observations of mating behaviour (Ota and Čokl, 1991) indicate that the antennal mechano-receptors of *N. viridula* may contribute to the detection of vibratory signals. Antennal campaniform sensilla and Johnston's organ with the central chordotonal organ were described by Jeram and Pabst (1996) and Jeram and Čokl (1996). On each antenna 12 campaniform sensilla were identified; six of them are grouped at the base of the first antennal segment. The Johnston's organ is located in the distal part of the pedicel in nymphs and in the distal part of the second pedicellite in adults. It is composed of 45 amphinematic scolopidia distributed around the periphery of the distal part of the third antennal segment (distal pedicellite). Scolopidia are attached separately in invaginations of cuticle between the pedicel and flagellum. Each scolopidium has three sensory cells and three enveloping cells (scolopale, attachment and accessory cell). Two sensory cells have short sensory cilia, attached distally to the extracellular tube, which has direct connection to the joint cuticle.

The third sensory cell with a longer sensory cilium attaches distally to the cuticle. Axons of 17 scolopidia join one antennal nerve, and 28 scolopidia of the opposite side join fibres of the other antennal nerve. The central organ described in the pedicel of *N. viridula* consists of seven mononematic scolopidia located proximally and centrally in regard to amphinematic scolopidia. Four of these are grouped into two scoloparia (each with two scolopidia). The four grouped and three separate scolopidia attach to the same place as those of the Johnston's organ. The axons of four scolopidia join one antennal nerve and those of the other three run together with fibres of the second antennal nerve.

Responses of receptor neurons to vibrations of the proximal flagellar segment revealed highest sensitivity around 50 Hz with threshold velocity sensitivity of 2×10^{-3} m/sec (Jeram, 1993, 1996; Jeram and Čokl, 1996). The phase-locked response pattern is characteristic for all of them. Back-fill staining of the *N. viridula* antennal nerve (Jeram, 1996) revealed mechano-sensory fibres with branches passing the antennal lobus and finally terminating at the ipsilateral side in the suboesophageal or prothoracic ganglion. A few of them project down to the abdominal region of the central ganglion. Such long axons were demonstrated in *Drosophila* to originate in campaniform sensilla of the pedicel (Strausfeld and Bacon, 1983).

TUNING OF VIBRATORY SIGNALS WITH TRANSMISSION MEDIA

Because of the resonant properties of the respiratory air bubble surrounding submerged *Corixidae*, frequency spectra of the emitted stridulatory signals (Theiss, 1982) and frequency sensitivities of the auditory system (Prager and Streng, 1982) are perfectly tuned. However, this is not so in plant-dwelling bugs. In the southern green stink bug, *N. viridula*, both subgenual organ receptor cells respond outside the dominant frequency range of the stink bug vibratory signals, recorded either as airborne sound (Čokl *et al.*, 1972, 1978; Kon *et al.*, 1988) or as vibration of a nonresonant loudspeaker membrane (Ryan *et al.*, 1995; Čokl *et al.*, 2000, 2001, 2004a; Pavlovčič and Čokl, 2001; McBrien and Millar, 2003; McBrien *et al.*, 2002; Čokl and Virant-Doberlet, 2003). Spectral properties of such recorded signals reflect frequency characteristics of body vibrations, depending directly on the repetition rate of muscle contraction (Kuštor, 1989). Representatives of the stink bug subfamily, Pentatominae, are entirely plant feeders (Panizzi, 1997; Panizzi *et al.*, 2000), mating on green plants which are likely to resonate when set in vibration by a singing bug. Recent recording of vibratory signals simultaneously with two laser vibrometers from the body and leaf below a singing *N. viridula*, revealed that several subdominant peaks, not present in spectra of signals recorded on a loudspeaker membrane appear in spectra of body and leaf recorded signals (Čokl *et al.*, 2005). Analyses of the resonant properties of a number of stink bug host plants (Čokl *et al.*, 2005) showed that the dominant resonant frequency peak lies well within the range of best sensitivity of the MFR subgenual organ receptor cell. A prominent subdominant resonant frequency peak lies between 80 and 100 Hz, and is tuned to the range of dominant frequencies of stink bug songs and to the peak sensitivity of the LFR neurons, coding precisely the temporal and frequency structure of vibratory signals in the frequency range below 100 Hz. Only the HFR cell responds optimally outside the frequency range of plant resonance and vibratory songs indicating that its role is preferentially oriented to detecting vibratory signals not directly connected with communication.

VIBRATORY SENSE ORGANS AND BEHAVIOUR

The distance of communication between mates on a plant depends on the intensity of the emitted signal, sensitivity of receptors and attenuation of signals during transmission through a medium. Behavioural experiments demonstrated that stink bugs communicate at distances well above 1 m when on the same plant. Long distance efficient communication is enabled by 20 dB to 40 dB difference between intensity of emitted signals recorded on the leaf immediately below the singing bug (3 to 31 mm/sec) (Čokl *et al.*, 2004b) and threshold-velocity sensitivity of leg vibratory receptors in the frequency range below 200 Hz (Figure 4.2). Attenuation of plant-transmitted vibratory signals was measured in several plants and can be as low as 0.3 dB/cm, as measured for 75 Hz vibratory signals transmitted through a banana leaf (Barth, 1998). In plant rod-like structures such as stems and stalks, we can expect standing wave conditions (Michelsen *et al.*, 1982) with the occurrence of readily repeated signal amplitude minima and maxima with increasing distance from the source. Investigations of amplitude variations with distance of artificially-induced pure-tone vibratory signals transmitted through *Cyperus* stem demonstrated that amplitude minima of 124 Hz signals do not fall more than 20 dB below the input value and the readily repeated amplitude maxima reach the input value at distances above 30 cm (Čokl, 1988). The amplitude of the main induced spectral component at 84 Hz varied above the input value at most measuring points exceeding values of the 124 Hz input signal.

There are no data about communication distances by broad band stridulatory signals transmitted through plants. The vibration velocity of broad band songs normal to the surface of the plants was between 0.1 and 1 mm/sec in most recordings from plant stems and leaves (Michelsen *et al.*, 1982). The threshold sensitivity of subgenual vibratory receptors in Cydnid bugs ranged at frequencies below 3000 Hz between 0.01 and 0.001 mm/sec (Devetak *et al.*, 1978). Attenuation of broad band stridulatory signals during transmission through plants has not been measured yet, so that the question of their value in long range communication remains open. It is possible that long distance mate location on a plant is enabled by low frequency components of the broad band signals and the stridulatory components are involved in species recognition at short distance or take part in another behavioural context.

COMPARISON WITH OTHER INSECTS

Debaisieux (1938) has demonstrated the most complete comparison of the morphology of leg scolopidial organs in insects. He described the subgenual organ and different joint chordotonal organs, emphasising significant differences between different groups. This work has not been followed up by comparative studies of functional properties, mainly because of great diversity of organ morphology and different technical problems connected to demands on stable surroundings for electrophysiological recordings. Also, the application of vibratory stimuli with defined parameters is very difficult. Although more data are now available because of the rapid development of relevant measuring and recording techniques, investigations are still focused on a small number of model species. So few complex morphological and physiological investigations on different species mean that a comprehensive comparison of subgenual organs between different insect groups is still not possible.

The subgenual organ of *N. viridula* has only two scolopidia and may be compared with those described in *Triecphora vulnerata* Germ. (Homoptera) with two scolopidia, and *Sialis lutaria* L. (Megaloptera) (Debaisieux, 1938) and *Chrysoperla carnea* (Neuroptera) Stephens (Devetak and Pabst, 1994) with three scolopidia, and *Panorpa communis* L. (Mecoptera) with only a single scolopidium (Debaisieux, 1938). Functional properties of the subgenual organ have been described only for *C. carnea* among these species (Devetak and Amon, 1997). The organ is composed of three scolopidia. The cell bodies of three cap cells form the velum, which is fixed to the leg wall and trachea so that it completely divides the blood channel into two parts. The organ shows best acceleration sensitivity (around 3 cm/sec^2) in the frequency range between 1.5 and 2.0 kHz (Devetak and Amon, 1997).

Data on biophysical characteristics of subgenual organs with a small number of scolopdia are lacking. To some extent the subgenual organ of *N. viridula* may be compared with that of the honeybee. The honeybee subgenual organ consists of about 40 scolopidia shaped as a hollow cone with its axis more or less parallel to the channel (Schön, 1911; McIndoo, 1922; Debaisieux, 1938). The organ is connected with the cuticle at two points, as in a termite (Richard, 1950) and an ant (Menzel and Tautz, 1994), with two connections between a membrane bag surrounding the organ and the membrane lining the tracheal walls. Biophysics of the organ was investigated by Kilpinen and Storm (1997). When the leg is accelerated, inertia causes the haemolymph and the subgenual organ to lag behind the movement of the rest of the leg. The magnitude of this phase lag determines the displacement of the organ relative to the leg and to the proximal end of the organ fixed to the cuticle. The oscillating tube model suggests that the sensory cells respond to displacements of the organ relative to the leg.

The morphology and function of the subgenual organ is best investigated in Orthoptera. The subgenual organ is present in all six legs. In the forelegs of most orthopteran groups it forms part of a complex tibial organ, which is highly specialised to detect airborne- and substrate-borne sound. Species of the family Haglidae (Mason, 1991), Stenopelmatidae (Ball and Field, 1981) and Tettigoniidae (Ball and Field, 1981; Lin *et al.*, 1993) have fully developed complex tibial organs with the subgenual organ, intermediate organ and crista acoustica. In the front legs of crickets (Gryllidae) the intermediate organ is lacking, in Raphidophoridae the complex organ consists of the subgenual and intermediate organ and in Acrididae *and* Blattidae only the subgenual and distal organs are present in the leg subgenual region (Schnorbus, 1971; Moran and Rowley, 1975). The distal organ may represent a predecessor of the intermediate organ (Lin *et al.*, 1995). The subgenual organ responds with highest sensitivity in the frequency range below 5 kHz with tuning of single cells to different frequencies in different groups. In *Troglophilus neglectus* (Raphidophoridae: Gryllacridoidea), for example, the best acceleration sensitivity (threshold between 3 and 20 cm/sec^2) of scolopidia of four functional types ranges between 0.3 and 1.4 kHz (Čokl *et al.*, 1995). In *Periplaneta* cockroaches, the subgenual organ responds also to low frequency airborne sound (Shaw, 1994) and differences in the shapes of auditory and vibratory threshold curves of the same receptor cells indicate that different mechanical structures transmit signals of both modalities to receptor cells (Čokl and Virant-Doberlet, 1997). The auditory–vibratory sensory system in bushcrickets is described in detail by Rössler, Jatho and Kalmring (Chapter 3).

One sensory cell of the *N. viridula* subgenual organ is tuned to frequencies around 200 Hz, which is well below the usual range of frequency sensitivity of organs investigated until now in other insect groups. More detailed biophysical investigations are needed to explain this low frequency sensitivity.

CONCLUDING REMARKS

Knowledge about sense organs involved in vibratory communication in general, and of bugs in particular, is still limited. Unlike data on song repertoire and related behaviour, most experimental data on receptors are based only on investigations on a few model species. Morphology of leg vibratory receptor organs in stridulating bugs is lacking. Observations of mating behaviour (Fish and Alcock, 1973; Harris and Todd, 1980; Borges *et al.*, 1987; Kon *et al.*, 1988) and data on pheromone emission modulated by female singing (Miklas *et al.*, 2003) demand detailed experimental investigations of multimodality at different levels of communication studies in Heteroptera.

ACKNOWLEDGEMENTS

The authors are grateful to friends and colleagues who share our interest in insect communication. We wish to thank both editors for helpful suggestions to improve the manuscript.

5 Use of Substrate Vibrations for Orientation: From Behaviour to Physiology

Meta Virant-Doberlet, Andrej Čokl and Maja Zorović

CONTENTS

INTRODUCTION

Substrate-borne signals play a crucial role in communication in many insect groups (Čokl and Virant-Doberlet, 2003; Virant-Doberlet and Čokl, 2004). Every walking, burrowing or sound-producing animal induces substrate vibrations that could be used in various inter- and intraspecific interactions. Many arthropods use vibratory signals in order to capture their prey, avoid predators or find their partners. For successful completion of these behaviour patterns, it is necessary not only to identify detected vibrations as signals emanating from a potential prey, predator or mate, but also to determine the location of the source of the vibrations.

    Observed behaviour, however, is a result of underlying neural mechanisms which in turn process the signals detected in the environment. In this chapter, we provide an overview of past and current research on mechanisms of vibratory directionality in insects, starting with the behavioural evidence which supports the idea that insects can accurately localise the vibration source. It is followed by discussion on possible mechanisms of localisation based on potential directional cues.

BEHAVIOURAL STUDIES OF LOCALISATION ABILITIES

While the role of vibratory signals as parts of specific-mate recognition systems has been widely studied, the ability of insects to localise a source of vibration has been less well documented. Table 5.1 provides a summary of major insect groups in which localisation based on vibratory information has been described. There are a number of studies which show that insects can localise a vibration source in the context of finding food (catching prey, finding a host, locating a fresh leaf or shoot) or locating a potential mate. However, we know of no studies which demonstrate that accurate localisation of source of vibration by insects plays an important role in avoiding predators. Substrate vibrations provide a warning signal for a singing cricket that a potential predator is approaching, though the common reaction is not a directional escape, but to stop singing (Dambach, 1989). Leafminer larvae react to vibrations generated by parasitoids, either by remaining still or in some cases by moving randomly (Meyhöfer et al., 1997a; Djemai et al., 2001). While on a water surface, mosquito larvae try to escape predators such as backswimmers (Notonectidae) and waterstriders (Gerridae) by a rapid diving response (Bleckmann, 1994).

Although anecdotal descriptions of behaviour or observational studies often suggest localisation based on vibratory cues, they rarely exclude other potential stimulis, such as chemical, visual or air-borne sound signals. Often it is not possible to determine whether insects use directional information which might be extracted from substrate vibrations, or whether detected vibratory signals induce increased locomotor activity and, as a result, insects find the source of vibrations by chance or by random search.

Accurate localisation of the vibration source can be particularly important for predators since their survival depends on their ability to catch prey. Insects which locate prey on the water surface orient towards the source of surface ripple waves generated by potential prey and localisation occurs by rotation in a series of discrete turning movements (Murphey, 1971a; Murphey and Mendenhall, 1973) (Figure 5.1). Terrestrial insect predators and parasitoids use vibratory signals generated during feeding or moving actions of the potential prey (Meyhöfer and Casas, 1999). Antlion larvae (Myrmeleontidae) wait for their prey at the bottom of funnel-shaped pit traps and their reaction to vibratory signals generated by walking prey is to throw the sand in its direction (Devetak, 1985; Mencinger, 1998) (Figure 5.1). Wasps parasitising immobile pupae use self-generated vibrations produced by tapping the substrate with their antennae and locating the position of a hidden host by analysing the reflected signals (vibratory sounding) (Wäckers et al., 1998).

When vibratory signals are used for locating other members of the group or a potential mate, the first step is usually to establish a duet (Strübing, 1958; Carne, 1962; Rupprecht, 1968; Hograefe, 1984; Ota and Čokl, 1991; Abbott and Stewart, 1993; Goulson et al., 1994; Čokl et al., 1999; Cocroft, 2001). Searching individuals alternate periods of emitting signals, and/or waiting for signals from conspecifics, with periods of walking. In Plecoptera, it has been shown that males took significantly less time to locate a female when pairs were engaged in continuous duets (Abbott and Stewart, 1993). It was also proposed that measuring the time males require for location after establishing a duet might be a possible cue for female sexual selection.

Males of the southern green stink bug, Nezara viridula (Pentatomidae), can localise accurately a vibration source, i.e. female on a plant (Ota and Čokl, 1991; Čokl et al., 1999) (Figure 5.2). When a male detects female calling signals, he responds with his own vibratory signals and starts walking on the plant. He approaches the source of the vibration with characteristic search behaviour which

TABLE 5.1
Behavioural Studies Demonstrating the Ability of Insects to Localise the Source of Vibrations

Insect Taxa	Behavioural Context	Localisation Task	Experimental Design	Ref.
Orthoptera				
Tettigoniidae	Mate location	Y branching point	Play-back	Latimer and Schatral (1983); Stiedel and Kalmring (1989)
		T branching point		De Luca and Morris (1998)
Gryllidae	Mate location	Y maze (soil)	Play-back	Weidemann and Keuper (1987)
Plecoptera				
Perlidae	Mate location	One dimension along the stem, branching	Observation	Rupprecht (1968)
Capniidae	Mate location	Two-dimensional surface (leaves)	Observation	Rupprecht (1968)
Pteronarcyidae	Mate location	Two-dimensional surface (cardboard)	Live couple	Abbott and Stewart (1993)
Hemiptera: Auchenorrhyncha				
Membracidae	Recruitment to new feeding site	One dimension along the stem, branching	Observation	Cocroft (2001)
	Mate location	Stem, branching	Observation, play-back	Hunt (1993)
Delphacidae	Mate location	One, two-dimensional surface	Observation	Strübing (1958)
		Y branching	Play-back	De Winter and Rollenhagen (1990)
Hemiptera: Heteroptera				
Gerridae	Prey location	Two-dimensional surface (water)	Play-back	Murphey (1971b)
	Mate location		Play-back	Wilcox (1972)
Vellidae	Prey location	Two-dimensional surface (water)	Observation	Jackson and Walls (1998)
Belostomatidae	Mate location	Two-dimensional surface (water)	Play-back	Kraus (1989)
Notonectidae	Prey location	Two-dimensional surface (water)	Play-back	Markl and Wiese (1969); Markl *et al.* (1973); Murphey and Mendenhall (1973)
Pentatomidae	Mate location	Y branching point	Play-back	Čokl *et al.* (1999)
	Prey location	Y branching point	Play-back	Pfannenstiel *et al.* (1995)
Neuroptera				
Myrmeleontidae	Prey location	Two-dimensional pit trap (sand)	Live prey	Devetak, (1985); Mencinger (1998)
Coleoptera				
Anobiidae	Mate location	Two-dimensional surface (wood)	Play-back	Goulson *et al.* (1994)

Continued

TABLE 5.1
Continued

Insect Taxa	Behavioural Context	Localisation Task	Experimental Design	Ref.
Gyrinidae	Prey location	Two-dimensional surface (water)	Play-back	Rudolph (1967)
			Live prey Observation	Reinig and Uhlemann (1973)
	Avoiding objects		Observation	Tucker (1969)
Tenebrionidae	Mate location	Two-dimensional surface (sand)	Observation	Hanrahan and Kirchner (1994)
Hymenoptera				
Tenthredinidae	Recruitment to new feeding site	One dimension along the stem, branching	Observation	Hograefe (1984)
Pergidae	Group cohesion	One dimension along the stem, branching	Observation	Carne, (1962)
Ichneumonidae	Host location (vibratory sounding)	One dimension, paper cylinder	Host mimic	Wäckers et al. (1998)
Eulophidae	Host location	Two-dimensional surface (leaf)	Live host, other cues excluded	Meyhöfer et al. (1997)
Braconiade	Host location	Two-dimensional surface (cloth, plexiglas)	Mobile/immobile host, Artificial signal	Lawrence (1981)
		Two-dimensional surface (agar)	Mobile/immobile host	Sokolowski and Turlings (1987)
Eucoilidae	Host location	Two-dimensional surface (agar)	Mobile/immobile host	Van Dijken and Van Alphen (1998)
Formicidae	Recruitment to food source	Y branching point	Play-back	Roces et al. (1993)
Sphecidae	Mate location	Two-dimensional surface (sand)	Play-back	Larsen et al. (1986)
Colletidae	Mate location	Two-dimensional surface (sand)	Play-back	Larsen et al. (1986)

includes stops at crossing points — waiting for the next female call and testing different possible paths with their legs. The same type of behaviour has been observed in males of a stonefly, *Perla marginata*, that searches for females on plant stems (Rupprecht, 1968). Strikingly similar male-search behaviour has also been described in a courtship of the webless plant-dwelling spider, *Cupiennius salei* (Rovner and Barth, 1981).

CUES FOR ORIENTATION

Localisation of the vibration source can be regarded as an interaction between the physical properties of the substrate and insect's anatomy, physiology and resulting behaviour. Although behavioural studies have demonstrated that insects can accurately localise the source

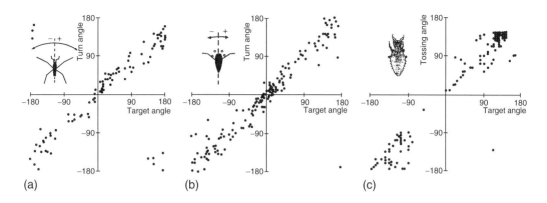

FIGURE 5.1 The relationship between target angle and turn angle in waterstrider, *Gerris remigis* (Heteroptera: Gerridae) (a), backswimmer, *Notonecta undulata* (Heteroptera: Notonectidae) (b) and between target angle and tossing angle in antlion, *Euroleon nostras* (Neuroptera: Myrmeleontidae) (c). Turns towards the right and left are given as positive and negative angles, respectively. Zero degree is directly ahead, 180° is directly behind. (Redrawn and modified from (a) Murphey, R. K. *Z. Vergl. Physiol.*, 72, 168–185, 1971b, Figure 1, p. 171. With permission from Springer-Verlag; (b) Murphey, R.K. and Mendenhall, B. *J. Comp. Physiol.*, 84, 19–30, 1973, Figure 2, p. 21. With permission from Springer-Verlag; (c) Mencinger, B. *Acta Zool. Fennica*, 209, 157–161, 1998, Figure 3a, p. 159. With permission from Finnish Zoological and Botanical Publishing Board.)

of vibrations, either on two-dimensional surfaces or at branching points, these experiments provide little information about the underlying mechanisms of localisation.

In insects, vibration receptors (campaniform sensilla, scolopidial organs including chordotonal organs and subgenual organ) are located primarily in all six legs (Chapter 3). An insect therefore has a spatial array of mechanoreceptors positioned on the substrate and a vibratory signal travelling through the substrate arrives at each leg at different times and with

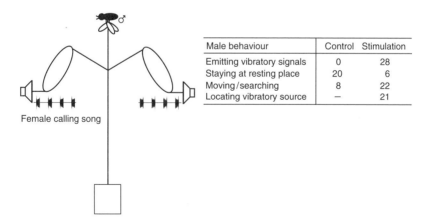

Male behaviour	Control	Stimulation
Emitting vibratory signals	0	28
Staying at resting place	20	6
Moving/searching	8	22
Locating vibratory source	–	21

Female calling song

FIGURE 5.2 Experimental arrangement used during directionality tests of males of *Nezara viridula* (Heteroptera: Pentatomidae) on a bean plant. Female calling signals were applied at the apex of the left or right leaf and male behaviour was recorded. Each male was tested in stimulus and no stimulus (control) conditions. Table on the right shows the results: in response to conspecific female vibratory signals males started to emit their own vibratory songs, they moved from their resting place and a great majority located the source of vibrations. A male was recorded as locating the signal if he reached the active loudspeaker or vibrated leaf and stayed there until the end of the trial. (Modified from Čokl, A., Virant-Doberlet, M., and McDowell, A. *Anim. Behav.*, 58, 1277–1283, 1999, Figure 1, p. 1278. With permission from Elsevier.)

different intensity (amplitude, velocity). Resulting differences in arrival times (Δt) and in amplitudes (Δd) are the most obvious directional cues, since each source direction is characterised by a unique temporal and amplitude pattern of leg stimulation. But the question is whether these directional cues are actually available to the insects.

The basic requirement to create time or amplitude differences that are large enough to be evaluated in the central nervous system is that the vibrations should travel through the substrate with relatively low propagation velocities or with relatively strong attenuation.

VIBRATIONS IN DIFFERENT SUBSTRATES

Vibration source can induce in a substrate (or on the surface of water) different types of vibrations simultaneously, which are characterised by different:

- Propagation velocities
- Loss of energy during the transmission (attenuation)
- Motions of the particles in relation to the direction of wave propagation

(Cremer and Heckl, 1973; Michelsen *et al.*, 1982; Markl, 1983; Gogala, 1985a; Bleckmann, 1994; Barth, 1998)

The most essential information concerning the physical properties of vibratory waves relevant for biological interactions (Table 5.2) is summarised as:

1. In sand, the surface Rayleigh waves propagate with low velocity and, in the frequency range of 0.1 to 5 kHz, attenuation is relatively low, increasing with increase in frequency (damping is particularly low in the frequency range of 300 to 400 Hz) and the slope of the attenuation curve decreases with distance (attenuation is stronger close to the source) (Brownell, 1977; Devetak, 1985; Aicher and Tautz, 1990; Brownell and van Hemmen, 2001).

2. On plants, bending waves are the most important type of vibrations for biological interactions. Bending waves propagate with low propagation velocity and with dispersion (velocity increases with increase in frequency). The velocity varies to some extent also with the mechanical properties of the plant (propagation velocity increases with increased stiffness and radius) (Michelsen *et al.*, 1982; Markl, 1983; Keuper and Kühne, 1983; Gogala, 1985a; Barth, 1998; Cocroft *et al.*, 2000; Miles *et al.*, 2001). Bending waves propagate with relatively little loss of energy and generally attenuation increases with increasing frequency (Baurecht and Barth, 1992; Barth, 1993). In structurally and mechanically complex and heterogeneous substrates, like plants, reflections create a frequency-dependent standing-wave pattern of vibration which results in complicated filtering properties (Michelsen *et al.*, 1982; Barth, 1993, 1998). The amplitude of vibratory signal does not decrease monotonically with the distance from the source, but oscillates (Michelsen *et al.*, 1982; Čokl, 1988). Signals may be more intense on the top of a plant or on a leaf than on a stem close to the source (Michelsen *et al.*, 1982; Keuper and Kühne, 1983; Amon and Čokl, 1990). At junctions, signals are not necessarily of higher amplitude at the branch closest to the source (Stritih *et al.*, 2000). Attenuation is also stronger during transmission through leaf lamina than through veins (Čokl *et al.*, 2004).

3. On water surfaces, waves propagate with very low propagation velocity and with dispersion. Attenuation is much greater than it is on plants, high frequencies are much more strongly damped than low frequencies and attenuation is stronger close to the source (Markl, 1983; Bleckmann, 1994; Barth, 1998).

TABLE 5.2
Propagation Velocities and Attenuation of Vibrations on Different Substrates

Substrate	Frequency (Hz)	Propagation Velocity (m/sec)	Attenuation (dB/cm)	Calculated Δt (msec) (1 cm/5 cm)	Ref.
Dry sand					
Rayleigh waves	300 to 400	40 to 50	0.23 to 0.40	0.2/1	Aicher and Tautz (1990); Brownell and van Hemmen (2001)
Plants					
Bending waves					
Vicia faba	200	39/36[a]		0.26/1.28	Michelsen *et al.* (1982)
	2000	122/120[a]		0.08/0.42	
Phragmites communis	200	78/75[a]		0.13/0.67	Michelsen *et al.* (1982)
	2000	246/220[a]		0.05/0.23	
Acer pseudo-platanus	200	88/95[a]		0.11/0.55	Michelsen *et al.* (1982)
	2000	279/–[a]		0.04/0.18	
Agave americana	30	4.4 to 35.7[b]	0.4[c]	2.27/11.23 to	Barth (1993, 1998)
	200	26 to 87[b]		0.28/1.40	
	2000	80 to 278[b]		0.39/1.82 to 0.12/0.58 0.13/0.63 to 0.04/0.18	
Musa sapentium	75		0.3[c]		Barth *et al.* (1988)
Water surface					
Surface waves					
	6	0.18[d]		55.56/277.78	Markl (1983); Barth (1998)
	100	0.55[d]		18.18/90.91	Barth (1998)
	10		1.67		Barth (1998)
	140		8.57		Markl (1983); Barth (1998)

[a] Calculated/measured.

[b] Measured at the apical third of the leaf and in its basal region.

[c] Average attenuation.

[d] Group propagation velocity.

DIFFERENCES IN TIME AND AMPLITUDE

Attenuation values and calculated differences in arrival time (Δt) (Table 5.2) indicate that values are large enough for both directional cues to be most probably available in insects' natural substrates, such as the water surface, sand, soil and plants. It has also been shown that arrival times and amplitude differences result in direction-dependent response patterns in vibratory ventral cord neurones (Čokl *et al.*, 1985). There are very few studies demonstrating which one of these cues (if any) insects actually use for orientation. Behavioural studies on scorpions (Brownell and Farley, 1979c) and spiders (Hergenröder and Barth, 1983), using two movable plates that vibrate legs either with different time delays or different amplitudes, have not yet been conducted on insects.

Water Surface

Because of low propagation velocities and strong attenuation of water surface waves, Δt and Δd are potentially valuable directional cues for insects living on the surface of water. In the backswimmer, *Notonecta*, the receptor closest to the wave source controls the direction of turn (Murphey, 1973) and the exact localisation is obtained by measurements of time intervals (Wiese, 1972, 1974). In waterstriders, localisation is obtained by determining which of the legs is closest to the source of ripples (Murphey, 1971b). In this case, which cue (Δt or Δd) was essential for localisation was not determined. It is very likely that they use Δt as a directional cue. Snyder (1998) described an artificial neural network which simulates directional rotational movements of a waterstider in which time differences between the arrivals of the stimuli were taken as a directional cue. The network was trained to rotate towards the source of vibrations and the ablations of vibration receptors were simulated. Behaviour simulated by the network corresponded well with behavioural data.

Sand

Sand appears to be a good medium for the conduction of biologically relevant vibratory signals due to low propagation velocity and low attenuation of Rayleigh waves. These properties suggest that Δt might be the more convenient directional cue. It has been shown that, for sand-dwelling scorpions, time delay between arrival of a vibratory signal at different legs is the most important directional cue (Brownell and Farley, 1979c). Larvae of the antlion, *Euroleon nostras*, can determine accurately the direction of the potential prey when it is still away from the rim of the sand pit (Mencinger, 1998) though there is no available information on localisation mechanisms.

Plants

On plants, insects face three different localisation tasks:

- Orientation on a two-dimensional surface like leaves
- Choice between the main stem and side branch or branches
- Forward or backward choice on a stem.

Heterogeneity of the substrate and the properties of bending waves impose some constraints on localisation based on Δt or Δd as directional cues since, for accurate localisation, a high degree of predictability in received information must be available. In particular, on plants there is often no reliable amplitude gradient (although attenuation may amount to up to10 dB at distances as short as 2 to 4 cm) (Michelsen *et al.*, 1982; Stritih *et al.*, 2000). Because of the dispersive nature of propagation, Δt depends on the frequency of the signal and, as a result, signals with higher frequencies create time delays which might be too short to be evaluated in the central nervous system. Although several behavioural studies have demonstrated that insects may successfully complete all the three localisation tasks as mentioned (see Table 5.1), there is no information about the underlying mechanism of localisation.

Observed search behaviour suggests that *N. viridula* could use both or either arrival time and amplitude differences as directional cues. However, amplitude gradients or amplitude differences between stems are often not reliable cues and, at least according to our current knowledge, comparing time difference might be more reliable. At junctions between a main stem and side branches, males of a stonefly, *Perla marginata*, showed the same type of search behaviour as *Nezara* (Rupprecht, 1968). Selective stimulation of one leg in males of this species resulted in a strong direction-dependent movement in the direction of the vibrated leg (Rupprecht, 1968). Orientation seems to depend on determination of which leg is closest to the source, *i.e.* the one that receives the signal first or with the highest amplitude. The mechanism underlying this behaviour has not been investigated.

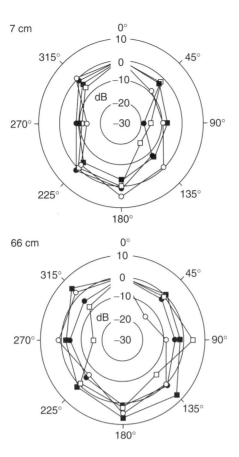

FIGURE 5.3 Characteristics of transversal signal transmission in *Cyperus alternifolius*. Artificial stimuli (duration 200 msec) were delivered to the stem 3 cm below them/leaves junction via a mini shaker being in contact with the stem over a cut end of aluminum cone. Signals were recorded at different distances from the point of vibration (7 cm and 66 cm are shown) with the microphone positioned around the stem at different angles according to the direction of vibration. Amplitude values are expressed in relative units (dB); reference (0 dB) is the amplitude determined at the particular distance at the same angle as applied vibration (0°). Single values at four different frequencies are shown: 100 Hz — filled circles, 200 Hz — open circles, 500 Hz — open squares, 1000 Hz — filled squares.

An insect standing on a stem does not have legs positioned in a two-dimensional circular arrangement as on a flat surface, but they are in a three-dimensional array around the stem. There are almost no two- or three-dimensional analyses of vibration transmission in plants available. The information that is available indicates that, in this situation, legs receive a complex spatial pattern of amplitude with increasing and decreasing values (Figure 5.3).

OTHER DIRECTIONAL CUES

Biological signals usually contain numerous frequency components and the vibratory signals of many different insects cover broad frequency ranges. As a consequence of frequency dispersion in plants, short broad band signals are transformed into frequency-modulated (FM) sweeps and their structures gradually change with distance from the source (Michelsen *et al.*, 1982; Gogala, 1985a). Time delay between the arrivals of components with different frequencies is in the range of several milliseconds even at short distances. For example, on *Vicia faba* 10 cm away from the source, the delay between 200 and 2000 Hz components would be 2 msec. With a proper underlying neuronal network, theoretically the insects could use this time delay to estimate the distance to the source and its direction. Insects have receptors which respond to different frequency ranges (campaniform sensilla and subgenual organs), but there is no behavioural or neurophysiological evidence indicating that insects might use dispersion time delay as a directional cue.

Directional information is available also in the mechanical response of the insect body to substrate vibration (Cocroft *et al.*, 2000; Miles *et al.*, 2001). Motion of the body along its long axis is a result of the rotational and translational mode of vibration and differs substantially depending on the stimulus direction.

Recently, the possibility has been raised that substrate vibrations might assist dancing honeybee foragers to attract dance-followers (Tautz *et al.*, 2001). At specific distances from the source of low frequency vibrations (200 to 300 Hz), comb cells reverse their phase of displacement across a single cell for a short period, effectively doubling the amplitude of the input signal. The distance from which the majority of remote dance-followers were recruited coincided with the location of the phase reversal phenomenon relative to the signal source.

ORIENTATION CUES IN SMALL INSECTS

Time and Amplitude Cues

The size of the insect is another essential factor in creating time or amplitude differences which are large enough to trigger accurate orientation. Localisation based on vibratory cues has been described in insects that vary considerably in size. For example, in *Tettigonia cantans*, distance between fore- and hind legs is 5 cm, while maximal leg span in *N. viridula* is 1 cm; in deathwatch beetle, *Xestobium rufovillosum*, distance between fore- and hindlegs is approximately 0.3 cm (Goulson *et al.*, 1994); in planthoppers, leafhoppers and parasitic wasps it can be even smaller. For a long time it was thought that, on solid substrates, small insects might be unable to extract directional information from vibratory signals; the main reason is that Δt and Δd are too small to be evaluated in the central nervous system. The shortest behavioural threshold time delay has been found in scorpions, which can detect time differences as small as 0.2 msec (Brownell and Farley, 1979c). However, calculations show that, for insects smaller than 1 cm, difference in arrival times between fore- and hindlegs would be well below this value (see Table 5.2). It is not known whether, in natural substrates, biologically relevant amplitude gradients can occur among points which are less than 1 cm apart. The lack of reliable amplitude cues in plants makes the use of Δd even more difficult. The exact mechanism underlying vibratory directionality in small insects still remains unclear and some additional potential directional cues have been proposed. These cues are available to large insects too.

It has been shown that, on a leaf, a consistent, strong attenuation of the signal even over a short distance takes place, especially when signals cross the veins (Magal *et al.*, 2000). Comparing the amplitude of the signal across a vein could provide the information about the relative position of the source of vibration. Such information may be used by parasitic wasps (Magal *et al.*, 2000) and possibly also by other insects. A male of the leafhopper, *Scaphoideus titanus* (leg span 0.2 cm), searching for a female on the lamina of a grapevine leaf stops during the initial search on veins and waits for a female response, moving away from veins only after he reaches the sector with the female (M. Virant-Doberlet, personal observation).

A directional mechanism which could amplify the effects of small time differences has been proposed for a treehopper, *Umbonia crassicornis* (leg span 0.5 cm) (Cocroft *et al.*, 2000; Miles *et al.*, 2001). A dynamic response of the body to vibration of the substrate results in a relatively large amplitude difference across the body and comparison of signal amplitude between receptors in the fore- and hind legs might provide directional information.

Behavioural Strategies

Solutions to complex localisation problems imposed by environmental and physical constraints might also result from behaviour. Maximal distance between the legs of *N. viridula* does not exceed 1 cm; however, during searching behaviour at branching points, where males have to choose among several paths, they test different stems by stretching their legs (and sometimes also antennae) between branches, extending the leg span to up to 2 cm (Ota and Čokl, 1991; Čokl *et al.*, 1999). This results in an increase in time delay (Čokl and Virant-Doberlet, 2003).

Some behavioural adaptations might not even require that directional information is extracted from the vibratory signal. Mate location in the leafhopper, *Graminella nigrifrons*, is facilitated by multiple cues (Hunt and Nault, 1991). In the first phase, males use a call-fly strategy to find a plant with a virgin female. After establishing a vibratory duet, male-search for a female on a plant is initially guided by positive photo taxis. This, in combination with perch-site preference by virgin females for the top of the plant, increases the likelihood of quick and successful location. Observations indicate that a call-fly strategy involving vibratory signalling might also be used by males of the alderflies *Sialis lutaria* and *S. fuliginosa* (Rupprecht, 1975).

The behaviour pattern observed in death-watch beetles suggests that males make large turns when they move beyond the range of the female response (Goulson *et al.*, 1994). Such a simple behavioural rule does not need any directional cue except that, after a duet has been established, attenuation of the signal drops below the receptor threshold. This orientation mechanism was not very efficient; males had to make a lot of turns and many failed to locate the female. The observed search path of the stonefly, *Capnia bifrons* (Rupprecht, 1968), also suggests that males might use amplitude of female response to determine whether he is approaching or leaving her.

Comparing the amplitude of the received signals and time delay between call and receipt of response at several locations (triangulation) could also enable determination of the direction of a vibratory source (Goulson *et al.*, 1994). Triangulation as a search pattern has been observed in the stonefly, *Pteronarcella badia* (Abbott and Stewart, 1993) (Figure 5.4). During their search, males made many turns with no clear directionality with regard to the position of the signalling female.

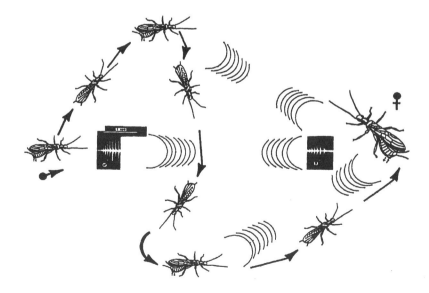

FIGURE 5.4 Diagrammatic presentation of triangulation search in *Pteronarcella badia* (Plecoptera). Insets show the actual male and female drumming signals. (From Abbott, J. C. and Stewart, K. W. *J. Insect Behav.*, 6, 467–481, 1993, Figure 7, p. 478. With permission from Kluwer/Springer-Verlag.)

DISTANCE DETERMINATION

As a part of the localisation process, animals should, in principle, determine not only a direction of the source of vibrations, but also its distance. However, there is almost no evidence of determination of distance in insects. Gyrinid whirligig beetles can avoid objects on the water surface by a form of echo location (Tucker, 1969). While swimming, they generate low-frequency waves which run ahead of them and with the antennae they can detect waves which are reflected from an obstacle. Even when swimming faster than the waves, travelling beetles can detect objects up to 1 cm away by means of the rising meniscus (Rudolph, 1967).

Several other available cues exist which could theoretically allow insects to estimate the distance to a source. On the water surface, such cues are curvature of a spreading concentric wave front, FM and frequency spectrum of the signal and frequency–amplitude content (since the decrease of signal amplitude and higher frequency content are more pronounced close to the source) (Bleckmann, 1994; Bleckmann et al., 1994).

In sand, amplitude gradient could be used to estimate the distance to a vibration source (Brownell and Farley, 1979c). The attenuation gradient across the leg span is large closer to the source while, at greater distances, different legs receive signals of approximately the same amplitude. Distance could also be determined from the time delay between the arrival of different types of wave (Brownell and Farley, 1979c).

On plants, the most obvious cue is the time interval between arrivals of different frequencies due to the dispersive propagation of bending waves (Michelsen et al., 1982). Insects may also roughly estimate the distance by comparing amplitudes of the signal in the three-dimensional array around the stem (Figure 5.3). Transversal amplitude oscillations depend on the frequency of the signal and distance from the source. With increasing distance, attenuation becomes smaller at angles perpendicular to the position of the source, and amplitude differences between angles become smaller.

CENTRAL MECHANISMS OF VIBRATION LOCALISATION IN INSECTS

In comparison with mechanisms of sound localisation in insects (reviewed by Gerhardt and Huber, 2002), processing of vibratory signals in the central nervous system has been very poorly studied and we are still far from understanding how information from vibration receptors is integrated and in turn controls orientation behaviour. In comparison with sound localisation in crickets, for example, localisation of a vibratory source is less amenable to experimental analysis. Most of the insects that rely on vibratory communication are very small, even compared with crickets, bushcrickets and grasshoppers. Also, because of very complicated modes of transmission through the substrate, it is extremely difficult to provide a predictable, experimentally manipulated stimulus. The task involves not only processing of the binaural cues in one segment, but also from several (usually six) spatially separated inputs located in three different thoracic ganglia, each having several receptor organs (subgenual organ; femoral, tibial, tarsal chordotonal organs; campaniform sensilla). The common feature of all vibratory systems in the ventral nerve cord studied up to now in bushcrickets (Tettigoniidae) (Kühne, 1982b), locusts (Acrididae) (Čokl et al., 1977; Kühne, 1982b; Bickmeyer et al., 1992), crickets (Gryllidae) (Kühne et al., 1984, 1985), cave crickets (Rhaphidophoridae) (Stritih et al., 2003) and bugs (Heteroptera) (Čokl and Amon, 1980; Zorović et al., 2003, 2004) is a relatively high number of interneurons with different morphologies and response properties. Most of these studies, however, describe only functional properties of interneurons responding to substrate vibrations. Only a few deal with possible integration and interactions of inputs from different legs.

MODEL NETWORKS

For insects, only one network model exists which represents the central nervous interactions of vibratory sensory inputs and refers to observed orientation behaviour. Artificial neuronal network model orientation behaviour in waterstriders (Snyder, 1998) does not use the same internal processing mechanism as the biological system, although it produced results qualitatively similar to real behaviour (Murphey, 1971b).

Using selective ablation of mesothoracic vibration receptors and observing orientation behaviour in *Notonecta*, Murphey (1973) proposed a model network which is compatible with observed behaviour. The model postulates inhibitory interactions among all sensory inputs. All inputs are equally weighted; the strongest inhibition is directed contralaterally; less strong is ipsilateral inhibition from front to back; and ipsilateral inhibition from back to front is the weakest.

PROCESSING OF VIBRATORY SIGNALS IN THE CENTRAL NERVOUS SYSTEM

Ironically, most of the available information about central processing of vibratory signals derives from research on locusts and crickets which primarily rely on air-borne sound communication and for which either behavioural vibratory directionality has not been established (locusts) or is relatively weak (crickets). In locusts, all ascending vibratory interneurons also receive inputs from the auditory system (Čokl *et al.*, 1977; Kühne, 1982b; Bickmeyer *et al.*, 1992), while in crickets several did not respond to air-borne sound (Kühne *et al.*, 1984, 1985).

The most important features regarding central integration of sensory inputs from the legs in the cricket, *Gryllus campestris*, are:

1. Most of the ascending vibratory interneurons received excitatory inputs from subgenual organs in all six legs (Dambach, 1972; Kühne *et al.*, 1984, 1985; Virant-Doberlet, 1989).
2. Inputs from one leg pair (in most cases midlegs) were dominant (Kühne *et al.*, 1984, 1985; Virant-Doberlet, 1989).
3. The influence of contralateral inhibition has been detected in the response of some interneurons (Kühne *et al.*, 1984, 1985; Virant-Doberlet, 1989).
4. Some interneurons showed a markedly different response pattern when the temporal pattern of leg stimulation has been changed (Virant-Doberlet, 1989) (Figure 5.5).

Interneurons connecting inputs from different legs with ascending interneurons have not been recorded in crickets. However, integration of the inputs from all six legs is certainly achieved via intersegmental interneurons and networks. Subgenual receptors project only to the ipsilateral half of their segmental ganglion (Eibl and Huber, 1979). Dendritic regions of the studied ascending interneurons were mainly restricted to one of the thoracic ganglia (Kühne *et al.*, 1984, 1985; Virant-Doberlet, 1989) and inputs from different legs differ by their latency of responses (Virant-Doberlet, 1989).

The ability of ventral cord neurons to code directional information has been tested in greater detail in *Locusta migratoria* (Čokl *et al.*, 1985). Legs were stimulated simultaneously or with a time delay between different leg pairs, simulating the position of a vibratory source in front or behind the animal. Results showed clear direction-dependent response patterns in some of the ascending interneurons (Figure 5.6). Some of these showed improved directional response when, in addition to a time delay, signal attenuation was also simulated. Directional information (in front–behind) encoded in the response pattern of ascending interneurons is obtained by integrating inputs from several legs or pairs of legs with an underlying inhibitory neuronal network. Some of the recorded interneurons were restricted to the thoracic ganglia and responded only to vibratory stimulation of a certain leg pair (Čokl *et al.*, 1985). These probably connect vibratory inputs from different legs with ascending interneurons.

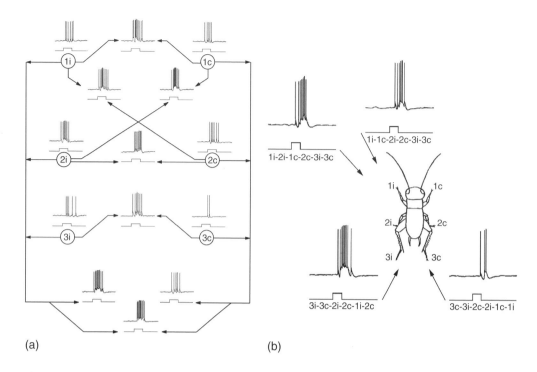

(a)   (b)

FIGURE 5.5 Responses of ascending interneuron to vibratory stimuli in *Gryllus campestris* (Orthoptera: Gryllidae). Responses were recorded intracellularly in the neck connectives. (a) Recorded responses of morphologically unidentified ventral cord neuron to stimulation of single legs and simultaneous stimulation of them in different combinations. i, c: ipsilateral/contralateral regarding the position of the axon of the recorded ascending interneuron. Stimulus frequency 500 Hz, duration 50 msec, intensity (acceleration 0.5 m/sec^2). (b) Recorded responses of the same interneuron to simulations of vibratory signals coming from different sides of the cricket. Arrows and time order of leg stimulation indicate the hypothetical direction of the stimuli. Stimulus frequency 500 Hz, duration 20 msec, acceleration 0.5 m/sec^2, time delay between each leg stimulation was 0.75 msec.

In reality, vibratory signals arrive at different spatially separated receptors at different times and with different amplitudes. Latency of neuronal response is inversely related to intensity of the stimuli (Gerhardt and Huber, 2002). In the cave cricket, *Troglophilus neglectus*, latency in different interneurons varies between 5 and 30 msec when intensity of vibratory stimuli differs for 5 dB (N. Stritih, personal communication). At the integration level such intensity-dependent time delays could enhance the time delay of arrival of the signal at different legs.

RECOGNITION VS. LOCALISATION

In order to catch a prey or find a conspecific mate, insects have to accomplish each task by recognising the emitter and locating the same. Under natural conditions, insects often have to choose between signals which are emitted from different, spatially separated sources. In order to orientate towards the correct one, the recognition and localisation tasks have to be integrated or co-ordinated. Integration processes have been examined in grasshoppers, crickets and bushcrickets in relation to processing directional information in sound communication (reviewed by Pollack, 1998; Gerhardt and Huber, 2002). Studies aimed at providing information on interactions

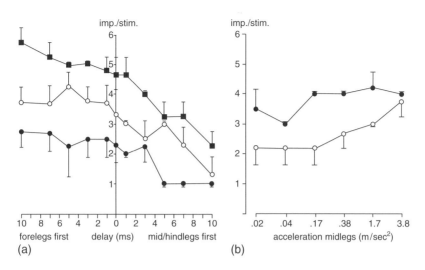

FIGURE 5.6 Responses of ascending interneuron in *Locusta migratoria* (Orthoptera: Acrididae) to vibratory stimuli presented to different leg pairs. (a) Response characteristics to stimulation of forelegs vs. mid- and hind-legs. Stimulus frequency 100 Hz, duration 20 msec, acceleration of 3.8 m/sec^2 (filled squares), 1.7 m/sec^2 (opened circles), 0.38 m/sec^2 (filled circles). (Modified from Čokl, A., Otto, C., and Kalmring, K. *J. Comp. Physiol. A*, 156, 45–52, 1985, Figure 2a, p. 48. With permission from Springer-Verlag.) (b) Response characteristics to stimulation of fore- and midlegs with a delay of 5 msec: midlegs vibrated before forelegs (filled circles), forelegs stimulated before midlegs (open circles). The acceleration of the stimulus applied to forelegs was kept constant (3.8 m/sec^2), the acceleration of the midleg stimulus is indicated on abscissa. Stimulus frequency 100 Hz, duration 20 msec. (Modified from Čokl, A., Otto, C., and Kalmring, K. *J. Comp. Physiol. A*, 156, 45–52, 1985, Figure 3b, p.49. With permission from Springer-Verlag.)

between recognition of vibratory source and localisation are lacking although some indirect evidence is available from results obtained on *Ribautodelphax* planthoppers by De Winter and Rollenhagen (1990). Although they tested the importance of vibratory signals in reproductive isolation, the results from two-way choice play-back experiments indicate that males recognise a species-specific female-call and also they are able to orient themselves towards the call. Taking into account the unpredictability of vibratory interactions when two signals are emitted simultaneously from different positions, it is not clear how planthoppers could extract the relevant information.

COMPARISON OF INSECTS WITH OTHER ARTHROPODS

Orientation towards a vibration source has been particularly well studied in spiders (Barth, 1998, 2002a, 2002b) and scorpions (Brownell and Farley, 1979a, 1979b, 1979c; Brownell and van Hemmen, 2001) and much insight into localisation mechanisms in insects derives from work done on other arthropods.

Spiders use substrate vibrations in two behavioural contexts while locating prey or a potential mate (Barth, 1998, 2002a, 2002b). Behavioural studies with two moveable platforms in which legs were vibrated with different time delay or different amplitudes showed that *C. salei* turns toward the leg which received a stronger or earlier stimulus (Hergenröder and Barth, 1983). The shortest time delay that resulted in directional response was 2 msec and the lowest amplitude difference was 10 dB (although a smaller Δd has not been tested) (Hergenröder and Barth, 1983; Barth, 1993, 1998, 2002a). The model network of possible central nervous interactions of the inputs from eight legs postulates inhibitory networks with ipsilateral and contralateral inhibition, acting only front to back, with differential weighting of the sensory inputs from each leg.

Fishing spiders capture their prey on the water surface and face the same localisation tasks as waterstriders and backswimmers. Spiders of genus *Dolomedes* use time interval to determine the direction of the wave source and curvature of a concentric wave for distance determination (Bleckmann and Barth, 1984; Bleckmann, 1994; Bleckmann *et al.*, 1994).

Plant-dwelling insects and spiders like *C. salei* have to deal with the same localisation problems caused by heterogeneity of the substrate and unpredictability of directional cues, and it is not yet known how spiders integrate the time and intensity differences which occur in their natural habitats (Barth, 2002a). The diagonal leg span of adult *Cupiennius* is 10 cm (Barth, 1993) and the resulting time delay for low-frequency courtship signals is in the range of the experimentally determined values which trigger directional response. However, signals emanating from walking prey contain higher frequencies with the resulting time delay below 1 msec and these values are even lower for young spiders with shorter leg spans (Barth, 1998).

Desert scorpions use substrate vibrations to locate their prey (Brownell, 1977; Brownell and Farley, 1979a, 1979b, 1979c). Behaviour studies with two moveable platforms showed that scorpions can use time delay or amplitude difference to localise the prey. However, when presented with conflicting cues (earlier, weaker/later, stronger), they invariably turned towards the side which moved first. This shows that, at least for scorpions, Δt is the critical cue for locating the source of vibrations and *Paruroctonus mesaensis* can detect time delays as short as 0.2 msec (Brownell and Farley, 1979c). A computational model of the neuronal mechanism underlying localisation of a vibratory source in a two-dimensional plane has been proposed (Stürzl *et al.*, 2000; Brownell and van Hemmen, 2001). The network consists of eight command neurons (one in each leg) that receive excitatory input from receptors in that leg and inhibitory input from the triad of receptors opposite to it. Localisation predicted by this model corresponds well with observed behaviour in animals with ablated receptors (Brownell and Farley, 1979c).

CONCLUDING REMARKS AND SUGGESTIONS FOR FURTHER STUDIES

Despite their small size, insects can localise the source of vibratory signals. It is, at least from a human perspective, an extremely difficult task since, in reality, insects probably have to deal with complex temporal, spatial and amplitude frequency-dependent patterns of stimulation of receptors in different legs. From this chapter, it is evident that we have hardly scratched the surface and the gaps in our knowledge and understanding are large. In order to bridge at least some of these gaps, some effort should be made to gather more information on the physical properties of natural substrates, particularly plants. Until we know more about available directional cues, localisation mechanisms will remain uncertain. In this context it is necessary to stress that behavioural experiments dealing with ability to locate the source of vibration and with mechanisms by which these are achieved should be performed on insects' natural substrates, since propagation velocity, attenuation and also frequency filtering are highly dependent on the substrate. Future research should be extended to a greater number of species and also to more detailed neurophysiological investigations of selected model species. Following the August Krogh principle (Sanford *et al.*, 2002) that "for many problems there is an animal on which it can be most conveniently studied", and taking into account the great number and variability of insects, it should not be too difficult to find such model species. However, localisation is a multi-level task. To get a proper insight, one has to analyse the behaviour, identify the properties of biologically relevant stimuli and link those to physical properties of the natural habitat and to receptor mechanisms, and subsequently to information processing in the central nervous system.

ACKNOWLEDGEMENTS

We wish to thank all our colleagues, past and present, who over the years shared with us the interest in research of vibratory communication. We are grateful to Dr. Bill Symondson and Dr. Mike Wilson, who helped to improve upon the manuscript. Last but not the least, we gratefully acknowledge the support and patience of the editors.

6 Size–Frequency Relationships in Insect Vibratory Signals

Reginald B. Cocroft and Paul De Luca

CONTENTS

INTRODUCTION

For insects that communicate using airborne sound, physical constraints on sound production and transmission lead to a relationship between body size and signal frequency. The efficiency with which sound is radiated depends on the size of the sound-producing structure. The lower the frequency, the larger the structure needed for efficient sound radiation (Michelsen and Nocke, 1974; Bennet-Clark, 1998a). This constraint places a lower limit, but not an upper limit, on the frequency of sounds used in communication. Once a sound has been radiated into the air, absorption of sound energy and scattering by objects in the sound path are both frequency-dependent, favouring lower frequency sounds for long-distance communication (Bradbury and Vehrencamp, 1998). As a compromise between these two constraints, animals often produce sounds whose frequencies are near the lower limit of efficient sound radiation (Bennet-Clark, 1998a).

The link between size and signal frequency has important evolutionary consequences. Because the size of the sound-producing structure is often closely related to overall body size, comparative studies usually reveal a negative relationship between body size and sound frequency among insects and other animals which use sound for long-distance communication (Ryan and Brenowitz, 1985; Bennet-Clark, 1998a; Gerhardt and Huber, 2002). Within species, ecological sources of selection on body size can influence the evolution of communication systems through their correlated effect on signal frequency (Ryan and Wilczynski, 1991). Social sources of selection may target frequency directly as a result of its reliable association with body size (Morton, 1977; Gerhardt and Huber, 2002).

In contrast to species which communicate using airborne sound, the relationship of body size to signal frequency has not been systematically investigated for species which communicate using substrate vibrations. Use of the vibratory channel is far more prevalent in insects than the use of airborne sound (Michelsen *et al.*, 1982; Markl, 1983; Claridge, 1985b; Gogala, 1985a; Henry, 1994; Stewart, 1997; Virant-Doberlet and Čokl, 2004), which probably occurs in hundreds of thousands of species (Cocroft and Rodríguez, 2005). One of the most striking differences between the signals of species communicating with vibrations and those communicating with sound is that the vibratory signals, on the whole, have much lower carrier frequencies (Cocroft and Rodríguez, 2005). It is unclear whether body size plays as important a role in the evolution of vibratory signals as it does in the evolution of airborne signals. There is recent evidence for the importance of frequency differences for mate recognition in species using pure tone signals for vibratory communication (Rodríguez et al., 2004). If divergence in signal frequency is decoupled from divergence in body size in such species, then vibratory signal frequency may be an evolutionarily labile trait, and one that could contribute to the rapid evolution of reproductive isolation among diverging populations.

Plant stems and leaves are among the most widely used substrates for insect communication. Plant borne vibratory signals are transmitted in the form of bending waves (Michelsen et al., 1982; Gogala, 1985a; Barth, 1997). As with sound waves, absorption of energy during propagation of bending waves is frequency-dependent, with greater losses at higher frequencies (Greenfield, 2002). Therefore, as with airborne sound, frequency-dependent attenuation should favour lower frequency signals for longer range communication. A second, less predictable influence on vibration transmission arises from the frequency filtering properties of plant stems and leaves. Although such filtering can favour lower frequency signals, it does not always do so, and there are too few studies of the vibration transmitting properties of plant tissue to permit broad generalisations (Michelsen et al., 1982; Čokl and Virant-Doberlet, 2003; Cocroft and Rodríguez, 2005).

The question is whether the mechanics of vibratory signal production impose a relationship between body size and signal frequency as with airborne sounds. Although use of substrate vibrations releases animals from some constraints on signal frequency, such as the acoustic short-circuit which makes radiation of low frequency sounds by small dipole sources inefficient (Gerhardt and Huber, 2002), the potential for other size-related constraints on signal frequency has not been explored. Little is known about the coupling of a vibratory signal between an insect and a plant stem (Michelsen et al., 1982). The details of vibratory signal production are also unknown in most cases, apart from observations of which body parts are involved (Virant-Doberlet and Čokl, 2004). Our lack of knowledge of the details of vibratory signal production and transmission prevent us from making specific predictions about the relationship of body size to frequency.

Here we take an empirical approach to the question of size–frequency relationships in vibratory signals. Comparing the spectral features of vibratory signals across different individuals and species, or using data drawn from different studies, presents two challenges. First, the distribution of energy across the different frequencies in a signal will be influenced by the properties of the substrate on which a signal is recorded (Michelsen et al., 1982). For signals using a narrow band of frequencies, the influence of substrate may be small or absent (Sattman and Cocroft, 2003). For signals containing a wider range of frequencies, differences in substrate filtering properties may introduce a significant amount of variation into a comparative dataset. Second, different investigators may use transducers which measure different components of a vibratory signal, and this will be reflected in the amplitude spectrum of the signal. For example, in a signal with a range of frequencies, acceleration amplitude will increase by 6 dB/octave relative to velocity amplitude. In this study we use methods which minimise substrate- and transducer-induced variation in signal amplitude spectra; or, for comparisons in which these sources of variation cannot be eliminated, we discuss their implications for interpretation of the results.

We investigated the relationship between body size and frequency at three levels: within a population; between closely related species; and across a wide range of species in different insect orders. For our investigations of size–frequency relationships within populations, we recorded a sample of individuals on a common substrate. For comparisons among closely related species, we use our own library of recordings of the signals of membracid treehoppers, and for the broader comparison we have drawn information from the literature.

METHODS

WITHIN-POPULATION VARIATION IN SIZE AND SIGNAL FREQUENCY

To investigate the relationship of body size to signal frequency within a population, mate advertisement signals were recorded from a sample of 35 males of the Neotropical treehopper *Umbonia crassicornis* (Hemiptera: Membracidae). Males were drawn from a greenhouse colony at the University of Missouri established with periodic collections from populations in southern Florida, USA, where this species has been naturalised. Signals were transduced using a laser vibrometer (Polytec CLV 1000 with CLV M030 decoder modules) at 5 mm/sec/V sensitivity. The laser head was positioned on a tripod approximately 10 to 15 cm from the stem and a small ($\sim$1 mm^2) piece of reflective tape was attached to the stem at the recording location to increase reflectance of the laser signal. Signals were digitised at 44.1 kHz using a National Instruments data acquisition board and a custom written data acquisition program in LabVIEW v. 7.0.

Substrate related variation was minimised by recording each male at the same location ($\pm$ 2 cm) on a 1 m tall potted host plant (Mimosaceae: *Albizia julibrissin*). Males were positioned with their dorso-ventral axis in the plane of the laser. A recorded male–female duet was played through a loudspeaker to induce signalling once a male was placed on the stem. Recordings were made with the host plant on a Kinetic Systems Vibraplane isolation table in a temperature-controlled room maintained at (23 ± 1)°C. After each male was recorded, its mass was measured using a Mettler Toledo AB545 balance, and its body length (front of vertex to tip of abdomen) was measured with an ocular micrometer using a Leica MZ75 microscope.

Signal frequency was measured for one signal from each male using a custom written program in MATLAB v. 6.5, with an FFT size of 8192 points. In *U. crassicornis* signals, the dominant frequency was typically the second harmonic; however, for some signals it was the fundamental. Accordingly, the principal energy in the second harmonic was measured for each signal to provide a more consistent measure across individuals.

The males recorded in this study were drawn from seven family groups. Families remain together through nymphal and early adult development, and thus siblings are similar in both genetic and environmental contributions to the phenotype. Accordingly, we used multiple regression, with family included as a nominal variable. The analysis was conducted using JMP IN 5.1, with size and frequency measurements log-transformed for statistical analysis.

We also provide data drawn from a study of size–frequency relationships in a species in the *Enchenopa binotata* species complex (Sattman and Cocroft, 2003). As in the present study, a sample of males was recorded with laser vibrometry. Males were recorded on more than one host plant stem. However, as shown in that study, the differences among individual host plants did not influence measurements of signal dominant frequency.

BETWEEN-SPECIES VARIATION IN SIZE AND SIGNAL FREQUENCY IN THE MEMBRACIDAE

To examine the relationship between body size and signal frequency among a set of closely related species, we used our own library of recordings of temperate and tropical species of Membracidae (see Table 6.1 for a list of species). We included one aetalionid, *Aetalion reticulatum*, which was placed within the Membracidae in the phylogenetic tree we used. Recordings of male mate advertisement signals were made on cut host plant stems, usually about 0.25 m in length, with the base placed in a florist's water tube held in a clamp. The transducer was placed within 5 to 10 cm of the signalling insect, reducing the changes in signal spectra imposed by transmission through a stem. In order to allow for use of phylogenetic comparative methods (see below), we included only those species in our recording library which were also present in an unpublished molecular phylogeny of membracids made available to us by C.P. Lin and R.L. Snyder (personal communication).

TABLE 6.1
Membracid Species Included in the Comparative Analysis

Acutalis tartarea	*Heteronotus trinodosus*	*Potnia brevicornis*
Aetalion reticulatum	*Hypsoprora coronata*	*Potnia dubia*
Alchisme apicalis	*Ischnocentrus inconspicua*	*Publilia concava*
Aphetea inconspicua	*Lycoderes* sp. Panama	*Smilia camelus*
Atymna querci	*Metheisa lucillodes*	*Stictocephala diceros*
Bajulata bajula	*Microcentrus perditus*	*Stictocephala lutea*
Bolbonota sp. Panama	*Micrutalis calva*	*Stylocentrus championi*
Campylenchia latipes	*Micrutalis* sp. Panama	*Thelia bimaculata*
Campylocentrus brunneus	*Nassunia bipunctata*	*Tolania* sp. Panama
Cladonota apicalis	*Nassunia* sp. Panama	*Trinarea sallei*
Cymbomorpha prasina	*Notocera bituberculata*	*Tropidaspis affinis*
Cyphonia clavata	*Ophiderma definita*	*Tylopelta americana*
Cyrtolobus vau	*Ophiderma salamandra*	*Umbonia crassicornis*
Darnis latior	*Oxyrachis tarandus*	*Umbonia spinosa*
Enchenopa binotata Panama	*Platycotis* sp. Panama	*Vanduzea arquata*
Enchophyllum melaleucum	*Polyglypta costata*	*Vanduzea mayana*
Glossonotus crataegi	*Poppea capricornis*	*Vanduzea segmentata*

Our recordings of membracid signals were made with three different transducers. In addition to the laser vibrometer described above, we used a Knowles BU-1771 accelerometer and an Astatic 91T ceramic phonograph cartridge. Different transducers measure different components of a vibratory signal, accelerometers measure its acceleration; laser vibrometers measure its velocity; and ceramic phonograph cartridges measure its displacement. To compare, measurements made with different transducers we converted all signal spectra to velocity units. We first played band-limited noise (80 Hz to 5 kHz) through a shaker and measured it simultaneously with the three transducers. We then used the ratios between the relevant amplitude spectra to adjust the spectrum of signals recorded with the accelerometer or phonograph cartridge. As a result, all of our measurements of the amplitude spectra of membracid signals are directly comparable.

For many of the membracids we have recorded, and especially for rarely encountered tropical species, signals are available for only one or a few individuals. Our measurement of signal frequency was therefore made from measurement of one individual per species. For purposes of this analysis, we assumed that between-species variation was greater than within-species variation. In species for which we have many recordings, this assumption appears to be met as frequency has a low coefficient of variation within populations (unpublished data). Size was likewise measured for one individual per species, where possible from the voucher specimen from which the signal was recorded. As an index of body size, the total length was measured from the front of the vertex to the tip of the folded wing as in McKamey and Deitz (1996), using either an ocular micrometer (as above) or digital photos with a scale included. Size measurements for a few species for which specimens were not available were obtained from Funkhouser (1917).

We measured the dominant frequency using MATLAB as above. Because every individual was recorded on a different substrate, these dominant frequency measurements are subject to substrate-induced variation. Accordingly, we also measured the lowest frequency in the signal containing appreciable energy (20 dB below the amplitude of the dominant frequency). We anticipated that a measurement reflecting the overall bandwidth of the signal would be less substrate-dependent than measurements reflecting the relative amplitude of different frequencies within that frequency band. All variables (body length, dominant frequency, low frequency) were log-transformed for statistical analysis.

Before proceeding with a statistical analysis of the correlation of size and frequency, we assessed the degree of phylogenetic autocorrelation in the data using the test for serial independence in PI (Phylogenetic Independence) v. 2.0 (Reeve and Abouheif, 2003). The molecular phylogenetic tree provided by C.P. Lin and R.L. Snyder (personal communication) was pruned to exclude species not used in our analysis. The test for serial independence revealed the presence of significant phylogenetic signal in dominant frequency ($p < .01$), but not in size ($p = .35$) or in the lowest frequency ($p = .31$). To adjust for the phylogenetic component of variation in dominant frequency, we used the CAIC program v. 2.0.0 (Purvis and Rambaut, 1995) to calculate phylogenetically independent contrasts between size and dominant frequency. Below we present the original data and the independent contrasts.

BETWEEN-SPECIES VARIATION IN SIZE AND SIGNAL FREQUENCY IN THE INSECTA

We examined the relationship of body size and signal frequency on a broad scale across several insect orders using data drawn from the literature (see Table 6.2 for a list of species and references). We obtained measures of dominant frequency and the lowest frequency present in the signal (20 dB below peak, as above) either from the measurements reported in the paper or from estimates based on amplitude spectra or spectrograms. Only one of the papers on insect vibratory communication that we examined included information on the size of the insects. For the rest of the species, we used approximate size information by obtaining a measurement of total length from values reported in the literature for the same sex and species recorded. Information was also obtained from the investigators who published the signal analyses; from values reported on websites; or from our measurements of museum specimens. Because there is no species-level phylogeny available for the Insecta as a whole, our examination of size–frequency relationships at this scale did not incorporate information on phylogenetic relationships.

The sample drawn from the literature includes signals with a variety of functions, including mate attraction, solicitation of maternal care and attraction of ant mutualists. We first examined the relationship of body size to frequency for all signals regardless of function using log-transformed variables. Then, because the above investigations of size–frequency relationships within populations and among closely related species were based on male mate advertisement signals, we conducted a second analysis based only on mating signals.

RESULTS

WITHIN-POPULATION VARIATION IN SIZE AND SIGNAL FREQUENCY

There was no relationship between body length and signal frequency (measured as the principal energy in the second harmonic) for a sample of 35 *U. crassicornis* males (Figure 6.1a; $F = 1.33$, $p > .25$); the same was true for body mass ($F = .04$, $p > .8$). There also was no effect of family on signal frequency (for length and mass $p > .3$). Analyses using dominant frequency, which for some individuals was the fundamental rather than the second harmonic, yielded similar results (*e.g.* for mass, $F = .001$, $p > .9$).

We also provide data from a previous study examining the relationship of size and mating signal frequency in a second species of membracid, the *E. binotata* complex member which uses *Ptelea trifoliata* host plants (Sattman and Cocroft, 2003). In that dataset there was no relationship between body length and the dominant frequency of the signal (Figure 6.1b; $N = 24$).

TABLE 6.2
Insect Species Used in the Comparative Analysis with References for Signal Descriptions

Coleoptera	Curculionidae	*Hylobius abietis*	Selander and Jansson (1977)
Diptera	Chloropidae	*Lipara lucens*	Ewing (1977)
—	Psychodidae	*Lutzomyia longipalpis*	Ward *et al.* (1988)
Hemiptera	Aleyrodidae	*Trialeurodes vaporariorum*	Kanmiya (1996a, 1996b, 1996c)
—	Alydidae	*Alydus calcaratus*	Gogala (1990)
—	—	*Riptortus clavatus*	Numata *et al.* (1989)
—	Aphodidae	*Aphodius ater*	Hirschberger and Rohrseitz (1995)
—	Cicadellidae	*Dalbulus* sp.	Heady *et al.* (1986)
—	—	*Empoasca fabae*	Shaw *et al.* (1974)
—	—	*Graminella nigrifron*	Heady and Nault (1991)
—	—	*Macrosteles fascifrons*	Purcell and Loher (1975)
—	Cydnidae	*Sehirus bicolor*	Michelsen *et al.* (1982)
—	Membracidae	*Enchenopa binotata*	Sattman and Cocroft (2003)
—	—	*Spissistilus festinus*	Hunt (1993)
—	—	*Umbonia crassicornis*	Cocroft (1999)
—	Pentatomidae	*Acrosternum hilare*	Čokl *et al.* (2001)
—	—	*Nezara viridula*	Čokl *et al.* (2000)
—	—	*Thyanta custator*	McBrien *et al.* (2002)
—	—	*Thyanta pallidovirens*	McBrien *et al.* (2002)
—	Plataspidae	*Coptosoma scutellatum*	Gogala (1990)
—	Reduviidae	*Triatoma infestans*	Roces and Manrique (1996)
—	Rhopalidae	*Corizus hyoscyami*	Gogala (1990)
—	Tettigarctidae	*Tettigarcta crinita*	M.F. Claridge, personal communication
Hymenoptera	Apidae	*Apis mellifera*	Michelsen *et al.* (1986a, 1986b)
—	Formicidae	*Atta sexdens*	Masters *et al.* (1983)
Lepidoptera	Lycaenidae	*Arawacus lincoides*	DeVries (1991)
—	—	*Chlorostrymon simaethis*	DeVries (1991)
—	—	*Jalmenus evagoras*	DeVries (1991)
—	—	*Leptotes cassius*	DeVries (1991)
—	—	*Lysandra bellargus*	DeVries (1991)
—	—	*Maculinea alcon*	DeVries (1991)
—	—	*Panthiades bitias*	DeVries (1991)
—	—	*Polyommatus icarus*	DeVries (1991)
—	—	*Rekoa palegon*	DeVries (1991)
—	—	*Strymon yojoa*	DeVries (1991)
—	—	*Thereus pedusa*	DeVries (1991)
—	Riodinidae	*Calospila cilissa*	DeVries (1991)
—	—	*Calospila emylius*	DeVries (1991)
—	—	*Juditha molpe*	DeVries (1991)
—	—	*Nymphidium mantus*	DeVries (1991)
—	—	*Synargis gela*	DeVries (1991)
—	—	*Synargis mycone*	DeVries (1991)
—	—	*Theope matuta*	DeVries (1991)
—	—	*Theope thestias*	DeVries (1991)
—	—	*Theope virgilius*	DeVries (1991)
—	—	*Thisbe irenea*	DeVries (1991)
Mecoptera	Panorpidae	*Panorpa* sp.	Rupprecht (1975)
Neuroptera	Chrysoperlidae	*Chrysoperla plorabunda*	Henry and Wells (1990)
—	Sialidae	*Sialis* sp.	Gogala (1985a)

Continued

TABLE 6.2
Continued

Orthoptera	Tettigoniidae	*Choeroparnops gigliotosi*	Morris *et al.* (1994)
—	—	*Conocephalus nigropleurum*	De Luca and Morris (1998)
—	—	*Copiphora brevirostris*	Morris *et al.* (1994)
—	—	*Docidocercus gigliotosi*	Morris *et al.* (1994)

BETWEEN-SPECIES VARIATION IN SIZE AND SIGNAL FREQUENCY IN THE MEMBRACIDAE

If we first examine the data using species as independent data points in a linear regression, there is a significant negative relationship between size (body length) and dominant frequency (Figure 6.2a; $N = 51$, $r^2 = .09$, slope $= -.60$; $p < .05$). There is also a significant negative relationship between size and the lowest frequency in the signal (Figure 6.2b; $N = 51$, $r^2 = .17$, slope $= -.69$, $p < .01$). Although there is considerable scatter in both figures, there were no small species that produced low frequency signals. Dominant frequency was the only variable

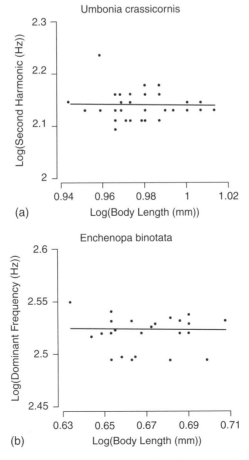

FIGURE 6.1 The relationship of body size and frequency in the vibratory mating signals of two species in the family Membracidae. (a) *U. crassicornis*; (b) *E. binotata* from *P. trifoliata*.

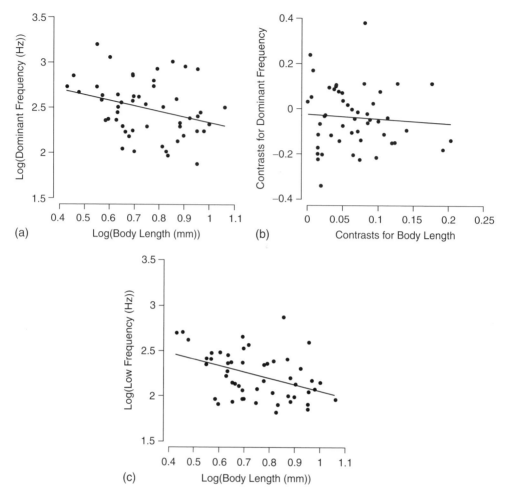

FIGURE 6.2 The relationship of body size and frequency among 51 species in the family Membracidae. (a) Relationship between size and dominant frequency; (b) phylogenetically independent contrasts between size and dominant frequency; (c) relationship between size and the lowest frequency in the signal.

which showed significant phylogenetic signal; if we examine the phylogenetically independent contrasts, the results are qualitatively similar with a marginally significant relationship between size and dominant frequency (Figure 6.2c; $N = 50$ contrasts, $r^2 = .07$, slope $= -.45$, $p = .06$).

BETWEEN-SPECIES VARIATION IN SIZE AND SIGNAL FREQUENCY IN THE INSECTA

If all of the signals from our literature survey are included regardless of signal function, linear regression reveals no relationship between body size and dominant frequency (Figure 6.3a; $N = 52$) or between body size and the lowest frequency in the signal (Figure 6.3b; $N = 53$). Note that these comparisons use species as data points in the absence of a species-level phylogeny for this broad comparison.

In contrast, if we consider only the species for which measurements were available for mate advertisement signals, there is a significant negative relationship between body size and dominant frequency (Figure 6.4a; $N = 28$, $r^2 = .28$, slope $= -.60$, $p < .01$) and between body size and the lowest frequency in the signal (Figure 6.4b; $N = 29$, $r^2 = .42$, slope $= -.71$, $p < .001$).

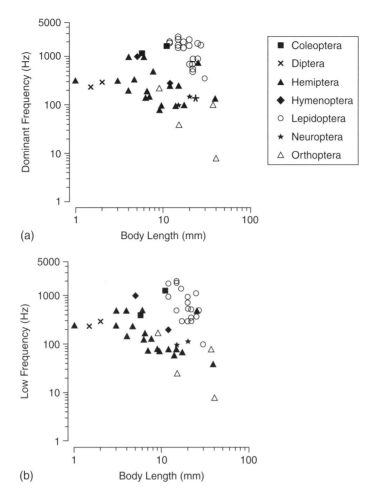

FIGURE 6.3 The relationship of body size and frequency in 53 species of insects from several orders using data drawn from the literature. (a) and (b) show the relationship of size and dominant and low frequency for all species, whether the signal described for a given species was used in mating, alarm or attraction of mutualists.

DISCUSSION

For communication systems using airborne sound, the close relationship between body size and frequency has shaped the evolution of signal function and diversity (Morton, 1977; Gerhardt and Huber, 2002). Here we asked if a similar relationship exists for the most widespread form of mechanical signalling, substrate-borne vibratory communication. We examined the relationship of body size to signal frequency within populations and across species.

We found rather different patterns in the within-population comparison than in the between-species comparisons. In the two species of insects for which we examined within-population variation (the membracid treehoppers *U. crassicornis* and *E. binotata*), there was no correlation between the size of the signaller and the frequency of the signal. In contrast, there was a negative relationship between body size and measurements of signal frequency among 51 species in the family Membracidae, when using species as independent data points and when using phylogenetically independent contrasts. When we expanded our comparison to variation in size and signal frequency across various insect orders, the results depended on whether or not we

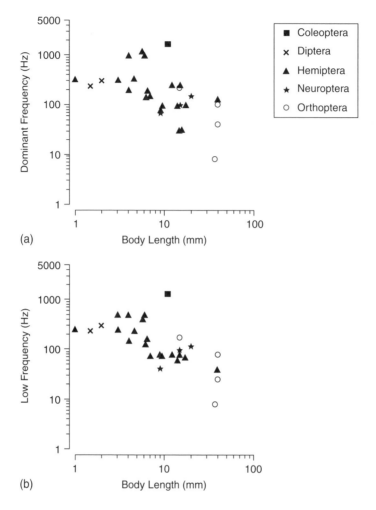

FIGURE 6.4 The relationship of size to dominant (a) and low (b) frequency in the insect vibratory signals shown in Figure 6.3, with the analysis restricted to the signals used in mating.

grouped together signals with different functions. Considering all signals together, including those used in mating, alarm communication and attracting ant mutualists, there was no relationship of size and frequency. However, if we restricted our analysis to mate advertisement signals, as we did for the signals of membracid treehoppers, there was a strong negative relationship between body size and both the dominant and the lowest frequency of the signal. Examination of Figures 6.2 to 6.4 reveals a lack of data points in the lower left-hand corner of the figure; although there is considerable variation in frequency for species of a similar size, there were no very small species which produced very low frequency signals.

The question is how should we interpret these results in terms of the reliability of spectral measurements of vibratory signals and in terms of the implications for vibratory signal production and evolution. For the within-species comparisons, the potential challenges for signal measurement were minimal or absent, maximising the probability of detecting a size–frequency relationship. However, we found no suggestion of such a relationship. For the two populations examined (one each for *U. crassicornis* and *E. binotata* from *Ptelea*), variation in signal frequency appears to be decoupled from variation in body size. Within-population variation in signal frequency can be important for female choice, where female preference may exert stabilising or directional selection on male signal frequency (reviewed for insects and anurans in Gerhardt and Huber, 2002). For *E. binotata* from *P. trifoliata*, females prefer frequencies near the mean frequency of males in the population (Cocroft and Rodríguez, 2005). Signal frequency differs substantially among species in the *E. binotata* complex (Rodríguez *et al.*, 2004), and females of each species examined have preferences centred on the mean frequency of males in their population (Rodríguez, Ramaswamy and Cocroft, unpublished data). If *E. binotata* from *Ptelea* is representative, however, female preference based on signal frequency is unlikely to lead to correlated changes in male size.

For the broad between-species comparisons reported here, the potential to mask any relationship between size and frequency might exist for the following reasons:

- The sources of variation in measurements of vibratory signal spectra discussed above
- The differences among substrates and transducers
- The limited availability of accurate size measurements

However, for the membracid data set and the overall insect data set, there was a significant negative correlation between body size and dominant frequency and the lowest frequency in of male mating signals. The correlation was higher between size and the lowest frequency in the signal, possibly because the overall bandwidth of the signals was less substrate-dependent than the relative amplitude of different frequencies within that frequency band. The r^2 values (reflecting the proportion of variation explained) were higher at the broader level of comparison, suggesting that the relationship between size and frequency is relatively loose, and unlikely to be detected unless there is a large range of values for both variables. The lack of a relationship for the analysis which included signals other than those used in mating, is primarily accounted for by the lycaenid and riodinid caterpillars, which produce signals to attract ant mutualists (DeVries, 1991). These signals are typically broadband and relatively high in frequency, perhaps as a result of selection to produce vibratory signals similar to those of the ants with which they are communicating (DeVries *et al.*, 1993).

Two factors might explain why there is an inverse relationship between body size and frequency in insect vibratory signals. First, an insect resting on six legs can be modelled as a mass on a set of springs (Tieu, 1996; Cocroft *et al.*, 2000; also see Aicher *et al.*, 1983). Other things being equal, the greater the mass, the lower the resonant frequency of a mass-and-spring system. It is not known whether insects use this resonance in signal production (or signal reception; see Aicher *et al.*, 1983), but if they do this could explain why larger insects produce lower frequency vibratory signals. Second, it is likely that at least some species use the thoracic muscles to generate vibratory signals (Gogala, 1985a; Cocroft and McNett, Chapter 23). Wingbeat frequency is inversely correlated with mass in insects (Dudley, 2000), and if the wing muscles are used to produce vibratory signals, this could generate a negative correlation between signal frequency and mass.

In this study, we did not examine the relationship of size with temporal variables or overall signal amplitude. For vibratorily communicating katydids, size was tightly correlated with tremulation rate, larger males signalled at a faster rate (De Luca and Morris, 1998). In communication systems using airborne sound, size has been found to correlate with a variety of temporal signal traits (Gerhardt and Huber, 2002). In general, larger animals can produce higher amplitude signals (see Markl (1983) for vibratory signals). At a finer level (*e.g.* among individuals within a population or between closely related species), amplitude comparisons will be difficult to make for insects communicating with plant-borne vibrations because the amplitude of plant-borne signals is highly substrate dependent (Čokl and Virant-Doberlet, 2003). The same insect will produce a higher amplitude signal on a thin stem than on a thick stem, and the amplitude of a signal recorded at different distances from a signaller does not decrease monotonically (Michelsen *et al.*, 1982).

For the membracid data set, there was evidence that close relatives tend to be similar in dominant frequency, though not in size. Ecology and behaviour are also similar within many membracid clades (Wood, 1979, 1984; Dietrich and Deitz, 1991; McKamey and Deitz, 1996), and it would be worthwhile to investigate whether phylogenetic aspects of signal variation reflect adaptation to similar ecological conditions such as low population densities or use of herbaceous vs. woody plants.

ACKNOWLEDGEMENTS

We thank L.M. Sullivan for making the recordings of *U. crassicornis* males, and M.F. Claridge, G.D. McNett and R.L. Rodríguez for comments on the manuscript. We also thank M.F. Claridge for providing recordings of *Tettigarcta*, and T.M. Jones, G.K. Morris and P. DeVries for providing size information on insects whose signals they have described in the literature. Financial support during preparation of this manuscript was provided by an NSERC graduate fellowship to P.A.D and NSF IBN 0318326 to R.B.C.

7 Acoustic Signals and Temperature

Allen F. Sanborn

CONTENTS

INTRODUCTION

Notable early studies of the effect of temperature on acoustic insects were the investigations into the cricket thermometer. The relationship between syllable repetition rate and temperature was first discussed in the 19th century. Brooks (1882) reported a letter to the *Salem Gazette* where W.G.B. determined that there were 72 "strokes" per minute at 60°F (15.6°C) in a cricket call and the rate increased or decreased four "strokes" for every degree increase or decrease in ambient temperature (T_a). She then made a series of observations which showed that the call rate of the local species (which was not identified) was also temperature dependent (Brooks, 1882). Dolbear (1897) took the observations a step further in 1897 and found out that one could determine T_a (in °F) by knowing the chirp rate (N) in chirps per minute using the formula

$$T_a = 50 + \frac{N - 4}{4} \tag{7.1}$$

These results were then confirmed by Bessey and Bessey (1898), who also observed that chirping rate in the tree cricket, *Oecanthus niveus* (De Geer), varied with T_a. The influence of temperature was demonstrated conclusively when they moved a calling individual inside a warm room on a cold night and that individual began to chirp at nearly twice the rate of the crickets outdoors and at a rate which conformed to individuals at the same temperature out of doors on different evenings (Bessey and Bessey, 1898). Bessey and Bessey were able to determine T_a through the chirp rate of *O. niveus*

111

and the formula

$$T_a = 60 + \frac{N - 92}{4.7} \qquad (7.2)$$

These results illustrate that the degree to which acoustic signals are altered by T_a varies with individual species. And so began the analysis of temperature and acoustic insects.

TEMPERATURE AND BIOLOGICAL PROCESSES

Temperature is a physical variable that affects all aspects of animal life. Animals are able to function because the various chemical reactions necessary for life are able to proceed at sufficient rates to maintain life processes. However, the rates of these chemical reactions are also temperature dependent. The ability of insects to produce acoustic signals is dependent on chemical reactions occurring at a specific rate, so that the mechanisms producing and receiving sound function properly. Therefore, temperature will affect acoustic insects at a subcellular level, but these effects will influence the abilities of populations to communicate effectively.

The rate of chemical reactions varies with temperature in a predictable fashion. The Q_{10} effect describes how reaction rates vary over a 10°C temperature change. The relationship is described by equation 7.3:

$$Q_{10} = \frac{k_2^{\frac{10}{(T_2 - T_1)}}}{k_1} \qquad (7.3)$$

where k_1 and k_2 are the rates of chemical reactions at temperatures T_1 and T_2 and should be about two in the normal temperature range of animal activity (Withers, 1992). This relationship between temperature and chemical reaction rates means that insect communication systems will be altered by changes in temperature. Acoustic insects must be able to deal with changes in temperature if they are going to communicate over a range of T_a. If a species cannot adjust to changes in T_a, there will be only a small range over which they can communicate and this will limit the usefulness of sound as a reproductive signal.

Acoustic insects have two potential strategies with respect to changes in T_a. They can allow their body temperature (T_b) to fluctuate with the environment to be a thermoconformer (historically a poikilotherm or cold-blooded animal) or they can regulate T_b independent of T_a and be a thermoregulator. Thermoconformers must be able to adjust receiver preferences as T_a alters the call structure through the Q_{10} effect in order to maintain a response to the altered signal. Thermoregulators can avoid any temperature effects on call structure or receiver preferences since these can maintain T_b in a narrow range so that there will be minimal variation in call parameters as T_a changes. The implementation of specific thermoregulatory strategies by particular groups of acoustic insects will be influenced by physical factors.

Animals exchange heat with the environment based on the physical mechanisms of heat transfer. The small body size and high surface-to-volume ratio of most insect species means that they will exchange heat with the environment quickly. This fact will have significant influence on the thermoregulatory strategy employed by the insects. Small insects (*e.g.* planthoppers) will have difficulty in maintaining a thermal gradient from T_a. In effect the physics of heat transfer will cause these animals to be thermoconformers and the receivers must be able to compensate for the changes in the acoustic signals which will result with changes in T_b as a result of changes in T_a. Larger insects (*e.g.* cicadas) will be able to maintain a thermal gradient from T_a and can avoid temperature effects on call structure since they can regulate their T_b (*e.g.* Villet *et al.*, 2003).

Another factor which will influence the choice of thermoregulatory strategy by an insect species is the activity period of the species. Solar radiation can be used as a potential heat source to regulate T_b by diurnally active insects. The amount of radiative heat gain from the sun has been shown to cause the difference between the ability of some insects to signal or not (Sanborn and Phillips, 1992). Crepuscular or nocturnally active animals will not have solar radiation as a potential heat source so they will be required to generate their own body heat for thermoregulation (be endothermic). This forces them to be thermoconformers. The physics of heat transfer again means that smaller animals are more likely to be thermoconformers.

A survey of acoustic insects shows that the different groups of acoustic insects employ both the thermoregulatory strategies. Animals like planthoppers and crickets are thermoconformers based on their size or time of activity. Larger insects, like katydids and cicadas, can generate sufficient heat to regulate their T_b during crepuscular or nocturnal activity and are thermoregulators. The thermal strategies and the effects of temperature on the acoustic behaviour in these groups of insects will follow.

ACOUSTIC INSECTS AND TEMPERATURE

CICADAS (HEMIPTERA, CICADIDAE)

Much work has been done relating to temperature effects on the acoustic components of the sound production mechanisms as well as the relationship of acoustic signals on temperature in cicadas. The relatively large size of most cicadas has meant that experiments could be performed on individual components of the sound production system which would be impossible in much smaller species. This provides an opportunity to investigate temperature effects on the components of the sound production system of cicadas which would be technically difficult or impossible in much smaller insects.

The nervous system is the ultimate control system for acoustic behaviour. Nakao (1952, 1958) suggested that decreasing T_a initiates calling activity in crepuscular cicadas. Further, it has been suggested that temperature sets the limits on the motor control of chorusing in cicadas (Crawford and Dadone, 1979) and many authors refer to song production being inhibited by low T_a (references in Sanborn, 1998). Measurements of the T_b range over which cicadas can call are limited, regardless of thermoregulatory strategy or mechanism of sound production (Table 7.1). Signalling could be coordinated over a greater T_b range than tymballing species and this had been suggested (Sanborn and Phillips, 1999) to be a possible reason for crepitating. However, further experiments showed that crepitating species have a similar T_b range to tymballing ones (Sanborn *et al.*, 2002b). The T_b range to coordinate acoustic activity ranges from 2.2 to 13.2°C in cicadas, with the extremes found in endothermic species (Table 7.1). The larger T_b ranges are all seen in endothermic species and include the production of the "warm-up" sounds which are characteristic of animals, elevating their T_b to the range necessary to produce a call (Sanborn *et al.*, 1995a; Sanborn, 2000, 2004; Villet *et al.*, 2003). The absolute T_b range of singing has been suggested to separate sibling species, temporally facilitating sympatry (Sanborn *et al.*, 2002a).

Analysis of nervous system activity demonstrates the expected changes in neuronal firing rate with changes in temperature. The firing rate in the tymbal nerve was temperature dependent in *Graptopsaltria nigrofuscata* (Motschulsky), *Oncotympana maculaticollis* (Motschulsky) and *Tanna japonensis* (Distant) (Wakabayashi and Hagiwara, 1953; Wakabayashi and Ikeda, 1961). The change in tymbal muscle activity occurred with a Q_{10} of 1.7 to 1.9 in the three species over a T_a of 24 to 32°C (Wakabayashi and Ikeda, 1961), which is an expected thermal-dependent rate change. The effect of temperature on nervous system function is a primary concern for acoustic insects. The stimulation of the sound production mechanism will change with changes in temperature, altering the temporal parameters of the acoustic signals.

TABLE 7.1
Body Temperature and Acoustic Activity in Cicadas

Species	T_b Range of Calling (°C)	Method of Thermo-regulation	Song Production Mechanism	Ref.
Platypedia putnami var. *lutea* Davis	6.7	Ectothermic	Crepitation	Sanborn *et al.* (2002b)
Okanagana hesperia (Uhler)	2.8	Ectothermic	Tymbal	Heath (1972)
Okanagana striatipes (Uhler)	4.3	Ectothermic	Tymbal	Sanborn *et al.* (2002a)
Okanagana utahensis Davis	5.3	Ectothermic	Tymbal	Sanborn *et al.* (2002a)
Okanagodes gracilis Davis	4.9	Ectothermic	Tymbal	Sanborn *et al.* (1992)
Magicicada cassinii (Fisher)	6.8	Ectothermic	Tymbal	Heath (1967)
Cacama valvata (Uhler)	9.5	Ectothermic	Tymbal	Heath *et al.* (1972)
Diceroprocta apache (Davis)	4.8	Ectothermic	Tymbal	Heath and Wilkin (1970)
Diceroprocta olympusa (Walker)	5.8	Ectothermic	Tymbal	Sanborn and Maté (2000)
Tibicen chloromerus (Walker)	8.1	Ectothermic	Tymbal	Sanborn (2000)
Albanycada albigera (Walker)	6.2	Ectothermic	Tymbal	Sanborn *et al.* (2004)
Guyalna bonaerensis (Berg)	11.2	Endothermic	Tymbal	Sanborn *et al.* (1995a)
Proarna bergi (Distant)	3.6	Endothermic	Tymbal	Sanborn *et al.* (1995b)
Proarna insignis Distant	3.6	Endothermic	Tymbal	Sanborn *et al.* (1995b)
Tibicen winnemanna (Davis)	13.2 (6.7 full song)	Endothermic	Tymbal	Sanborn (1997, 2000)
Tibicen cultriformis (Davis)	4.4	Endothermic	Tymbal	Sanborn (2004)
Pycna semiclara (Germar)	5.0	Endothermic	Tymbal	Villet *et al.* (2003)
Platypleura capensis (L.)	3.8	Endothermic	Tymbal	Sanborn *et al.* (2003b, 2004)
Platypleura hirtipennis (Germar)	8.1	Endothermic	Tymbal	Sanborn *et al.* (2003b, 2004)
Platypleura plumosa (Germar)	2.2	Endothermic	Tymbal	Sanborn *et al.* (2004)
Platypleura wahlbergi Stål	6.4	Endothermic	Tymbal	Sanborn *et al.* (2004)

The T_b range over which animals can coordinate acoustic activity is finite. Similar T_b ranges are found in ectothermic and endothermic species as well as the crepitating cicada *Platypedia putnami* var. *lutea*.

Changes to T_b alter the activity of the nervous system which then alters call temporal parameters. The slight increase (2°C) in T_b which occurs as a result of tymbal muscle activity in *Cystosoma saundersii* Westwood was sufficient to change the cycle period of the song produced (Josephson and Young, 1979). There is direct evidence in *Tibicen winnemanna* (Davis) (Sanborn, 1997), and indirect evidence in several other species (Davis, 1894a, 1894b; Myers, 1926; Beamer, 1928; Jacobs, 1953; Alexander, 1956) that song coordination is temperature dependent.

Although the speed of tymbal muscle contraction is influenced by the ultrastructure of the muscle (Josephson and Young, 1981, 1985, 1987), the contraction kinetics of the tymbal muscles depend on temperature (Aidley and White, 1969; Josephson and Young, 1979, 1985, 1987; Josephson, 1981; Young and Josephson, 1983b; Sanborn, 2001). Tymbal muscles contract more rapidly and with greater force as the temperature of the muscle increases (Figure 7.1). The temperature of the tymbal muscles also increases well above T_a during activity (Figure 7.2) (Josephson and Young, 1979, 1985; Sanborn, 2001). This condition is seen in ectotherms with (Josephson and Young, 1985) and without (Josephson and Young, 1979) access to solar radiation as well as endotherms (Sanborn, 2001), and is necessary for cicadas to produce the calling song (Sanborn, 2001).

The study of the endothermic cicada, *Tibicen winnemanna* (Davis) (Sanborn, 2000), showed that tymbal muscle temperature was related to the type of song produced by the individual (Figure 7.3). As T_b and tymbal muscle temperature increase during endogenous warming, the animals began to add components of the normal calling song. The full calling song was produced

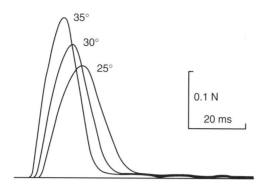

FIGURE 7.1 Tymbal muscle contraction kinetics of the cicada *Cystosoma saundersii* at three different temperatures. The tymbal muscle contracts more rapidly and with greater force as muscle temperature increases. The twitch at 35°C showed a 20% increase in muscle force while the completing the twitch in 73% of the time when compared to a twitch at 25°C. (From Josephson, R. K. and Young, D., *J. Exp. Biol.*, 80, 69–81, 1979. With permission from The Company of Biologists, Limited.)

when T_b and tymbal muscle temperatures averaged at 7.5 and 13.1°C, respectively, above T_a (Sanborn, 1997, 2000, 2001). The contraction kinetics for the tymbal muscles require that the muscle temperature be elevated well above T_a in order for the species specific calling song to be produced (Sanborn, 2001).

The investigations into the relationship between call parameters and temperature have led to contradictory results. Call frequency has been shown to be independent of T_a in *Magicicada* species (Marshall and Cooley, 2000) and *Pyca semiclara* (Germar) (Villet *et al.*, 2003). This would be expected since the frequency of cicada calls is dependent on the physics of the sound producing structures (Pringle, 1954; Bennet-Clark and Young, 1992).

The temporal parameters of *Cicada orni* L. (Quartau *et al.*, 2000), *Tettigetta brullei* (Fieber) (Popov *et al.*, 1997), *T. argentata* (Olivier), *T. josei* Boulard, and *Tympanistalna gastrica* (Stål)

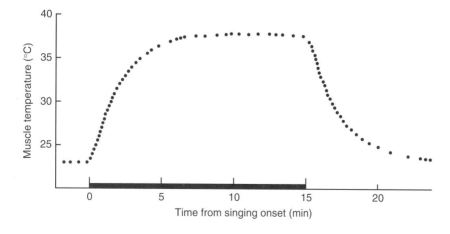

FIGURE 7.2 Tymbal muscle temperature during activity in the cicada *Cystosoma saundersii*. Tymbal muscle temperature increases during activity with the mean muscle temperature reaching a plateau at 13.6°C above ambient. The muscle cools to ambient after singing stops. Body temperature was only elevated an average of 2.0°C above ambient during these experiments. (From Josephson, R. K. and Young, D., *J. Exp. Biol.*, 80, 69–81, 1979. With permission from The Company of Biologists, Limited.)

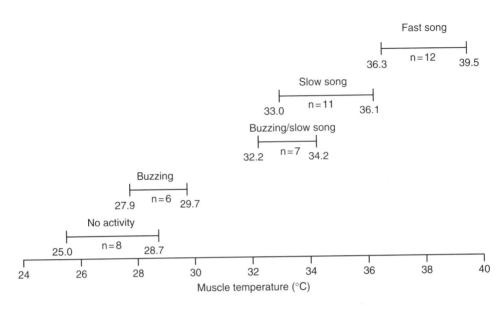

FIGURE 7.3 Tymbal muscle temperature and the acoustic output of *Tibicen winnemanna*. Silent animals (no activity) have muscle temperatures near ambient. Muscle temperature increases as animals begin to produce sound (buzzing) and continue to rise until the cicada is producing the calling song characteristic of the species (fast song). (Reprinted from Sanborn, A. F., *Comp. Biochem. Physiol. A. Mol. Integr. Physiol.*, 130(1), 9–19, 2001. With permission from Elsevier.)

(Fonseca and Revez, 2002) are related to T_a. In contrast, the temporal parameters of *Tibicen plebejus* (Scopoli) (Popov, 1975), *Cicadetta tibialis* (Panzer) (Gogala *et al.*, 1996) and the endothermic *Pycna semiclara* (Germar) (Villet *et al.*, 2003) are independent of T_a. This independence of song parameters from T_a has been suggested to be a benefit of endothermy (Sanborn *et al.*, 2003; Villet *et al.*, 2003). However, most cicadas are thermoregulators (Sanborn, 2002) so one would expect less of a relationship to T_a than with generally thermoconforming animals like crickets (Prestwich and Walker, 1981; Toms *et al.*, 1993).

Two species showing contradictory results have been investigated with respect to the relationship of T_b and temporal parameters in cicada songs. One of the studies (Sanborn and Maté, 2000) directly correlated T_b and temporal parameters, while the second analysis can be done indirectly by combining two separate studies (Sanborn, 1997, 2001). A direct correlation between T_b and temporal parameters showed that there was more variation within the call of *Diceroprocta olympusa* (Walker) than any affect T_b had on the temporal parameters of the call (Sanborn and Maté, 2000). In contrast, the acoustic output of *Tibicen winnemanna* was shown to be dependent on T_b (Beamer, 1928) and there was a direct correlation between the temporal parameters of the call and the type of call produced (and thus T_b) and tymbal muscle temperature (Sanborn, 2001). Both species thermoregulate, but *D. olympusa* is an ectotherm while *T. winnemanna* is an endotherm (Sanborn, 2000; Sanborn and Maté, 2000). The range of T_b over which the species call is similar is 5.8 and 6.7°C, respectively (Table 7.1), so perhaps the development of endothermy has led to a more restricted temporal pattern in *T. winnemanna* resulting in the observed effect of temperature on call parameters.

Two studies have shown a relationship between temperature and the sound pressure level (SPL) of cicada calls. The first was an investigation into the acoustic behaviour and T_b in *Tibicen winnemanna* (Sanborn, 1997). As the endothermic species (Sanborn, 2000) warms, the song intensity increases from an average of 79.64 dB when they initiate acoustic activity to 105.74 dB during full song production (Sanborn, 1997). Average T_b was 30.38 and 35.36°C at the initiation of activity and during full song production, respectively (Sanborn, 1997). This is probably the result of greater tension being produced at higher temperatures by the tymbal muscles (Josephson and Young, 1979, 1985; Josephson, 1981; Young and Josephson, 1983b; Sanborn, 2001) and tymbal tensor muscle which has been shown to influence the intensity of cicada calls (Hennig *et al.*, 1994).

A positive correlation between T_a and SPL was found in three species of *Tibicina* (Sueur and Sanborn, 2003). Minimal sound power was produced at 22 to 24°C, which appeared to be the minimal T_a for calling. The behaviour of the *Tibicina* in the field suggested that they are ectothermic thermoregulators. As T_a and presumably T_b and tymbal muscle temperature increased, the SPL also increased. This is probably a result of the increased tension produced in the tymbal and tymbal tensor muscles as described above. The SPL levels plateau at elevated T_a, which further suggests the changes to intensity were the result of thermally induced changes to the sound production system (Sueur and Sanborn, 2003).

PLANTHOPPERS (HOMOPTERA, DELPHACIDAE)

The smaller relatives of cicadas have been shown to have vibratory signals influenced by T_a. de Vrijer (1984) showed that the pulse rate and frequency of signals produced by *Javesella pellucida* (F.) correlated to T_a. Female click rates were also temperature dependent and corresponded to the click rate of a portion of the male call (de Vrijer, 1984). This would be expected due to the smaller body size of the planthoppers. The increased surface area to volume ratio would exchange heat at a more rapid rate making it difficult to maintain a thermal gradient and to thermoregulate. As a result, these animals are most likely thermoconformers and the observed relationship is a result of the Q_{10} effect. Support for this assumption is the rate of change in the duration and rate of signals produced by *J. pellucida*, which exhibits a Q_{10} of 1.7 to 2.1, a range predicted in temperature-dependent systems.

CRICKETS (ORTHOPTERA, GRYLLIDAE)

Field measurements of T_b in crickets suggest that most are thermoconformers (Prestwich and Walker, 1981; Toms *et al.*, 1993). This accounts for the cricket thermometers. The acoustic parameters which are the result of temperature-dependent systems (*e.g.* nerve and muscle) exhibit thermal dependency as well. There is a temperature dependency for pulse frequency, wing stroke rate, pulse length, interpulse interval and carrier frequency in crickets (Brooks, 1882; Dolbear, 1897; Bessey and Bessey, 1898; Alexander, 1957b; Walker, 1957, 1962a, 1962b, 1963, 1969a, 1969b, 1969c, 1998, 2000; Bennet-Clark, 1970b; Popov and Shuvalov, 1977; Sismondo, 1979; Prestwich and Walker, 1981; Doherty and Huber, 1983; Forrest, 1983; Doherty, 1985; Doherty and Callos, 1991; Pires and Hoy, 1992a, 1992b; Souroukis *et al.*, 1992; Toms, 1992; Toms *et al.*, 1993; Ciceran *et al.*, 1994; Olvido and Mousseau, 1995; Van Wyk and Ferguson, 1995; Hill, 1998; Gray and Cade, 2000; Martin *et al.*, 2000) (Figure 7.4). The temperature dependency of carrier frequency, which is not found in cicadas, is due to the mechanisms of sound production in crickets (Bennet-Clark, 1989).

Temperature has been demonstrated to influence when crickets call and the pulse rate of calls produced within a species. Rence and Loher (1975) showed that crickets entrain calling behaviour to diurnal changes in T_a. Crickets have also been shown to alter the timing of their calling behaviour when T_a inhibits normal calling activity. When low T_a inhibits nocturnal calling, crickets will change to a diurnal calling pattern (Loher and Wiedenmann, 1981; Walker, 1983). Walker (2000) further showed that variation in call pulse rates between spring and fall generations is the result of rearing temperature in *Gryllus rubens* Scudder.

Although there is a temperature dependency to many of the song components, some crickets can compensate for temperature effects. Walker (1962) showed that *Gryllus rubens* was able to

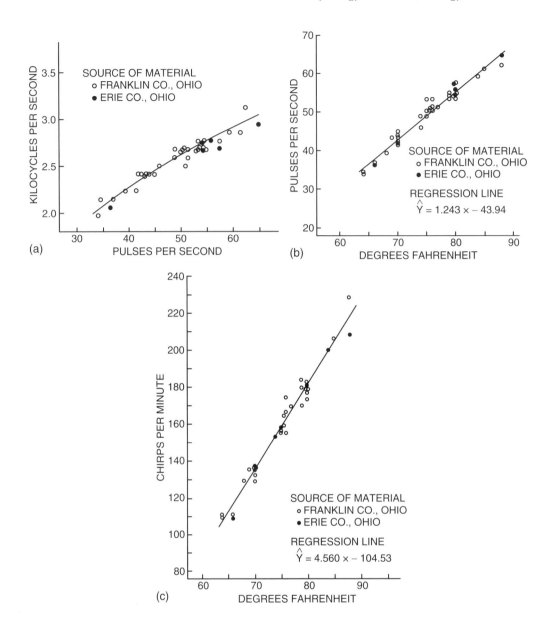

FIGURE 7.4 Relationship of acoustic parameters and temperature in the tree cricket *Oecanthus fultoni*. (a) Relationship between pulse rate and frequency. (b) Effect of temperature on pulse rate. The change in pulse rate will lead to changes in frequency with changes in temperature. (c) Effect of temperature on chirp rate. (From Walker, T. J., *Ann. Entomol. Soc. Amer.*, 55, 303–322, 1962b. With permission from Entomological Society of America.)

alter the number of tooth strikes per pulse and change the proportion of time the wings were closing in the wing cycle to maintain song parameters (Figure 7.5). Koch *et al.* (1988) showed that *Gryllus campestris* L. could alter the closing speed of the wings, the tooth impact rate, and the carrier frequency between 20 to 30°C to produce a song of constant characteristics. Martin *et al.* (2000) showed that the number of trills in calls of *Gryllus integer* Scudder does not vary with temperature, providing consistence to their song over a range of T_a. Some crickets are

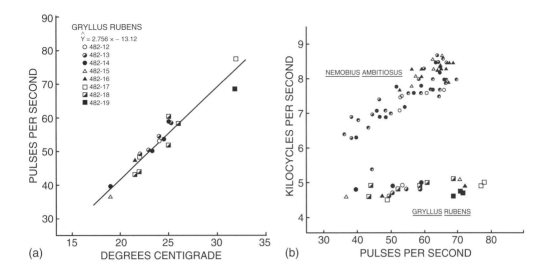

FIGURE 7.5 Compensation of song parameters with changing ambient temperature in *Gryllus rubens*. (a) The pulse rate changes with ambient temperature. (b) The compensatory ability of *G. rubens* (lower symbols) is illustrated by the constancy of frequency over variable pulse rates (and ambient temperatures). The compensation is a result of a decrease in the number of tooth strikes and a change in the proportion of time spent to open and close the wings during the wing cycle at higher ambient temperatures. *Pictonemobius* (= *Nemobius*) *ambitiosus* (Scudder) (upper symbols) is unable to compensate for changes in pulse rate and the frequency of its call varies with pulse rate (and ambient temperature). (From Walker, T. J., *Evolution*, 16, 407–428, 1962a. With permission from Society for the Study of Evolution.)

capable of compensating for some thermal effects on the sound production mechanism, presumably to be more attractive to females.

Chorusing crickets can also compensate for the differences in calling rate at different T_a (Walker, 1969a). Walker (1969a) played *Oecanthus fultoni* Walker recordings of males at different temperatures. Males were able to synchronise their chirp pattern if the stimulus chirp rate was within 10% of their own chirp rate. This suggests that individuals can compensate for slight environmental differences which would alter T_b and call rate. However, if the stimulus rate was outside this compensatory range, the responding males would follow the stimulus with a series of short pulses (at low chirp rates) or lengthen their chirps and match alternate chirps of the stimulus (Walker, 1969a). These tactics permit the animals to maintain a degree of synchrony even over a large T_a range.

Temperature also affects courtship songs in crickets. The courtship song of *Gryllus texensis* Cade and Otte was shown to correlate significantly with temperature (Gray and Eckhardt, 2001). The components of the song that were altered with temperature were the interphrase interval length and the high-frequency tick period (Gray and Eckhardt, 2001). These changes to temporal parameters would be another illustration of the Q_{10} effect.

KATYDIDS (ORTHOPTERA, TETTIGONIIDAE)

Temperature will influence the components of the katydid sound production system in a manner similar to that for cicadas. Although the speed of contraction will be determined by the ultrastructure of the muscle (particularly the amount of sarcoplasmic reticulum) (Josephson, 1975), individual muscles will still exhibit temperature-dependent contraction kinetics (Josephson, 1973, 1981, 1984). Temporal parameters of katydid song, such as wing stroke and tooth strike rates, have been shown to be temperature dependent (Frings and Frings, 1957; Shaw, 1968; Walker *et al.*, 1973;

Walker, 1975a, 1975b; Samways, 1976a; Gwynne and Bailey, 1988; Stiedl *et al.*, 1994) as expected in ectotherms. For example, the chirp rate of *Neoconocephalus ensiger* (Harris) correlates to T_a with a Q_{10} of about two (Frings and Frings, 1957). Even though temperature varies the acoustic parameters of some katydid calls, there are species who can alter the rate of their calls to remain in synchrony with the temporal patterns of calls produced by males at different temperatures (Samways, 1976a), as seen in some crickets.

Although some katydids appear to be ectotherms, others are endothermic like cicadas (Heath and Josephson, 1970; Josephson, 1973, 1984). There is a low correlation between song components and T_a in these species (Walker *et al.*, 1973). The high rate of wing movement necessary to generate their acoustic signals suggests that katydids would elevate T_b (or T_{muscle}) in order to be able to contract their wing muscles fast enough to produce their calling song.

Recordings of T_b in the field and in the laboratory illustrate that *Neoconocephalus robustus* (Scudder) actively regulates its T_b during acoustic activity with T_b measured as much as 15°C above ambient during activity (Heath and Josephson, 1970). Recordings of muscle electrical potentials show that the animals elevate T_b through a period of shivering thermogenesis where antagonistic wing muscles are activated in a synchronous fashion causing T_b to elevate without wing movement. When the animals reach a T_b sufficient to support calling (33.5°C), the electrical activity of the antagonistic wing musculature begins to occur out of phase. This stimulates the opening and closing of the wings which can result in sound production (Heath and Josephson, 1970) (Figure 7.6).

The wing musculature of *N. robustus* is synchronous (Josephson and Halverson, 1971). The high frequency of wing movements needed to generate the song can only be accomplished with elevated T_{muscle}. The increase in percentage of sarcoplasmic reticulum within the muscles (Josephson and Halverson, 1971) can account for the high contraction frequency, but sufficient contractile speed and tension must also be produced if sound is to be generated. The Q_{10} effect would prohibit rapid wing movements without the elevation of T_b (or T_{muscle}). In addition, Josephson (1973) showed that endothermy and the elevated T_{muscle} were necessary to produce the song in *Euconocephalus nasutus* (Thunberg). Measurements of the contraction kinetics demonstrate that *E. nasutus* muscles would enter tetany if muscle and ambient temperatures were equal. Only by elevating muscle temperature can the contractions occur rapidly enough to produce the song (Josephson, 1973). Similar changes to morphology are seen in rapidly contracting cicada muscles which also operate at temperatures well above ambient (Josephson and Young, 1985).

The endothermic species expend significantly greater energy than ectothermic species. It was estimated that *N. robustus* would expend 0.8% of its body mass in one hour of singing just to regulate its T_b (additional energy would be required to move the wings and generate sound energy). This raises the question why the species would expend that much energy to thermoregulate when ectothermy is a viable alternative. Heath and Josephson (1970) suggested that the endothermy used by *N. robustus* is a means to isolate the species from the cogener, *N. ensiger*, with which it shares a habitat. The elevated T_b of *N. robustus* permits the production of a loud, highly distinctive song [150 to 200 chirps sec^{-1} (Josephson and Halverson, 1971)] in comparison to the soft intermittent song (10 to 15 chirps sec^{-1}) of the presumed ectothermic *N. ensiger* (Heath and Josephson, 1970).

Grasshoppers and Locusts (Orthoptera, Acrididae)

There is limited work on the influence of temperature on the acoustic signals of grasshoppers although the group has been used in female preference studies (see below). The acoustic parameters of *Chorthippus parallelus* (Zetterstedt) and *Chorthippus montanus* (Charpentier) (von Helverson and von Helverson, 1981), *Chorthippus biguttulus* (L.) (von Helverson and von Helverson, 1983, 1994) and *Omocestus viridulus* L. (Skovmand and Pedersen, 1983) have been shown to be

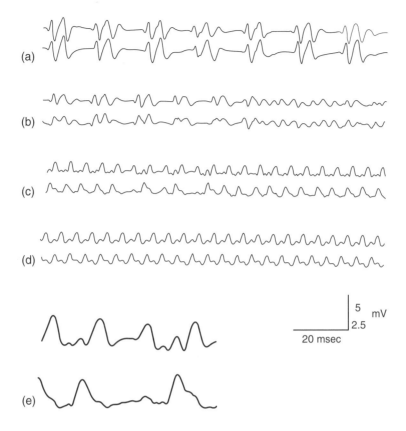

FIGURE 7.6 Changes to the activation pattern of katydid (*Neoconocephalus robustus*) sound muscles. During endogenous warm-up (a and b) there is synchronous activity in the muscles that open (tergosternal, lower trace) and close (subalar, upper trace) the wings. As the animals attain a body temperature necessary to produce a song, the activation pattern become asynchronous and sound production is initiated (c) after the tergosternal muscle (lower trace) failed to fire for one or more cycles (expanded in e). The muscles are activated out of phase when the animal is singing (last half of c and d). (From Heath, J. E. and Josephson, R. K., *Biol. Bull.*, 138(3), 272–285, 1970. With permission from The Marine Biological Laboratory.)

temperature dependent. The observed changes to acoustic parameters is another example of the Q_{10} effect on insect sound generating systems.

The tuning and physiology of the locust sound reception system has been shown to be temperature dependent (Oldfield, 1988). Characteristic sound frequency shifted by 1 to 2 octaves per 10°C with a shift of maximal sensitivity to higher frequencies at elevated T_a. This suggests that tuning is determined by cellular processes (Oldfield, 1988).

DROSOPHILA (DIPTERA, DROSOPHILIDAE)

The small size of *Drosophila* species means they will be obligate thermoconformers. This means their T_b will approximate T_a and they will be unable to maintain a gradient from T_a in order to regulate T_b and prevent thermal effects on their calls. Any alteration in T_a should result in changes to the acoustic parameters produced by the flies, so one would predict that temperature influences the call parameters.

Evidence to support the above assumption can be found in the literature. The interpulse interval, intrapulse frequency and carrier frequency of courtship songs have been shown to be temperature dependent in *Drosophila* (Shorey, 1962; Kyriacou and Hall, 1980; Hoikkala, 1985; Ritchie and Kyriacou, 1994; Ritchie and Gleason, 1995; Noor and Aquadro, 1998; Ritchie *et al.*, 2001). The temperature-dependent acoustic parameters are the result of changes in the activity in neural and muscular systems. Since the components of the sound production systems are influenced by temperature, it is not surprising that there is a temperature dependency to the acoustic parameters.

SOUND RECEPTION AND TEMPERATURE

Receivers must be able to recognise a signal in order to respond properly to that signal. It has been shown that the signals produced by acoustic insects can be altered by changes to T_b or T_a. Therefore, a "temperature coupling" (Gerhardt, 1978) between the signaller and receiver should occur if the communication channel is to function effectively. Otherwise the receiver may not respond correctly or at all to the signal being produced.

Several studies have shown a temperature coupling between female preference and male song production. Female planthoppers (de Vrijer, 1984), grasshoppers (von Helverson and von Helverson, 1981; Skovmand and Pedersen, 1983; Bauer and von Helverson, 1987) and crickets (Walker, 1957, 1963; Doherty and Huber, 1983; Doherty, 1985; Pires and Hoy, 1992a, 1992b) (Figure 7.7) have been shown to respond best to calls produced by males whose T_b is similar to the female. Evidence has also been provided that female cicadas, both ectothermic (Heath, 1967; Heath *et al.*, 1971; Sanborn *et al.*, 1992, 2002a) and endothermic (Sanborn *et al.*, 1995a; Sanborn, 2000; Villet *et al.*, 2003), have a T_b similar to calling males. The temperature coupling observed would insure that females respond to the male signal. Thermoregulating is an additional means to insure T_b of males and females are within the same range so this coupling can occur.

Although thermal coupling of acoustic signals makes logical sense, Ritchie *et al.* (2001) provide evidence that *Drosophila montana* Wheeler does not exhibit temperature coupling. They found female preferences only poorly match the predicted values based on the observed temperature-induced changes to song parameters. They suggest that the temperature-induced song variation may be irrelevant to mate choice if courtship occurs over a restricted temperature range in the field and if females have directional preferences for song traits (Ritchie *et al.*, 2001). Additional problems may occur if the tuning frequency of the auditory system is more temperature dependent than the carrier frequency (Oldfield, 1988). For example, the frequency shifts which occur in the tuning of the auditory receptors of *Valanga irregularis* (Walker) exhibit a Q_{10} of 2.93, a value greater than expected from temperature-dependent changes in the mechanical properties of the sound reception system (Oldfield, 1988).

One possible solution to temperature changing the acoustic signal is to have a receiver that can respond to a range of acoustic parameters. The females of the grasshopper, *Chorthippus biguttulus,* appear to be able to respond to a range of pulse rates that would correspond to males calling over an extended T_b range. von Helverson and von Helverson (1983, 1994) showed that the female nervous system at 34°C would be able to respond to songs produced by males with body temperatures of between 20 and 40°C.

It appears that temperature coupling is based on temperature influences on independent neural components for signalling and receiving (Bauer and von Helverson, 1987; Pires and Hoy, 1992a, 1992b). An elegant experiment by Bauer and von Helverson (1987) was able to produce a mismatch between preferences for signal time intervals in the head and the time intervals produced by the thorax by differentially heating the head and thorax. An additional experiment by Pires and Hoy (1992a, 1992b) showed that heating of both the head and thorax were necessary to change female preferences in crickets. These results suggest there is a separate neural recognition system in insects that is temperature dependent.

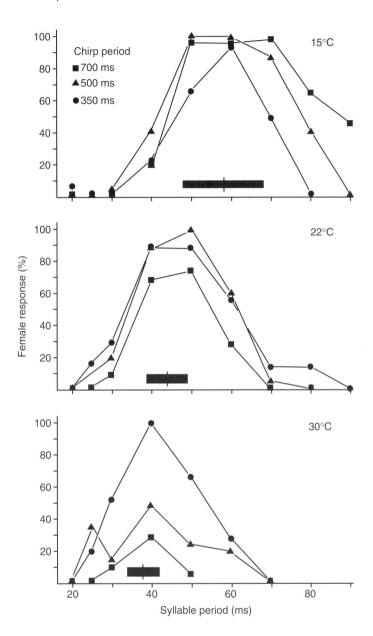

FIGURE 7.7 Female preference and temperature in the cricket *Gryllus bimaculatus*. Females prefer synthetic calls similar to those that would be produced by males at the same body temperature as the female. The vertical bars at the bottom of each graph correspond to the population means and standard deviations of syllable period in the natural calling song at the specific temperatures. (From Doherty, J. A., *J. Exp. Biol.*, 114, 17–35, 1985b. With permission from The Company of Biologists, Limited.)

SOUND PROPAGATION AND TEMPERATURE

A final issue that needs to be discussed is the influence temperature has on the propagation of acoustic signals. Temperature gradients across upper forest canopies at dawn and dusk reduce the amount of excess attenuation in forests by trapping sound energy under the canopy (Waser and Waser, 1977; Wiley and Richards, 1978; van Staaden and Römer, 1997; Römer, 1998). The broadcast area of acoustic signals has been shown to be greatest during these times (Figure 7.8)

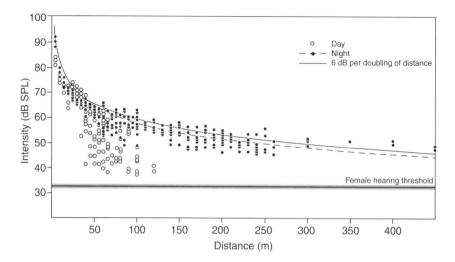

FIGURE 7.8 Decrease in sound attenuation due to thermal inversion. The differences in temperature above and below the canopy cause sound to be reflected back toward the ground increasing the range of acoustic signals. Excess attenuation and upward-refracting sound conditions during the day (open circles) limits the transmission distance of the call while calling at night increases the transmission distance more than 10-fold due to the downward-refracting temperature inversions. (From van Staaden, M. J. and Römer, H., *J. Exp. Biol.*, 200, 2597–2608, 1997. With permission from The Company of Biologists, Limited.)

(Henwood and Fabrick, 1979; van Staaden and Römer, 1997). Young (1981) hypothesised that dawn–dusk chorusing in cicadas is an adaptation for communicatory optimisation. Thus the influence of temperature on sound transmission may have been selected for the dawn–dusk chorusing activity seen in many acoustic animals. The increased sound transmission at dawn and dusk has also been hypothesised to be a selective advantage for endotherms (Sanborn *et al.*, 1995a, 1995b, 2003a, 2003b, 2004; Sanborn, 2000; Villet, *et al.*, 2003).

The propagation of vibratory signals is also influenced by temperature. Sound propagation and communicatory distance of the vibratory signal of the mole cricket, *Gryllotalpa major* Suassure, was shown to vary with temperature (Hill and Shadley, 2001).

CONCLUSIONS

Many of the temperature-dependent characteristics of insect sound, *e.g.* changes to pulse rate, can be traced to the effect of temperature on biological processes, or the Q_{10} effect. The influence of temperature on the rates of chemical reactions means that processes such as nervous activity and muscle contraction will be dependent on temperature. Since the sound production system is mediated through nerve and muscle, it is not surprising that insect sounds often vary with temperature. Although there is evidence that some insects can modify their signal to compensate for thermally induced changes, the only means by which the insects can avoid these changes to their signals is to try to regulate their T_b.

Larger insects in general regulate their T_b and minimise the effect of temperature on their calls. This strategy is not available for smaller insects due to the physics of heat transfer and the inability to maintain a thermal gradient. A similar problem occurs in many nocturnal insects and do not have a heat source to regulate T_b. Diurnally active animals can use the radiant energy of the sun to regulate their T_b while larger crepuscular and nocturnal species must use endogenous heat for thermoregulation if their call is to remain unaffected by T_a. The ability to thermoregulate uncouples the reproductive signal from environmental influence providing a more consistent signal.

The influence of temperature on acoustic signals needs to be considered whenever analyses are performed on these signals. The most reliable measurement a researcher could record would be the T_b of the calling insect. This is relatively easy to accomplish with the small, highly accurate, modern thermal probes. However, this is not always feasible due to the size (where additional precautions must be taken due to the ratio of the mass of the insect and the mass of the probes) or calling location of the insect. In these cases the T_a, but preferably perch temperature, should be recorded along with the signal, so that there is a standard by which the call can be compared with later analyses. Although there is voluminous literature on thermoregulation and temperature effects in insects, the physiological study of temperature on acoustic signals is still limited. The recent findings of endothermy as an almost exclusive thermoregulatory strategy within the widespread cicada tribe Platypleurini has been suggested to evolve as a means of maintaining consistent acoustic signals (Sanborn *et al.*, 2003a, 2003b, 2004; Villet *et al.*, 2003). There is much work that remains to be done on understanding the relationship between temperature and acoustic signals.

8 Insect Songs — The Evolution of Signal Complexity

Winston J. Bailey

CONTENTS

INTRODUCTION

The repeated evolution of sound producing structures, amplified by internal cavities and sound receivers developed as microphone-like tympana, is as inevitable for insects as homiothermy is for mammals. Insects are built that way. Almost any conceivable mechanism which can be developed for producing and detecting airborne sound and substrate transmitted standing waves would appear to have been achieved by this taxon (Bennet-Clark, 1984b; Bailey, 1991; Greenfield, 2002). The motion of a hard cuticle amplified by an underlying fluid-filled cavity, and for sound detection, the vibration of an air-backed tympanum with its dedicated mechano-receptor array (Rössler, Jatho and Kalmring, Chapter 3), is a nearly universal mechanism for sound production in insects. Biologically relevant sounds made by insects may be used for defence, often as aposematic signals, or more usually for mate attraction. In these cases, signalling has evolved as a mate recognition system which is distinct to each species and often to each race or population.

For this reason, the production of sound has been studied either as a tool to distinguish between species as a biological criterion to support morphology (Sueur, Chapter 15), or as a mechanism to better understand the workings of the insect's complex nervous system (summary in Gerhardt and Huber, 2002). But in many ways a more challenging approach has been the focus on ways in which sender and receiver have evolved to optimise reproductive effort — the recognition of like species and the reproductive advantage of increased offspring survival and better genes through female choice and male competition (Endler, 1992; Arak and Enquist, 1995; Endler and Basolo, 1998; Greenfield, 2002). For most signalling systems the sender evolves in parallel with the receiver where it is the relative investment in reproduction of each sex that governs which sex produces the signal and which searches. So in these contexts, the complexity of the signal should be influenced by increased opportunities of each sex for mating and any extremes through run-away selection balanced by the costs of producing the signal.

In this chapter the conflicting interaction between the sexes as to who leads whom in the evolution of signalling and searching strategies is discussed, and attempts made to examine the different selective pressures on each sex which increase mating opportunity and survival. At one extreme, very simple signals consist of a series of brief syllables, usually produced by one wing or tymbal movement, while at the other, males may produce long, complex and in some cases "ornate" acoustic or vibratory signals. The author's personal interest in the Orthoptera, and particularly the ensiferan Tettigoniidae (bushcrickets and katydids), undoubtedly dominates many of the examples provided, but from this perspective the chapter may draw common threads of research which are applicable to most insect taxa. The chapter focuses almost entirely on the temporal structure of insect signals, but it may be argued that variation in the carrier frequency of the calls may be under similar selective pressure and forms an additional level of complexity. At the conclusion of the discussion, the variation in carrier frequency is commented upon and, given the extraordinarily complex calls of many membracid treehoppers, the opportunity for research in this area is all too obvious (see Cocroft and McNett, Chapter 23).

Two related scenarios in regard to call length and complexity have been examined. First, where signal length is under selection through female choice, there is an assumption that males will produce long, complex, or loud signals which reflect fitter males. The question is whether there is evidence for selection through female call preference among acoustic insects and whether there is a case that male signals have evolved through female preference alone. Second, if call length is costly, perhaps through development of elaborate signalling structures, the requirement of metabolic energy for producing the call or the risks from possible predation involved, then what is the minimum length of a signal and what strategies can each sex adopt in terms of signalling and searching to optimise reproductive effort? In other words, if selection operates to increase signal length, then counter selection should reduce call length and, because of conflicting interests of each sex, strategies of calling and searching should evolve to optimise the mating advantage of each sex. Figure 8.1 summarises some of these paths of selection resulting in complex calling patterns at one extreme and brief calling at the other.

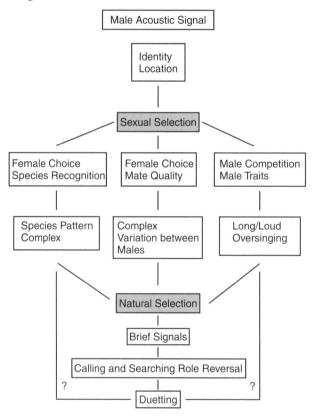

FIGURE 8.1 Selection on acoustic signals through sexual selection illustrating the potential for long and complex calls. Counter selection through energy costs and predation results in signal reduction and duetting, but the evolution of duetting behaviour may possibly increase selection on call length and also male competition.

LONG AND COMPLEX CALLING

Long signals are not necessarily complex. For example, many cicadas, crickets and bushcrickets call for extremely long periods of time, either alone or in a chorus with calls that have little structure except a repeated syllable. Such stereotyped calling inevitably has high redundancy, providing advantage within the noise of the habitat and the calls of other insects. The pattern of the calling male may contain species' information and where species are syntopic, females are usually able to distinguish between males based on syllable repetition rates or the internal temporal structure of the signal (Bailey and Robinson, 1970; Claridge, 1985b; Ritchie, 1996). In these cases, call pattern can be considered to be under stabilising selection (Hedrick and Weber (1998) in *Gryllus integer* and Stumpner and Helversen (1994) in species of *Chorthippus*). Some continuously calling insects divide their calls into chirps and trills, where part of the call is more attractive to females — perhaps a first step toward complexity. In *Teleogryllus oceanicus*, for example, females prefer a narrow range of chirp intervals and even calls that contain chirps alone (Pollack and Hoy, 1981; Pollack, 1982).

For sexual selection to be invoked as a primary driver for call complexity, there are some basic assumptions. First, males producing more complex signals should provide females with increased benefit, either as direct benefit through resources or indirect benefit of better genes. Second, there should be an association between signal length and complexity, with the male trait indicating quality, such as male size, mating status or age. In other words the trait should be linked with the benefit. Finally any increase in signal complexity should be costly to the male and any female preference function should be a balance of trade-off between immediate benefits and the future benefits of offspring mating success.

For most insects an increase in call length is reflected in female preference; females are attracted to longer, louder signals which are easy to detect, but evidence that call length is under selection through female choice is far from secure. Female *Scudderia curvicauda* prefer long to short bouts of calling (Tuckerman *et al.*, 1993), as do females of *Gryllus interger* (Hedrick and Weber, 1998) and those of *Chorthippus biguttulus* (Helversen and Helversen, 1993). However, Gerhardt and Huber (2002) point out that while there have been extensive studies of insects using short calls (<0.5 sec) with preference based on syllable/chirp number, there seems to be no general or reliable trend. In some species there is no preference at all, while in others there is distinct stabilising selection on species' structure, and again in others there is strong directional selection. For example, stabilising selection has been shown in grasshoppers (Butlin *et al.*, 1985), crickets (Gray and Cade, 1999), bladder cicadas (Doolan and Young, 1989), and the bushcricket *Ephippiger ephippiger* (Ritchie, 1996). In the bushcricket, *Phaneroptera nana*, females responded to pulses with durations beyond the natural range shown by males and with greater number of pulses per call (Tauber *et al.*, 2001). In cricket species such as *Laupala cerasina*, female preference was not only strongly directional for pulse duration but also appeared open-ended (Shaw, 1996a).

Signals that are over 0.5 sec may be classed as long and there is some evidence that call length can be associated with body size or mating status. For example, as indicated above, duetting females of the bushcricket, *Scudderia cauvicauda*, respond more reliably to long phrases than to short phrases and this trait is associated with body size (Tuckerman *et al.*, 1993). For some bushcrickets male body size and mass is linked with spermatophore size, where the spermatophylax represents a costly nuptial gift for the female, but for *S. cauvicauda* there was no clear link between body size and the mass of the donated spermatophylax (Tuckerman *et al.*, 1993). While long signals may result in increased mating advantage for some species, a more proximate explanation is that there is more signal for the searching insect, which may be important where searching is costly. Nevertheless, experiments such as those of Tuckerman *et al.* (1993) on *S. cauvicauda* and similar experiments by Galliart and Shaw (1996) with *Ambylocorypha* may present one opportunity to test female preference as a selective process for call structure; for duetting insects it is often the male that both initiates the call and searches for the responding female. In these duetting species the female merely indicates her readiness to mate and her location to the male by producing a series of simple clicks; her willingness to reply and the number of replies become a simple metric of preference.

Insect calls are highly complex. Recently, Walker (2004) suggested that katydids such as *Amblycorypha longinicta* and *A. arenicola* produce some of the most complex of insect calls (and see also descriptions by Spooner (1968)). These calls are not only complex in their temporal design but also show extreme variation in intensity and each call lasting for several seconds (4 to 60 sec), for many species. The call may contain multiple syllable types, where each sequence is produced in a "more or less predictable order" (Walker, 2004) (Figure 8.2a). Heller (1988, 1990) described exceptional complex calling in many of the European Barbistini and a recent examination of an Australian bushcricket belonging to the genus *Caedicia* show similar complexity and, importantly in the context of this discussion, all signalling in all genera includes male–female duetting. Parts of these complex calls are under selection through species' recognition, for example females of *Amblycorypha* (Figure 8.2a) and *Caedicia* sp. 12 (Figure 8.2b) show preference for critical elements that appear species' distinct characters (Hammond and Bailey, 2003; Walker, 2004). Thus, *A. longinicta* and *A. arenicola* females produce responding ticks in response to the "type 4" calls described for these species and experimental removal of these elements significantly reduced

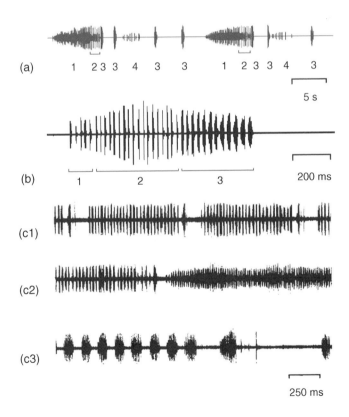

FIGURE 8.2 Call complexity in the tettigoniid Orthoptera. (a) *Ambylocorypha longinicta* showing some 40 sec of the song with a repeated phrase of the call consisting of four call types. (Reproduced with permission of Walker, T. J., *J. Orthoptera Res.*, 13(2), 169–183, 2004.) (b) One of the calls of *Caedicia* species 10 showing three call types. (Redrawn from Bailey and Hammond, 2004.) (c) A long call of *Polysarcus scutatus* divided, for the purposes of this figure, into three sections. (Redrawn from Heller, K.-G., *Bioakustik der Europäischen Laubheuschrecken. Ökologie in Forschung und Anwendung*, Verlag Josef Margraf, Weikersheim, 1988.)

level of response. Earlier studies by Galliart and Shaw (1992) of *Amblycorypha parvipennis* found that females moved to louder calls with longer phrases (see discussion above), but also to males that overlapped the calls with those of their competitors a shorter period of time as possible; the calls lacked interference. Further, as with *Scudderia cauvicauda* (*cf.* Tuckerman *et al.*, 1993), these authors were able to show that females preferred to mate with louder and larger males (see also Gray, 1997, in crickets). But perhaps the most complex of calls within the *Tettigoniidae* are produced by species of *Polysarcus* (Phaneropterinae) (Figure 8.2c) where male *P. scutatus* have calls of extraordinary length and complexity (Heller, 1988).

Sections of the calls of species from these genera function in male competition. Walker speculates on the function of some of the complex call elements in *Amblycorypha* but, typical of many phaneropterines, the calls contain elements which vary in relative intensity. Spooner (1968) shows that these quiet call elements are used once a duet is established and the female is close to the male. He speculates that this is a tactic to avoid other males taking over the courting female. Hammond and Bailey (2003) demonstrate that duets are vulnerable to take-overs by satellites (see also Bailey and Field, 2000, in *Elephantodeta nobilis*) and that one function of the loud calls of *Caedicia* is to mask the calls of the duetting female, so protecting her from identification by other males. Even more remarkably, in *Caedicia*, and perhaps also in *Amblycorypha*, some elements of the call have no other function than to act as spiteful disruption to the calls of competing males. For example, *Caedicia* sp. 12 will produce buzz-like elements over the critical parts of the male call used by females to determine conspecifics (Bailey, McLeay and Gordon, submitted ms), and still other "female-like" calls are produced after an alpha male's call as if to distract the searching male from its duetting female partner (Figure 8.3). Sueur (2003) describes similar aggressive behaviour in the cicada *Tibicina*, where males appear to intrude on the acoustic "space" of neighbouring males with aggressive calls. Importantly for such a system to have evolved, as in the case of many phaneropterine tettigoniids, satellite behaviour occurs in this species.

In summary, the long and often complex calls of duetting phaneropterine bushcrickets potentially contain elements that are subject to sexual selection through female choice, but also there are elements which are clearly used in male competition where males overlap the calls of other males or even obfuscate the calls of replying females so that intruding satellite males will not locate them, a situation that is best described as mate guarding.

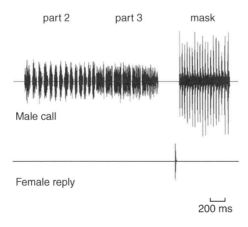

FIGURE 8.3 *Caedicia* species 12 producing a masking signal of significantly higher amplitude to parts two and three of its own song. The female reply comes just before the mask (Hammond, T. J. and Bailey, W. J., *Behaviour*, 140, 79–95, 2003).

MINIMAL SIGNAL LENGTH

Counter selection on call length comes from two primary sources. First, energy invested in the production of the call may be a trade-off with energy required for current or future mating, for example the replenishment of costly accessory gland products such as the spermatophylax of many ensiferan Orthoptera (Gwynne, 1984; Thornhill and Gwynne, 1986; Simmons *et al.*, 1999). Second, the risks incurred through predation from bats (Belwood, 1990) and parasitoid flies (Cade, 1979; Heller, 1992; Allen, 1995; Rotenberry *et al.*, 1996) may offer selection to minimise the exposure of the advertising signal. In both cases there will be strong selection on the calling sex, which is invariably the male. But at least theoretically, while female preference may be for higher than mean values of the species' signal, such selection invariably holds the male call beneath these values (Kotiaho *et al.*, 1998).

Evidence of a trade-off among tettigoniids between metabolic costs of sound production and mating effort is invariably dependent on both postemergence development, usually referred to as the refractory period, and food availability and quality (Simmons and Gwynne, 1991). For most insects there is a time between emergence and calling where spermatogenesis and the development of accessory gland material represent a premium over calling (Simmons, 2001). During this time, the propensity to call and the call length are usually shorter than when males are fully mature. In extreme cases, such as *Kawanaphila nartee* (Zaprochilinae), where calling and searching roles are reversed through a lack of nutrients within the habitat, females may compete for males and during this time males may provide the most minimal of signals (Gwynne and Bailey, 1999).

Are metabolic costs involved in changing signal components such as length, type or loudness? While Prestwich *et al.* (1989) suggest that the metabolic costs of calling are trivial for insects, particularly compared with frogs where energy conversion to acoustic power is less efficient, there are a number of examples where food-limited insects call less and with signals of lower amplitude. For example food-deprived bushcrickets, such as *Requena verticalis*, call less following mating, with the production of a spermatophore, and are influenced by the availability of food (Bailey *et al.*, 1993). *Ephippiger ephippiger* males on a lowered diet produce fewer chirps per day and at significantly reduced amplitude compared with males on a normal diet (Ritchie *et al.*, 1998). Similarly, *Gryllus lineaticeps* on high diet tend to sing more than their siblings on lower diets (Wagner and Hoback, 1999). One assumption is that insects can replenish their food resources with ease, and while this may be so for species in nutrient-rich environments, other species are specialist feeders and may be restricted in diet by volume and quality (*Requena verticalis*; Simmons *et al.*, 1992).

The second selection pressure is predation. Here it is assumed that the call length will decrease with rising incidence of predation. The most thoroughly researched predation event is that from parasitoid flies belonging to the Tachinidae. Commonly, ormine flies locate the calling insect and either lay eggs or larvae on or close to the host (Cade, 1975, 1979; Allen, 1998; Zuk *et al.*, 1998; Gray and Cade, 1999; Allen and Hunt, 2000). Indirect evidence can be inferred from the current calling activity of closely related species. For example, Heller (1992) suggested that the difference between searching and calling roles of two Mediterranean bushcrickets is a response to predation history, in part through parasitoid flies. More direct evidence comes from changes in call structure and female preference function in the Australian cricket, *Teleogryllus oceanicus*. In regions where ormine flies are abundant, not only is calling time altered, but male call length is decreased (Zuk *et al.*, 1993a; Rotenberry *et al.*, 1996).

Evidence for a reduction in call length is more indirect for predation by bats, and here most evidence relates to tropical bushcrickets. The large picture in regard to ensiferan Orthoptera suggests that there are many more short or intermittent calling species in tropical rain forests inhabited by foliage gleaning bats compared to more temperate or open habitats (Rentz, 1975; Belwood, 1990; Morris *et al.*, 1994). Foliage gleaning bats are more attracted to mist-nets loaded with calling bushcrickets than nets without bushcrickets (Belwood and Morris, 1987). The effects of call length were tested by attracting foliage gleaning bats to different species producing different calls rates. Bats were more attracted to a species producing a call rate of 60/min compared with one calling infrequently at less that one call per minute (Belwood and Morris, 1987). Gleaning bats also prefer the long calls of crickets (Bailey and Haythornthwaite, 1998).

Belwood (1990) made the observation that many nocturnal species of tropical bushcricket call within a very narrow time window, with some calling during the early hours of the morning, possibly to escape the flight times of bats. By comparison, species that are continuous callers are more frequent in open grass areas where bat activity is far less (*e.g. Neoconocephalus* (Greenfield and Roizen, 1993)). Interestingly, the continuously calling forest-dwelling pseudophylline, *Ischnomela pulchripennis*, calls from refuges such as bromeliads where bats are unable to intrude (Belwood, 1990). Perhaps the most compelling evidence for call structure change is the switch by tropical forest bushcricket species from acoustic to vibratory signalling strategies. Many species combine airborne high frequency signals with substrate tremulation (Belwood, 1990; Morris *et al.*, 1994) and, in the pseudophylline species *Docidocerus gigliotosi*, the call is extremely brief, consisting of a diplosyllable while the tremulation signal consists of a long series of syllables (Belwood and Morris, 1987; Belwood, 1990).

There exists a minimum signal length for information transfer. Once the signal reaches minimal values the call may no longer achieve the signaller's two primary goals of identity and location. (See Beecher (1989), Bradbury and Vehrencamp (1998), and Gerhardt and Huber (2002) for a broader discussion of the selective forces on acoustic signals that provide information for recognition.) Where signals fall into this category, the signaller's relative fitness will presumably increase with higher chances of mutuality and decrease where there is deceit, or where alternative tactics are employed. Further, the signaller's fitness will increase with its ability to eavesdrop or use its signal in an aggressive or spiteful manner (Wiley, 1993). Critical, however, in any consideration on signal effectiveness is the decision by the receiver to respond, which is often considered as its internal state. But such a decision may be influenced by more complex factors such as exposure to predation during searching or phonotaxis and also the ability to locate the sender. Gerhardt and Huber (2002) summarise the situation: *"if individuals cannot be distinguished statistically by their signals, then in the long run there will be no benefit in investing time and energy in discriminating between signals, the signal differences (must) reliably reflect on average differences in state"*. Further, the response and recognition of signals is compounded by memory of previous events, including signals (Greenfield *et al.*, 1989). While this is frequently considered for vertebrate signallers, there has been little experimental evidence that this may be important for insects.

The notion of acoustic memory in insects is a discussion beyond the scope of this chapter and such events have received surprisingly little attention, yet with intermittent callers or those producing extremely short signals, spatial memory would seem intuitive. There is simply not sufficient signal to locate one's mate in a complex habitat. Interestingly, there have been two recent papers that identify behaviour patterns of acoustically searching females of nocturnal species that are linked with visual patterns associated with the caller (Helversen and Wendler, 2000; Bailey *et al.*, 2003). The evidence so far strongly favours the idea that visual cues are an integral part of the phonotaxis process for nocturnal insects.

One way to view a theoretically minimum signal length is to consider the ability of an insect to resolve repeated patterns. For example, lower limits of pattern resolution have been described for the grasshopper, *Chortippus bigutulus*, which produces distinctly structured sounds from each leg, but the combined stridulation of both legs results in a near continuous noise interrupted by gaps (Helversen, 1984). The behavioural and neural story of each sex's response to conspecific sounds is far from clear, and while the shape of each pulse is important, females fail to respond to songs if gaps of longer than 2 to 4 msec are inserted. These are presumably close to a recognition threshold (Ronacher and Stumpner, 1988; Stumpner *et al.*, 1991). Evidence at a neural level, however, shows that neurons such as AN12 produce bursts with song pause durations of 20 to 25 msec and are as influenced by syllable duration. For many copiphorine bushcrickets, the calls consist of an almost continuous series of syllables produced by the wing's closure, and in some species the inter-syllable interval can be as low as 5 msec. For example, in extremely fast callers such as *Neoconocephalus robustus*, the syllable repetition rate is close to 150/sec, producing an interval between syllables of 7 msec (Josephson and Halverson, 1971). In the genus *Hexacentrus*, repetition rates reach over 200/sec (Heller, 1986), suggesting an interval resolution lower than 5 msec.

An initial survey of calls less than 0.5 sec of bushcrickets fails to provide any convincing pattern in regard to minimum syllable of pulse interval, largely because there is often extreme variation within one taxon, and some species produce near pure tone calls, which may provide as much information to the receiver as the interval between the syllables. The expectation was that there would be a lower call interval that matched the theoretically lower limit of "gap" resolution. However, two basic short signal strategies did emerge from this survey. First, males may produce a single syllable of rising amplitude formed by a single wing closure (Figure 8.4a) and, while the action of the plectrum on the file creates a series of internal tooth impacts, these impacts are not probably resolvable by the insect nervous system as they are below the theoretical limit provided by other studies. In this case the signal ends with a sharp and final loud pulse of sound; the receiver is warned of the oncoming signal by the rising phase and there is then an opportunity for recognition with final louder pulse. The signal may be repeated several seconds later and the interval between repetitions may be random; essentially the element is a syllable with a defined species' shape. The key feature of the signal is that the receiver is set up for the explosive end of the syllable.

By contrast, other species may produce distinct syllables in pairs or triplets (Figure 8.4b). In these examples, the initial signal warns the receiver that the caller is on-air and the interval between this and the proceeding signal has the capacity to provide a species' distinct signal; the latency of the diplosyllable contains temporal information. Tropical species that are presumed to be under selection from predation by foliage gleaning bats are more likely to have such extreme calling strategies. Morris *et al.* (1994) illustrate a number of cases that fit this style of calling. For example the pseudophylline *Myopophyllum speciosum* produces a doublet sound of extremely high carrier

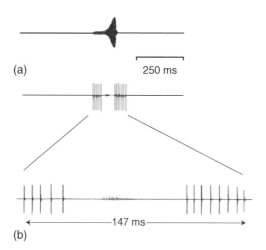

FIGURE 8.4 Two short call types. (a) A single syllable with a rising amplitude employed by the duetting phaneropterine bushcricket *Poecilimon affinis*. (Redrawn after Heller, K.-G., Evolution of song pattern in east Mediterranean Phaneropterinae: Constraints by the communication system, In *The Tettigoniidae: Biology, Systematics and Evolution*, Bailey, W. J. and Rentz, D. C. F., Eds., Crawford House Press, Bathurst and Springer, Berlin, pp. 130–151, 1990.) (b) A diplosyllable of the tropical pseudophylline *Myopophyllum speciosum* drawn to the same scale as A and below the detail of the 147 msec syllable. (Redrawn after Morris, G. K., Mason, A. C., and Wall, P., *J. Zool. London*, 233, 129–163, 1994.)

frequency (>75 kHz). These authors report that the interval between calls was over 8 sec and, moreover, the song period was highly variable giving a duty cycle of 1.7%, which is extremely low for any calling insect.

SHORT SIGNALS, THE DUET AND A RETURN TO LONG COMPLEX CALLS

Theoretically, the briefest piece of acoustic information is a single click formed by a transient created by a single tooth impact and the only context where such a brief signal may contain information is as a reply to another signal. This distinguishes it from other transient sounds within the habitat including those made by potential predators moving within the vegetation. Thus, for many insects the phonoresponse within a duet contains information for the receiver solely on the basis of its latency (Bailey, 2003). While male–male duets are possible among acoustically signalling insects, the commonest is for females to respond to the calls of males where the reply falls within a precise species' defined window (Heller and Helversen, 1986; Zimmermann *et al.*, 1989; Heller, 1990; Robinson, 1990a, 1990b; reviewed by Bailey, 2003). In such cases, information provided by the female creates a brief pair-bond, which results in the male searching for the female. Call latency of the duet is therefore critical for mate attraction and males do not respond to female-created transient signals outside the species range (*e.g.* Zimmermann *et al.*, 1989). But factors that determine this latency, as perceived by the receiving male, are not only the ability of the female to reply within the species' interval, but also the distance between sender and receiver. For very brief species' latencies, sometimes as low as 15 msec (Robinson, 1990), as the distance between the sexes increases, so the travelling time for sound from the male to the female and back again effectively increases the latency to a point where a female, over a certain distance, cannot make an effective response. This is clearly adaptive, in that it may be too costly for males to search females outside this range.

Females frequently take the less costly role of producing extremely short signals and require the male to search. Although this is not always so, in many species the role of searching is shared by both sexes (Bailey, 2003; Walker, 2004). Selection may now operate through female choice on the male call, so male signals that initiate a duet become increasingly ornate. Supporting this hypothesis is the observation that the most complex signals are exhibited by duetting species (Spooner, 1968; Heller, 1988, 1990; Tuckerman *et al.*, 1993; Hammond and Bailey, 2003; Walker, 2004). In this context, while there are few biological data on the longest of calling species, *Polysacus scutatus*, Heller (1990) does provide behavioural information on the related *P. dendticaudus*, where at the beginning of the first phrase the male moves around whilst calling. But after about 2 min the male stops walking and then increases its calling rate and after several seconds produces loud distinct syllables. Females that are ready to mate will respond to these loud syllables and males then move to them. A secondary prediction is that, as selection through female choice increases with the associated increase in male call complexity because of the higher male costs, males should evolve defensive (or offensive) calling tactics. Complex signals have more than one component with a section of the call containing species' information, as well as masking signals that can best be described as mate guarding tactics.

VARIATION IN CARRIER FREQUENCY

The signals of many insects contain frequency modulations that appear, in some cases, to be merely the consequence of the changing acoustic dynamics of the stridulatory apparatus and associated resonating structures. For example Simmons and Ritchie (1996) describe shifts in the call frequency of the cricket *Gryllus campestris* dependent on wing morphology where females preferred pure tones of low carrier frequency that were characteristic of symmetrical harps. An argument suggests that wing structural symmetry is under selection through female choice. In certain species of bushcricket the calls are delivered in more than one frequency by using different parts of the stridulatory file and different associated resonant structures. One of many examples of this call type is *Tympanophora similis* (Schatral and Yeoh, 1990). Again in species such as *Metrioptera sphangnorum* males produce different songs at different frequencies, a low audible call associated with a second call at high intensities and in the ultrasonic range (Morris, 1970). Morris suggests that the high frequency call of this species could be used in male–male aggression. Undoubtedly the Mozarts of insect calls are the treehoppers (see Cocroft and McNett, Chapter 23) and as indicated in their chapter, the opportunities to examine function and selection on such extreme call types represents a new challenge.

SUMMARY AND CONCLUSIONS

A common assumption is that insect acoustic signals are under sexual selection, which includes, as a component of female choice, species' identity and preference for traits that reflect male quality. If selection operates in this way then we may expect an increase in signal length and complexity of the male call. Counter selection should come through increased expenditure of metabolic energy used in calling, most noticeable as a trade-off, with energy diverted to spermatogenesis and the production of accessory gland material and predation. Predation events, particularly among the ensiferan Orthoptera, are most noticeable by parasitic flies and bats where, in tropical and subtropical regions, bats include foliage gleaning species. There is evidence from studies on crickets that predation from parasitic flies not only influences call length but also female call preference. Evidence for counter selection from bats is more indirect, and there are few experimental studies that support this hypothesis. But foliage gleaning bats are more inclined to prey on long-calling species of tettigoniids than on short callers, and short calling species appear to be more prevalent in areas containing such a bat fauna. Further, many of these species adopt alternative signalling strategies, which include tremulation and so the question is whether these species adopt this strategy to avoid bats.

The assessment of the length of signal capable of retaining sufficient information is open for discussion. Short signals may contain information within their structure, largely as changes in intensity or shape, or as doublets and triplets. In the latter case information is temporal and may be contained in the latency between each syllable. A case can be made that extremely short signals, consisting of a single transient, can only be used within the context of a duet. Here, as long as the female reply falls within a distinct time window, conspecific males will search for the responding sex, and call latency contains information on species and location. However, at least among certain families of bushcrickets, selection from the female phonoresponse may result in extremes of male call complexity where calls may have multiple functions, including female attraction as well as male competition. Where this occurs, females may shift roles by increasing their call length or the number of replies and also commence searching.

The duet among calling Orthoptera would appear a rich vein for research on female preference as the experimental paradigm is simple and, given that many duetting species have very ornate signals, such experiments would begin to redress the lack of empirical evidence for selection through female choice among acoustic insects.

9 Song Evolution and Speciation in Bushcrickets

K.-G. Heller

CONTENTS

INTRODUCTION

Bushcrickets comprise one of the most obvious groups of acoustically active insects. Together with cicadas, they are the loudest singers. In temperate regions tettigoniid species are active day and night and compete acoustically, mainly during daylight, with cicadas. In tropical rain forests they, together with true crickets (Gryllidae), are acoustically dominant at night. The range of the carrier frequencies of their songs is much wider than that of their singing competitors, ranging from 0.6 kHz to more than 100 kHz. Males sing and females respond by approaching the singing male phonotactically or by producing special acoustic response signals. Typically every species has a species-specific song pattern. The first nonverbal description of these species-specific differences was given 150 years ago by Yersin (1854) in Europe, followed by Scudder (1868) in North America, both using musical notation. In the last century, many detailed studies of the song patterns followed, mainly on the tettigoniid fauna of North America and Europe (Alexander, 1956; Heller, 1988; Ragge and Reynolds, 1998; Walker and Moore, 2004). From his studies, Alexander (1957a) established the rule that songs are always different in species occurring simultaneously at one locality. Complementing this observation, behavioural studies showed that females recognise the song of conspecific males (acridids: Perdeck, 1957; gryllids: Walker, 1957; tettigoniids: Bailey and Robinson, 1971). Perdeck (1957) also formulated a hypothesis about the origin of song differences between species. Working with acridids of the genus *Chorthippus,* he demonstrated that in these morphologically similar species, the species-specific song pattern represents the main isolating barrier. Other premating and postmating barriers seem to be very weak.

From similar results, Walker (1964) argued that there might be a high percentage of cryptic species not recognised by morphological methods. After having detected differences in song, usually differences in morphology are also found, but some species may differ in song only. For example, no morphological differences have been found between some of the species mentioned by Walker in 1964 (Alexander, 1960) and described later (Walker *et al.*, 2003). From these results it could be concluded that in acoustic Orthoptera, speciation usually starts with differences in song as one of the first steps in the mating process. Perdeck (1957) assumed that the origin of song differences occurred as an adaptation to different acoustic environments in isolated populations, and that the differences possibly enlarged by character displacement or reinforcement. However, even today, identification of unknown specimens and description of new species is done quite often without knowledge of the song. The most important characters used for these purposes are found in morphology. Like many other insects, bushcricket species differ morphologically mainly in male genitalia (Harz, 1969). Obviously, not only the song becomes different during speciation, but also the genitalia. Both are part of a species' mate recognition system and are used for an exchange of information of different kinds and purposes between the mating partners, for example about identity and quality. Both character complexes are assumed to evolve under sexual selection (Eberhard 1985, 1996; Searcy and Andersson, 1986; Gray and Cade, 2000; Gerhardt and Huber, 2002; Sirot, 2003) and often play a primary role in speciation.

But what kinds of characters are likely to evolve first and establish a starting point for speciation? Is it typically the acoustic component or do other elements of premating behaviour change first, which may be connected with morphological changes? Are there any rules for this sequence, and are there similarities or differences between bushcrickets and other groups of singing Orthoptera or singing insects in this respect?

Here we should like to present new data and combine these with published examples of species with differing acoustic and morphological characters. These examples may help to understand what can happen during speciation and which characters may change first. The results may give the taxonomist indications what to look for and for which taxonomic level.

Walker (1974) tried to find species in the last step of speciation 30 years ago. He looked for species with acoustical character displacement. If song is important for species recognition in areas where two closely related species with similar songs occur together, larger differences in song might be expected than where only one of the species occurs. Surprisingly to him, Walker (1974) did not find many examples for this phenomenon among acoustic insects and not a single one in bushcrickets. He gave a long list of reasons why it is found so rarely; one of them was insufficient data. In the meantime, some obvious examples of acoustic character displacement were found among the crickets of Hawaii (Otte, 1989), but for bushcrickets still no such examples have been published.

However, the theoretical background has changed. For characters under sexual selection large differences between species would not be surprising even without contact with other species (Panhuis *et al.*, 2001). For theoretical reasons, reproductive character displacement or reinforcement may be restricted only to relatively special conditions (Turelli *et al.*, 2001; Marshall *et al.*, 2002). Therefore, other methods may be more promising for the analysis of character change during speciation:

Comparison of closely related but allopatric forms (populations or subspecies or species): Comparing characters in allopatric forms, which are important for recognition in sympatric species, may indicate which traits differentiate first and may form the starting point for species-specific differences. In one of the few studies of this type, Den Hollander and Barrientos (1994) have shown that in the tettigoniid, *Pterophylla beltrani*, morphological characters (mainly male genitalia) evolved more rapidly than acoustic signals. They concluded to have found a species in the process of splitting into different species. We will present some more examples from tettigoniids by bringing together separately published data from acoustics and morphology.

Tracking change of song and genitalia in groups where a phylogenetic tree is available: This powerful method depends on the availability and reliability of a phylogenetic tree. At the moment, there are still very few examples, but data will probably increase rapidly mainly due to molecular genetic methods. We shall try to review all data of this type, but it is often difficult (see below).

MATERIAL AND METHODS

Specimens recorded and examined for this work are deposited in the collection of K.-G. Heller (CH, followed by the specimen code). At http://www.dorsa.de the localities of specimens (except those from Turkey) can be seen on a web-based Geographical Information System (GIS) map. Digitised sound recordings are available at the taxonomic database Systax (http://www.biologie.uni-ulm.de/systax). Details of recorded specimens are given in the appendix (accompanying compact disc [CD]). Authors and years of description of other species mentioned in the text are in Otte *et al.* (2004).

For sound recording in the field a Uher tape recorder 4200 IC and a Sony WM3 tape cassette recorder were used with Uher M 645 and M 517 microphones (frequency response flat up to 20 kHz and 15 kHz, respectively). For sound recording in the laboratory a Racal store 4 D tape recorder with Brüel and Kjaer 4133 and 4135 microphones (frequency response flat up to 40 resp. 70 kHz) were used. After digitising calls, oscillograms (after filtering) and sound analysis were made using the programs Turbolab, Amadeus (Apple) and CoolEdit. Wing movements were registered by an opto-electronic device (Helversen and Elsner, 1977; modified as in Heller, 1988).

Song terminology:

- *Calling song*: Song produced by an isolated male.
- *Syllable*: The sound produced by one complete up (opening) and down (closing) stroke of the forewings.
- *Hemisyllable*: The sound produced by one unidirectional movement (opening or closing) of the forewings.
- *Duty cycle*: Ratio of the durations of sound emission and silent interval.

RESULTS

Species occurring syntopically at one locality show distinct differences in calling songs in frequency composition and amplitude modulation. Frequency composition depends mainly on the size of body and tegmina of the sound-producing animal and is quite often typical for a genus or a group of species of similar size. Hearing sensitivity typically matches the spectrum of the conspecific song, but the spectral information cannot be used to discriminate between closely related forms. Related species of one genus differ typically in amplitude modulation (Samways, 1976a, 1976b; Heller, 1988; Ragge and Reynolds, 1998; Walker and Moore, 2004). For example, several species of *Metrioptera* are found together in large parts of Europe without any obvious intraspecific variation in song (Figure 9.1). Although the species have slightly different habitat requirements, at many localities they occur together within acoustic range.

COMPARING CHANGE OF SONG AND MORPHOLOGY OF MALE GENITALIA IN ISOLATED, ALLOPATRIC POPULATIONS OF CLOSELY RELATED FORMS

Differences in song or morphology between isolated, allopatric populations of members of one clade are typically smaller than the differences between widespread sympatric species. These are,

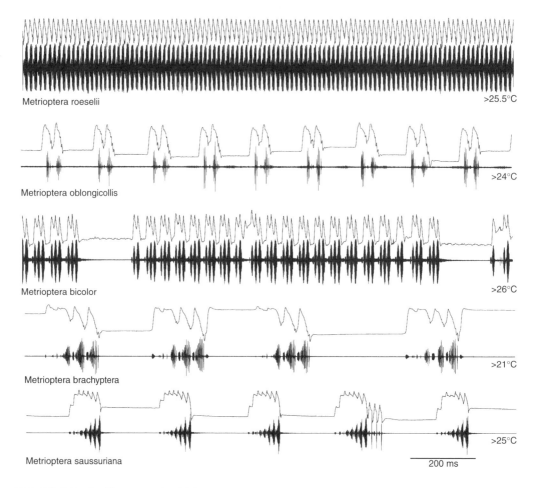

FIGURE 9.1 Oscillograms of stridulatory movements and song (synchronous registration of left tegmen movement [here and in Figure 9.2, Figure 9.4 and Figure 9.5 upward deflection represents opening, downward closing] and sound) of some species of the genus *Metrioptera* (localities of recorded specimen on CD).

however, important in models of allopatric speciation and there is a large literature on this (Otte and Endler, 1989). Here, we shall concentrate on the comparison between differences in song and male genitalia. In the four parts of this section we shall:

- Describe the situation in the subgenus *Parnassiana* (genus *Platycleis*) in detail.
- List examples where only one trait (song or morphology) shows differences and the other seems to be more or less stable.
- Mention rare examples where allopatric populations of one species differ in song and morphology.

The Subgenus *Parnassiana* (Tettigoniidae: Platycleidini: *Platycleis*)

Parnassiana occurs with about 15 taxa (12 monotypic species and one species with three subspecies) on high mountains in Greece. A further species (*P. vicheti*), endemic to Southern France, probably belongs in another (sub)genus. All described forms occur allopatrically at high altitudes on different mountains (Figure 9.2). The species may be differentiated mainly by their genitalia (Figure 9.3), although a few differences in coloration and pronotum shape are also important (mainly for subspecies recognition; see Willemse, 1985). Calling songs are known for

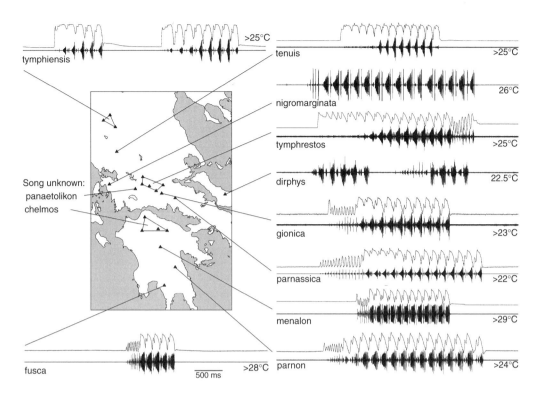

FIGURE 9.2 Distribution map and oscillograms of song of the species of the subgenus *Parnassiana* (genus *Platycleis*) and of stridulatory movements (in eight species only; synchronous registration of left tegmen movement and sound). Localities of recorded specimen on CD, distribution data from Willemse, F., *Catalogue of the Orthoptera of Greece, Fauna Graeciae*, Hellenic Zoological Society, Athens, 1984, p. 1, i–xii, 1–275; Willemse, L. and Willemse, F., *Entomol. Ber. Amsterdam*, 47, 105–107, 1987; Heller, K.-G. and Willemse, F., *Entomol. Ber.*, 49, 144–156, 1989.

10 species (Figure 9.3; Heller, 1988; Heller and Willemse, 1989). In all species the song contains two groups of elements: sequences of very short syllables (microsyllables) which can be positioned at the beginning (Figure 9.2: *P. parnassica*) or end (Figure 9.2: *P. tymphrestos*) of a syllable group, and sequences where short and long syllables (macrosyllables) alternate. This pattern is unknown in any other European tettigoniid. Series of microsyllables are produced by many other species related to *Platycleis* (examples in Heller, 1988; see also Figure 9.1: *Metrioptera saussuriana*) and they are possibly directed towards other males (Samways, 1976). From a comparison with related species it can be assumed that the alternate sequence with macrosyllables carries species-specific information. However, in all *Parnassiana* this sequence is quite similar. All syllable types are repeated at about the same rate in all species, so there is no evidence for distinct song differences, although the amount of song variation is still insufficiently known for all species. It is known for example that the duration of the alternate syllable series and the frequency of microsyllable series can be very variable within one species (Heller, 1988), indicating that these parameters are not useful for species recognition, but they seem to differ reliably between other species. The differences in male genitalia, however, are so pronounced that the forms have been described as different species. These data suggest that in this group song does not form the starting point for speciation. Some observations on Mt. Tsoumerka may even provide evidence that two species (*P. tenuis* and an unidentified one) with quite similar songs but different genitalia, occur on the same location (Heller and Willemse, 1989). The same situation as

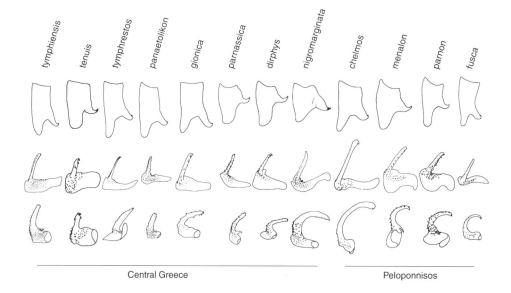

FIGURE 9.3   Male genitalia of the species of the subgenus *Parnassiana* (genus *Platycleis*). First row: left cercus, dorsal view. Second row: titillator, anterior view. Third row: titillator, lateral view. (Figures from Willemse, F., *Entomol. Ber. Amsterdam*, 40, 103–112, 1980; Willemse, L. and Willemse, F., *Entomol. Ber. Amsterdam*, 47, 105–107, 1987; Heller, K.-G. and Willemse, F., *Entomol. Ber.*, 49, 144–156, 1989.) Bottom row: distribution.

in *Parnassiana* occurs in some other groups where songs are similar but genitalia are different in allopatric populations.

Allopatric Forms Differing in Genitalic Morphology, but not in Song

Many published examples describe allopatric forms which differ in genitalic morphology, but not in song. Probably the most striking example is found among the Greek species of *Eupholidoptera*. Among the about 21 completely allo- or parapatric species, the song pattern of all those studied (14) is the same, with the possible exception of the east Aegean *E. prasina*, which may have a faster syllable rate (Heller, 1988; Willemse and Heller, 2001). In male genitalia, however, very large differences between most species are observed (Willemse, 1985; Willemse and Heller, 2001).

Psorodonotus and the subgenus *Tessellana* of *Platycleis* contain distinctly less subspecies or species, respectively, than *Eupholidoptera* in Europe, but present otherwise the same picture. Within each genus the three subspecies of *Psorodonotus* and four species of *Tessellana* have nearly identical songs (Heller, 1988), but differ in morphology of genitalia (Harz, 1969). Other examples are *Acrometopa macropoda*, *Platycleis albopunctata*, *Zeuneriana*, *Pholidoptera*, *Pterolepis spoliata*, *Callimenus* and *Uromenus* (Heller, 1988).

Ragge and Reynolds (1998) discussed two species complexes within the genus *Metrioptera* — the *M. saussuriana* and *M. roeselii* complexes. The widespread *M. saussuriana* and the local *M. buysonii*, which do not differ in song, show small but stable differences in genitalia. They occur close together in the Pyrenees (maps in Voisin, 2003). The Italian *M. caprai* shows only small differences in genitalia compared with *M. saussuriana*. The song of some populations of this species is very similar to that of *M. saussuriana*, but others have a higher syllable number per syllable group. In the *M. roeselii* complex several local allied forms have been described in southern Europe. Some show relatively large differences in genitalia, but the song patterns of all of them are the same (Ragge and Reynolds, 1998; Fontana, 2001).

In the *ampliatus* group of *Poecilimon*, songs of the two parapatric species, *P. ampliatus* and *P. ebneri*, are quite similar, while the male cerci show distinct differences. *P. ampliatus* differs further from *P. ebneri* in the possession of a large dorsal gland which the female feeds on during mating (Heller and Lehmann, 2004).

A more complicated situation is found in *Ephippiger*. There has been a long discussion about the taxonomic status of species related to *Ephippiger ephippiger* (Kidd and Ritchie, 2000). Obviously local forms do exist, but their status is unclear. They differ mainly in male genitalia (Fontana, Kidd in preparation; Kidd and Ritchie, 2000) and song with high variability concerning syllable numbers per syllable group. In the western part of the range of *E. ephippiger* three mainly allopatric archetypes (Kidd and Ritchie, 2000) may be found, but they hybridise where they meet. Obviously neither differences in song nor in genitalia are large enough to prevent mixing of the gene pools. Over the whole range of the species, the pattern of song variation and its relationship to morphological variation is not sufficiently known.

Another well-known example from North America is the genus *Orchelimum*. Two parapatric species, *O. nigripes* and *O. pulchellum*, differ slightly in genitalia, but not in song (Walker, 1974). Hybridisation between these forms has been documented in two contact zones (Shapiro, 1998). It should be noted that *Orchelimum* species are long-winged animals capable of relatively fast dispersal while all other examples mentioned above concern short-winged animals.

Belocephalus contains five species falling into two groups (see Walker and Moore, 2004). Members of both groups may occur sympatrically. The three are morphologically quite similar; allopatric species of the *B. sabalis* group all have the same song, but one species differs clearly in life-history traits, while the other two are geographically separated. The other two allopatric species of the *B. subapterus* group are morphologically similar, but differ mainly in song and belong to the species treated in the following subsection. However, within *B. subapterus* there are two geographically separated forms (with hybrid zone) differing in male genitalia only.

Allopatric Forms Differing in Song, but not in Genitalic Morphology

In a few examples among allopatric forms, only the song seems to have changed. Species will be mentioned here where only the song and the stridulatory organs together have changed, or at least the song much more than genitalia.

The most distinctive examples are found in *Decticus verrucivorus*. This well-known and widespread species shows some variability in wing length and body size (Samways and Harz, 1982). In the song, which has been recorded at many localities in Europe (Ragge and Reynolds, 1998), no significant differences have been found, except in Spain. Specimens from the Iberian Peninsula produced a quite distinct song, which was used to establish a subspecies, *Decticus verrucivorus assiduus* (Ingrisch *et al.*, 1992). No differences were found in genitalia. Recently, in another form of this group from south Italy, a song pattern was detected which also differs slightly from that of the nominate subspecies (Fontana *et al.*, 1999).

Two other examples of song change with no, or only minor, divergence in genitalic morphology are found in *Poecilimon*. Heller (1984) described the subspecies *P. obesus artedentatus*, now usually considered as a species (Willemse and Heller, 1992), differentiated clearly in song and morphology of the stridulatory file. This is a possible example of reproductive character displacement because *P. artedentatus* occurs in an area where another species of *Poecilimon*, *P. nobilis*, is found with a song very similar to that of *P. obesus*. The second example concerns *Poecilimon paros*. This island species was discovered by its song, which differs distinctly from that of its allopatric sister, *P. hamatus* (Heller and Reinhold, 1992). A small difference in the shape of the male cerci has now been found, which can be used for morphological separation of both species. A similar discovery was made recently in Turkey, where the new species *P. martinae* differs from its sister species, *P. inflatus*, mainly in song, but also in morphology of the male cerci (Heller, 2004).

In east Asia, two species of *Hexacentrus*, *H. japonicus* and *H. unicolour*, are very similar in morphology, but differ in song (Inagaki *et al.*, 1986). They are largely allopatric, but overlap in a small area in Japan where they prefer separate habitats (Inagaki *et al.*, 1986). Morphologically they can be differentiated by small differences in male genitalia (Inagaki and Sugimoto, 1994) and the structure of the stridulatory file (Inagaki *et al.*, 1990). In the laboratory interspecific hybrids can be obtained (F1 and F2) (Inagaki and Sugimoto, 1994).

At this point those rare examples should be mentioned where not only allo- but also sympatric species differ only by song. Among bushcrickets only some species of North American *Amblycorypha* belong to this category (Walker *et al.*, 2003). Walker (1964) expected that such cases of cryptic species would be relatively common, but he also predicted that, after discovery of a cryptic species by its song, morphological differences would later be found (see *Poecilimon* example above). This may also apply, for example, to the different song forms of *Phyllomimus inversus* in Malaysia, for which the colour of the "knees" of the hind legs may be diagnostic (Heller, 1995).

Unusually Variable Species

Besides species in which allopatric forms differ mainly in genitalia or song, there are some examples where allopatric forms differ in morphology and song but are still considered members of one species.

Allo- and parapatric populations of several species of *Pterophylla* have been analysed in North America. In *P. camellifolia*, Shaw and Carlson (1969) made one of the first studies of this type, demonstrating song differences together with differences in male genitalia. Because of presumed hybridisation, they considered both parapatric forms as members of one species. Den Hollander and Barrientos (1994) found acoustic and morphometric differences between allopatric populations of *P. beltrani* in Mexico. Morphometric differences (most importantly in male cerci) seem to have diverged more rapidly than song. In hybridisation experiments these authors demonstrated that there are neither strong premating nor postmating reproductive barriers. They compared *P. beltrani* also to the allopatric *P. robertsi* (Barrientos and Den Hollander, 1994). As expected for different species, they found differences in morphology and song. They found some overlap in the measured song characters between both species, but not in morphological characters. Hybridisation experiments indicated strong postmating barriers (successful matings and egg deposition, but no hatching of offspring), although in a second series of tests some offspring were obtained (Barrientos-Lozano, 1998).

TRACKING SONG EVOLUTION BY THE USE OF PHYLOGENETIC TREES

As examples of this powerful method we shall:

1. Use a phylogenetic tree based on morphological characters (Çiplak, 2000) and add song data to this tree,
2. Present data from a combined approach which used DNA together with song and morphology (Lehmann, 1998)
3. Review other published phylogenetic data.

Genus *Parapholidoptera* (Tettigoniidae)

The genus *Parapholidoptera* was revised by Çiplak (2000). On the basis of morphological characters, he was able to establish a cladogram (Figure 9.4) indicating a separation of the genus into two groups, the larger one with several subgroups. Comprising 17 species, the genus occurs in Turkey and some surrounding countries. The species of each group are allo- to parapatric, as can be judged from the distribution data, but the ranges of the groups overlap slightly (see Appendix for the first record of a syntopic occurrence of *P. distincta-signata*). For six species from groups and most subgroups, sound records are available. Mapping the song characters on the tree (Figure 9.4) shows that within the *castaneoviridis* group (with the possible exception of *salmani*) the basic song pattern does not change. Species from all subgroups produce a multi-syllable song with syllable groups separated by silent intervals of several seconds. Within each syllable group the amplitude increased at the beginning and remained stable or decreased slightly towards the end. In the syllable number per syllable group there is some variation between species (from seven to nine in *spinulosa*, *antaliae* and *signata* to 11 to 13 in *castaneoviridis*), which could be the basis for species-specific song recognition. It is unknown whether this song pattern represents the ancestral condition for the genus since the song of the most outgroup species is not known. However, one of the possible outgroups, *Uvarovistia satunini*, also shows a song pattern relatively similar to that of members of the *Parapholidoptera castaneoviridis* group (Figure 9.4). The songs of the two members of the second group which have been recorded differ distinctly. These consist of fast sequences of short syllable groups (about three syllables per group), produced for quite long time periods (up to 30 sec recorded). Both species differed in syllable duration and syllable group repetition rate. The second character was variable in *P. distincta*. The situation in the first group may indicate a slowly divergent drift in song pattern in allopatric species, as found in the structure of male genitalia for

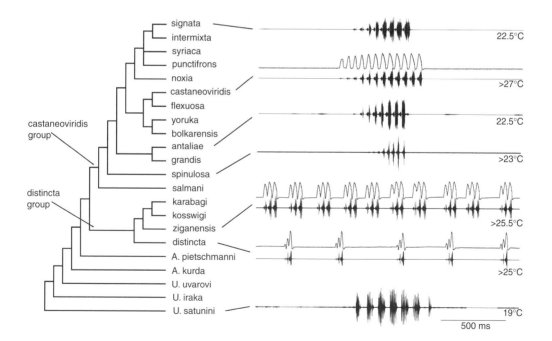

FIGURE 9.4 Phylogenetic tree of the genus *Parapholidoptera* (outgroups genera *Apholidoptera* and *Uvarovistia*) together with oscillograms of stridulatory movements (in three species only) and song of seven species (synchronous registration of left tegmen movement and sound) (phylogenetic tree according to Çiplak, 2000; for localities of specimen, see Appendix).

most species (Çiplak, 2000). Some species, however, show unexpected large differences in male cerci or titillators. In the second group the differences in song pattern are more pronounced than in genitalia.

Poecilimon propinquus Group (Phaneropteridae, Barbitistinae, *Poecilimon*)

Evolution of song and male genitalia was studied in detail by Lehmann (1998) in a group of species of the genus *Poecilimon*. In the first application of DNA sequencing for tettigoniid phylogeny, he used differences in the base composition of the cytochrome oxidase gene to establish a phylogenetic tree independent of morphological and song characters. The eight species of the *Poecilimon propinquus* group which he studied occur allo- or parapatrically in Greece. Some species form quite narrow contact zones, often at ecological borders, where within less than 100 m one species is replaced by another. All species except two can be differentiated most easily by the structure of the male cerci. According to the phylogenetic tree these species-specific differences have evolved independently and partly convergently. Distinct differences in song pattern between species have been found (Figure 9.5). The apparent plesiomorphic pattern, a song consisting of isolated syllables, is retained in three species. Twice independently it has changed to polysyllabic syllable groups. Lehmann (1998) tested whether reproductive character displacement could be involved in the changes in morphology and song, but found no evidence for that. Therefore, he concluded that sexual selection probably was the most important factor for the evolution of the differences.

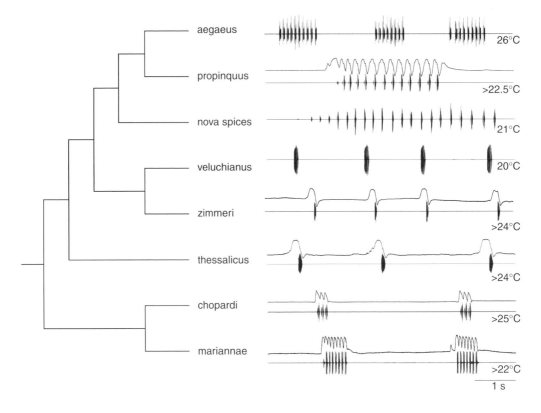

FIGURE 9.5 Phylogenetic tree of the *propinquus* group of the genus *Poecilimon* together with oscillograms of stridulatory movements (in six species only) and song (synchronous registration of left tegmen movement and sound) (phylogenetic tree according to Lehmann, 1998; for localities of recorded specimen, see Appendix).

Other Published Phylogenetic Data

At the moment the most significant limitation for the application of this method is the very limited number of phylogenetic trees available. To my knowledge, there are only about ten groups of species for which cladograms have been published, and not all of them can be used to study song evolution. Two cladograms refer to groups with few species. In their study of *Ephippiger*, Oudman *et al.* (1989) analysed the relationships of a few taxa by enzyme electrophoresis. As mentioned above, there is great variability in song and genital morphology in this genus and conclusions about evolution from a restricted data set are premature. Barendse (1990) studied all four species of the Australian genus, *Mygalopsis,* using also allozyme data. The species differ in song and morphology of male cerci (Bailey, 1979). Two species, *M. marki* and *M. pauperculus*, are parapatric, but the differences between them are of about the same order of magnitude as between sympatric species. Surprisingly, *P. marki* consists of two genetically distinct, allopatric subspecies without noticeable differences in song and general morphology and only slight differences in number of teeth on the stridulatory file.

Several published phylogenetic trees are based mainly on the evaluation of data from song pattern and genitalic morphology (*Poecilimon*: Heller, 1984; *Pycnogaster*: Pfau, 1988; Pfau and Pfau, 1995; *Platystolus*: Pfau, 1996; *Calliphona*: Pfau and Pfau, 2002). These represent valid hypotheses about the phylogeny of the groups considered. However, these are of limited value for the question analysed here because the characters whose evolution has to be examined were used to establish their evolutionary relationships.

With caution some conclusions can be drawn from these data. In *Poecilimon*, morphologically similar allo- and parapatric forms (mainly considered as subspecies) show similar songs (*e.g.* *P. jonicus* group, *P. brunneri* group; Heller, 1984). The same situation is found in *Pycnogaster* (Pfau, 1988; Pfau and Pfau, 1995; Heller, 2003). All species show basically the same song pattern, but differ in syllable repetition rate and the presence or absence of opening hemisyllables. In morphology, differences in male genitalia as well as in other characters are found. All species are allo- and parapatric; sympatric occurrences are not known. The differences in song are much larger within *Platystolus* (ten species) than with *Pycnogaster* (nine species), even though they both inhabit the Iberian Peninsula. The genus differs from *Pycnogaster* also because some species are sympatric (map in Pfau, 1996) and because many species use complicated acoustic male–female duets. According to the phylogenetic tree, in the subgenus *Neocallicrania* related and even partly sympatric species have similar songs, while in the subgenus *Platystolus* sister species differ strongly in song (but not genitalia). An interpretation of these surprising differences has to await an independently established cladogram.

An especially interesting situation is found in the genus *Calliphona*. Three species occur in allopatry on different Canary Islands. Two species distributed on three islands have similar songs, while the third exhibits a song as different as typically is found for sympatric species (Pfau and Pfau, 2002). Contrary to older hypotheses (Holzapfel and Cantrall, 1972), the authors favour the hypothesis that the islands were settled twice independently from Africa. This would explain the large song differences, which could be assumed to have been inherited from sympatric ancestors.

The two most recent phylogenetic trees were established for *Anterastes* (Çiplak, 2004) and *Panacanthus* (Montealegre and Morris, 2004). For *Anterastes*, too few song data are available to combine song and morphological characters. The cladogram for *Panacanthus* (seven species) is mainly based on morphological characters, but also includes song data from four species. All species occur allopatrically in the northwest of South America. Their songs consist of short groups of isolated syllables. Only in one species was a stable combination found of two syllables produced in short groups. There is no variation in basic song pattern. However, there was variation in spectral composition from nearly pure tone, resonant songs to broad band signals. According to the tree, the broad band forms evolved from pure tone signals. Montealegre and Morris (2004) discussed possible advantages of the derived type. First they mentioned advantages in inter-male competition for females. Bailey (1976) observed similar differences in spectral characteristics in the songs of several species of *Ruspolia* in Uganda, where the differences seem to be related to habitat structure.

CONCLUSIONS

From the data discussed here it is clear that there are no strict rules about what happens during the first steps of the differentiation of allopatric populations. In most examples, however, changes in calling song seem to appear more slowly than changes in morphology. This impression is supported by the comparison of allopatric populations of closely related forms as well as by analyses of phylogenetic trees. In isolated populations, modifications of male genitalia are typically observed first. In these the song pattern often remains unchanged, and even in morphologically widely divergent forms, calling songs are often quite similar. This rapid evolutionary change in male genitalia is a widespread phenomenon and not restricted to bushcrickets. There is much evidence that it is an effect of sexual selection (Eberhard, 1985).

But why does song pattern not change equally fast or equally often? Genitalia and song may be part of a species' mate recognition system and can be used to judge the quality of a mating partner, typically the male. In other groups of animals, songs do change quite often in allopatric, closely related forms.

Typical examples are found among birds, although the ranges of the forms involved may be much larger than in bushcrickets (Martens and Nazarenko, 1993; Martens and Steil, 1997; Price and Lanyon, 2004). But rapid song changes are not restricted to vertebrates. Among insects, they are also commonly found, as for example in *Drosophila* (Ritchie and Gleason, 1995) and many other groups of insects (Sueur, Chapter 15). Even in another group of Orthoptera, the acridid grasshoppers, many examples of rapidly diverging songs are known (Helversen, 1986; Ragge, 1987; Dagley *et al.*, 1994; Vedenina and Helversen, 2003), but where morphology of genitalia is little changed (Harz, 1975), and sometimes not even mtDNA sequences (Frieder Mayer, personal communication).

What are the reasons for these taxon-specific differences? Comparing different groups of insects, one important difference in the function of song is quite obvious. In bushcrickets, song is used for long-distance communication between animals sitting in bushes or trees. Their maximal hearing or communication range may be between 5 and 40 m or even larger, depending on the frequency content and the loudness of the signal (Römer and Lewald, 1992). Either males or both sexes produce their songs until male and female meet. After establishing physical contact, they stop acoustic communication. Then they use their antennae, mouthparts or genitalia to touch and check each other. They also use information about their partner's body weight for mating decisions, but they typically do not sing. Accordingly, tettigoniids have no special courtship songs except for some unusual species (*e.g. Pycnogaster* [Pfau, 1988; Pfau and Pfau, 1995]). During mating, the tettigoniid male transfers a spermatophore amounting to up to 40% of his body weight (Gwynne, 2001).

Acridid grasshoppers and *Drosophila*, on the other side, typically sing even when males and females are in close contact. In *Drosophila*, the song can even be heard by females only from a very small distance, up to a few mm. Acridid grasshoppers are able to communicate acoustically over larger distances, but they usually sing sitting on the ground where sound propagation is restricted. Species with loud songs can communicate over a distance of about 1 m, under optimal conditions of up to 3 m only (Lang, 2000). The male continues to sing after male and female are in close contact. Usually a male tries to mate after some courtship (rarely immediately after meeting), but continues to sing if the female rejects. The same type of behaviour is also seen, for example, in many Hemiptera (this volume).

This difference in song production after contact between the groups has some important consequences. In a long-distance communication system the effects of the acoustic environment on sound propagation are much more important than in a close range system (Römer, 1993). Fine temporal elements of the song may be lost due to reverberations and scattering. Sounds of conspecifics and other species may mask the signal partly or completely. In acoustic long-range systems with many animals singing simultaneously, it will be more difficult to extract information about the quality of a signaller than if a female hears only one or a few males. It may also be difficult, especially at high densities, to connect the properties of a special acoustic signal with the particular male the female contacts. In accordance with these considerations, song patterns in bushcrickets are generally quite simple. In many genera similar types of amplitude modulation are found. Most common are trills, chirps of various duration and double syllable patterns (see Figure 9.1), although the recognition of the latter can be based on the combined duration of both syllables of a pair (Schul, 1998). So it would not be surprising if song patterns in tettigoniids are under stabilising selection, as known for crickets (Ferreira and Ferguson, 2002). They may change mainly under selection pressure coming from outside the species, from the environment and from other calling species. There are still very few exceptions to the rule (Alexander, 1957a) that syntopic species always differ in song. Other song characters, like frequency content, are possibly more often under directional selection, if they confer direct advantages for sender or receiver. Low frequencies in calling songs may be difficult for small animals to produce, but they generally have advantages in aerial sound propagation (although they provide problems for localisation). A continuing conflict between costs and benefits is expected with surprisingly low frequencies in some species (Stumpner, 2002; but see Gwynne and Bailey, 1988). After a male and female tettigoniid ready to mate have met, estimating information on the partner's condition becomes important. This may be generally true in mating, but especially in bushcrickets where the spermatophore is a costly investment for the male and a valuable resource for the female. Cues may be received and transferred via male genitalic contact. The structure of the genitalia can be considered as an extreme close-range recognition or communication system and one of the last possible steps for prezygotic isolation. If sexual selection is important, close-range signals which are directly linked to an individual partner could be much more reliable than long-distance signals for which the origin is uncertain.

In acridids which continue to signal acoustically after having established physical contact and use sounds for close range recognition, many species have elaborate courtship songs, which are typically more or even much more complex than the calling songs (Helversen, 1986; Ragge and Reynolds, 1998). In this group of Orthoptera not only stabilising selection on song patterns is observed (Helversen and Helversen, 1983; Butlin *et al.*, 1985), but also clear cases of directional selection on parts of song which may indicate male quality (Stumpner and Helversen, 1992; Klappert and Reinhold, 2003). At least at high population density, song seems not to be necessary for male and female to find each other, but for mating, singing males are clearly preferred by females compared with mute ones (Kriegbaum and Helversen, 1989), and even males with slightly disturbed songs were often rejected (Kriegbaum, 1989).

The song patterns in many Phaneropteridae and Ephippigerinae differ from those of other bush-crickets because in both groups females respond acoustically to the male song. For example, the male songs often have lower duty cycles (Heller and Helversen, 1993) and contain trigger elements, after which the females respond (Heller, 1990). Duetting behaviour creates opportunities for alternative male tactics (Bailey, 2003) and song-pattern evolution is different from that in "normal" bushcrickets. Possibly some cases of song-pattern changes with or without accompanying changes in male genitalia are related to this type of communication, although in some of those species (*Poecilimon propinquus* group) the females have lost the ability to respond acoustically.

The differing speed of evolution in song pattern and genitalia also has some consequences for taxonomy. According to the data presented above, in tettigoniids different song patterns typically indicate that the speciation process is complete and that the forms in question should be treated as full species. However, populations or forms with slightly different genitalia but the same song pattern are more difficult to evaluate. They may often not be reproductively isolated, their gene pools are still not separated and they can be treated as subspecies according to the biological species concept. The occurrence of hybrids as in *Ephippiger* (Kidd and Ritchie, 2000), *Orchelimum* (Shapiro, 1998), *Belocephalus* (Walker and Moore, 2004) and the *Poecilimon propinquus* group (Lehmann, 1998) supports this view, although in strictly allopatric forms this can only be tested by experiments. The presence of the same song pattern indicates that at the moment they will not be able to coexist at the same place. Of course, under the phylogenetic species concept all forms differing in any character can be treated as species, especially in short-winged insects like many bushcrickets. However, the strict application of this concept will not only result in a simple taxonomic inflation as in primates (Isaac *et al.*, 2004), but in a hyperinflation combined with the loss of important information about phylogenetic relationships.

ACKNOWLEDGEMENTS

The chapter was originally planned as part of a comparison of the acoustic behaviour of acridids and tettigoniids by Otto von Helversen and myself, but the acridid part still awaits its completion. My special thanks go to Klaus Reinhold, Bielefeld and Andreas Stumpner, Göttingen for helpful comments on an earlier version of the manuscript.

APPENDIX

Genus *Metrioptera*
M. bicolor (Philippi, 1830): GERMANY: Baden-Württemberg, Utzenfeld near Todtnau, Schwarzwald (47°48′N, 7°54′E), 8 × 1979, coll. Heller and Volleth (CH3230)

M. brachyptera (Linnaeus, 1761): GERMANY: Bavaria, Oberailsfeld, Fränkische Schweiz (49°49′N, 11°20′E), 25 ix 1981, coll. Heller and Volleth (CH3840)

M. oblongicollis (Brunner v. Wattenwyl, 1882): YUGOSLAVIA: Serbien, Grdelicka gorge south of Nis (42°51′N, 22°7′E), 6–7 viii 1981, coll. Heller and Volleth (CH3845)

M. roeselii Hagenbach, 1822: GERMANY: Baden-Württemberg, Utzenfeld near Todtnau, Schwarzwald (47°48′N, 7°54′E), 8 × 1979, coll. Heller and Volleth (CH3235)

M. saussuriana (Frey-Gessner, 1872): SPAIN: Lerida, Valle de Aran: Baqueia (42°41′N, 0°56′E), 22 viii 1984, coll. Heller (CH5175).

Genus *Parapholidoptera*:
P. antaliae Nadig, 1991: TURKEY: Antalya, above Gazipasa (near Ilõca) (36°24′N, 32°23′E), 800 m, 10 vii 2002, coll. K.-G. Heller (1 *M*, 2FF: CH5393, CH5710-1; Figure 9.4); TURKEY: Antalya, Gazipasa (36°17′N, 32°18′E), 17 iv 1987, coll. M. Gebhardt (1 F:CH5395); TURKEY: Antalya, Günlükly (=Kücükörpe) southwest of Antalya, 8 iv 1987, coll. M. Gebhardt (2 *M*, 2F: CH5396-9)

P. castaneoviridis (Brunner v. Wattenwyl, 1882): TURKEY: Bursa, Ulufer Cayiy südl. d. Ulu Dagh (40°5′N, 29°4′E), 400 m, 21–22 vii 1983, coll. Heller (1 *M*: CH4350; Figure 9.4); TURKEY: Edirne, 3 km west of Havsa (41°34′N, 26°47′E), 20–21 vii 1983, coll. Heller (2 *M*: CH4348-9; recorded)

P. distincta (Uvarov, 1921): TURKEY: Artvin, Yalnizcam pass (41°4′N, 42°18′E), 9 viii 1987, coll. K. Reinhold (1 *M*: CH4572); TURKEY: Artvin/Kars, Yalnizcam pass, eastern slope (41°4′N, 42°19′E), 2200–2400 m, 10–11 viii 1983, coll. Heller (3 *M*, 1 F: CH5985-8; recorded); TURKEY: Erzurum, pass northwest of Ispir (Tatos Daghlari) (40°35′N, 40°53′E), 2150 m, 4 viii 1983, coll. Heller (1 *M*, CH5989; Figure 9.4); TURKEY: Kars, 10 km north of lake Cildir (42°8′N, 43°6′E), 11 viii 1988, coll. K. Reinhold (1 *M*: CH4573); TURKEY: Kars, pass south of Sarikamis (40°16′N, 42°39′E), 2000 m, 9 viii 1983, coll. K.-G. Heller (1 *M*:CH5993; recorded).

P. signata (Brunner v. Wattenwyl, 1861): TURKEY: Erzurum, valley of river Coruh, 30 km northeast of Ispir (40°39′N, 41°16′E), 950 m, 5 viii 1983, coll. Heller (2 *M*: CH5980-1); TURKEY: Erzurum, pass northwest of Ispir (Tatos Daghlari), near village of Cayirözü (40°35′N, 40°53′E), 1800–1950 m, 4 viii 1983, coll. Heller (2 *M*, 2 F: CH5976-9; recorded); TURKEY: Kars, pass south of Sarikamis (40°16′N, 42°39′E), 2000 m, 9 viii 1983, coll. K.-G. Heller (1 F: CH5982); TURKEY: Mersin, below Güzeloluk (22 km from Erdemli) (36°44′N, 34°9′E), 1000 m, 6 vii 2002, coll. K.-G. Heller (1 *M*, 1 F: CH5394, CH5709; Figure 9.4).

P. spinulosa Karabag, 1956 TURKEY: Kastamonu, Ilgaz mts. (41°3′N, 33°40′E), 1800–2000 m, 15–16 viii 1983, coll. Heller (1 *M*, 1 F: CH5983-4; Figure 9.4).

P. ziganensis Karabag, 1963: TURKEY: Giresun, valley near Tandere (Teknecik) (40°31′N, 38°23′E), 1500 m, 1 viii 1983, coll. Heller (1 F: CH5992); TURKEY: Giresun, Egribel pass (40°27′N, 38°23′E), 2100–2200 m, 1 viii 1983, coll. Heller (2 *M*: CH5990-1; Figure 9.4).

Genus *Uvarovistia*

U. satunini (Uvarov, 1916): TURKEY: Erzurum, pass gorge northwest of Ispir (40°30′N, 40°54′E), 1400 m, 4–5 viii 1983, coll. Heller (3 *M*, 2 F: CH0576, CH5994-7; Figure 9.4).

Genus *Poecilimon*

P. aegaeus Werner, 1932: GREECE: Evia, near Oktonia southeast of Kimi (38°33′N, 24°10′E), 100 m, 10 vi 1999, coll. Heller and Volleth (CH4750)

P. chopardi Ramme, 1933: GREECE: Florina, pass west of Florina (40°47′N, 21°16′E), 1200 m, 30 vii 1982, coll. Heller (CH1775)

P. mariannae Willemse and Heller, 1992: GREECE: Fthiotis, 6 km south of Metalio north of Lamia (39°1′N, 22°22′E), 400 m, 21 vi 1980, coll. Heller (CH1783)

P. n. sp.: GREECE: Fthiotis, Domokos (39°8′N, 22°17′E), 4 vi 1998, coll. K.-G. Heller and M. Volleth (CH4437)

P. propinquus Brunner v. Wattenwyl, 1878: GREECE: Korinthia, Akrokorinth (37°56′N, 22°55′E), 30 iii 1983, coll. E. Blümm (CH0331)

P. thessalicus Brunner v. Wattenwyl, 1891: GREECE: Pieria, Mt. Olymp (40°4′N, 22°20′E), 23 viii 1981, coll. O.v.Helversen (CH1720)

P. veluchianus Ramme, 1933: GREECE: Fthiotis, Vitoli near Makrakomi (38°58′N, 22°1′E), 27 v–28 vi 1987, coll. K.-G. Heller (CHX132)

P. zimmeri Ramme, 1933: GREECE: Fokis, Mt. Oiti, Gorgopotamos spring (38°46′N, 22°18′E), 1500 m, 18–19 viii 1981, coll. Heller (CH1748).

10 Acoustic Communication in Neuropterid Insects

Charles S. Henry

CONTENTS

INTRODUCTION

Insects are noisemakers. Their hardened exoskeletons click and tap or grind and crunch with nearly every movement, much like mechanical toys made of plastic. Insects and other arthropods are predisposed by their ground plan for acoustic communication, especially through stridulation (the rubbing of body parts together), and they have independently evolved an extraordinary array of sound-producing devices (Ewing, 1989, p. 16). Their noises, songs and music often appeal to our own auditory sensitivities, and have been the subject of much description, analysis and experimentation (reviewed in the current volume and in many other works, *e.g.* Pierce, 1948; Alexander, 1960; Busnel, 1963; Otte, 1977; Lewis, 1983; Ewing, 1989, Bailey, 1991; Bailey and Ridsill-Smith, 1991; Gerhardt and Huber, 2002; Greenfield, 2002). Yet clearly audible songs are

153

not evenly distributed across the 28 orders of insects. Orders such as Orthoptera and Hemiptera are replete with singing species, but others are nearly mute. The superorder Neuropterida, comprising the orders Raphidioptera (snakeflies), Megaloptera (dobson- and alderflies), and Neuroptera (lacewings), is a relatively silent taxon.

Although insects are preadapted for airborne sound production, several constraints affect the audibility of their acoustic signals and strongly influence the evolution and phylogenetic distribution of such signals. The best-known factor is body size. Insects need to be small due to surface-area-to-volume considerations which reduce the effectiveness of the exoskeleton and the tracheal system at larger body sizes. Smaller sizes in turn reduce the efficiency of energy transfer from the body to the air during sound production (Bennet-Clark, 1998). Therefore, communication using airborne sound should be limited to relatively large insects, as exemplified by crickets, katydids, grasshoppers (Orthoptera, Greenfield, 1997) and cicadas (Hemiptera, Bennet-Clark and Daws, 1999).

A less recognised constraint on sound production is the sclerotisation of an insect's body. Hard-bodied or heavily armoured insects can produce loud airborne sounds quite easily, whereas soft-bodied insects, even the large ones, will be less able to sing, especially via stridulation (Greenfield, 2002, p. 127). Moreover, body plans tend to characterise entire insect orders, so certain orders will include more loud singers than others. Leathery front wings are a focus of sound production in ensiferan Orthoptera, but soft-bodied stoneflies (Plecoptera) or caddisflies (Trichoptera) lack strongly sclerotised structures which might be coopted for sound production. For those insects, a better alternative to airborne sound for intraspecific communication is substrate-borne vibration. This "silent" acoustic mode has several advantages for small insects as well, notably long-distance propagation of signals through plant stems and leaves (Ossiannilsson, 1949; Michelsen *et al.*, 1982; Markl, 1983; Gogala, 1985a; Stewart, 1997; Čokl and Doberlet, 2003). As predicted, vibratory signals are particularly common in weakly sclerotised and small-bodied representatives of the clades Plecoptera, Psocoptera, Hemiptera, Neuropterida, Diptera and Trichoptera (Stewart, 1997).

Predominantly, neuropterid insects are not only small in size, but also soft-bodied. Perhaps because of these dual constraints, and like other insect groups of similar size and body plan, neuropterids communicate largely through their substrates (Devetak, 1998). In this chapter, I describe what is known of air- and substrate-borne acoustic signals in Neuropterida, taking a phylogenetic perspective when possible.

NEUROPTERIDA

Neuropterida is a relatively small taxon of three archaic insect orders united by a few weak morphological synapomorphies (Kristensen, 1999). Recent DNA sequence data nonetheless support the clade (Whiting, 2002; Haring and Aspöck, 2004). It is thought to be the sister clade of the Coleoptera, and together the two taxa constitute the monophyletic "neuropteroid complex" or "Neuropterodea" of Endopterygota (Holometabola).

The globally distributed neuropterid order Megaloptera includes some 300 species in two families. Raphidioptera is a small Holarctic order of two families and about 200 species. Cosmopolitan Neuroptera is much larger, at 6000 described species. Neuroptera is also the most neatly delimited of the three orders: although its 17 families are extraordinarily disparate in morphology and habits, all share a suite of specialised suctorial mouthpart structures in their larvae. In contrast, neither Megaloptera nor Raphidioptera possess compelling morphological synapomorphies, and their phylogenetic positions within Neuropterida are uncertain (reviewed in Aspöck *et al.*, 2001). Nevertheless, monophyly of Megaloptera and Raphidioptera has been affirmed by DNA sequence data (Haring and Aspöck, 2004). The most recent morphological and molecular studies (Figure 10.1) support a clade of Megaloptera + Neuroptera, with Raphidioptera as its outgroup (Aspöck *et al.*, 2001; Aspöck, 2002; Haring and Aspöck, 2004).

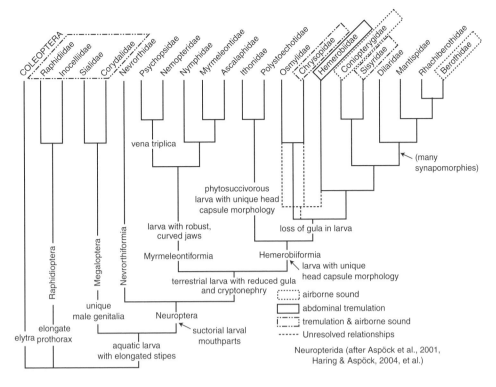

FIGURE 10.1 Phylogenetic hypothesis for the relationships of the 21 families of the superorder Neuropterida, indicating taxa within which acoustic communication has been described.

TYPES OF ACOUSTIC COMMUNICATION IN NEUROPTERIDA

As mentioned, the neuropterids are not noted singers. Acoustic or vibratory signals, in the context of sexual behaviour, have been confirmed in just nine of 21 families (Figure 10.1). In most of these, communication is by substrate-borne vibratory signals, although putative stridulatory devices have been described in several genera of green lacewings (Chrysopidae: Adams, 1962; Riek, 1967; Brooks, 1987). Because vibratory signals are not easily detected, acoustic communication will likely be found in additional neuropterid taxa when appropriate instrumentation is used.

Vibratory signals in neuropterids may be produced percussively or by tremulation. Percussive (drumming) signals result from some part of the insect's body, usually the abdomen or wings in neuropterids, which actually strike the substrate such that compression waves are propagated through the material. In contrast, tremulation produces bending waves in the substrate (Michelsen *et al.*, 1982). As the insect vibrates its abdomen vigorously in the vertical plane, the lightweight stem or leaf upon which it is standing is shaken up and down at low frequencies ranging typically between 10 and 150 Hz. Tremulation produces a more constrained, consistent signal than percussion. Whereas the frequency structure (tone or pitch) of a percussive signal depends upon the resonant properties of the substrate, that of a tremulation signal is largely independent of substrate characteristics and instead accurately reflects the frequency of vibration of the abdomen itself. As we shall see, tremulation is much more common than percussion for signal production in neuropterids. Several species of green lacewings have evolved modified regions of the wings which are hammered against the substrate during courtship and mating (see later discussion), while other chrysopids and some sialids (Megaloptera) will strike the substrate with the vibrating abdomen rather than (or in addition to) tremulating. Even when audible (airborne) sound results from percussive behaviour, as in some lacewings, it is never very loud. The substrate-borne component of such signals probably carries most of the information.

Structures and behaviour specifically dedicated to the production of airborne sound are very rare in neuropterids, again being limited to Chrysopidae. These will be reviewed in a section devoted to green lacewing acoustic communication.

ACOUSTIC COMMUNICATION IN MEGALOPTERA AND RAPHIDIOPTERA

Tremulation is found in all four families of Megaloptera and Raphidioptera (Figure 10.1). In every case, it includes "bouts" or volleys of abdominal vibrations which initiate and accompany courtship, sometimes in both sexes. These bouts may be accompanied by fluttering of the wings, which can produce faint but audible airborne sounds. In those species in which both sexes "sing", duetting between males and females is usually present.

MEGALOPTERA: SIALIDAE AND CORYDALIDAE

Courtship and mating within Megaloptera have been described for just one genus of Sialidae (alderflies) and five of Corydalidae (dobsonflies and fishflies). A detailed overview of megalopteran mating systems can be found in Henry (1997).

Sialidae

Best studied of all Megaloptera are two European alderflies, *Sialis lutaria* (Linnaeus) and *Sialis fuliginosa* F. Pictet. In *S. fuliginosa*, Killington first described tremulation as a mutual "twitching of abdomens upward at intervals" in both sexes (Killington, 1932, p. 67). Much later, Rupprecht (1975) distinguished two types of tremulation signals in both species, which he described as "rhythmic" vs. "prolonged, unstructured".

"Rhythmic" vibrations are relatively short volleys of low frequency abdominal tremulation repeated at regular intervals by both sexes. These signals "allow mutual approach and recognition of species and sex" (Rupprecht, 1975, p. 305). Volley characteristics are similar in both species. The duration of each volley is several hundred milliseconds, and carrier frequency gradually decreases (*i.e.* is modulated) within each volley from a mean of 200 Hz at volley onset to about 120 Hz at the finish. Volley period ranges from 250 msec to 2 sec, depending on the behavioural context. Both species use only rhythmic tremulation signals during heterosexual duets. However, duetting in *Sialis* is not precise: the stimulus produced by one courting insect does not trigger a predictably timed response from its partner.

Rupprecht's "prolonged, unstructured" vibrations were found only in males of the two *Sialis* species he studied. These signals do not show consistent temporal or frequency characteristics. They appear to reflect a general state of sexual receptivity in males, and often segue into periods of organised volley production, courtship and copulation (Rupprecht, 1975, Figure 4).

Rupprecht also noted percussive signals in *Sialis*, produced by "tapping of abdomen and wings on the ground" (Rupprecht, 1975, p. 305). These he interpreted as providing encouraging feedback from one partner to the other during the early stages of courtship. Drumming occurs only in males of *S. lutaria*, but in both sexes of *S. fuliginosa*. Females of the latter species can drum very rapidly, at a rate of nearly 20 strikes/sec (Rupprecht, 1975, Figure 7).

If acoustic signals have evolved in the context of premating reproductive isolation, one expects them to be significantly different in closely related, sympatric species, whether because of stabilising selection on different mate recognition systems (Paterson, 1985; Butlin, 1995) or because of reinforcement and reproductive character displacement (Butlin, 1989; Howard, 1993; Liou and Price, 1994). Yet few consistent differences are found between the vibratory "songs" of *S. lutaria* and *S. fuliginosa*, suggesting that alternative evolutionary dynamics have been at work.

Corydalidae

The most complete description of acoustic communication in corydalids is Parfin's (1952) on the North and Central American species *Corydalus cornutus* (Linnaeus). Additional relevant anecdotes are mentioned in New and Theischinger (1993). A number of other excellent studies focus on nonacoustic aspects of reproductive behaviour, including intrasexual selection, mate guarding, nuptial gifts, courtship feeding, spermatophore investment and sperm cooperation (Hayashi, 1993, 1996; Contreras-Ramos, 1999).

Parfin (1952, pp. 429–432) described how the male of *C. cornutus* approached the female, laid his long mandibles across her wings for several minutes, withdrew to a position next to the female, and then "wriggled his soft abdomen for about half a minute with three to four series of several rapid quivers each" prior to final approach and copulation. From this, it is clear that corydalids can tremulate in the same manner as sialids, producing discrete volleys of abdominal vibration during courtship. However, tremulation in corydalids seems to be limited to males, and possibly to *Corydalus* alone; for example, in his careful study of premating behaviour in *Platyneuromus* sp., Contreras-Ramos (1999) reported no such vibratory signalling, only wing fluttering. No recordings or analyses of corydalid male "songs" have been published.

RAPHIDIOPTERA: RAPHIDIIDAE AND INOCELLIIDAE

Courtship and mating in this order have been described for three genera of Raphidiidae and one genus of Inocelliidae (reviewed in Henry, 1997). Unfortunately, observations of inocelliids have generally been merged with those of raphidiids, so it is not possible here to treat acoustic communication in the two families separately.

It has been recognised for years that abdominal vibration accompanies courtship in snakeflies (Eglin, 1939; Zabel, 1941: *Raphidia ophiopsis* Linnaeus and *Inocellia crassicornis* (Schummel); Woglum and McGregor, 1958: *Agulla bractea* Carpenter; Acker, 1966: three species of *Agulla*). The most detailed descriptions are found in Kovarik *et al.* (1991) in their study of North American *Agulla bicolor* (Albarda) (as *Raphidia bicolor* Albarda). Courtship in snakeflies is similar to that of alderflies (Sialidae). Males and females approach each other and vigorously vibrate their abdomens for sustained periods. As in sialids, the courting individuals establish duets, such that "prolonged vibrations by a persistent male often elicit a similar response from the courted female" (Kovarik *et al.*, 1991, p. 362). Acker (1966) had earlier made note of the same kind of sustained abdominal vibration in *Agulla astuta* (Banks), *A. adnixa* (Hagen) and *A. bicolor* (Albarda), additionally observing that signals in females are less intense than in males. Audible wing fluttering usually accompanies courtship activity. No formal descriptions of snakefly tremulation exist, but Kovarik's statement that "the female often responds with short vigorous vibrations" implies that abdominal vibration in *A. bicolor* is temporally organised into discrete volleys, as in *Sialis* (Rupprecht, 1975).

PERSPECTIVE

Tremulation using abdominal vibration is a fundamental component of mating behaviour in Megaloptera and Raphidioptera. The archaic (plesiomorphic) nature of both orders implies that abdominal tremulation is part of the ground plan of Neuropterida, and that male–female duetting using these vibratory signals is included within that ground plan. The challenge is therefore to assess retention, loss and perhaps secondary gain of tremulation in the principal neuropterid order, Neuroptera.

ACOUSTIC COMMUNICATION IN NEUROPTERA

Communication using some type of sound or substrate vibration has been confirmed in just five of the 17 families of the order Neuroptera, all within the suborder Hemerobiiformia (Figure 10.1).

Abdominal tremulation is found in Sisyridae (spongilla flies), Hemerobiidae (brown lacewings) and Chrysopidae (green lacewings), while Coniopterygidae (dusty-wings) and Berothidae (beaded lacewings) simply vibrate or flutter their wings during courtship. Only in Chrysopidae has acoustic signalling become sophisticated and diverse. Additional aspects of courtship and mating systems throughout the order are compiled in Henry (1997).

SISYRIDAE AND CONIOPTERYGIDAE

A recent cladistic analysis based on 36 morphological features argues for a sister-group relationship between these two families of small to minute Neuropterans (Aspöck *et al.*, 2001; Aspöck, 2002). However, earlier work came to very different conclusions (summarised in New, 1991), and molecular data have so far failed to support the hypothesised relationship (Haring and Aspöck, 2004). These two families will be considered together because both include species which communicate acoustically.

Sisyridae

Spongilla-flies are small, peculiar Neuropterans whose larvae are subaquatic predators of freshwater sponges. Although constituting only 50 described species, Sisyridae is nonetheless cosmopolitan in distribution (Pupedis, 1980, 1985). Reproductive behaviour has been observed in a handful of species. In North American *Climacia areolaris* (Hagen), courtship includes horizontal extension and rapid vibration of the wings on one side of the male's body, whereby he "fans" the head of the female and perhaps produces a faint sound (Brown, 1952). In contrast, Holarctic *Sisyra nigra* (Retzius) (*S. fuscata* [Fabricius]) foregoes wing fanning in favour of abdominal tremulation (Killington, 1936; Rupprecht, 1995). Here, males and females oscillate their abdomens erratically during courtship at frequencies between 100 and 450 Hz. Additionally, published sonograms show some evidence of percussive drumming during tremulation (Rupprecht, 1995, see figure). Tremulation in spongilla flies does not seem to be organised into discrete volleys of vibration.

Coniopterygidae

The reproductive behaviour of the "dusty-wings" or "wax flies" (Plant, 1991) has been little studied, probably because they are so small. More than 300 species have been described, but courtship and mating have been seen in only seven (Withycombe, 1922; Collyer, 1951; Henry, 1976; Johnson and Morrison, 1979).

In his recent review of the subject, Devetak (1998) mentions "vibratory signals" in Coniopterygidae, but only precopulatory wing fluttering was actually documented in the cited study (Johnson and Morrison, 1979). Johnson and Morrison reported wing fluttering in males and females of the three California species they examined. Furthermore, fluttering appeared to be temporally structured into 1 to 2 sec "calls", delivered intermittently. It is therefore not out of the question that abdominal vibration accompanies or even causes wing fluttering, but was simply overlooked in such tiny insects.

PERSPECTIVE

The phylogenetic positions of Sisyridae and Coniopterygidae have not been determined. Both families have been perceived at various times either as highly specialised and of relatively recent origin (Aspöck *et al.*, 2001; Aspöck, 2002), or as ancient, plesiomorphic lineages (Withycombe, 1925; Klingstedt, 1937; Hughes-Schrader, 1975; Gaumont, 1976; Henry, 1982b; New, 1991). Of interest in this regard is the "staggered parallel, female above" copulation position seen in coniopterygine Coniopterygidae (Johnson and Morrison, 1979), which is shared only with the archaic Megaloptera (Sialidae) and Raphidioptera (both families). If Sisyridae and

Coniopterygidae are one another's closest relatives (Aspöck *et al.*, 2001) and also of ancient origin, then the presence in the clade of abdominal vibration and wing fluttering during courtship and mating could be interpreted as retention of those traits, along with copulation position, from the ground plan of Neuropterida.

BEROTHIDAE

Only wing fluttering has been reported from the "beaded lacewings". This is a small but widespread family of about 60 species, belonging to a clade, Mantispoidea, which includes Rhachiberothidae, Mantispidae (mantis flies) and Dilaridae (pleasing lacewings) (Willmann, 1990; Aspöck, 2002). MacLeod and Adams (1967) describe courting individuals of North American *Lomamyia* spp. lifting their wings to a horizontal position and alternately "vibrating" these at one another in a brief duet before copulating, which is reminiscent of "wing fanning" behaviour in the sisyrid *Climacia areolaris* (Brown, 1952). Within the Mantispoidea, it is not known whether such behaviour is unique to Berothidae or also present in one or more of the other three families. The phylogenetic position of Mantispoidea within the Neuroptera is not well understood, making interpretation of character origin and evolution difficult. However, wing fluttering in Berothidae appears to be simple and therefore perhaps evolutionarily labile; as such, the trait might not have larger phylogenetic significance. Thus, the occurrence of wing fluttering in disparate taxa such as Berothidae, Corydalidae, Raphidiidae, Sisyridae and Coniopterygidae (see above) could be incidental, that is, a consequence of convergent evolution.

HEMEROBIIDAE

No published records of acoustic communication in the brown lacewings exist, even though the family includes some 550 species (Oswald, 1993). However, I have personally observed receptive males of *Hemerobius* sp. vibrating their abdomens in a temporally structured manner, much like Chrysopidae. Until very recently, hemerobiids and chrysopids have been treated as sister taxa, so the simplest hypothesis explaining the presence of tremulation in brown lacewings is the inheritance of that trait from a most recent common ancestor with green lacewings. However, Aspöck *et al.* (2001) challenged that view on morphological grounds, arguing instead for a closer relationship of Chrysopidae with Osmylidae (but see Haring and Aspöck, 2004, for a molecular alternative). In any case, relationships among the families of Hemerobiiformia are sufficiently uncertain to preclude accurate tracing of signal evolution within the suborder.

CHRYSOPIDAE

The green lacewings number about 1200 species, making it one of the largest families of Neuroptera. Here, acoustic communication has blossomed, such that a wide variety of low-intensity sounds and vibrations are used by different taxa.

STRIDULATION

Airborne sound production via stridulation has been inferred in Chrysopidae from morphology but never demonstrated in living individuals. Adams (1962) was the first to notice rows of striae on the lateral margins of the second and third abdominal sternites in both sexes of *Meleoma schwartzi* (Banks), together with a row of wart-like tubercles on the inner surface of each hind femur. He concluded that vertical movements of the abdomen would cause the striae to scrape against the femoral tubercles, producing a sound. Tauber (1969) confirmed these observations and reported similar structures in two related species, *Meleoma pinalena* (Banks) and *M. adamsi* Tauber. Subsequently, Brooks (1987) conducted a comprehensive search for putative stridulatory structures in the family, finding and describing them in *Meleoma* Fitch (three of the 22 known species, both

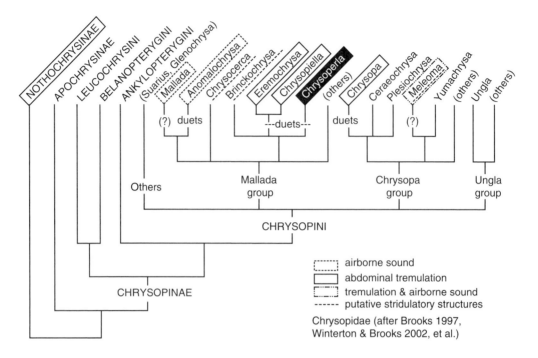

FIGURE 10.2 Phylogenetic hypothesis for the relationships of the subfamilies and tribes of Chrysopidae, including the major genera of tribe *Chrysopini*. Taxa with some type of acoustic communication are indicated. Relationships within *Chrysopini* are those hypothesised by Dr. Stephen Brooks (personal communication, July 2004; see Acknowledgments).

sexes in each), *Brinckochrysa* Tjeder (all 13 species, both sexes), and *Chrysocerca* Weele (one of five species, males only). He suggested independent evolutionary origins for these structures in each of the three genera, based upon fundamental differences in their anatomy and placement. That conclusion is also supported by phylogeny (Figure 10.2) — the three genera are not particularly close relatives and they do not trace their origin back to a more distant common ancestor possessing a "stridulatory" apparatus.

Circumstantial evidence for stridulation in green lacewings is compelling, even though sounds have not yet been recorded. Based on morphology, such signals are likely to be faint, of relatively high frequency and probably sexually dimorphic because the femoral pegs are usually larger in males than females and the form of the abdominal striae differs between the sexes (Brooks, 1987). The presence of tremulation in several close relatives of "stridulating" genera is also significant because abdominal vibration could have provided the initial behavioural basis for stridulation (Adams, 1962).

A stridulatory function has also been ascribed to interacting patches of microtrichia, located at the point where the folded forewings contact the metanotum (Riek, 1967; Eichele and Villiger, 1974). However, these "organs" are widely distributed in other Neuropteran families, Trichoptera and micro-Lepidoptera, and are more likely to function in wing positioning than sound production (Henry, 1979; Brooks, 1987).

WING FLUTTERING

Generalised fluttering or rattling of wings during courtship and mating has been reported in several species of *Meleoma* and *Eremochrysa*, including *Eremochrysa*/subgenus *Chrysopiella* (Duelli and Johnson, 1982). This behaviour seems to be a byproduct of particularly vigorous abdominal vibration characterising *Meleoma emuncta* (Fitch), *M. arizonensis* (Banks) and *M. furcata* (Banks) (Tauber, 1969; P. Duelli, personal communication). The effect is to produce a soft but distinctly audible rustling sound of unknown significance.

A specialised type of wing fluttering is found in *Anomalochrysa maclachlani* Blackburn, one of 22 species in this endemic Hawaiian genus. Both sexes flick the wings anteriorly so as to produce a clicking sound which is clearly audible several meters from the caller (Tauber *et al.*, 1991). Courting individuals will often produce trains of 50 or more clicks. The clicks are repeated at a steadily increasing rate during each train, from one click every 2 sec to three clicks per second. The male and female will alternately exchange clicks in a duet which usually continues until copulation. Simultaneous clicking, which could mask the partner's signal, is rare. The mechanism of sound production has not been determined, but may involve "a sclerotised structure at the base of the forewings" (Tauber *et al.*, 1991, p. 1024).

PERCUSSION

Mallada basalis (Walker), distributed widely across the Indo-Pacific region, produces audible buzzing noises by striking the modified costal margins of its vibrating hind wings forcibly against the substrate (Duelli and Johnson, 1982). Only the male has the thickened, hammer-like pterostigma on the wing required for signalling. Males call vigorously in the presence of females, and are capable of producing the loudest airborne sounds of any neuropterid.

Percussive sounds can also be found in lacewings which tremulate, *e.g.* two species in the *carnea* group of *Chrysoperla* (Henry *et al.*, 2002, 2003). In such cases, the tremulating abdomen will periodically strike the substrate, usually toward the end of a song consisting of a single long volley (*Chrysoperla agilis*; Henry *et al.*) or many short volleys (*C. pallida*; Henry *et al.*). The sound which we hear is a low-intensity ticking or rattling noise, detectable over a range of perhaps 25 cm.

TREMULATION

The predominant form of acoustic communication in green lacewings is tremulation via abdominal vibration. Tremulation signals in chrysopids are remarkably similar to those in *Megaloptera* and *Raphidioptera*, but more diverse and often more complex. Five genera are known to include tremulating species: *Nothochrysa*, *Eremochrysa* (including *Chrysopiella*), *Chrysoperla*, *Chrysopa* and *Meleoma*.

Of the three subfamilies of Chrysopidae, Nothochrysinae is the most plesiomorphic, lacking the ultrasonic "ear" found in all other green lacewings. Nonetheless, tremulation is present even in this archaic group, based on Toschi's (1965) observations of *Nothochrysa californica* Banks. Here, abdominal vibration is apparently confined to males during the "approach" and "contact" phases of courtship. No details of volley structure or frequency are given.

Chrysopa and *Meleoma* share membership in the "*Chrysopa* group" of Chrysopinae (Figure 10.2; S. Brooks, personal communication), and they share simple tremulation as well. In *Meleoma emuncta*, *M. kennethi* Tauber and *M. hageni* Banks, abdominal vibration has been demonstrated only in courting males (Tauber, 1969; Duelli and Johnson, 1982). On the other hand, *Chrysopa* definitely exhibits tremulation in both sexes (Smith, 1922; Principi, 1949), and partners will duet using nearly identical, nonoverlapping signals during courtship (Henry, 1982a). The two best-studied species, North American *Chrysopa oculata* Say and *C. chi* Fitch, produce trains of simple volleys. Volleys of *C. chi* are twice as long (160 vs. 82 msec) and three times as far apart (800 vs. 250 msec) as those of *C. oculata*, and are also lower pitched (77 vs. 109 Hz). Of more significance for premating reproductive isolation, courting partners of *C. chi* trade volley-for-volley during duets, while the unit of exchange (the "shortest repeated unit" or SRU) in duets of *C. oculata* is a cluster of multiple volleys (Henry, 1982a). If coordination of signals during duets is important to success in copulation in these two species (as seems likely), then this basic difference in the mode of duetting will effectively preclude heterospecific interactions.

Eremochrysa (including *Chrysopiella*) and *Chrysoperla* are two closely related members of the *Mallada* species group (Figure 10.2). "Polite" male–female duetting using identical signals is characteristic of both genera. Although *Eremochrysa* seems to have retained simple tremulation signals (*e.g. E. minora* [Banks]; see Henry and Johnson, 1989), *Chrysoperla* has become the tremulation champion of the green lacewings. Tremulation signals have been found whenever looked for in this genus, in representatives of each of the four recognised "species groups" (Brooks, 1994). As in *Chrysopa* and *Eremochrysa*, their low-frequency signals are sexually monomorphic and organised into repeating volleys of abdominal vibration, which in turn are grouped into identical SRUs which serve as the currency of exchange during heterosexual duets (Figure 10.3). A SRU consists of just one, several or many volleys and may include more than one type of volley, *e.g.* of different durations or pitches (Figure 10.4, Figure 10.5C). From a theoretical perspective, the temporal and frequency characteristics of the songs are well suited to the general biomechanical properties of plant stems (Michelsen *et al.*, 1982; Čokl and Doberlet, 2003). However, playback experiments comparing signal propagation in grass vs. conifer stems using two lacewing species ecologically associated with those substrates suggest that songs may not show truly fine-tuned bioacoustic adaptations to specific plant types (Henry and Wells, 2004).

Courting partners of *Chrysoperla* lacewings synchronise their songs to one another with remarkable accuracy. Because songs of males and females are identical within a species, an individual need only recognise its own song in the signal of its partner to establish a duet leading to copulation. Individuals will listen to their partners carefully, quickly and mutually adjusting the tempo of their songs. In fact, if changes in a stimulus signal are presented incrementally, it is possible to more than double or halve the SRU interval of a responding male or female of *C. plorabunda* (Figure 10.5, unpublished data). However, song phenotypes with parameters outside the acceptable range for the species elicit little or no response. Sexual interactions of two species singing different songs will terminate quickly, based on the inability of the partners to synchronise their signals and establish the mandatory duet (Wells and Henry, 1992).

CRYPTIC SPECIES

Lacewing species characterised by sophisticated duetting behaviour, such as those in *Chrysoperla*, are often very difficult to delimit morphologically. Accordingly, we now know that the easily recognised and widespread species *Chrysoperla carnea* (*sensu lato*) is really a complex of many cryptic "biological" species, separable chiefly by unique tremulation signals. Under natural conditions, courtship duetting prevents mating between individuals with different songs, even though most "song species" of the *carnea* group are potentially interfertile (Wells, 1993; Wells and Henry, 1994). To date, at least 15 cryptic species of the *carnea* group have been recognised, five in North America and ten in Eurasia (Figure 10.3). Each species has a broad geographic range (Wells and Henry, 1998), yet intraspecific song variation over that range is remarkably small. Sympatry and even syntopy of species is extensive, such that several can be collected at a given site on a given day. Yet hybrids between sympatric song species, which are easily recognised by their intermediate song phenotypes, have never been found in the field.

So far, cryptic, duetting song species have been described only for the *carnea* group of *Chrysoperla*. However, it is very likely that hidden taxonomic diversity will soon be found in other subsections of the genus. For example, complex signals and duets have been recorded from four species of the cosmopolitan *pudica* group: *C. comanche* (Banks) (Henry, 1989), *C. pudica* (Navás), *C. mutata* (McLachlan) and *C. congrua* (Walker). In addition, both *C. comans* (Tjeder) and *C. volcanicola* Hölzel *et al.* of the Afro-Asian *comans* group produce remarkably sophisticated tremulation songs and could easily harbour cryptic song species within their diagnostic limits (unpublished data). Representatives of the fourth species group, *nyerina*, remain unstudied.

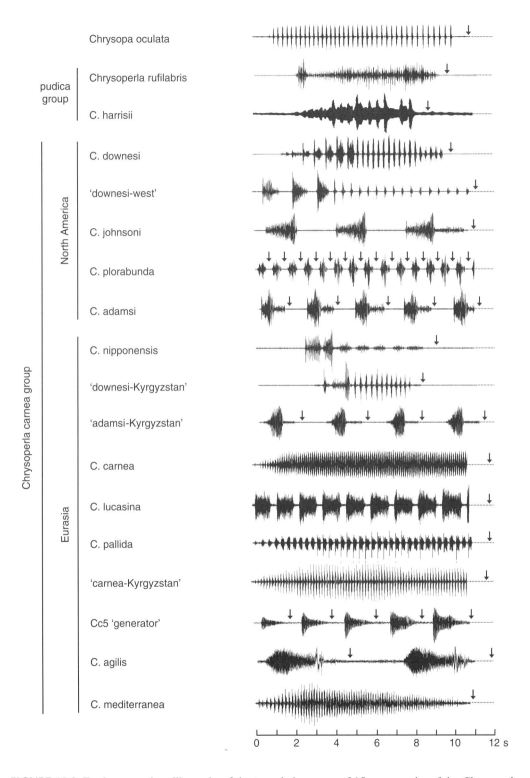

FIGURE 10.3 Twelve-second oscillographs of the tremulation songs of 15 song species of the *Chrysoperla carnea* group, two species from the closely related *pudica* group and one species from the more distant genus *Chrysopa*. Arrows indicate where the partner would insert its signal during a heterosexual duet.

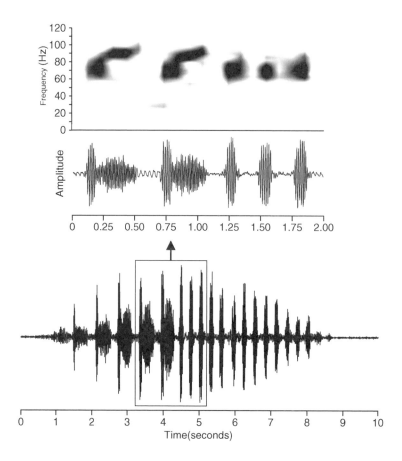

FIGURE 10.4 Tremulation song of *Chrysoperla downesi*, a species having a long, multivolley courtship signal. The oscillograph at the bottom of the figure shows a single SRU, which is exchanged in alternating fashion with the SRU of the duetting partner. The box contains a 2-sec detail of part of the SRU, showing the temporal and frequency structure of the two distinct volley types.

Once cryptic species have been properly delimited by songs, subtle morphological and ecological differences among them usually become apparent (Henry *et al.*, 2001, 2002a). Eventually, identification of such species will be possible using more traditional taxonomic tools.

SPECIATION

Precise duetting between heterosexual partners singing the same song has set the stage for rampant speciation within *Chrysoperla*. Mutual and rapid adjustment of song tempo during the initial phase of the duet assures that acoustically compatible individuals will sing in lockstep. This unusually well-coordinated premating interaction between conspecifics has a potent reproductive isolating effect between species singing different songs. Consequently, any random but measurable change in one type of song — say, due to a mutation in a song-controlling gene — could precipitate a speciation event. It is thought that occasional changes such as these have probably generated the swarms of cryptic species in *Chrysoperla*. The driving force is sexual selection, in this case mate choice by both sexes, which quickly segregates the new song types reproductively from existing ones. Such speciation could even have been sympatric rather than allopatric. It has also been

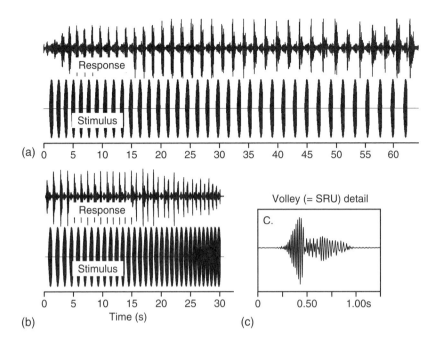

FIGURE 10.5 Duetting responses at 25°C (upper oscillographs in (a) and (b)) by an individual of *Chrysoperla plorabunda* to a stimulus signal (lower oscillographs in (a) and (b)) which gradually increases (a) or decreases (b) in SRU period. This species has a single-volley SRU (c). SRU duration of both stimulus signals was held constant at the species mean, 636 msec (25°C). Initial SRU period for both stimuli was 1230 msec (also the species mean at 25°C). In (a), the period of the stimulus increased to a maximum of 2280 msec; in (b), the period of the stimulus decreased to a minimum of 610 msec. Note that the individual is able to maintain perfect synchrony with the stimulus, regardless of whether the period of the latter markedly increases or decreases.

very recent: genetic distances and nucleotide sequence divergences among the song species are vanishingly small (Wells, 1994; Henry *et al.*, 1999).

A prediction of this model of speciation is that the genetic architecture of song phenotype should be simple, consisting of relatively few genes of large effect. Indeed, the results of a hybridisation experiment between *C. downesi* and *C. plorabunda* showed that song features segregated in a manner consistent with simple architecture (Henry, 1985). Furthermore, in a Bayesian Castle-Wright analysis of song data from *C. plorabunda* × *C. johnsoni*, Henry *et al.* (2002b) showed that as few as one "genetic element" might be responsible for a crucial duetting difference between those two species. If these conclusions continue to receive support from future studies, *Chrysoperla* will be seen as a rare exemplar of rapid, sympatric speciation caused by sexual selection acting on chance mutational differences in mating signals.

OVERVIEW

From the human perspective, Neuropterida is a silent taxon of insects. That can be blamed on their soft bodies and small size, which together conspire to limit their abilities to produce loud sounds. However, many members of the superorder have instead exploited low-frequency acoustic channels which are always available in plant stems and leaves. These are private lines of communication, not so easily accessible to eavesdropping predators, parasites or even conspecific sexual competitors. To sing "silently" in this way, the neuropterids use simple vertical oscillations of their abdomens to produce tremulation signals in plant tissues.

Substrate-borne tremulation signals have a deep evolutionary history within Neuropterida. Abdominal vibration during courtship and mating is found not only in the ancient, relict orders Megaloptera and Raphidioptera, but also in several plesiomorphic taxa of Neuroptera. Unfortunately, courtship and mating behaviour are poorly known in most families of Neuroptera, so it is currently impossible to trace the evolution of tremulation with precision or confidence. However, it is very likely that abdominal tremulation is part of the ground plan of the neuropterids, and in modified form it has contributed to recent explosive speciation within at least one genus of Chrysopidae, *Chrysoperla*.

Tremulation may also have been indirectly responsible for the origin of airborne sound production in several other lacewing genera through the evolution of putative files and scrapers which apparently use the same vertical jerking motion of the abdomen. Even if such sounds exist, though, they must be very soft. A few other lacewing species produce faintly audible sounds percussively by drumming their abdomens or wing margins against the substrate. Endemic Hawaiian lacewings employ a unique wing-flicking behaviour to produce audible clicks.

A complete understanding of the taxonomic distribution of acoustic communication in Neuropterida will undoubtedly shed some light on the phylogenetic relationships of the families which are currently controversial. However, as part of the dynamic mating systems of species, sexual signals usually undergo rapid evolutionary change which can quickly overwrite phylogenetic information through convergence, parallelism and reversal. Therefore, mating signals are of greater value for the light they can shed on evolutionary processes, particularly speciation. The tremulation songs of *Chrysoperla* reflect rapid, repeated and behaviour-based speciation processes which challenge the entrenched dogma of more gradual allopatric speciation. Future work along these lines will certainly yield new surprises.

ACKNOWLEDGEMENTS

The research by the author described in this chapter was supported in part by the Research Foundation of the University of Connecticut. I thank Drs. Marta M. Wells (University of Connecticut and Yale University, U.S.A.), Peter Duelli (Swiss Federal Research Institute WSL, Switzerland), Stephen J. Brooks (The Natural History Museum, London) and James B. Johnson (University of Idaho, U.S.A.) for fruitful collaboration and friendship over many years of work together. For valid species names, I am indebted to Dr. John Oswald's web-based searchable index of species names (Oswald, 2003). I also thank Dr. Cynthia S. Jones, Suegene Noh and David Lubertazzi (University of Connecticut, U.S.A.), who critically read and improved the manuscript.

11 Inheritance of Male Sound Characteristics in *Drosophila* Species

Anneli Hoikkala

CONTENTS

INTRODUCTION

Drosophila males and females produce a variety of acoustic, chemical and visual cues during courtship, acoustic signals being an essential part of the courtship in most species. Male courtship song produced by wing vibration was first described in *Drosophila melanogaster* (Shorey, 1962). The males of this species produce two kinds of songs, a pulse song (a train of single cycle pulses with about 20 to 50 msec inter-pulse interval [IPI]) and a sine song (a modified sine wave with a carrier frequency of 170 Hz) (Cowling and Burnet, 1981). The pulse song has a rhythmic *modulation* in the IPI with a period of approximately 1 min (Kyriacou and Hall, 1980).

The courtship songs of *D. melanogaster* are quite simple compared with those of most other *Drosophila* species. In many species the songs consist of trains of polycyclic sound pulses, which may form different song types and which also may have a rhythmic structure. In some Hawaiian species the males produce high-frequency songs resembling cricket song more than fly song (Hoy *et al.*, 1988). Males may also produce different song types at different stages of courtship, which gives the female information on the male's location as well as on the stage of courtship (Crossley, 1986). The song is usually produced before mating, but males may also continue singing during copulation or produce only copulation song (Hoikkala and Crossley, 2000). Examples of different song types as well as the traits usually analysed in the songs are shown (Figure 11.1). The song traits pulse length (PL), interpulse interval (IPI) and pulse train length (PTL) are measured and CN (number of cycles in a sound pulse) and PN (number of pulses in a pulse train) are calculated from the oscillograms of the songs, while the carrier frequency is usually measured from the Fourier spectra. The difference between a pulse song and a sine song may not always be clear (Cowling and Burnet, 1981), and sometimes it is also difficult to trace the homology of song traits in different species. Many song traits are sensitive to environmental

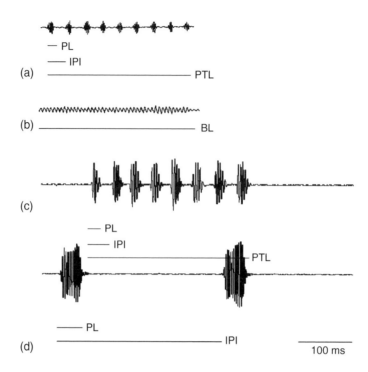

FIGURE 11.1 (a) A train of a pulse song of *D. birchii* with polycyclic sound pulses (CN 4 to 5 cycles/pulse, PL around 12 msec, IPI around 30 msec and frequency around 400 Hz). Number of pulses in the pulse train (PN) is nine and the length of the pulse train (PTL) 250 msec. (b) A sine song of *D. serrata* with a carrier frequency of about 140 Hz and a burst length (BL) of 230 msec. (c) A pulse train of *D. montana* consisting of eight sound pulses (PN) with PTL of 380 msec. Among the pulse characters CN is around 5 cycles/pulse, PL around 20 msec and IPI around 35 msec. The carrier frequency of the song is about 300 Hz. (d) The first two pulses of a pulse train of *D. littoralis* consisting of long sound pulses (PL around 40 msec, CN around 12 cycles/pulse, IPI around 300 msec and frequency around 320 Hz).

temperature, which has to be taken in account when recording the songs (Hoikkala, 1985; Noor and Aquadro, 1998).

In addition to males, females also may produce acoustic cues during courtship. Ewing and Bennet-Clark (1968) reported that the "buzz" is essentially the same in all species of the *melanogaster* and *obscura* groups. Closer inspection of female songs has revealed species differences, *e.g.* in the *virilis* group (Satokangas *et al.*, 1994) and in the *bipectinata* complex of the *melanogaster* subgroup (Crossley, 1986). Female songs are generally more irregular than the male songs.

In *D. melanogaster*, antennal hearing organs mediate the detection of conspecific songs in females and the arista tips of male and female antennae are moderately tuned to frequencies around 425 Hz (Göpfert and Robert, 2002). Sensory feed-back seems to play an important role also in shaping the courtship song in *Drosophila*, as males with auditory mutations are not able to produce normal song (Tauber and Eberl, 2001). Tauber and Eberl (2003) have reviewed acoustic communication in *Drosophila* focusing on the proximate, neural aspects of sound production and hearing.

In the present review I shall first give an overview on song variation between and within various *Drosophila* species and then go through the studies on the inheritance of male song characters using classical crossing experiments, quantitative genetic methods and gene transfer techniques. Finally there will be a short discussion on major questions concerning the evolution of species-specific courtship songs in *Drosophila* species.

MALE COURTSHIP SONGS IN DIFFERENT *DROSOPHILA* SPECIES GROUPS

Male courtship songs have been described for more than 110 *Drosophila* species belonging to *Sophophora*, *Drosophila* and *Idiomyia* subgenus. Even though the songs may differ drastically between closely related species, there are some trends to be seen in song evolution within the species groups. In this review, male courtship songs are described most thoroughly for the species groups for which there are data also on the inheritance of song characters (*melanogaster*, *obscura* and *virilis* groups) and in the Hawaiian picture-winged species group, where the song evolution has led to the most elaborate courtship songs.

In the *melanogaster* group species (genus *Sophophora*), males produce a simple pulse song (mono- or polycyclic) and sound bursts (a sine song) that occasionally may have a pulse structure. Males of the *melanogaster* subgroup (except *Drosophila yakuba*) produce both a pulse and a sine song, and alternate between these songs during courtship (Cowling and Burnet, 1981). In the *ananassae* subgroup, males either produce two types of pulse song, long song at the early and short song at the late stages of the courtship (Crossley, 1986), or only one kind of pulse song with a variable burst length (Yamada *et al.*, 2002). In the *montium* subgroup, males of the species of the *auraria* complex produce a simple pulse song, a large proportion of the song being produced during copulation (Tomaru and Oguma, 1994a). *Drosophila birchii* and *Drosophila serrata*, endemic Australian species belonging to the same subgroup, have a pure copulatory courtship, *i.e.* they produce sound bursts with or without pulse structure only during and after copulation, even though they have an ability to produce also a pulse-structured precopulatory song (Hoikkala and Crossley, 2000).

In the *obscura* group, male songs are more complicated than in the *melanogaster* group. Males of the two most thoroughly studied species, *Drosophila persimilis* and *Drosophila pseudoobscura*, produce two kinds of songs consisting of polycyclic sound pulses, a high repetition rate (HRR) song, IPI < 70 msec) and a LRR song (low repetition rate song, IPI > 230 msec) (Ewing and Bennet-Clark, 1968; Noor and Aquadro, 1998). The IPI of the HRR song of *D. persimilis* has a periodicity of 1.3 sec (Noor and Aquadro, 1998). According to Ewing and Bennet-Clark (1968) *D. ambiqua* males produce two kinds of songs resembling those of *D. pseudoobscura* males, while *Drosophila affinis*, *Drosophila algonquin* and *Drosophila athabasca* males produce only one kind of song consisting of single or polycyclic sound pulses and *Drosophila subobscura* males do not produce sounds at all.

In the *virilis* group, the songs of all species consist of polycyclic sound pulses arranged in pulse trains with shorter or longer IPIs (Hoikkala and Lumme, 1987). In songs of the *virilis* subgroup, sound pulses are arranged in dense pulse trains with no pauses between the pulses, with both the PLs and IPIs being about 20 msec. In the *montana* subgroup, only *D. borealis* has this kind of song, whereas the songs of all other species have clear pauses between the successive sound pulses. In most of these species the IPIs are 30 to 80 msec long, while in *Drosophila littoralis* they are about 300 msec long. The males of most species of the group also produce secondary song, which is more irregular than the primary song and is not an essential part of the courtship (Suvanto *et al.*, 1994).

Hawaiian picture-winged species are a good example of a species group with a great diversity of songs. The flies of this group produce songs by different kinds of wing vibrations and by abdomen "purring". They also produce high-frequency song by an unknown mechanism (Hoy *et al.*, 1988). In the *planitibia* subgroup the males either produce wing and abdomen songs or no songs at all. The wing songs fall into three categories, pulse song, rhythmic phrase song and sound bursts (Hoikkala *et al.*, 1994). In one intermediate species, *Drosophila neoperkinsi*, males produce all three songs, while in other species males use different combinations of the songs. Males of the most derived species (*e.g. Drosophila silvestris* and *Drosophila heteroneura*) produce only sound bursts, but they have added abdomen purring in their song repertoire. In the species of the *pilimana*, *vesciseta* and *grimshawi* subgroups, the males may semaphore or vibrate their wings producing no audible sound. After this, the males fold their wings back over the thorax and abdomen and make small "scissoring" movements producing loud high-frequency clicking sounds (up to 15,000 Hz), which differ from all other described *Drosophila* sounds both in their spectral and temporal structures (Hoikkala *et al.*, 1989).

Male songs have been described also for several species of the *willistoni*, *funebris*, *immigrans*, *melanica*, *quinaria* and *repleta* groups. In the *willistoni* group the major song types are a pulse song, rasps (short sound bursts) and trembles (rapid vibrations of the whole body) (Ritchie and Gleason, 1995). In the *funebris* group, males of most species produce single cycle sound pulses near the minimum IPI (approximately 10 msec) (Ewing, 1979a, 1979b). Ewing argues that this has produced an evolutionary bottleneck with regard to the song pattern, and so the minimum IPI songs have been arranged into bursts with species-specific inter-burst intervals (IBIs) and males also have evolved secondary song with longer IPIs. In the *immigrans* group, males of the three studied species of the *hypocausta* subgroup produce three or four different song types or no song at all (Asada *et al.*, 1992). Also in the *melanica* group, males can produce up to four qualitatively different song types (Ewing, 1970), while in the *quinaria* group, males produce only one song type consisting either of single polycyclic sound pulses with long IPIs or of groups of pulses arranged in dense pulse trains (Neems *et al.*, 1997). The *repleta* group is a large species group consisting of several subgroups and species complexes. Ewing and Miyan (1986) have suggested that the archetypical song in this group was composed of two distinct components produced at the early or late stages of courtship. During speciation processes, some species lost the first song, others the latter song and in many species the latter song became less regular and more complex.

VARIATION IN MALE SONGS BETWEEN CONSPECIFIC POPULATIONS

Variation in male courtship songs within species has been studied using laboratory strains as well as the progeny of wild-caught females. The problem with many of these studies is that a few laboratory strains do not necessarily represent a wild population from the same geographic region as the songs may change during laboratory maintenance (Aspi, 2000).

Variation in song traits has been studied most thoroughly in *D. melanogaster*. Kawanishi and Watanabe (1980) found variation in IPI to be much smaller in *D. melanogaster* than in its sibling species *Drosophila simulans*. Ritchie *et al.* (1994) reported variation in IPI not to be significant among replicated *D. melanogaster* laboratory stocks of different geographic origin. Subsequently, Colegrave *et al.* (2000) found unusually short IPI in the songs of some African *D. melanogaster* populations.

Noor and Aquadro (1998) studied song variation among 20 *D. pseudoobcura* (*obscura* group) strains from four locations in North America and found significant between population variation in the IPI of the HRR song. In another study by Noor *et al.* (2000), two partly sexually isolated subspecies, *D. p. pseudoobscura* in western North America and *D. p. bogotana* in Colombia, showed no divergence in their song characters, even though these subspecies show moderate pre- and postzygotic isolation.

Variation in male courtship songs has been studied also in the *virilis* group. Hoikkala (1985) analysed male songs of 42 fresh isofemale strains (derived from wild-caught females) of *D. littoralis* from three localities in Finland and of several laboratory strains from Europe and the Caucasus. The songs of the old laboratory strains differed from each other more than the songs of the fresh strains, but there was no sign of geographical variation in any song trait. A recent study on genetic (48 microsatellites) and phenotypic (male song traits) variation among 15 fresh and 30 older laboratory strains of *Drosophila virilis* revealed significant geographic variation in male song characters, only PL and IPI being constant among the strains from different localities (unpublished data).

TABLE 11.1
***Drosophila* Species for which the Male Songs Have Been Described and a List of References where the Song Descriptions Can Be Found**

Subgenus *Sophophora*
  Group *melanogaster*
    Subgroup *melanogaster*   Cowling and Burnet, 1981
      erecta
      mauritiana
      melanogaster
      simulans
      teissieri
      yakuba
    Subgroup *ananassae*   Crossley, 1986; Yamada *et al.*, 2002
      ananassae
      bipectinata
      malerkotliana
      pallidosa
      parabipectinata
      pseudoananassae
    Subgroup *montium*   Tomaru and Oguma, 1994a, 1994b;
      auraria   Hoikkala and Crossley, 2000
      biauraria
      birchii
      quadraria
      serrata
      subauraria
      triauraria
  Group *obscura*   Ewing and Bennet-Clark, 1968;
    affinis   Noor and Aquadro, 1998
    algonquin
    ambiqua
    athabasca
    persimilis
    pseudoobsura
    subobscura
  Group *willistoni*   Ewing, 1970; Ritchie and Gleason, 1995
    carmondy
    equinoxialis
    insularis
    paulistorum superspecies
    pavlovskiana
    tropicalis
    willistoni

Subgenus *Drosophila*
  Group *funebris*   Ewing, 1979a, 1979b
    funebris
    limpiensis
    macrospina
    multispina
    subfunebris

Continued

TABLE 11.1
Continued

Group *immigrans*	Asada *et al.*, 1992
hypocausta	
neohypocausta	
siamana	
Group *melanica*	Ewing, 1970
euronotus	
melanica	
melanura	
micromelanica	
nigromelanica	
paramelanica	
Group *quinaria*	Neems *et al.*, 1997
kuntzei	
limbata	
phalerata	
transversa	
Group *repleta*	Ewing and Miyan, 1986;
Subgroup *fasciola*	Alonso-Pimentel *et al.*, 1995;
coroica	Costa and Sene, 2002
ellisoni	
fascioloides	
moju	
onca	
rosinae	
Subgroup *hydei*	
eohydei	
hydei	
neohydei	
Subgroup *mercatorum*	
mercatorum	
paranaensis	
peninsularis	
Subgroup *mulleri*	
aldrichi	
anceps	
arizonensis	
buzzatii	
eremophila	
leonis	
longicornis	
martensis	
meridiana	
mettleri	
micromettleri	
microspiracula	
mojavensis	
mulleri	
navojoa	

Continued

TABLE 11.1
Continued

  ritae
  stalkeri
 Subgroup *repleta*
  canapalpa
  limensis
  melanopalpa
  repleta
Group *virilis*   Hoikkala and Lumme, 1987; Suvanto *et al.*, 1994
 Subgroup *virilis*
  americana americana
  americana texana
  lummei
  novamexica
  virilis
 Subgroup *montana*
  borealis
  ezoana
  flavomontana
  kanekoi
  lacicola
  littoralis
  montana

Subgenus *idiomyia*
 Group picture-wing
 Subgroup *planitibia*   Hoy *et al.*, 1988; Hoikkala *et al.*, 1994
  cyrtoloma
  differens
  hanaulae
  hemipeza
  heteroneura
  ingens
  melanocephala
  neoperkinsi
  oahuensis
  obscuripes
  planitibia
  silvestris
 Subgroup *pilimana*   Hoikkala *et al.*, 1989
  glabriapex
  fasciculisetae
  lineosetae
 Subgroup *vesciseta*
  digressa
 Subgroup *grimshaw*
  disjuncta
  affinisdisjuncta

STUDIES ON THE INHERITANCE OF MALE SONG CHARACTERS USING CLASSICAL CROSSING EXPERIMENTS AND QTL TECHNIQUES

Classical crosses in studies of the genetic basis of variation in male song traits involve the production of F1, F2 or backcross hybrids between the parental species/strains. Special crossing schemes may help to find out how large a proportion of the variation is additive and dominant, what is the direction of dominance, how much interaction (epistasis) there is between nonallelic loci and what is the number of genes affecting the studied traits. Quantitative trait loci (QTL) methods also enable localisation of the genes (QTLs) affecting the studied traits on different chromosomal regions.

Studies on the songs of the hybrids between *D. melanogaster* and *D. simulans* (Kawanishi and Watanabe, 1981; Kyriacou and Hall, 1986) have shown the species-specific IPI to be autosomally inherited, while the differences in this song trait between *D. melanogaster* and *Drosophila mauritiana* are determined by genes distributed evenly throughout the genome (Pugh and Ritchie, 1996). A QTL study on a species difference in IPI between *D. simulans* and *Drosophila sechellia* (Gleason and Ritchie, 2004) showed the presence of dominant alleles or epistasis and revealed six autosomal QTLs explaining a total of 40.7% of the phenotypic variance. Within species variation in IPI in *D. melanogaster* populations seems to be additive and autosomal (Cowling, 1980; Ritchie *et al.*, 1994). Chromosomal replacement studies between the African and the cosmopolitan strains of this species showed the between strain differences in IPI to be largely due to genes on the third chromosome, with significant interactions involving other chromosomes (Colegrave *et al.*, 2000). A QTL study on recombinant inbred lines of *D. melanogaster* revealed three significant QTLs explaining 54% of the genetic variation (Gleason *et al.*, 2002). None of these three QTLs overlapped with the six QTLs found to have an effect on interspecific variation in IPI (Gleason and Ritchie, 2004). Inheritance of male song characters in the *melanogaster* group has also been studied in *Drosophila auraria* and *Drosophila biauraria* (montium subgroup), where the song data suggest autosomal control of species-specific IPI (Tomaru and Oguma, 1994b).

The courtship songs of *D. pseudoobscura* and *D. persimilis* of the obscura group differ from each other quantitatively in IPI and intrapulse frequency (FRE) of the HRR song and qualitatively in the amount of the LRR song produced. Ewing (1969) suggested that the presence of the LRR song and the frequency of the HRR song are controlled by genes on the X chromosome, while the pulse repetition rate of the HRR song (reverse of the IPI) is inherited independently of the former characters. Noor and Aquadro (1998), however, found the FRE and IPI of the HRR song to be determined by both X chromosomal and autosomal genes. Williams *et al.* (2001) dissected the genetic basis of the species differences in these two song characters using 15 molecular markers and found them to be associated with at least two or three genomic regions both on the X chromosome and the autosomes. The QTLs were associated with nonrecombining portions of the X and second chromosomes and so it is possible that they cover a large part of the genome.

In the *D. virilis* group, the genetic basis of song variation has been studied both between and within species. In the *virilis* subgroup species with structurally similar songs, the only song traits varying profoundly between conspecific strains or species are PN and PTL (Hoikkala and Lumme, 1987). A diallel cross between all species and subspecies of the subgroup (*D. virilis*, *Drosophila lummei*, *D. a. americana*, *D. a. texana* and *Drosophila novamexicana*) showed PN and PTL to be autosomal and truly polygenic with the direction of dominance for longer pulse trains (higher PN and longer PTL) and for denser sound pulses (Hoikkala and Lumme, 1987). Huttunen and Aspi (2003) performed a biometrical analysis (16 crosses over three generations) of song differences between two strains representing extreme phenotypes in PN and PTL in *D. virilis*, a species of the *virilis* subgroup with the shortest pulse trains. Here the joint scaling test revealed significant additive and dominance components and the direction of dominance was towards shorter and denser sound pulses. A QTL analysis between the two above-mentioned *D. virilis* strains using 26 microsatellite markers revealed two QTLs on the third chromosome, both of them explaining slightly above 10% of the variation in PN and less than 10% in PTL (Huttunen *et al.*, 2004).

In the species of the *montana* subgroup of the *virilis* group, male songs show clear species-specific characters, and in these species the X chromosomal song genes have been found to play a major role in determining the IPI and PL. Hoikkala and Lumme (1987) suggested that an X chromosomal change increasing variation in IPI has occurred during the separation of the *virilis* and *montana* subgroups. The long IPI allowed variation also in PL. The gene (or a group of genes) with a major effect on song differences between *D. virilis* and *D. littoralis* has been localised on the proximal end of the X chromosome, but a large inversion in *D. littoralis* prevented more precise localisation by classical crossing methods (Hoikkala *et al.*, 2000; Päällysaho *et al.*, 2001). The role of the X chromosomal song genes was studied further by crossing females of *D. virilis* (*virilis* subgroup) and of *Drosophila flavomontana* (*montana* subgroup) with males of four species of the group (Päällysaho *et al.*, 2003). The study confirmed a central role of the X chromosomal factors in determining species-specific song characters and in controlling the action of autosomal genes, at least those located on the third and fourth chromosome. Ewing (1969) has argued that X-linked genes are more likely to be involved in behavioural differences between closely related species because favourable partial recessives would immediately be exposed to selection, allowing more rapid evolution, and this study gives some support to his idea.

The genetic basis of song variation within a species has been studied most thoroughly in the *Drosophila montana* subgroup. Diallel crosses between four inbred *D. montana* strains revealed additive genetic variation in four out of five song traits with the carrier frequency of the song showing unidirectional dominance towards a higher frequency (Suvanto *et al.*, 2000). Another study partly using the same inbred strains (Aspi, 2000) showed evidence of additivity for the pulse train characters (PN, PTL and IPI) and additivity, dominance and epistasis for the pulse characters (PL, CN, FRE). In FRE the inbreeding depression was 14% suggesting that this song trait is associated with fitness.

A CANDIDATE GENE APPROACH

Several mutations are known to alter the pattern of male courtship songs in *D. melanogaster*, and the genes where these mutations have taken place can be regarded as candidate genes affecting natural song variation. The most thoroughly studied mutations are *per*^{0}, *per*^{L} and *per*^{S} mutations in *period* locus, *dissonance* (*diss*) mutation in *no-on- transient-A* (*nonA*) locus, *cacophony* (*cac*) mutation in *DmcalA* locus and several mutant alleles in *fruitless* locus. Mutant alleles in *period* locus change several kinds of behavioural rhythms in the flies, including the 1-min courtship song rhythm of IPIs of *D. melanogaster* song (Kyriacou and Hall, 1980). In *dissonance* mutants, the song pattern is severely distorted due to extreme polycyclicity and loudness; the sound pulses on a given train become progressively more polycyclic (Kulkarni *et al.*, 1988). This gene is also accompanied by visual defects *cacophony* causes similar changes in song and visual system (Kulkarni and Hall, 1987), which suggests that the *nonA* and *DmcalA* might be involved in the same biochemical pathway. The *fruitless* locus has several alleles that alter the pulse structure of the courtship song with some alleles of the gene completely blocking male song (Villella *et al.*, 1997). There are also several other mutant genes known to affect male courtship song in *D. melanogaster* (Peixoto, 2002). Peixoto and Hall (1998) suggest that the genes involved in ion-channel function (like many of the above-mentioned genes) might be a source of genetic variation in the fly's courtship song.

To study whether the genes that influence song could affect variation in different song traits within and between the species, two kinds of a candidate gene approaches have been used. First, candidate song genes have been used as markers in QTL analysis to determine if any of them coincide with QTLs. Second, orthologous genes from other *Drosophila* species have been transformed into *D. melanogaster* to investigate whether they carry species-specific information on songs. In a QTL study on *D. melanogaster* recombinant inbred strains only one candidate song gene (*tipE*) coincided with a significant QTL (Gleason *et al.*, 2002), while in a study on hybrids between

D. simulans and *D. sechellia* three candidate genes (*mle*, *cro* and *fru*) were coincident with significant QTLs (Gleason and Ritchie, 2004). A gene localisation study on the songs of the hybrids between *D. virilis* and *D. littoralis* showed *nonA* and *DmcalA* (but not *per*) to be located on the same end of the X chromosome as the major X linked gene(s) causing species differences in male song in the *virilis* group species (Päällysaho *et al.*, 2001). The role of these candidate genes in song variation in the *virilis* group is, however, still quite obscure.

The first transformation experiment with candidate song genes involved transformation of *D. simulans per* gene to *D. melanogaster* (Wheeler *et al.*, 1991). This experiment confirmed that the central fragment of the *per* gene controls species-specific song rhythmicity. Campesan *et al.* (2001) applied transformation of *nonA* gene between *D. virilis* and *D. melanogaster* to investigate whether *nonA* encodes species-specific information for song characters. The courtship song of transformant males showed several features characteristic of the corresponding *D. virilis* signal, indicating that *nonA* may encode species-specific information. Huttunen *et al.* sequenced the coding region of the *nonA* gene of *D. littoralis* and compared it with that of *D. virilis* and *D. melanogaster* (Huttunen *et al.*, 2002a) and also studied sequence variation in a repetitive region of the gene in all species of the *virilis* group (Huttunen *et al.*, 2002b). The data did not support the hypothesis that the *nonA* coding region affects interspecific song differences in the *virilis* group, but it is still possible that species-specific information lies in the regulatory region of the gene.

THE EVOLUTION OF SPECIES-SPECIFIC COURTSHIP SONGS

The divergence of male courtship songs in closely related species raises several questions. First, is the genetic basis of variation in the song traits within species qualitatively similar to that between species, or does speciation involve novel genetic processes such as fixation of alleles with a large effect and genomic resetting? Second, what kinds of selection pressures have affected songs during their evolution? And third, what are the physiological constraints in sound production and reception and how do they restrict song evolution?

In the *melanogaster* group, variation both at the intra- and interspecific level in the song IPI seems to be mainly caused by autosomal genes, but the genes affecting interspecifc variation are different from those contributing to intraspecific variation (Gleason and Ritchie, 2004). Gleason and Ritchie (2004) proposed that the type I architecture of many small-effect genes (Templeton, 1981) combined with bidirectional allelic effects found for the species difference in IPI is most compatible with a history of gradual divergence without strong selection. The *montana* subgroup of the *virilis* group offers another kind of song evolution. Here an X chromosomal gene(s) plays a major role in determining the large species differences, *e.g.* in IPI and PL, and this gene also seems to control the action autosomal song genes (Päällysaho *et al.*, 2003). It still remains to be studied whether this gene also affects song variation within the species.

Recent studies on song variation between populations in different *Drosophila* species show that songs may vary over a species' distribution range. Stabilising and directional selection leave different signs on the genetic architecture of song traits, and so it is possible to trace whether song differences have evolved as a consequence of genetic drift or whether songs have been affected by directional selection pressures (character displacement, sexual selection) during their evolution. There is no convincing evidence so far of the effects of character displacement on song evolution in *Drosophila* species, but there is some evidence on the effects of sexual selection. In *D. montana* a diallel study has revealed strong directional dominance for shorter sound pulses and a higher carrier frequency (Suvanto *et al.*, 2000). Here the direction of dominance was the same as the direction of female preferences for the song, and so female song preferences might be a driving force in song evolution (Ritchie *et al.*, 1998). The question of whether coevolution of male song traits and female preferences for these traits require coordinated changes in both sexes remains to be studied.

Evolution of male courtship songs can be hindered by the restricted ability of males to produce songs or of females to receive them and to recognise different song traits. Ewing (1979a) has argued that in the *funebris* group, the flies' inability to produce songs with much shorter than 10 msec IPI has caused a bottleneck in song evolution. Also, Göpfert and Robert (2002) have shown that arista tips of the antennae in *D. melanogaster* are moderately tuned to frequencies around 425 Hz. The restrictions can, however, be broken by evolving new ways to produce and perceive sounds, as has happened in Hawaiian picture-winged *Drosophila* species where the males can produce high-frequency sound pulses with 5 to 6 msec IPIs and the females can recognise the songs of up to 15,000 Hz (Hoikkala *et al.*, 1989).

ACKNOWLEDGEMENTS

I am grateful to Jenny Gleason and Jouni Aspi for critical reading of the earlier draft of the manuscript. This work has been supported by The Finnish Academy (project 50591).

12 Vibratory Communication and Mate Searching Behaviour in Stoneflies

Kenneth W. Stewart and John B. Sandberg

CONTENTS

INTRODUCTION

The order Plecoptera is composed of the northern hemisphere suborder Arctoperlaria and the southern hemisphere suborder Antarctoperlaria (Illies, 1966; Zwick, 1973, 1980). Research over the past four decades on about 150 species from Europe, New Zealand and North America has established that the typical mating system of Arctoperlaria involves the sequence of:

1. Encounter site aggregation of sexes near the time of emergence (Stewart, 1994)
2. Species–specific calling by males with vibratory signals, initially during a ranging, nonoriented search for females
3. Duet establishment when females within communicable range answer with vibratory signals then become stationary
4. A localised search by the moving male orienting toward the stationary female until location and mating are accomplished

The duetting portion of this system generally conforms to the Plecoptera model described by Bailey (2003). Further, research has established that the intersexual vibratory duetting portion of this mating system in stoneflies is one of the most diverse and complex known in insects (Rupprecht, 1968, 1969, 1976; Zeigler and Stewart, 1977; Stewart and Maketon, 1991; Stewart, 1994, 1997, 2001).

In contrast, observations over many years by stonefly specialists in Australia, New Zealand and South America and experiments with several New Zealand species, including *Stenoperla*, have failed to detect drumming in any species of Antarctoperlaria, suggesting that members of that suborder have never adopted vibratory communication. Stewart (2001) hypothesised that they may have evolved highly specific encounter site aggregation behaviour patterns, possibly supplemented by as yet undiscovered intersexual communication modes that bring the sexes sufficiently close enough together to accommodate mate-finding.

Newport (1851) first used the term drumming to describe stonefly communication; his and other early observations by MacNamara (1926), Brinck (1949), Briggs (1897) and others (Stewart 2001) were qualitative descriptions of the abdominal tapping behaviour noted in various species. The first quantified studies (Rupprecht, 1968, 1969; Zeigler and Stewart, 1977; Szczytko and Stewart, 1979) established that:

1. Signals were produced by percussion with either the unmodified or specialised distal, ventral portion of the abdomen.
2. Duets were either "two-way" (male call–female answer) or "three-way" (male call–female answer–male reply) sequences.
3. Male calls were more diverse and complex than either female answers or male replies, and signals of sexes and duet pattern were species-specific fixed action patterns.
4. During duetting, males searched for stationary females.

Further studies from the 1980s to the present, mostly by K.W. Stewart and students, using pairing experiments and playback experiments with computer-modified male calls to live females and other experimental protocols, have expanded our knowledge mainly in discovery of:

1. Additional methods of signal production by males, other than percussion.
2. Additional complex male signal patterns, seemingly derived from ancestral monophasic signals by changes in number of beats, rhythm, amplitude modulation and phasing or grouping of signal beats.
3. Duetting patterns additional to call–answer sequences, including grouped and "symphonic" male–female exchanges.
4. Conspecific characters and informational content of male calls as recognised by females (Zeigler and Stewart, 1986; Stewart and Maketon, 1990).
5. That communication distances may be up to 8 m on a suitable substrate (dry woody stems).
6. That vibratory calls on a resonant substrate can be communicated to females sitting also on a resonant substrate over short distances of 10 to 200 cm (Stewart and Zeigler, 1984).

Since the early quantitative studies of Rupprecht (1968) and Zeigler and Stewart (1977) there have been major advances in technology, ranging from recording on cassette tapes and analysing signals with oscilloscopes to the current digital recording on minidiscs and analysing with computer sound software. Sandberg and Stewart (2003) addressed the question of whether this change in technology had substantially affected the accuracy of signal parameter description, and therefore capability to compare favourably historical descriptions of particular species with current ones from the same or different populations. They found that signals of five species recorded and analysed with old and current technology were favourably comparable, and suggested that precise duplication of signals of a species over time cannot be expected due to extraneous sources of variation, such as temperature and light conditions at recording, instrument calibration integrity, population (dialect) variation, age of adults recorded (Zeigler and Stewart, 1985) and variation in number of signals of pairs successfully recorded.

SIGNALLING METHODS

Males produce their vibratory calls by four basic methods: percussion, scratching (scraping), rubbing and tremulation. Percussion, as the name implies, involves tapping on substrate with the apical-ventral portion of the abdomen; the body contact surface may be either the unmodified sternum or variously modified abdominal appendages or surfaces such as vesicles, lobes, knobs or hammers (Stewart and Maketon, 1991). Males without such specialised structures all produce percussion or tremulation calls. Scratching (scraping) involves a short drag of the abdomen as it contacts the substrate, producing a raspy sound on resonant substrates, and rubbing involves a prolonged percussion stroke always with the textured ventral surface of a knob or hammer. The textured surface may be ridged or papillous. We have described this unique vibration-producing behaviour as an abdominal-substrate stridulation, and on a resonant substrate it produces a squeaking sound. So far as presently known, females normally produce their answering signals by percussion (or only in Chloroperlidae by tremulation), without specialised structures other than possibly the subgenital plate.

SIGNAL COMPLEXITY AND PATTERNS

Male call signals and patterns vary greatly among different taxa in diversity and complexity (Stewart, 1997, 2001), probably because they contain the parameters that convey to females specific male identity. The basic types of calls include:

1. Simple, percussive monophasic calls with approximately even interbeat time intervals and little amplitude modulation (Figure 12.1).
2. Calls of slightly changed and intermediate complexity having variable length, number of beats and rhythm of interbeat time intervals (examples in Figure 12.2 and Figure 12.3). These are common types of calls, represented by numerous species in all nine arcto-perlarian families with two to 40 drumbeats and 13 to 3350 msec interbeat intervals.
3. Diphasic percussive calls having two phases, each with several beats of different number and different interbeat intervals and rhythms (examples in Figure 12.4 and Figure 12.5).
4. Grouped calls with percussive bursts, each of few beats and variable intraburst time intervals (examples in Figure 12.6, Figure 12.7 and Figure 12.8).
5. Scratch or rub calls of one to seven strokes of variable duration, amplitude rhythm and inter-stroke intervals (examples in Figure 12.9 and Figure 12.10).
6. Tremulation calls of vibrations produced in plants by pushup – or forward–backward rocking motion of the whole body (Alexander and Stewart, 1997), without striking the substrate (example in Figure 12.11).

Stewart and Maketon (1990) showed experimentally with five species representing ancestral and derived duet patterns that the critical information content of male calls that females respond to are:

1. A critical minimum threshold of beat or rub number
2. A discriminant window of beat intervals
3. Other specific parameters such as the bi-beats of *Calliperla luctuosa*

Female response signals are always relatively simple, percussive, typically monophasic and usually of few beats (Figure 12.1 to Figure 12.6 and Figure 12.8 to Figure 12.10), probably because they need to convey to males only that they have recognised his specificity and to signal their now stationary location. Notable exceptions in female answers are:

1. The interspersed grouped answers within calls of some species with grouped calls, leading to a symphonic interchange (Figure 12.8)
2. Tremulation answers to males with tremulation calls (Figure 12.11)

The shortness of female answers is not only attributable to the fact that they do not need to convey complex information, as in the male, but also probably to reduce their vulnerability to potential vibration-detecting predators such as spiders. Females remain stationary only for a certain time,

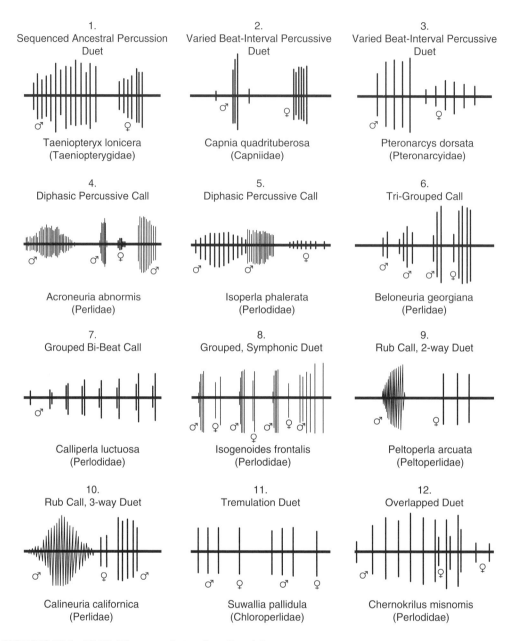

FIGURES 12.1–12.12 Diagrams of stonefly call and duet patterns.

while they presumably are able to measure male search fitness. The only prolonged female answers occur in females of large Perlidae that are larger than the prey threshold of most spiders.

DUET PATTERNS

Duets are typically either repeated sequences of "two-way" conversations (male call–female answer) (Figure 12.1, Figure 12.6 and Figure 12.10), or "three-way" ones (male call–female answer–male reply), or symphonies of grouped calls interspersed by grouped answers (Figure 12.8). In the sequenced duets, the female answer may follow the call after a variable latent time period (Figure 12.4, Figure 12.5 and Figure 12.10) or be overlapped with the latter portion of the

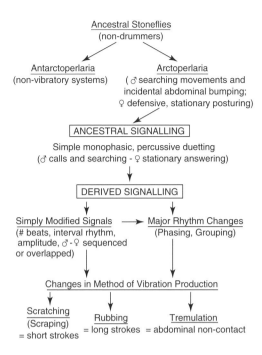

FIGURE 12.13 Drumming evolution paradigm (after Stewart 2001).

call (Figure 12.12). A unique pattern, discovered in only one species, *Paraperla wilsoni* Ricker (Stewart *et al.*, 1995), involves a long male call of 101 mode beats, with the female answer of seven mode beats fully embedded within the latter two-thirds of the call.

EVOLUTION OF STONEFLY DRUMMING

There is obviously no fossil record for vibratory communication, so formulation of an evolutionary scenario of the behaviour must rely on the patterns revealed from extant species. Out-group comparisons with the Grylloblattodea, other orthopteroid groups and the stonefly suborder Antarctoperlaria (Maketon, 1988) indicate that ancestral stoneflies were non-drummers, and that the Arctoperlaria have adopted and refined vibratory communication as an integral part of their mating system. This is further reinforced by the fact that special ventral abdominal structures have not appeared in males of the Antarctoperlaria. The first "drumming" in Arctoperlaria presumably resulted from the accidental bumping of the unmodified abdomen of males while searching for females, thus evoking the defensive response of females becoming motionless and stationary. Selection by enhanced finding success of females progressively reinforced the male bumping into a behavioural action, and a similar sequence in females followed, until relatively simple percussive, monophasic duetting became the typical ancestral communication (Figure 12.1, Figure 12.13). This, along with active males searching and stationary female response, became the ancestral mating system.

Out-group comparisons among Arctoperlaria families (Maketon and Stewart, 1988; Maketon *et al.*, 1988) indicate that species-specificity and behavioural isolation were then derived from the ancestral system, and are represented in extant species patterns as:

1. Ancestral (retained)
2. Simple modifications of ancestral number of beats, overlapping of male–female duet, altered beat interval rhythms and amplitude modulation in both male and female signals

3. Major rhythm changes in male calls by phasing or grouping of beats.
4. Changes in signal production methods, from percussion to scraping or abdominal-substrate stridulation (rubbing), or in rare cases tremulation (Figure 12.13)

It is assumed here that changes of male signals leading to specificity required concurrent changes in the females' capacity to recognise them. The result from study of a large sample of extant stonefly species is the exciting, highly derived and complex system of vibratory communication in Arctoperlaria, representing one of the most advanced such systems known in insects.

FAMILY PATTERNS

Table 12.1 shows a generalised summary of known drumming pattern diversity by arctoperlarian groups and families, relative to absence or presence of specialised drumming structures, signalling methods and pattern polarity. Families appear to fit the paradigm of Figure 12.13. in five basic ways:

1. The euholognathan families, Capniidae and Taeniopterygidae, and the systellognathan family, Pteronarcyidae, have largely retained the ancestral pattern of percussive, monophasic signals, with only enough change in beat number and interval rhythm and partial transition toward specialised vesicles, to derive species specificity (behavioural isolation). The several species of Taenionema, Doddsia and Oemopteryx that we have tested suggest the possibility that some Taeniopterygidae do not drum.
2. The euholognathan families Leuctridae and Nemouridae have retained percussion and derived specialised vesicles along with major modifications in percussion rhythm such as phased male calls or call burst groups.
3. The Perlodidae have largely retained percussion, and in some species the more ancestral model of monophasy, but numerous species have derived specialised grouped calls with interspersed grouped answers leading to symphonic duetting.
4. The most specialised calls that involve abdominal contact are made by males rubbing the substrate with a ventral knob (Peltoperlidae) or hammer (Perlidae); calls of the various species we have studied produce one to seven rubs, always answered by females with percussion. *Acroneuria evoluta* Klapálek may represent an evolutionary link in this percussion-to-rubbing method since some males produce both phases of their calls by percussion, but other males produce phase one by rubbing and phase two by percussion (Maketon and Stewart, 1984).
5. It appears that only some species of the family Chloroperlidae have derived tremulation. *Siphonoperla montana* Pictet and *Siphonoperla torrentuim* Pictet produce signals on plants by rapid, noncontact vertical movements of the abdomen (Rupprecht, 1981), and *Suwallia pallidula* (Banks) males produce a three-stroke tremulation signal by forward–backward rocking motions of the whole body (Alexander and Stewart, 1997). Over the years, we have tested several species of *Sweltsa* and other chloroperlid genera and never detected drumming through recording, possibly missing the possibility of tremulation.

MATE-SEARCHING

We began approaching the question of searching behaviour in relation to drumming in the later 1990s. We had learned quite a lot about the encounter site aggregation, calling and duetting components of the arctoperlarian mating system, but very little about the important steps of ranging and localised search, whereby a male utilises the vibratory information from a female to find her.

TABLE 12.1
Summary of Drumming Pattern Diversity of the 9 Arctoperlarian Families

Group/Family	Number Species	Specialised ♂ Abdominal Drumming Structures	Signalling Method	Evolutionary Scale Range of ♂ Calls
Euholognatha				
1. Capniidae	4	none or vesicles	percussion	ancestral to slightly derived change of beat number and intervals
2. Leuctridae	6	vesicles	percussion	ancestral to derived diphasic bursts or groups
3. Nemouridae	4	vesicles	percussion	all derived; bi-grouped, diphasic or diphasic bursts
4. Taeniopterygidae	7	none or vesicles	percussion	ancestral to slightly derived change of beat number and intervals
Systellognatha				
5. Chloroperlidae	6	none or lobes	percussion and tremulation	ancestral to diphasic
6. Peltoperlidae	13	knobs	percussion and rubbing	ancestral to diphasic
7. Perlidae	21	none or hammers	percussion or rubbing	ancestral, diphasic or grouped (transitional percussion and rubbing in *Acroneuria evoluta*)
8. Perlodidae	54	none or lobes	percussion and scratching	ancestral to derived bi-grouped, grouped or bi-beat grouped
9. Pteronarcyidae	8	none	percussion	ancestral to slightly derived change of beat number and intervals

Experiments, observations and videotaping were conducted with four species, selected as examples of different encounter site conventions: ground scramblers on flat or contoured substrates, and bushtoppers and treetoppers on linear or branched live or dead plant parts. The assumption was that the currently arrived-at system of any species should combine effective intersexual signalling with an efficient search modality.

First experiments with *Pteronarcella badia* (Hagen) (Abbott and Stewart, 1993) on a flat, gridded surface showed that the average find time for pairs engaging in strong, continuous duetting was significantly shorter than those for nonduetting or anomalously duetting pairs, and that the male localised search pattern was a triangulation, aided by the vibratory cues of answering females. Alexander and Stewart (1996a) videotaped a ground scrambling population of *Claassenia sabulosa* (Banks) with red light at night, and found that males drummed and searched near the shoreline of the Gunnison River, Colorado, on stones that protruded above the shallow marginal water surface. They circled the stones, then scrambled across the highest and ridged surface of the individual rock encounter sites that were female emergence sites. Alexander and Stewart (1996b) videotaped pairs of the treetopper *Perlinella drymo* (Newman) on a dead, branched tree limb and found that duetting increased male searching activity, influenced their directional movements and decreased the required find time for females, in contrast to nonduetting pairs. Males moving progressively up the limb toward a female placed high on one of its branches never called at branch bifurcations, but instead called at a position up the main stem or on the branch just above each fork. Therefore he was able to determine whether an answer came from out on that branch or from behind him, meaning she was further up on the limb. This process was repeated at each fork until he arrived at the branch leading directly to the answering female. The male in most trails found females within 10 min. Finally, Alexander and Stewart (1997) videotaped the tremulating chloroperlid *Suwallia pallidula* Banks in the field, and found that males flew during the afternoon on sunny days and landed on riparian alder where they began calling on stems and leaves with their typical three-stroke signal. Females answered with a typical one-stroke signal, moved to the petiole–leaf junction and became stationary. This position ensured that the male did not have to search entire leaf surfaces, so he searched only the petioles and leaf bases as they were encountered, while intermittently duetting with the female until she was found. Ensuing copulation occurred on the upper surface of leaves. Therefore, all four species conformed to the proposal that search patterns relate to the type of substrate at encounter sites (Stewart, 1994). The patterns were described as "fly–tremulate–search" for *Suwallia*, "rock to rock" for *Claassenia*, and "fly–run–search" for *Perlinella*.

Vibratory signalling has become a viable evolutionary strategy for Arctoperlaria, offering effective information for a searching caller (male) to locate a responding answerer (female). Male calls contain specific information that allows a female to use an apparently selectively arrived-at neuronal capacity to respond only to a conspecific mate, and possibly to a more fit conspecific mate. Females of *Pteronarcella badia* can discern the search-time fitness of particular males by remaining stationary only for a selectively allocated predator-safe time. If this time is not met, she moves to another position potentially to respond to calls of other males (Abbott and Stewart, 1993). Although there is little research to support it, directional location of a vibratory signal producer is probably determined by the differential time delays in reception of vibratory waves by the sensors of a receiver's planted tarsi. In scorpions, direction can be determined by integrating time delays as small as 0.2 msec, received by the subgenual organs of the different legs (Brownell and Farley, 1979c).

13 Communication by Substrate-Borne Vibrations in Cave Planthoppers

Hannelore Hoch and Andreas Wessel

CONTENTS

INTRODUCTION

With an estimated number of about 50,000 described species worldwide, Auchenorrhyncha (Fulgoromorpha and Cicadomorpha) is the largest hemimetabolous insect taxon which includes exclusively phytophageous species. As such, most Auchenorrhyncha are associated with the green parts of plants in grasslands and forests (Hoch, 2002). Auchenorrhyncha appear to be unlikely candidates for a permanent life underground. Nevertheless, five (out of 18) families of the Fulgoromorpha (planthoppers), a total of more than 50 subterranean species are now known from different parts of the world (Table 13.1). So far no cave-adapted Cicadomorpha (leafhoppers) have been reported.

In the cave planthoppers, adaptation to similar environments has given rise to the evolution of strikingly similar external appearance in different parts of the world, constituting a prime example of convergent evolution. Like many other obligately cavernicolous arthropods, cave planthoppers are characterised by a set of morphological features which have been acquired during the course of adaptation to cave life. Most represent reductive evolutionary trends, *e.g.* reduction and loss of compound eyes and ocelli, tegmina, wings and body pigment (Figure 13.1 to Figure 13.4).

TABLE 13.1
Cavernicolous *Auchenorrhyncha*

Family	Geographical distribution	Reference
Hypochthonellidae	Zimbabwe	China and Fennah (1952)
Delphacidae	New Caledonia	Fennah (1980a)
Kinnaridae	Mexico	Fennah (1973b)
	Jamaica	Fennah (1980b)
Meenoplidae	Australia	Fennah (1973b), Hoch (1990, 1993)
	New Caledonia	Hoch (1996)
	Western Samoa	Hoch and Asche (1988)
	Canary Islands	Remane and Hoch (1988), Hoch and Asche (1993)
	Cape Verde Islands	Hoch *et al.* (1999)
Cixiidae	Madagascar	Synave, (1953)
	Canary Islands	Remane and Hoch (1988), Hoch and Asche (1993)
	Azores	Hoch (1991)
	Mexico	Fennah (1973b), Hoch (1988a)
	Hawaii	Fennah (1973a), Hoch and Howarth (1999)
	Galápagos	Hoch and Izquierdo (1996)
	Argentina	Remes Lenicov (1992)
	Australia	Hoch and Howarth (1989a, 1989b)
	New Zealand	Fennah (1975)
	Reunion Island	Hoch *et al.* (2003)
	[Baleares]	Racovitza (1907) (unconfirmed record)

Other morphological characters observed in cave-dwelling planthoppers, although less obvious, may be of more adaptive value in the underground environment, *e.g.* specialised spine configurations on the hind tibiae and tarsi, serving for enhanced walking on wet or rocky surfaces (Hoch, 2002).

Why did all these species abandon environments abundant in food and light for a life in the seemingly hostile underground where permanent darkness and other adverse conditions such as high relative humidity close to saturation (Howarth, 1983) and sometimes abnormally high carbon dioxide concentrations (Howarth and Stone, 1990) prevail? Which physiological and behavioural adaptations did their epigean ancestors possess which enabled them to survive and even complete their life cycle underground?

FIGURE 13.1 *Oliarus polyphemus*, adult female, body length *ca.* 4 mm, Hawaii Island. (From Hoch, H. and Howarth, F. G., 1993. With permission.)

 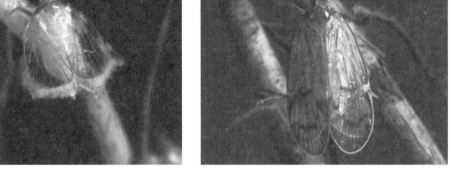

FIGURE 13.2 *Solonaima baylissa*, adult female, body length *ca.* 4.5 mm, Australia. (From Hoch, H., 2002. With permission.)

FIGURE 13.3 *Solonaima pholetor*, adult female, body length *ca.* 8 mm, Australia. (From Hoch, H., 2000. With permission.)

The past 30 years have seen a paradigm shift in our understanding of the evolution of terrestrial troglobites, especially in the tropics. Discoveries of terrestrial troglobites (mainly arthropods) from young oceanic tropical islands have challenged the long held belief that obligate cavernicoles were necessarily relicts, driven to inhospitable environments by deteriorating ecological conditions on the surface, *e.g.* during glaciation. According to this relict hypothesis (Barr 1968), these troglophilic populations acquired cave adaptations subsequent to the extinction or extirpation of

FIGURE 13.4 *Tachycixius lavatubus*, adult male, Canary Islands, Tenerife. (From Hoch, H., 2002. With permission.)

closely related epigean populations. Following his discoveries of a highly diverse cavernicolous fauna in lava tubes on Hawaii, Howarth (1986) suggested an alternative hypothesis. Here, the exploitation of a novel food resource is assumed to be the driving force in cave colonisation and evolution of terrestrial troglobites (the adaptive shift hypothesis). Howarth (1986) argues that if there were enough food, suitable habitats to lay eggs, and most crucially, if they had the ability to locate a mate and reproduce underground obligately cavernicolous species may derive from accidental invasions into subterranean spaces. In the planthoppers, nymphal habitat points to conceivable preadaptation.

All cavernicolous Fulgoromorph species belong to taxa in which even immature stages of epigean species live close to the soil, *e.g.* under the dead bark of rotting logs, in leaf litter or moss or even within the soil, feeding on roots or perhaps on fungi (Remane and Hoch, 1988). From this level of ecological preadaptation, it appears to be a small evolutionary step also for adults to switch to a permanent life underground (Hoch, 2002; Howarth and Hoch, 2004). Here, we shall focus on the characteristics of the mating behaviour of planthoppers which facilitate mate location and recognition in a permanently dark environment.

INTRASPECIFIC COMMUNICATION IN SURFACE-DWELLING PLANTHOPPERS

Mobile, sexually reproducing organisms have developed specific behaviour patterns which serve to bring together conspecific males and females for mating. These patterns may consist of visual, chemical, tactile or acoustic signals, or a combination thereof. In the Auchenorrhyncha (leaf- and planthoppers), it has been shown that mate recognition is primarily based on substrate-borne vibration signals (Ossiannilsson, 1949; Ichikawa, 1976). Little is known about the visual signals that may play a role in courtship, but observations in some epigean species suggest that wing fluttering of the male, which often accompanies acoustic signalling (*e.g.* Strübing, 1960; Ichikawa, 1976; Booij, 1982b; Drosopoulos, 1985) may provide additional stimulus for the female. Hitherto, there has been no confirmed evidence of chemical communication signals by pheromones in the courtship and mating of Auchenorrhyncha.

INTRASPECIFIC COMMUNICATION IN CAVE-DWELLING PLANTHOPPERS

Research on the intraspecific communication in obligately cavernicolous planthoppers in the Cixiidae has been conducted in Hawaii Island (*Oliarus*), Queensland, Australia (*Solonaima*) and the Tenerife, Canary Islands (*Tachycixius*).

RECORDING TECHNIQUES

Recordings were made in the field and in the laboratory. We used the *magneto-dynamic* (MD) system developed by Strübing and Rollenhagen (1988), which is comfortably portable and allowed an experimental setup inside a cave (Figure 13.5). In Hawaii, field recordings were made in a lava tube of Kilauea volcano (Pahoa Cave) to test the efficiency of vibratory communication in the natural habitat of *Oliarus polyphemus*. Roots of the endemic host tree, *Metrosideros polymorpha* (Myrtaceae), form dense curtains inside the lava tube (Figure 13.6), and particularly early instar nymphs (I–IV) are found feeding on these roots, while older nymphs (V) and adults are often observed on the rock surfaces of the cave (Hoch and Howarth, 1993). Tested substrates were:

(a) Living roots of Metrosideros polymorpha
(b) Rock surface

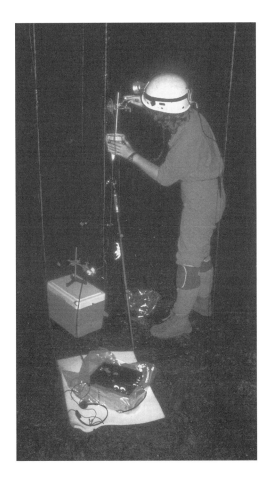

FIGURE 13.5 Recording the communication signals of cave-dwelling planthoppers in Pahoa Cave, Hawaii Island. (M. Asche, original photograph. With permission.)

FIGURE 13.6 Interior of a Hawaiian lava tube with root curtains (*Metrosideros polymorpha*). (From Hoch, H., 2002. With permission.)

A modified loudspeaker (a pin attached to the membrane) guaranteed the transmission of purely vibratory signals. Signals were recorded by a tape recorder (Sony TCD 5M Professional), subsequently digitised (40,000 data points/sec) and measured with MacLab/4s (AD Instruments) using Chart v. 3.5.4/s.

THE SPECIFIC MATE-RECOGNITION SYSTEM OF CAVERNICOLOUS PLANTHOPPERS

Our experiments revealed that cave-dwelling planthoppers retain the intraspecific communication system of their epigean relatives through communication by substrate-borne vibratory signals (Hoch and Howarth, 1993). The field experiments showed that communication by substrate-borne vibrations is extraordinarily efficient in cave environments. Living root tissue is a very well-suited substrate for the transmission of low-frequency vibrations (waterpipe principle), as is living plant tissue in general (Michelsen *et al.*, 1982). Signals travelled along root curtains and could be detected at distances up to 2.50 m from the source of vibration. In contrast, rocky substrates proved to have very poor transmission capacities: even intense banging with a forceps was not detectable by the MD system at a short distance (less than 10 cm) from the source (Hoch and Howarth, 1993).

HAWAII: *O. POLYPHEMUS* FENNAH — SPECIES COMPLEX

In Hawaii, the colonisation of caves has occurred repeatedly within *Oliarus*:

On the island of Molokai (one adaptive shift)
On Maui (three adaptive shifts)
On Hawaii Island (at least three adaptive shifts) (Hoch and Howarth, 1999)

Morphologically nearly identical populations of one of the evolutionary lineages on Hawaii Island which have invaded caves, *O. polyphemus* (Figure 13.1), are found in numerous lava tubes of four of the major volcanoes, Hualalai, Mauna Kea, Mauna Loa and Kilauea, ranging in age from less than a hundred to several thousand years.

A comparative analysis of male and female courtship call patterns in 11 populations from lava tubes in the Hualalai, Mauna Loa and Kilauea volcanic systems revealed the following results (Hoch and Howarth, 1993; Wessel and Hoch, 1999).

Courtship Behaviour

In all *O. polyphemus* populations studied, the following courtship pattern was observed. In the majority of male–female interactions, the female emitted spontaneous calls, that is, the female performed the initial step of courtship behaviour. In cases where these initial calls remained unanswered, the female eventually stopped calling. In cases where a nearby male answered these calls, the female did not change location and continued to emit calls at regular intervals. The male then usually approached the calling female, while responding to the female calls at irregular intervals. In all observed cases, courtship lasted for *ca.* 1 h before copulation commenced, while copulations persisted from 36 to 57 min. While in copula, neither male nor female emitted calls. Remarkably, in none of the observed male–female interactions displaying the described courtship pattern was a song-active female found to reject a responding male or to show avoiding or rejection behaviour when the male attempted to mate (Hoch and Howarth, 1993). Drosopoulos (1985) reported that there is specificity in precopulatory behaviour even in closely related species of the planthopper genus *Muellerianella*.

Call Structure

The time vs. amplitude pattern of the courtship signals of *O. polyphemus* individuals consisted of more or less homogenous pulse trains (Figure 13.7). Both sexes displayed similar call structures.

Both observations are in strong contrast with observations in epigean planthoppers which usually display highly complex time–amplitude patterns, especially in the male call, while females appear to be less differentiated (*e.g.* Claridge, 1985b; de Vrijer, 1986).

Within each population variation of courtship calls among individuals was found to be greater than individual variation of calls, although overall variation was comparatively small (Figure 13.7).

Remarkably, variation between populations was higher than within single populations (Figure 13.8; see song examples of *O. polyphemus* populations from Kaumana Cave, Pahoa Cave and McKenzie Park Cave on CD). Consequently, Hoch and Howarth (1993) assumed that *O. polyphemus*, previously considered to be a single species widely distributed on the island of Hawaii, was rather a complex of reproductively isolated populations, *i.e.* separate species. In the meantime, we showed that *O. polyphemus* is indeed a species complex *in statu nascendi* (Wessel and Hoch, 1999). Evidence is accumulating (Wessel and Hoch, in preparation) in support of the model suggested by Hoch and Howarth (1993) and Hoch (1999) which links evolutionary divergence to the succession of vegetation in dynamic, that is, volcanically active environments.

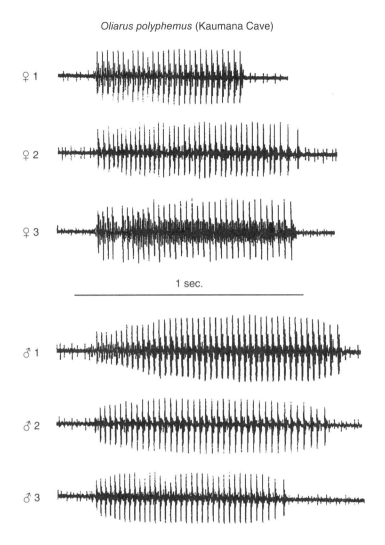

Oliarus polyphemus (Kaumana Cave)

FIGURE 13.7 Variation of *Oliarus polyphemus* male and female calls from Kaumana Cave population, Hawaii Island. (From Hoch, H. and Howarth, 1993. With permission.)

AUSTRALIA: *SOLONAIMA* KIRKALDY— SPECIES

In Australia, the (monophyletic) genus *Solonaima* is represented with nine epigean and six cavernicolous species in Queensland and New South Wales (Hoch, 1988b; Hoch and Howarth, 1989b; Erbe and Hoch, 2004). Cave-dwelling *Solonaima* occur in several million-year-old limestone caves in the Chillagoe and Mitchell Palmer Karst as well as in young lava tubes in Undara. The cave-dwelling *Solonaima* display varying degrees of troglomorphy, with external morphologies ranging from a virtually epigean appearance with only slightly reduced compound eyes, yet fully developed wings and body pigmentation, to greatly modified taxa, blind, flight- and pigmentless. In *Solonaima*, the degree of troglomorphy is positively correlated with the physical features of their respective environments rather than with the age of the caves. *S. baylissa* (Figure 13.2), which is strongly troglomorphic, is restricted to the deep cave zone of lava tubes in the 190,000-year-old Undara lava flow, while the only slightly troglomorphic species, *S. sullivani,* is found in limestone cave at Mitchell Palmer,

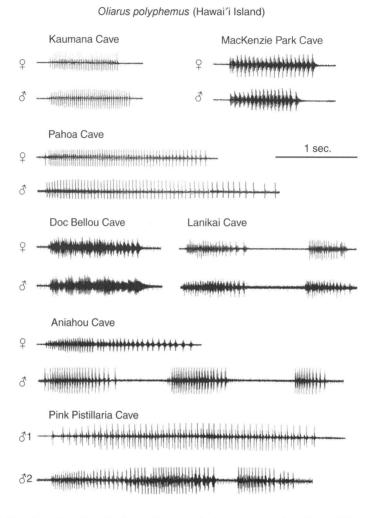

FIGURE 13.8 Female and male calls from *Oliarus polyphemus* populations from different lava tubes on Hawaii Island. (From Hoch, H. and Howarth, 1993. With permission.)

estimated to have been available for colonisation for at least 5 million years (Hoch and Howarth, 1989b).

Here, we compared the courtship behaviour of a facultatively cavernicolous species, *Solonaima pholetor* (Figure 13.3), which displays only slight modifications in external morphology, and the troglobitic, *S. baylissa*, which is highly troglomorphic.

Courtship Behaviour

From both cavernicolous *Solonaima* species, observations and recordings were obtained under field conditions. For each species, only a few male–female interactions were observed. Available information is too sparse to recognise a specific behaviour pattern.

Call Structure

In the facultative cave species, *S. pholetor*, an extremely simplified call structure was observed: calls of male and females consisted merely of single "clicks" (Figure 13.9 and song example on CD).

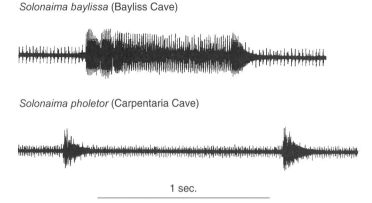

Solonaima baylissa (Bayliss Cave)

Solonaima pholetor (Carpentaria Cave)

1 sec.

FIGURE 13.9 Structure of male courtship call of *Solonaima baylissa* (Queensland, Australia: Bayliss Cave) (above) and *Solonaima pholetor* (Queensland, Australia: Carpentaria Cave, Chillagoe) (below). (From Hoch, H., 2000. With permission.)

Amazingly, these signals obviously contain sufficient information to lead to successful courtship and copulation. In contrast, males of the obligate cave species studied, the pale, blind and flightless *S. baylissa*, display a much more differentiated calling signal, which consists of three distinguishable elements — two initial chirps, a longer trill and a final pulse resembling the sound of a fog-horn (Figure 13.9, and song example on CD) (Hoch, 2000).

CANARY ISLANDS: *TACHYCIXIUS LAVATUBUS* REMANE AND HOCH

In the Canary Islands, several evolutionary lineages have colonised subterranean habitats on Tenerife, La Palma and El Hierro (Hoch and Asche, 1993). On the island of Tenerife, populations of *T. lavatubus* were found to occur in lava tubes in three disjunct areas, *e.g.* in the western part of the island, the north and the northeast (Hoch, 1994).

Recordings of courtship signals were obtained from *T. lavatubus* individuals from populations in the respective areas. Although very few male–female interactions were observed either in the field or the laboratory and recorded signals might not have been truly representative, a high degree of variation among the populations from different caves was observed (Hoch and Asche, unpublished). Unfortunately, recorded signals were erased from the tapes due to inadequate tape quality and/or storage, rendering preliminary results irreproducible. They are nevertheless mentioned here to point out the possible existence of another species complex comparable to that of *O. polyphemus* in Hawaii which now awaits further investigation.

EVOLUTIONARY INFERENCE AND POSSIBLE ADAPTIVE VALUE OF OBSERVED MODIFICATION OF BEHAVIOUR DURING CAVE ADAPTATION

ALTERATION OF GENERAL COURTSHIP PATTERN DURING CAVE ADAPTATION

Here, we refer to observations on the *O. polyphemus* species complex in Hawaii.

It is remarkable that, in nearly all cases observed, females initiated calling and remained the more song-active partner throughout the courtship phase. No female was ever observed to reject a responding male which had been successful in locating her (Hoch and Howarth, 1993). This behaviour is contrary to the courtship and mating behaviour observed in numerous epigean planthopper species (*e.g.* in the family Delphacidae, except for *Muellerianella brevipennis*

(Drosopoulos, 1985). There, males usually initiate calling, and in the genus *Javesella* males were observed to emit a few successive calls, then to stop calling for a variable amount of time, walk off the grass stem if there is no response by a female and, when placed back on the grass stem, produce a new series of calling signals. Under natural conditions, this strategy of acoustic exploration of environments with complex architecture (*e.g.* dense vegetation, grass tussocks, *etc.*) is apparently applied by males to locate receptive females (de Vrijer, 1986). Unfortunately, little is known about the courtship pattern in epigean *Oliarus* species which are closely related to the cavernicolous species. This limited knowledge makes it impossible to confirm that the switch from male to female initial calling occurs during cave adaptation. It is conceivable that the behaviour observed in *O. polyphemus* is a strategy to economise mate location under the specific conditions of the cave environment. The fact that song-active females readily accepted responding males for copulation indicates that song activity in females is usually directly correlated with receptiveness or, in the epigean delphacid *Muellerianella brevipennis*, to ensure pair forming (Drosopoulos, 1985). The female is actively advertising an opportunity for males in search of a mate. Consequently, males are made aware of receptive females only and, by this investment of less energy in exploring the environment acoustically in search for a receptive female, may instead be able to cover a wider area (Hoch, 2000). In the natural habitat, signals of a *Oliarus* female are perceptible by males within a radius of at least 2.5 m. The first male to reach the female will most likely perform a successful mating. The apparent low motility of the female during the courtship phase may facilitate the male's attempts to locate her. Both low song activity in the male and low motility of the female may also serve to minimise predation risk (Hoch and Howarth, 1993) by other troglobitic arthropods, *e.g.* spiders and crickets which also colonise the lava tubes (Howarth, 1981).

Reduction of Signal Complexity in Cave-Dwelling Planthoppers

An interesting phenomenon observed in the call structure of cavernicolous planthoppers is the reduction of complexity of single calls as compared with epigean taxa (see other contributions on Auchenorrhyncha in this volume), although substrate-borne vibratory signals are apparently the most important, if not the only, component, of the specific mate-recognition system of cave-dwelling planthoppers. Alternative hypotheses to rationalise this phenomenon have been discussed (Hoch, 2000). Preference was given to the ethological-release hypothesis which assumes that, when organisms are released from the competitive pressure of related species, selection on highly specific sounds may be less intense (Booij, 1982b). In Hawaii, the members of the *O. polyphemus* species complex occur largely in allopatry, and in any given cave there is only one cavernicolous planthopper species present (with one exception — see Hoch and Howarth, 1993). Consequently, competitive pressure by sympatric allies enhancing the maintenance of high signal complexity in order to minimise or avoid interspecific mating (which would probably yield less viable offspring) does not exist. Thus, maintaining a signal with minimal complexity may have been sufficient to serve the purpose of mate recognition and location under the specific conditions of the cave environment.

In *Solonaima* from Australia, the degree of reduction of signal complexity differs between the two cavernicolous species studied. The strongly troglomorphic, obligate cave species, *S. baylissa,* maintains a remarkably higher degree of call complexity than the less troglomorphic, facultative cave species, *S. pholetor*. In the cave where *S. pholetor* is found, it is the only cave-dwelling cixiid and perhaps the only insect using substrate-borne vibration as a means of intraspecific communication. The *S. baylissa* population studied, however, occurs in sympatry with another cavernicolous cixiid, *Undarana* species (Hoch and Howarth, 1989a). Although *Undarana* is not a close relative of *Solonaima*, maintaining signal complexity might be necessary to ensure specific mate recognition in an environment where mate recognition is apparently totally dependent on acoustic clues (Hoch, 2000).

SUMMARISING REMARKS AND PERSPECTIVES

From the results of behavioural studies on cave-dwelling planthoppers from Hawaii, Australia and the Canary Islands we are tempted to conclude that the maintenance of the intraspecific communication system based on substrate-borne vibrations in taxa with subterranean immature stages was crucial for the major ecological shift from surface to subterranean habitats. Whether or not the alterations of courtship and mating behaviour specifically observed in cave-dwelling *Oliarus* from Hawaii (female vs. male calling activity) are directly linked to the process of cave adaptation, however, is yet unknown. Although studies on intraspecific communication in cave-dwelling planthoppers have contributed to our knowledge of the reproductive biology of these insects, many questions remain open. In-depth studies on intraspecific communication in other cave-dwelling Fulgoromorpha may hold surprises and give new insights into the evolution and adaptation of communication systems in the course of major ecological shifts.

Very little is known about sensory structures and their physiological properties either in cave-dwelling planthoppers or even in Auchenorrhyncha in general (Čokl and Doberlet, 2003). Another promising field for further research is the role of sexual selection and female choice in the process of species formation in cave planthoppers. The only information on their prospective mating partner available to cave planthoppers must be encoded in the vibratory signals (Hoch, 2000). Other influences on choice which are critical to ensure mating success in epigean species, such as optical cues, are absent in troglobitic taxa. Obligately cavernicolous planthoppers offer prime opportunities to study sexually selected traits in a communication system with a minimal set of signal types, as compared with epigean taxa. Suitable laboratory facilities which are mandatory for rearing cave planthoppers by simulating the natural conditions of the cave environment (provision of root supply, permanently high relative humidity and darkness) are now becoming available for these studies at the *Museum für Naturkunde*, Berlin.

Considering that much of the existing knowledge stems from very few cavernicolous taxa and given the small proportion of investigated karst and volcanic areas containing suitable habitat, especially in the tropics, we predict that exploration of the respective areas will bring to light many more exciting discoveries.

ACKNOWLEDGEMENTS

We would like to express our sincere gratitude to the editors for inviting us to contribute to this volume. We thank all the people who have supported our work over the years in many ways and provided logistic and emotional support. Special thanks go to our colleagues and dear friends, Frank Howarth, Bishop Museum, Honolulu and Fred Stone, University of Hawaii, Hilo, for introducing us to the exciting underground biota, for sharing their wealth of knowledge, their support in the field, the many stimulating discussions and their friendship which never — not even in the tightest passages — failed us. We are also indebted to Manfred Asche, *Museum für Naturkunde*, Berlin, for his help and good companionship in the field, for his patience during long recording sessions and for valuable advice and comments, not only on this manuscript. Financial support was provided by the German Research Council and by The Nature Conservancy of Hawaii. This article is a contribution to the BEFRI project (Biodiversity and Evolution of Fulgoromorpha — a global Research Initiative (http://bach.snv.jussieu.fr/befri).

14 Partitioning of Acoustic Transmission Channels in Grasshopper Communities

Maria Bukhvalova

Most works on bioacoustics of insects have been concerned with signals and the structure of the communication system, either in a single species or in groups of closely related ones. Investigations of acoustic interactions between sympatric species dwelling in the same biotope are scarce. Still, information on the subject is available both for groups of sympatric species belonging to the same taxonomic unit, *e.g.* singing cicadas, and for communities of sound-producing animals as a whole.

Wolda (1993) described interspecific acoustic interference for singing cicadas (Cicadidae) in Panama. The breeding seasons of two of three sympatric species forming a "dusk chorus" completely overlap. The season of acoustic activity of the third species began only a day or two after the last singing male of one of the first two species had disappeared. Two other species, *Selymbrya achyetios* Ramos and *Pacarina puella* Davis, also differ in breeding seasons when they are sympatric (approximately December to February and March to April, respectively). In a locality where *P. puella* alone was present, it was found during the longer period from December to April. It is quite possible that the reason for the shifting of singing season of this species was acoustic interference with *S. achyetios*.

Acoustic segregation between different higher taxa also takes place. Investigators in a Bornean lowland rain forest showed that they were able to attribute most songs accurately either to vertebrates (mammals, birds and frogs) or to different insect groups (crickets, katydids and cicadas), even when the singer was not visible. This was possible because each such group had its own "acoustic appearance" (Riede, 1996). Acoustic activity of singing cicadas in the studied area was diurnal while crickets started singing at dusk, followed by the strictly nocturnal katydids. Thus, there is a pronounced temporal segregation between these taxa.

The situation observed in other groups of animals is more or less similar. Thus, among 15 Central Amazonian sympatric species of frogs breeding synchronously on the mats of floating vegetation, 14 produced mating calls differing distinctly from each other in dominant frequency and temporal pattern. Two species with similar calls demonstrate spatial segregation of calling sites. One of these calls from the peripheral parts of floating meadows at water level among aquatic plants distributed low over the water surface. The other perches on high grass growing nearer to the edge of a flooded forest (Hödl, 1977).

Thus, competition and niche segregation between species in ecological communities involve not only territory, food and other similar resources, but also acoustic transmission channels. Partitioning of communication channels in grasshopper assemblages will be discussed here.

In the Palaearctic region, Gomphocerine grasshoppers are diverse and numerous in meadows, glades and other open habitats. Generally, all Gomphocerinae are active during the daytime, so one can hear signals of about ten species singing simultaneously in the same biotope and sounding like an unceasing chorus to the human ear. The structure of such a chorus is one subject of the author's study.

It is evident that calling signals of each species must differ in at least one physical parameter from signals of all other sympatric simultaneously singing ones. Otherwise, recognition of a conspecific signal by a potential mate would be impossible. Signals of Gomphocerinae have wide-band noise frequency spectra, widely overlapping in different species. There is no evidence of use of frequency characteristics for recognition of conspecific songs in these insects (Meyer and Elsner, 1997). For this reason, only temporal pattern in the comparative analysis of signals in assemblages of grasshoppers will be considered here.

The temporal pattern of Gomphocerine signals is very elaborate and diverse. Still, from a collection of oscillograms of songs of sufficient species, it can be seen that certain signals are very similar in general structure and in temporal pattern of syllables. As a result, most of the song patterns fall into one of a small number of basic categories. The classification and terminology of signals used in this chapter is similar to that of Ragge and Reynolds (1998), but it covers Gomphocerine grasshoppers only. Calling signals of Gomphocerinae consist either: (1) of single (or repeated, but with irregular pauses) echeme or (2) of those repeated with regular intervals (echeme-sequence, according to Ragge and Reynolds, 1998) (Figure 14.1). In (2), the song may last for a minute or more. In (1), normally producing single echemes, a similar situation may arise very occasionally in males with unusually high acoustic activity. In several species normal song as a rule consists of a sequence including two or three echemes, as in *Chorthippus biguttulus* L. and *Ch. maritimus* Mistsh. (Figure 14.4). Nevertheless, they frequently produce single echemes. On the other hand, songs with more than three to four echemes are very rare. For this reason we attribute such signals to category (1) of single or irregularly repeated echemes.

Within each category two major groups may be separated. In the signals belonging to the first group (1), each syllable consists of a rather prolonged low-amplitude initial part and several shorter high-amplitude fragments separated by distinct gaps. Species producing single echemes (2) are numerous, including *Stenobothrus stigmaticus* Ramb., *S. fischeri* Eversmann and *Ch. vagans* Eversmann (Figure 14.4). Syllables in the songs of *Ch. parallelus* Zett. (Figure 14.4), *Ch. montanus*

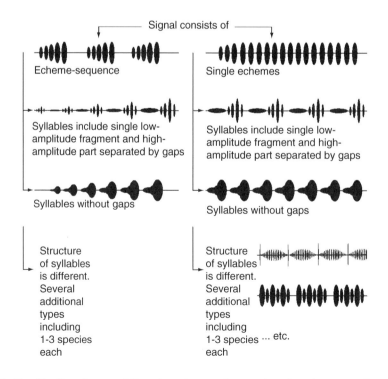

FIGURE 14.1 Classification of songs of Gomphocerine grasshoppers based on characters of temporal pattern.

Charp. (Figure 14.2), *Euchorthippus* species (Figure 14.4) and in a number of other species have the same structure, but the signal is an echeme-sequence.

The second group includes signals consisting of syllables without distinct gaps. Only in certain cases are low- and high-amplitude parts of the syllable separated by low-amplitude gaps. Single echemes consisting of such continuous syllables are intrinsic to *Omocestus viridulus* L., *O. haemorrhoidalis* Ch. (Figure 14.2), *etc. Chrysochraon dispar* Germ., *Euthystira brachyptera* Ocsk., *Stenobothrus nigromaculatus* H.-S. and a few other species produce echeme-sequences with the same syllable pattern.

As a result, four main types of signals may be separated. In addition, in a number of species the structure of syllables is quite different. For such species, several additional types, usually including one to three species each, may be recognised. The number of such "nontypical" species is not great, however. Among signals of more than 70 species of West European grasshoppers described by Ragge and Reynolds (1998), about 60% can be attributed to one of the four main types. Similarly, 24 of 35 species from Russia and adjacent territories (about 65%) produce signals belonging to one or another of these types (Bukhvalova and Vedenina, 1998; Vedenina and Bukhvalova, 2001).

It should be noted that the only purpose of this classification is to reveal the characters by which the signals of sympatric species differ from each other. It does not display phylogenetic relationships. Moreover, occasionally species from different genera produce quite similar songs, whereas closely related ones differ distinctly in temporal patterns.

Eleven grasshoppers assemblages in the Moscow Area, the steppe zone of the Lower Volga Region, the mountains of Northern Caucasus (Ossetia), Southern Siberia (Tuva) and the Russian Far East (Khabarovsk Province) were studied. In each community, signals of sympatric species were analysed and classified according to the scheme described above (Figure 14.1). All recordings were made at a shade air temperature of 26 to 31°C. Differences between the signals belonging to different types are obvious. Reliable characters for discrimination between the signals of different species within the same type are not so easy to find.

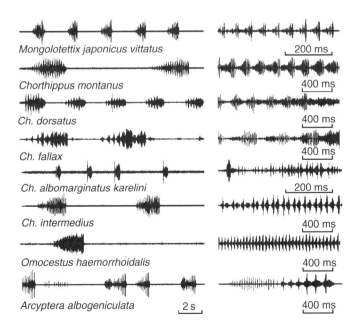

FIGURE 14.2 Oscillograms of calling songs of Gomphocerinae from a grasshopper assemblage in the floodland of Tes-Khem River in Southern Tuva (Southern Siberia). Two oscillograms of each signal are given at different speeds.

Echeme duration, as well as the echeme repetition period in species producing echeme-sequences, are variable and may overlap widely in sympatric species. The number of gaps (if present) in a syllable is also a variable character. Moreover, in certain species with similar patterns of syllables, the number of gaps is the same.

Comparison of the syllable repetition period (SRP) in signals belonging to the same type showed that the songs of sympatric species differ distinctly in this feature. This appeared to be quite similar in all biotopes studied. For this reason only one typical example will be described here, namely, the community of eight sympatric species in the floodland of Tes-Khem River in Southern Tuva (Southern Siberia) (Figure 14.2 and Figure 14.3).

The signals are subdivided into groups according to the classification given above (Figure 14.3). Species producing single echemes consisting of syllables with gaps, as well as species producing songs of any additional types, were absent from this community. The real distributions of values of SRP for each species are represented as a histogram. The following are general rules that appear to be observed in all communities. First, signals within the same type always have different SRP, only slightly overlapping in edge values. Second, the SRP in signals belonging to different types can overlap almost entirely. Only occasionally did we not find significant differences, for example in *Mongolotettix japonicus vittatus* Uvarov and in the second part of signal of *Arcyptera albogeniculata* Ikonn.

When the signal consists of two different parts, the situation is more complicated. Thus, in *A. albogeniculata*, the signal includes two successions of syllables (Figure 14.2). Sometimes these follow each other without a break forming a unified song. In other examples the two parts are subdivided by a pause lasting up to several seconds. Occasionally, a male omits any of the parts and the reduced signal consists of only one sequence of syllables. As a result, two sequences of syllables occasionally sound like different signals not connected to each other. In SRP these parts do not overlap either with each other or with signals of other species (Figure 14.3).

In *Chorthippus dorsatus* Zett., the first (main) part of the signal has a typical temporal pattern (Figure 14.2), whereas the second one is a monotonous fragment with indistinct inner structure. By contrast, in *A. albogeniculata* the pattern of song is constant and the second part is a necessary component of each signal. The SRP in the first part of the signal of *Ch. dorsatus* completely overlaps with that of *Ch. montanus* (Figure 14.3). Syllables in these two species also have the same

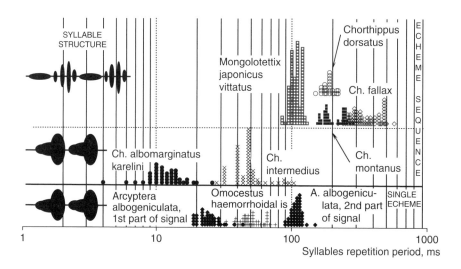

FIGURE 14.3 Histograms of distribution of syllable repetition period in eight sympatric species of Gomphocerinae from a grasshopper assemblage in the floodland of Tes-Khem River in Southern Tuva (Southern Siberia). Time scale is logarithmic.

pattern (Figure 14.2). Apparently, the song of *Ch. dorsatus* is exceptional among similar signals of sympatric species in the presence of additional fragments. The same phenomenon has been observed in a grasshopper community in the Moscow Area where this species coexists with *Ch. parallelus*, having almost the same SRP (for oscillogram of signal of the latter, see Figure 14.4).

Species producing signals of the same type with the same SRP were never found to be sympatric. For example, signals of *Ch. vagans* and *Ch. macrocerus* Fischer-Waldheim belong to the same type, namely, single echemes consisting of syllables with gaps (Figure 14.4). The SRP in both species are similar. According to our observations in the Lower Volga Region (Rostov Area), these species can live in the same locality, but they always inhabit different biotopes. *Ch. macrocerus* was found in steppes and other open habitats (on roadsides among the fields, *etc.*), but *Ch. vagans* lived exclusively in glades in forest plantations and on the edges of forest shelter belts. Thus, the two species replaced each other in different communities, whereas the remaining species producing signals of this type remained the same (Figure 14.5).

Several other examples of mutual replacement also were observed in the steppe zone of European Russia. *Ch. parallelus* was found here in all open habitats, but on saline land near Volgograd, it was replaced by *Eremippus costatus* Tarb., dwelling in the thickets of *Artemisia* (*Seriphidium*) (Figure 14.6). Signals of both species have the same type of temporal pattern and similar SRP (Figure 14.4).

Signals of *Ch. macrocerus* and *Stenobothrus miramae* Dirsh are also very similar, both in temporal structure and quantitative parameters (Figure 14.4). *S. miramae* was found in the Crimea, in steppes and anthropogenic landscapes, and also in the steppes of the Orenburg Region (South Urals). However, it was not found elsewhere in the steppe zone of south-east European Russia (between Crimea and South Urals). At the same time, in all localities studied in this area, *Ch. macrocerus* was present. Both species have very wide ranges of environmental tolerance. Two places where *S. miramae* was found are quite different in climatic conditions. The range of *Ch. macrocerus* includes central and southern parts of European Russia (northwards as far as Kursk), Caucasus (in Ossetia we have found this species both in the plains and on subalpine meadows about 2000 m above sea level), Transcaucasia, Northern and Western Kazakhstan,

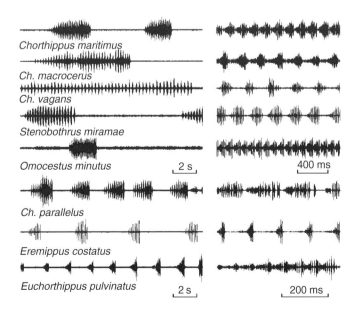

FIGURE 14.4 Oscillograms of calling songs of Gomphocerinae from steppe zone of European Russia. Two oscillograms of each signal are given at different speeds.

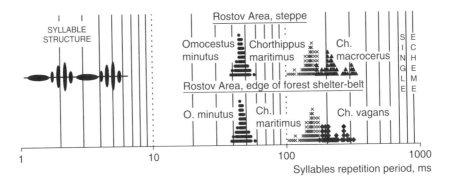

FIGURE 14.5 Histograms of distribution of syllable repetition period in four species of Gomphocerinae producing single echemes consisting of syllables with gaps. Time scale is logarithmic.

Turkmenistan (Kopet-Dagh Mountains), Iran and Afghanistan. It may be that the reasons for the allopatry of these species do not lie in a difference in ecological conditions between localities, but in the similarity of their calling signals.

Thus, we can note the third general rule observed in grasshoppers communities. The free range of SRP can be occupied in different communities by different species, but the place of this range on the time scale remains unchanged. The probable explanation of this phenomenon may be as follows. Difference in species composition between communities in different biotopes or localities is not great. As a rule, one can observe the appearance and loss of a small number of species in every biotope in comparison with nearby ones. For this reason, every species appearing in the community can occupy only the free range on the SRP scale of its signal. Consequently, in different biotopes and localities, species composition varies to some extent but the ranges of SRP that can be occupied by the signals remain the same.

As a result, all the complex of temporal parameters of the signal, including the general structure (single echemes or echeme-sequence), temporal pattern of syllables and the range of SRP determine for the song of each species its own "place" in the acoustic environment of the community, so-called acoustic niche, which is a part of the ecological niche as a whole (Zhantiev, 1981). Acoustic niche may be defined as the range of acoustic parameters of environment in which successful communication of the species is possible.

It can be assumed that the above-mentioned characters are most important for recognition of conspecific signals in Gomphocerinae. Ethological experiments with retranslation of model songs

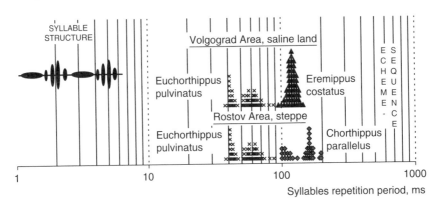

FIGURE 14.6 Histograms of distribution of syllable repetition period in three species of Gomphocerinae producing echeme-sequences consisting of syllables with gaps. Time scale is logarithmic.

to receptive females corroborate this supposition. Females of *Ch. biguttulus*, *Ch. parallelus* and *Ch. dorsatus* demonstrated significantly more higher levels of response reactions to artificial signals with gaps than to ones without gaps (Vedenina and Zhantiev, 1990; von Helversen and von Helversen, 1998). Apparently, the insects are capable of distinguishing between the two types of syllable patterns. As demonstrated in recognition tests on *Ch. biguttulus* (von Helversen and von Helversen, 1998), *Ch. parallelus* (Dagley *et al.*, 1994) and *O. viridulus* (Eiriksson, 1993), SRP also plays substantial role in the identification of conspecific song.

As a whole, the results of the comparative investigation of calling signals in grasshopper communities demonstrate that acoustic environment is at least as important for Gomphocerinae as other ecological conditions. For this reason, competition between species is not only for food and territory, but also for acoustic transmission channels. Sometimes, the absence of a free channel appears to be the main factor limiting the distribution of species and determining the species composition of the community.

15 Insect Species and Their Songs

Jérôme Sueur

CONTENTS

INTRODUCTION

In a comment entitled "Exploding species", Price (1996) wrote with some enthusiasm about the capacity of song analysis to discover new bird species. Price refers particularly to the Palaearctic leaf warblers (genus *Phylloscopus*) for which several sibling species have been discovered through analysis of acoustic characters. At the end of his essay, Price (1996) suggested that the number of acoustic sibling species could also be important in arthropods. The diversity of acoustic communication among arthropods is indeed so high — as attested by the contributions to this book — that the evoked species blast would certainly be more deafening for arthropods than for birds. Establishing an interdisciplinary link between biology and physics, bioacoustics is often viewed as a branch of ethology, mainly consisting of qualitative and quantitative behaviour descriptions and biomechanical and neurobiological investigations (Gerhardt and Huber, 2002; Greenfield, 2002). As questioned by Price, can bioacoustics also be used for systematics? What information can acoustic analysis contribute to the state and the dynamics of species diversity? Is the neologism "phonotaxonomy" coined by Ragge and Reynolds (1998) relevant?

    The first task of animal systematics is to identify taxa at the species level in the framework of an evolutionary theory, namely a species concept. Following Claridge *et al.* (1997a), the biological species concept, which applies only to biparental sexually reproducing organisms, may be considered in a broad sense, combining the points of view of both Mayr (1942) and Paterson (1985). This approach states that "different species are characterised by distinct specific mate recognition systems (SMRs) which result in the reproductive isolation observed between different sympatric species taxa in the field" (Claridge *et al.*, 1997a). In biparental sexually reproducing insects using sound to communicate, the structure of the acoustic signals, and of the underlying behaviour and mechanisms that produce and receive them, form part of the SMRS, ensuring syngamy. They are often stereotyped and, consequently, they provide potentially useful species-specific features or diagnostic characters for species recognition and delimitation. The use of acoustic characters has already been considered for animal species identification (Leroy, 1978, 1980) and more especially for identification of hemipteran (Claridge, 1985a, 1985b; Claridge and de Vrijer, 1994; Den Hollander, 1995; Claridge *et al.*, 1997b) and orthopteran insects (Ragge and Reynolds, 1998), but no very recent overview has summarised how such behavioural characters may be useful for studying species in different insect groups.

DIAGNOSTIC CHARACTERS: WHERE ARE THE SPECIES-SPECIFIC ACOUSTIC MARKERS?

Taxonomy, which mainly involves species description and synonymisation, requires recognising as many species-specific markers as possible. Acoustic markers may be sought in morpho-anatomy of sound transducers, selection of the transmission channel, associated behaviour and the sound signals themselves. In all cases, these characters should be considered and treated like any other characters. They must then adhere to three main rules: (1) be heritable, (2) show a low degree of intraspecific variation but (3) also display a high degree of interspecific variation.

MORPHOLOGY AND MECHANISM OF SOUND TRANSDUCERS

Sound-producing structures of the emitter ensure the conversion of a mechanical vibration (friction, percussion, tymbal activation, wing vibration and air expulsion) into a wave signal transmitted through the environment. This signal then reaches the sensory organs of a receiver (antennae, tympanal and other chordotonal organs), which convert it into a neuronal message. The producing and receiving structures, acting as sound transducers, thus constitute the starting and ending points of a species-specific acoustic communication chain. It is therefore not surprising that their structure, mechanism and physiology show species-specific properties. Stridulating structures have often been used for taxonomic purposes in Orthoptera (Bennet-Clark, 1970a, 1970b; Nickle and Walker, 1974; Ragge and Reynolds, 1988, 1998; Çiplak and Heller, 2001). These components of sound-producing structures in cicadas show a typical shape used for species and higher category taxonomy (Duffels, 1993; Boulard, Chapter 25). External acoustic structures of larger size than the insect, which allow small species to produce loud and low frequency calls, can also show species-specific features, as is well known for mole-cricket burrows (Bennet-Clark, 1970b, 1999) and cricket leaf-baffles (Forrest, 1982). Mechanically, differences can be found between closely related species, as in the case of three sympatric grasshopper species (*Chortippus dorsatus*, *C. dichrous* and *C. loratus*) having species-specific ranges of leg movement when stridulating (Stumpner and von Helversen, 1994). There have been fewer investigations using a systematic approach involving the acoustic receiving organs. Some recent studies reveal that this is an extraordinary field to explore. Revising the structure of tympanal organs in geometrid moths and discussing their relative importance in the subfamily classification of Lepidoptera, Cook and Scoble (1992) claimed that hearing organs should be routinely examined by taxonomists. Parasitoid flies of the family Tachinidae have complex tympanal organs on the prothorax, which enable females to find their hosts in the dark (Robert, 2001). The examination of the structure of these peculiar sense organs shows significant interspecies differences that could be used for identification as well as for phylogenetic reconstruction (D. Robert, personal communication). A recent study suggested that all Brachyceran flies should be able to hear with their vibrating antennae (Robert and Göpfert, 2002). Although showing similar general morphology and, probably, vibration mechanism, the antennae of higher flies are extremely diverse in form. The detailed description of their anatomy in conjunction with the analysis of their particular vibrating properties could provide new insights into fly systematics.

ACOUSTIC NICHES

Communication is constrained by various environmental factors, such as habitat structure, climatic conditions, surrounding noise and other animal activities (including predators) (Römer, 1993). As a result, insect species use specifically adapted times and sites when communicating. Ecological and phenological factors linked to sound communication constitute an acoustic niche with species-specific particularities that can be included in the species diagnosis.

The description of acoustic niches implies precise calling time period and habitat/microhabitat selection. For example, the species of the Mediterranean cicada genus, *Tibicina*, occupy distinctive habitats and are acoustically active at specific periods of the year. These specificities, together with geographical variables, probably contribute to acoustic segregation between species (Sueur and Puissant, 2002). More precisely, for many insects the microhabitat used when communicating, *i.e.* the calling and hearing sites, is often species-specific. Tropical cicada species have been documented to select specific heights in the forest, leading to a vertical stratification (Sueur, 2002a). Similarly, phytophagous species using substrate-borne waves often show strong host plant selection, leading to acoustic isolation by horizontal stratification. Specific host plant relations such as those observed in Hemiptera are indeed of direct relevance to the reproductive isolation of closely related species using sound to communicate (Drosopoulos, 1985; Claridge and de Vrijer, 1994; Rodriguez *et al.*, 2004). When exploiting the same plant as a transmission channel, species have been known to occupy preferentially different parts of the plant (root, stems, leaves and trunk). For example, epigean species of the Hawaiian planthopper genus, *Oliarus*, communicate on aerial parts of a native tree, whereas cave-dwelling species of the same genus communicate through the roots (Hoch and Howarth, 1993; Hoch and Wessel, Chapter 13).

In addition to spatial selection, species often show stereotyped calling periods that may contribute to species diagnosis. A most striking example is found in the cicada genus *Magicicada*. This well-known North American genus is divided into seven species which fall into three morphologically and behaviourally distinct groups. Within each group, species are very similar, particularly in regard to their acoustics, but differ greatly by their life cycle duration of 13 or 17 years (Alexander and Moore, 1962; Cooley *et al.*, 2003). This supra-annual temporal separation leads to an acoustical periodicity that constitutes the main diagnostic character differentiating the species. Species may also have seasonally distinct acoustic periods. For example, two North American crickets have similar calling songs: *Gryllus veletis* stridulates during spring and *G. pennsylvanicus* through autumn (Alexander and Bigelow, 1960; Walker, 1974). Species normally communicate only for a specific period of the 24 h cycle. Heller and von Helversen (1993) reported that in some species of *Poecilimon* bushcrickets, males restrict their calls to darkness or call during both day and night or during the day only. Fine time selection is also documented in tropical cicadas, where species use specific time windows leading to a time-sharing of sound activity (Gogala and Riede, 1995; Riede and Kroker, 1995; Sueur, 2002a).

SINGING POSTURES AND STRATEGIES

Behaviour associated with sound emission and sound reception may also be species-specific and may thus be used as diagnostic characters. Calling and receiving body posture may differ from one taxon to another. Some species of the cicada, *Cicadetta*, raise their abdomens and lift their wings when calling, whereas other species of the same genus do not adopt such an attitude. When searching for a mate, the calling sex may use different calling strategies. Two main tactics have been recognised: to call continuously from a near-permanent calling site (call-and-stay strategy), or move from one calling site to another one (call-and-fly strategy) (Gwynne, 1987; Sueur, 2002a). Although both strategies occasionally occur in a single species, they often differ between sympatric species, as in some Mexican cicadas (Sueur, 2002a). Sound emission can be characteristic of one or both sexes. In the latter, a duet may be engaged. The time interval between male and female signals can then be species-specific and thus important for species recognition processes, as in several bush-crickets (Heller and von Helversen, 1986; Robinson, 1990a; Bailey, 2003) and cicadas (Gwynne, 1987; Lane, 1995; Cooley and Marshall, 2001). Heller and von Helversen (1993) reported two different communication systems in *Poecilimon* bushcrickets. In some species, the mute female approaches the male phonotactically, while in other species the female responds acoustically to male song and can thus induce the male to approach her. Such differences have also been observed in cicada courtship (Sueur and Aubin, 2004). Alternatively, the signal repertoire may be species-specific. For example, the bug, *Nezara viridula*, uses courtship calls at short distances just before copulation, whereas *N. antennata* does not (Kon *et al.*, 1988). Finally, differences in calling strategies have been observed at the subspecific level: the grasshopper *Chorthippus parallelus* sings less, but moves more while searching for females than does *C. p. erythropus* (Neems and Butlin, 1993).

ACOUSTIC SIGNALS

Acoustic signals produced by insects have a genetic basis and there is little evidence of learning (Hoy, 1974; Ewing, 1989; De Winter, 1992; Tomaru and Oguma, 1994b; Shaw, 1996b; Gray and Cade, 2000; Williams *et al.*, 2001; Henry *et al.*, 2002a, 2002b; Hoikkala, Chapter 11). The species markers embedded in acoustic signals can be analysed in the three dimensions describing mechanical vibrations: amplitude, time and frequency. These three dimensions constitute a space of variability where each species occupies a typical volume (Figure 15.1). Species may be defined and differentiated by any combination of these three dimensions, as illustrated in comparative acoustic studies (King, 1999; Stumpner and Meyer, 2001; Sueur and Aubin, 2002). The main point is then to define the species volume, *i.e.* the species variability and limits. This requires properly estimating the intraspecific (intraindividual, interindividual, intrapopulation, interpopulation) and interspecific acoustic variability and subsequently recognising which acoustic variables show the lowest intravariability and the highest intervariability.

  As reviewed by Villet (1995), many factors may be sources of intraspecific variation at any of the three stages of the communication chain, *i.e.* production, transmission and reception of signals. These factors can be divided into two sets: the natural factors inherent to the species and their environment, and the artificial factors which are mainly due to recording and analytical procedures (Table 15.1). The latter can be minimised by good experimental design that includes accurate

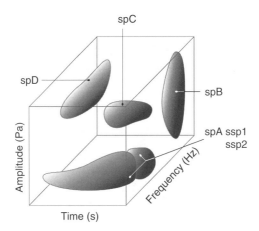

FIGURE 15.1 Schematic representation of species in the 3D acoustic space (time vs. frequency vs. amplitude). Inspired from the Bradbury and Vehrencamp (1998) graph describing signal space and limits of repertoire size.

TABLE 15.1
Intraspecific Acoustic Variation

	Main Effects	Main References
	Natural Factors	
Intrinsic Factors		
Species recognition	Stabilisation	Searcy and Andersson (1986), Shaw and Herlihy (2000), Ferreira and Ferguson (2002), Gerhardt and Huber (2002)
Sexual selection	Diversification	
Geographic isolation	Drift	Claridge (1990, 1993), Claridge and Morgan, (1993), Čokl *et al.* (2000), Zuk *et al.* (2001)
Reproductive character displacement	Shift	Walker (1974), Gray and Cade (2000), Marshall and Cooley (2000)
Developmental conditions (season)	Modification of temporal pattern	Whitesell and Walker (1978), Walker (1998, 2000)
Age	Modification of all parameters (maturation)	Hoch and Howarth (1993), Ciceran *et al.* (1994), Bertram (2000), Moulin *et al.* (2001)
Size	Negative correlation with frequency	Bennet-Clark (1998a), Fletcher (2004)
Physiological state	Increase/decrease of activity	Heller and von Helversen (1993)
Extrinsic Factors		
Ambient temperature and solar radiation	Modification of temporal rates, frequencies or intensity	Sanborn, Chapter 7
Food availability	Decrease of calling activity and signal amplitude	Bailey, Chapter 8
Habitat structure	Temporal delay, echo, amplitude filtration, frequency filtration, *etc.*	Wiley and Richards (1978), Richards and Wiley (1980), Padgham (2004)
Presence of congeners	Increase or decrease of call rate when synchronising or alternating signals	Greenfield (1994, 2002)
Presence of noncongeners	Increase or decrease of call rate when synchronising or alternating signals, shift in diel periodicity when avoiding masking	Schatral (1990), Greenfield (1988)
Presence of predators	Cease activity or decrease call rate, diversification	Spangler (1984), Zuk *et al.* (1993), Gray and Cade (2003)
	Artificial Factors	
Signal acquisition	Amplitude filtering, frequency filtering, aliasing, undersampling	Gerhardt (1998), Clements (1998)
Signal analysis	Frequency/time resolution	Beecher (1988), Beeman (1998), Jones *et al.* (2001)

A distinction is made between natural factors which are inherent to species biology and artificial factors inherent to acoustic analysis.

calibration and tests for the homogeneity of variables. Some of the natural factors may introduce variability which does not reflect genetic differences among species. Such extrinsic factors must be recognised and corrected, as is routinely done by regression analyses for temperature effects (*e.g.* Walker, 2000).

Practically, intra- and interspecific variability of acoustic parameters can be assessed by most uni- and multivariate analyses used in biometry (Sokal and Rohlf, 1995). Hence, variation can be estimated by within and between individual coefficients of variation and their ratio (Sueur and Aubin, 2002), repeatability indexes (Miklas *et al.*, 2003a), cluster analysis (Zuk *et al.*, 2001), principal component analysis (Claridge and de Vrijer, 1994; Henry *et al.*, 1999a; Sueur and Aubin, 2003) or discriminant function analysis (Ferreira and Ferguson, 2002; Henry *et al.*, 2002a, 2003). More specifically linked to acoustic data, waveform envelopes and frequency spectra may be compared by correlation and coherence procedures, whereas spectrogram similarity can be assessed objectively by cross-correlation algorithms (Khanna *et al.*, 1997).

EXPERIMENTAL TAXONOMY: ARE YOU ONE OF OURS?

Are the species-specific markers identified by the analysis of variation effectively used as such by sexual partners? Do individuals actually rely on these features to ensure species identity? In other words, do taxonomists and insects use the same characters for species identification?

Acoustic differences do not automatically indicate the presence of different species. Indeed, two signals that look distinct in shape may be similar functionally. Some allopatric populations of the planthopper *Nilaparvata bakeri* show obvious divergence in their vibratory signals, but mate choice experiments showed no preference for homogametic matings between these different calls, suggesting that they all belong to the same species (Claridge and Morgan, 1993). Similarly, the small spectral differences found between the tremulation songs of closely related lacewing species of the genus *Chrysoperla* are not considered relevant for a taxonomic differentiation because individuals apparently cannot process fine frequency differences (Henry *et al.*, 1999b). Conversely, two apparently similar signals may trigger different receiver responses, as illustrated by males of the planthopper *Ribautodelphax* that can distinguish between female conspecific and heterospecific calls, even if the simple structure of the female signal does not seem to offer sufficiently stable cues for effective species discrimination (De Winter and Rollenhagen, 1990). The next step of the analysis is therefore to understand the encoding–decoding processes of the species-specific information and then to estimate a possible mismatch between the phenotype (potential species-specific marker) and the biological function (to ensure syngamy). To achieve this, cross-mating tests and playback experiments may be conducted (Claridge and de Vrijer, 1994).

Cross-mating tests consist of bringing together individuals belonging to distinct populations where species identity is uncertain. Successful mating may indicate that populations belong to the same species. Nonetheless, a failure to mate by individuals originating from different populations does not automatically indicate the existence of an acoustic mating barrier. Conversely, if hybridisation is successful in an artificial environment, this does not mean that it would occur under natural conditions (Claridge and de Vrijer, 1994). For example, F1, F2 and backcross generations of two species of Hawaiian cricket (*Laupala paranigra* and *L. kohalensis*) can be obtained in the laboratory, despite the fact that they have clearly distinct calling songs (Shaw, 1996b). Simple cross-mating tests may be completed with choice experiments, in which an individual has to choose between a congener and an individual coming from another population. To gain information on the role of the acoustic channel, mating tests may also be conducted with animal made mute or deaf surgically or genetically (*e.g.* Tomaru *et al.*, 1995, 2000).

Another way to test the function of acoustic markers is by carrying out playback experiments (Hopp and Morton, 1998; McGregor, 2000). This experimental approach consists of manipulating the acoustic markers and inferring their importance in triggering species-specific responses from the receiver. Signal manipulations range from the complete deletion of the markers to negative and positive shifts from their natural values. Practically, the amplitude, temporal or frequency parameters of the reference signal is independently manipulated, or artificial signals are generated *de novo* by progressively incorporating parameters of the natural signal. Allospecific signals can also be used, especially those produced by sympatric and syntopic species for which the risk of unproductive courting is high. All these stimuli are broadcast to receivers, behavioural responses are measured qualitatively (level of display) and quantitatively (time of display) and the results are interpreted in terms of good functioning or breaking of species-specific communication. Experiments can be based on no-choice or two-choice (or even more) paradigms. As discussed by Doherty (1985a), no-choice and two-choice presentations each have advantages and limitations. An animal who does not reply to a signal in no-choice experiments could show a preference for the same signal when that signal is presented in competition with another one (Pollack, 1986; Ryan and Rand, 1993; McLennan and Ryan, 1997). Hence, two-choice experiments seem necessary when studying the recognition mechanisms of species sharing the same habitat. Playback experiments should also consider both long- and short-range signals. These two categories of signals can indeed rely on distinct encoding–decoding processes that may have different functions in the species recognition system (Doolan and Young, 1989; Guerra and Morris, 2002). Ultimately, playback experiments are not sufficient on their own to prove species isolation. In fact, postmating isolation can occur whatever the isolating properties of premating behaviour. Differences have been observed between the calling songs of two sister cricket species (*Allonemobius fasciatus* and *A. socius*), which were originally distinguished by allozyme assay (Benedix and Howard, 1991; Mousseau and Howard, 1998). However, phonotaxis experiments have shown that females of both species were insensitive to these interspecific differences (Doherty and Howard, 1996), the isolation being achieved by postmating barriers (Gregory and Howard, 1994).

To conclude, experiments should ideally combine both cross-mating and playback experiments, with no-choice and two-choice experiments, for both short- and long-range signals.

TAXONOMIST'S DECISION: "SHALL I SPLIT OR SHALL I FUSE"?

Once acoustic variability has been estimated and, if possible, tested experimentally, the time has come for the systematist to make a decision: to validate a species, to split it into different new species or to proceed to synonymisation. Acoustic observations have to be reconciled with nonacoustic information such as ecological, phenological, morphological or molecular data. If we consider a 2-dimensional reference system with one axis representing acoustic variables and the other nonacoustic variables, we can then schematise six main cases that the systematist may encounter (Figure 15.2). These six cases are detailed in the following sections, the first three (a to c) corresponding to invalid species and the three others (d to f) to valid species. Ragge and Reynolds (1998) already reported and discussed several taxonomic problems solved by acoustics in Western European Orthoptera taxonomy. I will not refer to these examples here, but each of them can be classified as one of the six types of cases here detailed.

NON-ACOUSTIC FISSION TRAP

The first case (a) corresponds to a low variability of the acoustic parameters associated with high variability of nonacoustic parameters. The lack of an acoustic analysis or poor data sampling can lead to recognition of "invalid" species, here depicted at the extreme of nonacoustic variation. For example, the cicada *Lyristes lyricen* from North America has different morphological varieties that could have been interpreted as subspecies or species. However, there is no clear geographic isolation, no barriers to introgression among the varieties and virtually no differences in the calling song (Moore, 1993). Immatures and adults of the Nearctic lacewing *Chrysoperla mediterranea*

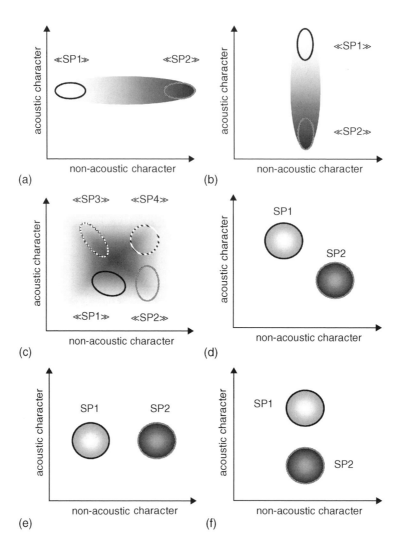

FIGURE 15.2 Variability of the acoustic characters and species validity. *x*-axis: set of nonacoustic variables such as morphological, ecological, phenological, or molecular variables. *y*-axis: set of acoustic variables. The suspected "species" of cases (a), (b) and (c) are not valid whereas those of cases (d), (e) and (f) are valid. See text for an explanation of each case. (Modified from Sueur, J., Aubin, T., and Bourgoin, T., *Mém. Soc. Ent. Fr.*, 6, 45–62, 2002. With permission.)

show high morphological polymorphism associated with a diverse larval ecophysiology. This variability could suggest the existence of different taxa. *C. mediterranea* is nevertheless regarded as a single valid species because the substrate-borne vibrations produced during mating are stable all across the species distribution (Henry *et al.*, 1999a). *Tibicina tomentosa* and *T. baetica* were two cicada species originally recognised from France and the Iberian Peninsular on the basis of morphological differences. However, these differences are not significant and the insects have similar acoustic niches, acoustic behaviour and calling signals. These similarities have been considered enough to put them in synonymy (Sueur and Puissant, 2003). Similarly, the name *T. atra* referring to a cicada species from Iberia has been recognised as invalid since its morphological and acoustic diagnostic characters previously thought to be unique were in fact the same as those of *T. argentata* from France (Sueur *et al.*, 2004).

Acoustic Fission Trap

This case (b) is the reverse of the previous one. Acoustic variability is here combined with a nonacoustic stability. Here, the consequence of inadequate data sampling would be to describe false sibling species (*cf.* case f). In Asia, several populations of the planthopper *N. bakeri* exhibit differences in their calling signals, although all use the same host plant. These acoustical divergences could indicate a species complex. However, cross-mating and mate choice experiments have shown that the populations are not sexually isolated and that they all belong to one species (Claridge and Morgan, 1993; Claridge and de Vrijer, 1994).

Multi-Factorial Fission Trap

The multifactorial fission trap (case c) is a combination of cases (a) and (b), both types of parameters varying. If not studied over a sufficient area, the variation can be considered as discrete. Locally, population characters could be then interpreted as peculiar adaptations, which could lead to naming new species erroneously. *C. johnsoni*, a lacewing living in the western part of North America, shows an acoustical variability of its calling signal and a colour polymorphism, which at first suggest the occurrence of a complex of species. Studying this acoustic and nonacoustic variation all over the species distribution, Henry (1993) reported that these two types of variation were not correlated, that the frequency differences observed between the populations were not statistically significant and that the temporal differences followed a cline without any important disruptions. Furthermore, there were neither discrete borders nor obvious hybridisation zones between the different suspected species. *C. johnsoni* is thus best regarded as a single biological species. The morphological and acoustical study of three allopatric populations of the Mexican bushcricket, *Pterophylla beltrani*, revealed large differences in size, wing venation, colour and a limited divergence in acoustic signals. Yet low genetic incompatibilities suggested that the populations might simply represent different subspecies, not different species (Den Hollander and Barrientos, 1990). In France, *Cicadetta brevipennis*, previously *C. montana* (Gogala and Trilar, 2004), and *C. petryi* were considered to be two valid species differing by their thoracic colouration and the duration of their calling songs (Boulard, 1995). However, a recent study covering a larger area showed that these characters were continuous and overlapping in such a way that the species have recently been synonymised (Puissant, 2001). Also in Hemiptera, differences in pheromones and songs between allopatric populations all around the world associated with data on asymmetric mating suggested that the bug, *N. viridula*, contains more than one species. Avoiding the artificial variability introduced in previous studies by using different plants as transmission channels, Čokl *et al.* (2000) and subsequently Miklas *et al.* (2003a, 2003b) reassessed acoustic variability between populations. They found temporal and, in some cases, frequency specificity at the population level. However, cross-mating and playback experiments showed that these differences did not sexually isolate the different populations, suggesting that they are all part of the same species.

"Standard" Species

"Standard" species differ in a set of acoustic and nonacoustic characters (d). The degree of intraspecific variability is low whereas the interspecific variability is high, making taxonomic identification relatively easy. For example, *Fidicinoides picea* and *F. pronoe* are two Neotropical cicadas that differ in their morphology, anatomy, acoustic niche, acoustic strategy and acoustic signals (Sueur, 2002a).

Some species show very similar character states (the distance between the two circles is short). Crickets of the complex, *A. fasciatus*, are very similar but even so can be distinguished by a set of electrophoretic, acoustic, habitat and distribution characters (Howard and Furth, 1986). Males of the two sympatric grasshoppers, *C. biguttulus* and *C. mollis*, differ slightly in wing venation, but produce signals with different frequency patterns, which are received by a tuned hearing system showing species-specific biomechanical and neurological characteristics (Meyer and Elsner, 1997). The Mexican *Pterophylla* bushcrickets are difficult to recognise and their taxonomy was confused until it was clarified by combining acoustic signals, morphometrics, mate choice and hybridisation experiments together (Barrientos and Den Hollander, 1994; Barrientos-Lozano, 1998). Similarly, combining morphological, enzymatic and acoustic traits with geographical information gives support to the notion of distinct subspecies within the bushcricket *Ephippiger ephippiger* (Kidd and Ritchie, 2000).

Acoustic behaviour can also confirm the existence of species or distinguish new ones exhibiting greater differences in acoustics than in other variables. Remarkably, this has been the case for widely distributed species. The full description of the acoustic behaviour of the green lacewing, *C. lucasina*, validated its taxonomic status and led to the recognition of new morphological attributes that could be useful in museum collections (Henry *et al.*, 1996). Two new species of the Hawaiian cricket, *Laupala*, were first detected by acoustic analysis and then confirmed by morpho-anatomy (body size, genitalia shape) and molecular (mt DNA) characters (Shaw, 2000). Conversely, in the case of *Magicicada neotredecim*, a periodical cicada species described in 2000, slight morphological and molecular differences with sister species were known for a long time before acoustic analysis confirmed the existence of a new species (Marshall and Cooley, 2000). The validation of *M. neotredecim* entailed evolutionary and phylogenetic analyses that go far beyond simple taxonomic interest (Cooley *et al.*, 2003).

PHYLOGENETIC CONSTRAINTS AND CONVERGENCES

Species of the fifth category (e) emit similar songs but differ in other variables. In a few cases, acoustic similarity has been reported between sympatric species (Blondheim and Shulov, 1972; Dadour and Johnson, 1983; Wilcox and Spence, 1986). In these closely related species, the acoustic resemblance is probably due to phylogenetic constraints (plesiomorphic traits), species not having diverged as much in acoustic characters as they have in other parameters. The signals produced during mating are then not crucial for species isolation. Acoustic similarity could also be the result of convergent evolution (homoplasic traits). Then, acoustic similarity is expected to be found among allochronic, allopatric and allotopic species. Similarities between allochronic species have been reported in some crickets and cicadas (see Section 15.2.2). A remarkable case of convergence has been demonstrated between two allopatric lacewings from North America and Kyrgyzstan, respectively (Henry *et al.*, 1999b). Indeed, these two *Chrysoperla* species, which clearly belong to two different clades, produce similar substrate-borne songs with only slight frequencies differences that do not make the signals functionally different.

HIDDEN BIODIVERSITY

The last case (f) refers to species that differ greatly in their acoustic behaviour but very slightly or not at all in other parameters. When the main nonacoustic variable is morphological or anatomical, species are usually named "sibling" or "cryptic" species. This hidden biodiversity, so qualified because species are virtually identical when looking at museum specimens, was reported early in Orthoptera (Thomas and Alexander, 1957; Alexander and Moore, 1958; Perdeck, 1958; Alexander and Bigelow, 1960; Walker, 1964; Bennet-Clark, 1970a; Walker *et al.*, 1973) and Diptera (Waldron, 1964; Ewing and Bennet-Clark, 1968; Bennet-Clark and Ewing, 1969; Bennet-Clark and Leroy, 1978). More recently acoustic sibling species have been discovered or confirmed not only in Orthoptera (Otte, 1994; Stumpner and von Helversen, 1994; Broza *et al.*, 1998; Cade and Otte, 2000; Walker *et al.*, 2003) and Diptera (Brogdon, 1994, 1998; Ritchie and Gleason, 1995; Noor and Aquadro, 1998), but also in Neuroptera (Henry, 1993, Chapter 10; Henry *et al.*, 1993, 2002a, 2003) and Hemiptera (Claridge *et al.*, 1985a, 1988; Claridge and Nixon, 1986; Boulard, 1988; Popov, 1997, 1998; Puissant and Boulard, 2000; Gogala and Trilar, 2004). These analyses increased considerably the number of species, especially in the case of the *Laupala* crickets from the Hawaiian archipelago (Otte, 1994; Shaw, 2000). Some of them revealed the presence of widely distributed species. The description of *G. texensis* defines a cricket species that was known for a long time and intensely studied in Texas. *G. texensis* was indeed previously referred to as *G. integer*, a closely related species but with different calling song and distributed in the west of the United States (Cade and Otte, 2000). *G. texensis* is even more similar to another species, *G. rubens*, also found in southeast United States. *G. texensis* and *G. rubens* differ slightly in their calling and courtship songs, but the role of these acoustic signals in premating isolation is in debate (Walker, 1998, 2000; Fitzpatrick and Cade, 2001; Higgins and Waugaman, 2004). The validation of *G. texensis* based on acoustic characters has clarified the taxonomy of the genus *Gryllus*, stimulating phylogenetic and speciation analyses (Gray and Cade, 2000; Huang *et al.*, 2000). Evolutionary or ecological studies can thus precede taxonomic descriptions. Hence, some species are still waiting to be officially described, *e.g.* the *N. lugens* complex (Claridge *et al.*, 1985b, 1988) and the cave planthoppers of Hawaii (Hoch and Howarth, 1993).

CONCLUSION — LISTENING TO THE DIVERSITY OF LIFE

Thus, bioacoustics seems to fulfil its pledges to insect systematics. Acoustic analysis has indeed been used in almost all insect groups producing sound and, by estimating phenotypic species variability and demarcations, has revealed many unsuspected species and solved several confusing taxonomic problems. Acoustic descriptions should be routinely included in species diagnosis (Tischeskin, 2000, 2002). All these taxonomic studies constitute the starting point of other insect research, especially studies on speciation modalities (Gray and Cade, 2000; Marshall and Cooley, 2000), evolution of behaviour (Henry *et al.*, 1999b; Desutter-Grandcolas, 2003; Desutter and Robillard, 2003), and automatic identification for species monitoring (Riede, 1993, 1998; Chesmore, 2001). These approaches, which work at the level of diversity patterns and processes, together expand the study of insect diversity. Combining two Greek words, *bios* and *akouien*, meaning *life* and *hearing*, respectively, bioacoustics can then be considered as the "listening to life" discipline, and even more as the "listening to the diversity of life" discipline when used in a systematic context. Let us therefore keep an eye and an ear on what species tell us.

ACKNOWLEDGEMENTS

I warmly thank Mike Claridge and Sakis Drosopoulos for their invitation to write this chapter and for their help during the preparation of the manuscript. More especially I would like to thank Mike Claridge and Charles Henry whose works have been a permanent source of inspiration during the preparation of this text. I am deeply grateful to Thierry Aubin and Thierry Bourgoin for the support and advice they have provided to me during my own studies associating bioacoustics and systematics. I am indebted to Charles Henry and Winston Bailey for having considerably improved the manuscript. I have been funded by the "Fondation Fyssen" (Paris, France) in the course of this work.

16 Is Migration Responsible for the Peculiar Geographical Distribution and Speciation Based on Acoustic Divergence of Two Cicadas in the Aegean Archipelago?

Sakis Drosopoulos, Elias Eliopoulos and Penelope Tsakalou

CONTENTS

INTRODUCTION

Lyristes (= *Tibicen*) *plebejus* (Scopoli) is the largest cicada in the Mediterranean region, extending north to central Europe (Austria, Germany, Czech Republic, Slovakia, Hungary and Poland) and East to Armenia, Georgia and Iran. Other species of the genus are distributed in the Far East (Nast, 1972). In addition, two new species have been described recently (Boulard, 1988) from Western Turkey, not far from the east Aegean Islands: *L. gemellus* from near the coast at Kemalpasa and *L. isodol*, a mountain species from the rocky slopes of Saimbeyli.

The first author, familiar only with the acoustic signals of the circummediterranean *Cicada orni* L., when hearing for the first time the sound of a second species, *C. barbara* Stål, believed that such a sound was emitted by a cicada of another genus, such as *Tibicina*. This experience occurred in southern Portugal where *C. barbara* is sympatric with *C. orni* (see Quartau and Simões, Chapter 17). This example of calling song differences of morphologically very similar species could be used as a classic example in introducing students to bioacoustics.

In Greece *L. plebejus* together with *C. orni* are the commonest species, in many places producing communal choruses. In continental Greece very high population densities of both are commonly observed, which is not generally the case in other regions such as in the Iberian Peninsula. However, where *L. plebejus* occurs in the Aegean Islands it is rarely common compared with the mainland of Greece and the Ionian Islands, including Kythira (Figure 16.1).

FIGURE 16.1 Distribution of *Lyristes* species in the Aegean Sea. Areas of absence are highlighted in dark grey, while the islands where this species is differentiated or two species occur are highlighted in light grey.

Historically, *L. plebejus*, rather than *C. orni*, was known to ancient Greeks who could already discriminate correctly between males and females. Socrates referring to his wife Xanthippe, said: "How happy male cicadas are that their wives never speak!" Worth mentioning here also is the fact that the first author and J.A. Quartau (personal communication), in their boyhood and independently, used *L. plebejus* for playing by way of inserting a thin stem of grass in the abdomen of a male and then letting it fly. The distance of the flight, even down the slope of a hill, was no more than 100 to 150 m, and this was repeated several times. This observation, together with others observing flight capabilities of this species, give some evidence that *L. plebejus,* due to its large size, cannot fly for long distances despite its well developed wings.

Behaviourally, *L. plebejus* is not at all shy and sometimes stops singing only when it is caught. However, in places where populations are not so dense it may stop singing when approached, as in many other cicadas. Despite this, it is the least shy species of all the 30 or so cicada species estimated to occur in Greece (Drosopoulos, unpublished). Gogala (personal communication) has not observed such behaviour of this species in its "type locality", Vipava (Wippach, West of Ljubljana). As in other cicadas, at high population densities males were observed to sing on any branch or trunk of a tree or shrub, walls, iron sticks and even electric poles. Comparatively, *C. orni* is more shy and normally is more difficult to collect when population densities are low.

In Greece *L. plebejus* is common, not only in natural areas on various trees and shrubs, but also in cultivated orchards such as olive, almond, pear, *etc.* These observations are based also on the exuvia found at places of its last mould, *e.g.* on the bark of trees and shrubs and also on walls. However, in big cities like Athens, this species may be heard in green areas like Kiphissia, but never in the sparsely green centre where, however, *C. orni* is often common.

The phenology of the adults of *L. plebejus* is relatively short, starting in warm places in mid-June and extending about 40 days. By contrast, *C. orni* may begin to sing early in June and can be heard (as in the centre of Athens) until the end of October. At places where the two species occur at high population densities, some damage can occur caused by the ovipositor of females when inserting their eggs, especially into the branches of young trees (Drosopoulos, unpublished). There are no reports on the damage caused by their larvae sucking the roots of cultivated orchards.

This chapter would not have been written if the first author had not heard and recorded differences in the songs of some populations of *L. plebejus* on the Island of Rhodes during his faunistic investigations in Greece. Similar differences were noticed on the Island of Chios later, during an expedition by the authors for recording both this species and *C. orni*. Moreover, in the three islands of Samos, Fourni and Ikaria, *C. orni* has also been heard making different songs. Thus, recordings of several males and sampling for morphological and DNA analysis from several localities in mainland Greece and the Aegean Islands, including Crete, were made. In addition recordings were made at Gordes (about 60 km NE of the city of Smyrna, western Turkey) during the summer of 2003 by P.C. Simões. The present chapter gives information on the distribution and the calling songs of populations of *Lyristes* species, where they are obviously differentiated.

GEOGRAPHIC DISTRIBUTION

Since the song of *L. plebejus* is very distinct on the mainland of Greece, collecting and recording were not necessary until a peculiar sound, apparently from this species, was heard in the island of Rhodes at the end of June, 1996. However, the size and morphology of these specimens did not differ greatly from those on the mainland. The population density was quite high at all localities investigated on Rhodes during our two expeditions compared with the other eastern Aegean Islands. Surprisingly, on this island the common species of *Cicada* was found only at one locality and in much lower numbers than *L. plebejus*. In contrast, on Chios, where a similar sound of a *Lyristes* was heard, the population densities of the two species were reversed. In all other islands investigated, Kos, Samos, Ikaria and Lesbos (Figure 16.1), where these two genera were found, *Lyristes* was local, rare and in low numbers in comparison to *Cicada*. In other Aegean islands *Lyristes* was absent. More striking was the absence in Crete where special expeditions were made during 1999 and 2000. After this, investigations were made on the island of Kythira, between South Peloponnesus and Crete, and on Karpathos, between Rhodes and Crete. Surprisingly, while on Karpathos no *Lyristes* species were recorded, in Kythira the two genera, occurring also in South Peloponnesus, were found everywhere in high population densities. These investigations prompted us to investigate further north of the Cyclades in order to have a more complete idea of the geographic distribution of *Lyristes* species. Thus, in the isolated island of Skyros (Cyclades) this species was not found, while in Skopelos in the Sporades the species was present and common.

These investigations enabled us to construct, for the first time, a distribution of *Lyristes* around the Aegean Sea (Figure 16.1). Whereas *Cicada* species were distributed all over the Aegean Archipelago, *Lyristes* were absent from Crete, the Cyclades and Karpathos.

ACOUSTIC DIFFERENTIATION

A detailed analysis of the acoustic songs of most of the Mediterranean *Cicada* species is presented here by Quartau and Simões (Chapter 17).

Recordings of *Lyristes* were made at all localities where they were found (Figure 16.1). Here we analyse only the calling songs of males easily distinguishable by ear at four localities in Greece (near Thebes — Central Greece, Kythira, Rhodes, and Chios islands) and one in Turkey (Gordes). Representative songs of the recorded specimens from a few of the localities visited are shown in Figure 16.2 and Figure 16.3. In all, the complete song (Figure 16.2) is composed of an energetic

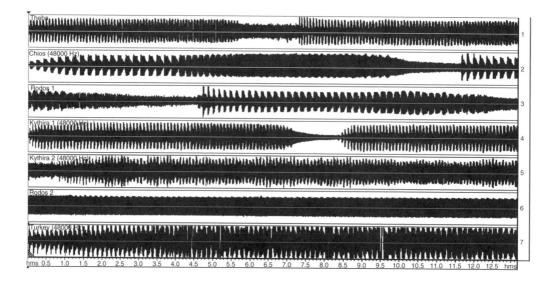

FIGURE 16.2 *Lyristes* songs recorded at 1: Thebes, 2: Chios, 3: Rhodes (type 1), 4: Kythira (type 1), 5: Kythira (type 2), 6: Rhodes (type 2), and 7: Turkey. Complete phrases are shown.

(loud) part and a second quieter part. Echemes from the energetic subphrase are shown in Figure 16.3. Frequency analyses were also made (Figure 16.4). As shown in Figure 16.2 and Figure 16.3, there is a clear difference between the male of Rhodes 2 and all males from the remaining localities. The other male of Rhodes 1 is again different from all other localities.

The Kythira 2 and the Turkish males probably belong to the same species, possibly one of those described by Boulard (1988 and Chapter 25) as *L. gemellus*. If this hypothesis is correct, then it is clear why the morphology of the species reported by Boulard (1988) is very similar to specimens collected from southern Peloponnesus and other localities of continental Greece (Tsakalou and Drosopoulos, unpublished data).

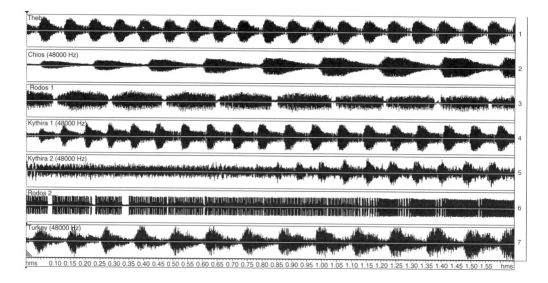

FIGURE 16.3 *Lyristes* songs recorded at 1: Thebes, 2: Chios, 3: Rhodes (type 1), 4: Kythira (type 1), 5: Kythira (type 2), 6: Rhodes (type 2), and 7: Turkey. Echemes are shown.

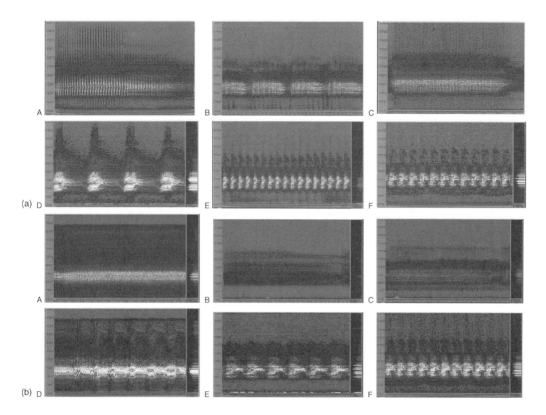

FIGURE 16.4 (a) FFT frequency analysis on *Lyristes* signals recorded at Rhodes (type 1) (A, D), Thebes (B, E), and Kythira (type 1) (C, F). Upper row (A, B, C) sonograms of complete phrases, (D, E, F), echeme analysis. (b) FFT frequency analysis on *Lyristes* signals recorded at Rhodes (type 2) (A, D), Turkey (B, E), and Kythira (type 2) (C, F). Upper row (A, B, C) sonograms of complete phrases, (D, E, F), echeme analysis.

In Thebes and Kythira 1 localities we have probably representatives of *L. plebejus*, but recordings of more males from the type locality (Vipava, Slovenia) are necessary before such identification can be guaranteed with certainty. In addition, careful analysis of other recordings published so far on this species should be made, *e.g.* from the Iberian Peninsula, southern France, and the southern parts of the former U.S.S.R. (Popov, 1975; Claridge *et al.*, 1979). At present we cannot comment on the Chios recordings with certainty, but it is very likely that this is another new species.

Frequency analyses of complete phrases and corresponding echemes (Figure 16.4a–d) support the conclusions about differentiation derived from the oscillogram analyses (Figure 16.1 to Figure 16.3). Furthermore, detailed acoustic analyses enable one to differentiate such *Lyristes* species, not only by the structure of the calling song, but also by frequency characteristics, as in species of genus *Tibicina* (Quartau and Simões, 2003).

Since in Rhodes and Thebes many specimens were recorded, we present here a few more in order to determine whether there is intraspecific or interspecific variation. Figure 16.5a and b show that there is some differentiation in specimens recorded at Thebes, but not to an extent likely to indicate different biological species. However in Rhodes, where two species are likely to occur, Rhodes 1 and Rhodes 5 (Figure 16.6a and b), there is evidence of even more introgression (Figure 16.6b) in the echemes of specimens Rhodes 3 and Rhodes 4.

This might explain why at some localities visited for a second time by the first author and several students during the same season (early July 1996 and 1999), the recordings were of

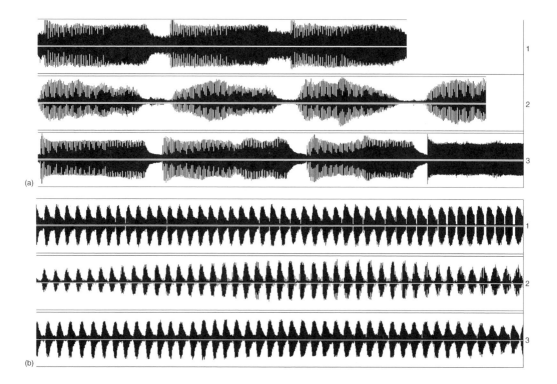

FIGURE 16.5 (a) *Lyristes* songs recorded at Thebes. Complete phrases are shown. (b) *Lyristes* songs recorded at Thebes. Echemes are shown.

various types. More materials recorded in 2000 by the first and last authors have not yet been analysed.

Thus, the present preliminary acoustic analyses on this genus surprisingly suggest the existence of a complex of closely related species in *Lyristes*. Already Boulard (1988) has shown that *Lyristes* in western Turkey includes more species than just *L. plebejus*. Based on his acoustic descriptions of these new species the present acoustic analysis allows us to conclude that:

(a) In Rhodes there are two new species.
(b) In Chios is another new species.
(c) *L. gemellus* is a very probable new record for Greece, found so far in Kythira.
(d) *L. plebejus* is probably represented in all other localities in Greece.

DISCUSSION

Studies on cicadas in the Mediterranean area require intensive field work which has to be strictly focused on them and carried out during a particular season that may shift from year to year, depending mainly on temperatures in early summer. For most cicadas, including *Lyristes*, this period is short and for each species adults do not emerge all at the same time. Particularly in *Cicada* species calling songs may start in mid-June and last until the end of October or even the beginning of November. The presence of one or several allochronically singing species certainly has not yet been established, though there are small differences in comparisons between early and late seasonal songs. The present investigations have demonstrated that there is a specific male calling song of *Cicada* species without clear indication that there is introgression between

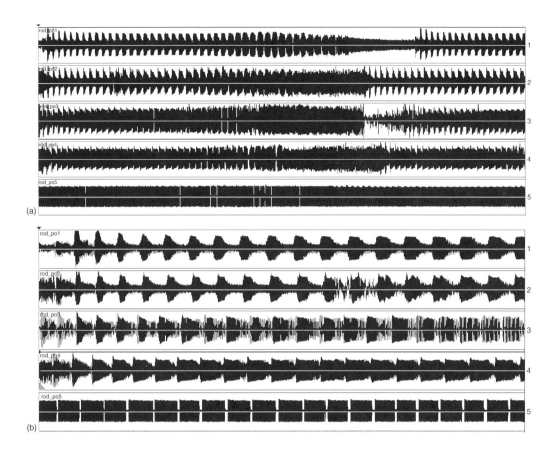

FIGURE 16.6 (a) *Lyristes* songs recorded at Rhodes. Complete phrases are shown. (b) *Lyristes* songs recorded at Rhodes. Echemes are shown.

them. *C. orni* and *C. barbara* can co-occur, for instance, in a few sites in Portugal, while *C. orni* and *C. mordoganensis* do not co-occur in the areas investigated in Greece and Turkey. This may be related to the extremely high population densities in the islands of Samos at least compared with the low ones in the Iberian Peninsula. In all other Aegean islands where *Cicada* species were found, populations were of moderate densities, while on the mainland and Kythira Island they occurred at high population densities. In addition the only *Cicada* found on Crete again occurred in high numbers across the whole island, whereas on Rhodes we were lucky to find a few specimens of *Cicada* at one locality, being almost sure until then that it was absent from this island. This again maybe due to the fact that *Lyristes* were very common there, while absent in Crete. Again this contrasts with the island of Kythira (South of Peloponnesus) where *Lyristes* and *Cicada* are both very common.

Presumably *C. cretensis* is a species that differentiated by geographic isolation on Crete, as also have many other insects and plants. The same model of speciation might have occurred in *C. mordoganensis* in the three islands that are close to each other (Samos, Fourni and Ikaria) from where it later dispersed to western Turkey. The other way round (first in Turkey and then in these islands) is less likely because this species does not occur in any other Greek island close to the west coast of Turkey. Because of the mountainous terrain, the annual precipitation is much higher than in the other islands in the vicinity and the Cyclades. Indeed, these three islands are richer floristically than any of the Cyclades islands.

Obviously, *Cicada* species, which are smaller and lighter than *Lyristes*, are capable of migrating between the mainland of Greece and Turkey and the islands of the Aegean Archipelago. However, due to the high population densities, even some if any migration of *C. orni* to the islands occupied by *C. cretensis* and *C. mardoganensis* may be of minor importance to prevent speciation procedure at present (see Drosopoulos, Chapter 19, and Strübing and Drosopoulos, Chapter 20).

Unlike *Cicada* species, the geographic distribution of *Lyristes* is peculiar since it is absent from all the small Cycladic islands and even also on Crete. Kühnelt (1986) in his review article on biogeography of the Balkan Peninsula reported that similar distributions occur in several other animal groups, including insects. A very striking example is the total absence of snakes of the family Viperidae in Crete. However, several geological explanations have been put forward for such unusual distributions. The present study suggests that dispersal from adjacent islands to Crete is very difficult for *Cicada* species and possibly impossible for *Lyristes*. How can it be that on Kythira both genera are common, but on Karpathos only *Cicada* is present? The answer may be that for *Lyristes* the distance from Rhodes is too great to reach Karpathos. It must be emphasised here that the species of the two genera have the same food requirements and adults can be found together even on the same trees, although they might prefer slightly different habitats as found in Portugal (Sueur *et al.*, 2004). That the flight (migration) capacity might be responsible for the absence of *Lyristes* in certain islands is supported also by the presence of this species in the chain of Sporades islands. Indeed, Skiathos island is closer to the mainland of Greece than Kythira and Antikythira are to Crete.

Speciation in *Lyristes* is difficult to explain at present; more field-work, further material, recordings and analyses are required. However, it is clear that some species, closely related to the common and widely distributed *C. orni* and *L. plebejus*, occur in various islands (Figure 16.1).

Bioacoustic studies have demonstrated their importance here in separating species that morphologically do not posses clear differences. Finally, it should be mentioned also that, due to the short duration of the adults and the long cycle of the nymphs, studies on cicadas are difficult as rearing has been successful only in a few rare cases. As far as we know, introgression is for the first time shown in Cicadas by the intermediate sounds emitted by hybrids of the two species in Rhodes. Similar songs are recorded by Drosopoulos (1985) in hybrids produced by crossing two closely related species of *Muellerianella* (Homoptera-Delphacidae)

ACKNOWLEDGEMENTS

The authors are indebted to Prof. M. Gogala for his great generosity in providing the sounds of *Lyristes plebejus* from its type locality and very useful comments on the manuscript, and to P.C. Simões for providing the recordings made at Gordes.

17 Acoustic Evolutionary Divergence in Cicadas: The Species of *Cicada* L. in Southern Europe

José A. Quartau and Paula C. Simões

CONTENTS

INTRODUCTION

The acoustically conspicuous species of genus *Cicada* L., typical of the Mediterranean area, is a group of insects of great acoustic and microevolutionary interest.

As particularly emphasised by Paterson (1985), sexually reproducing species may be defined as a set of organisms with a common specific mate-recognition system (SMRS). Thus, the defining properties of most species are their unique SMRSs and, therefore, the crucial event for the divergence and origin of a new species is the evolution of such new and unique mate-recognition systems.

As commonly referred to in the literature (*e.g.* Pringle, 1954; Claridge, 1985b; Bennet-Clark, 1998; Quartau *et al.*, 1999 and Simões *et al.*, 2000), most cicadas are distinguished mainly by the ability of males to produce loud and distinctive airborne acoustic signals during pair formation and courtship by means of a tymbal mechanism. Therefore, the structure of the acoustic signal is an important component of the SMRS and is probably the major factor responsible for the reproductive isolation between heterospecific cicadas, as elegantly demonstrated for North American *Magicicada* species by Alexander and Moore (1958). Moreover, examples of closely related cicadas occurring either in sympatry or allopatry would stand as unique case studies to investigate microevolutionary processes involved in speciation. This is indeed the case of genus *Cicada* which involves a complex of seven closely related species distributed mainly along the Mediterranean basin (Quartau and Simões, submitted): *C. cretensis* Quartau and Simões, submitted; *C. orni* Linnaeus, 1758; *C. cerisyi* Guérin-Méneville, 1844; *Cicada barbara* Stål, 1866; *C. permagna* Haupt, 1917; *C. lodosi* Boulard, 1979 and *C. mordoganensis*, Boulard, 1979.

The purpose of the present chapter deals with such closely related species on which we have accumulated a large set of original data (acoustic, morphological, biogeographic, ecologic and genetic) (*e.g.* Quartau and Simões, submitted; Quartau, 1988; Quartau and Fonseca, 1988;

Quartau, 1995; Quartau *et al.*, 1999, 2000a, 2000b, 2001, 2004; Seabra *et al.*, 2000, 2002; Simões *et al.*, 2000; Sueur *et al.*, 2004). This is a comparative analysis of the calling songs of the species present at both the extremes of Mediterranean Europe, where new species might have been produced and diverged as the result of strong geographical isolation in southern refugia during the glacial periods of the Pleistocene and possibly late Pliocene: (i) in the west (*e.g.* Portugal, Spain and Morocco) where *C. orni* and *C. barbara* are both present, either in different (allopatry) or in overlapping (sympatry) areas; and (ii) in the eastern part (*e.g.* Greece and Turkey), with *C. orni*, *C. mordoganensis*, *C. lodosi*, and *C. cretensis*. Two nominal species of doubtful affinities, since the calling songs are not yet properly known, were not considered in the present study: *C. permagna* (only given from a site in Turkey, but not found by us during recent field work) and *C. cerisyi* (referred to Egypt and Libya only).

SPECIES OF *CICADA* IN WESTERN MEDITERRANEAN EUROPE

Western Mediterranean Europe is an interesting geographical area where two species of *Cicada* are common and found generally from mid-June until October. Their habitats consist usually of "maquis" and open woodland, where males sing generally on trees such as *Olea europaea*, *Pinus pinaster*, *Quercus* species and the introduced *Eucalyptus globulus* (Sueur *et al.*, 2004) (Figure 17.1).

In this western area only two species occur: *Cicada orni* and *C. barbara*, which are closely related and often impossible to separate on grounds of external morphology alone (Figure 17.2 and Figure 17.3). Nevertheless, there are a few differences, especially if large series are analysed. In *C. orni* the general body colour is more olive-greenish and there are four well-defined spots on the forewings (Figure 17.4), on veins m-cu, m, r-m and r (Boulard and Mondon, 1995). By contrast, *C. barbara* is generally a brownish colour, and usually has only two well-defined wing spots on the forewings (Figure 17.5). With respect to the tymbals, there are no detectable differences between the two. However, when considering the male genitalia, separation is much easier. Indeed the analysis of this structure alone may enable species identification in most examples (Quartau, 1988). The dorsal spine of the pygophore is much bigger in *C. orni* than in *C. barbara* (Figure 17.6a and b).

These species overlap in some parts of their distributions, but are also widely allopatric. *C. barbara* is known to occur only in the south-western part of Europe and in the north of Africa, where *C. orni* is in general absent: Algeria, Italy (Sardinia and Sicily), Libya,

FIGURE 17.1 The typical western Mediterranean habitats where the *Cicada* species occur. (Montegordo, Portugal; Photo J.A. Quartau. With permission.)

FIGURE 17.2 *Cicada barbara*. (Crato, Portugal; Photo J.A. Quartau. With permission.)

FIGURE 17.3 *Cicada orni*. (Sousel, Portugal; Photo J.A. Quartau. With permission.)

FIGURE 17.4 Forewing spots of *C. orni*. (Tapada da Ajuda, Portugal; Photo J.A. Quartau. With permission.)

FIGURE 17.5 Forewing spots of *C. barbara*. (Praia da Rocha, Portuga; Photo J.A. Quartau. With permission.)

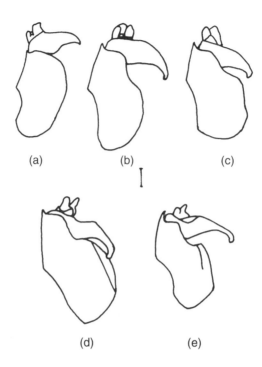

FIGURE 17.6 Terminal abdominal segments of different *Cicada* species (lateral view). (a) *C. barbara* (Crato; Portugal); (b) *C. orni* (Crato, Portugal); (c) *C. mordoganensis* (Samos; Greece); (d) *C. lodosi* (Alanya; Turkey); (e) *C. cretensis* (Crete; Greece). Scale = 0.8 mm.

Morocco, Portugal, Spain and Tunisia (Nast, 1972; Boulard, 1981, 1995; Quartau and Fonseca, 1988). On the other hand, *C. orni* has a much wider distribution in Europe (it also spreads into western Asia), where it extends from Portugal in the west to Greece and Turkey and the Near East, as well as around the Black Sea (Albania, Armenia, Azerbaijan, Cyprus, Egypt, Georgia, Israel, Jordan, Lebanon, Romania, Russia, Slovakia, Slovenia, Switzerland, Tunisia, Turkmenistan, Ukraine and Yugoslavia) (Nast, 1972; Popov, 1975). In a few areas populations of *C. barbara* and *C. orni* are truly sympatric, as in some Portuguese localities (*e.g.* in Alto Alentejo–Crato, Monforte, Sousel and in Estremadura–Serra da Arrábida) (Quartau *et al.*, 2001; Sueur *et al.*, 2004).

The best characters which readily separate *C. barbara* and *C. orni* are the calling songs (Figure 17.7.1 and Figure 17.7.5). If these are close in the spectral characteristics, in amplitude modulated patterns they are very distinct: in *C. barbara* the signal is continuous and monotonous, while in *C. orni* it is discontinuous with echemes lasting from 0.05 to 0.20 s (mean 0.09 s) and intervals between echemes varying from 0.06 and 0.20 s (mean 0.10 s) for the Portuguese sample (Quartau *et al.*, 1999). In frequency characters they are very similar, with ranges mostly between about 1 and 18 kHz and with peak frequency between 4 and 8 kHz (Figure 17.7).

Calling songs of males of *C. orni* from several localities in western and south-eastern Europe constitute a relatively homogeneous group. However, there is some tendency of songs from

FIGURE 17.7 Calling song profiles of species within genus *Cicada*. (a) Oscillogram (amplitude vs. time), (b) sonogram or spectrogram (frequency vs. time) and (c) mean amplitude spectrum (frequency vs. amplitude). (1) *C. orni* (Crato, Portugal); (2) *C. orni* (Athens, Greece); (3) *C. cretensis* (Crete; Greece); (4) *C. mordoganensis* (Samos; Greece); (5) *C. barbara* (Seville; Spain); (6) *C. lodosi* (Koyciegez; Turkey).

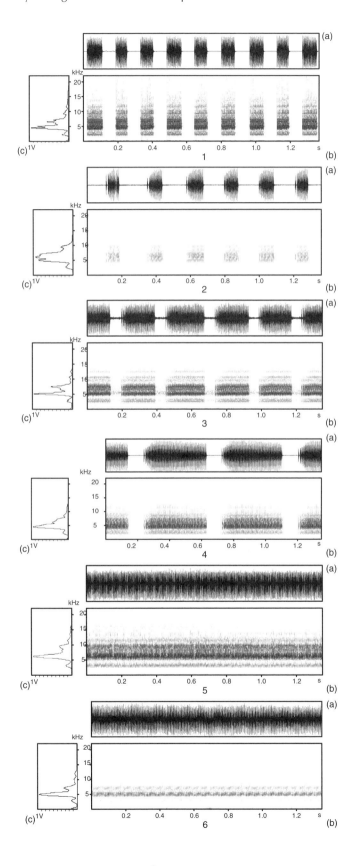

south-eastern Europe (Greece and Turkey) to group apart from those of the west (Iberian Peninsula and France) (Quartau *et al.*, 1999; Pinto-Juma *et al.*, 2005) (Figure 17.8). The interval between echemes (and consequently the number of echemes per second) was the acoustic signal parameter that contributed most to this separation. The calls proved to have interecheme intervals of much greater duration in the south-eastern than in the western populations (*cf.* Figure 17.7.1 and 17.7.2 for western and eastern populations, respectively). Besides this variable, males from the south-eastern area showed some significant differences in relation to the remaining studied regions in almost every acoustic variable (Pinto-Juma *et al.*, 2005). It seems, therefore, that there is geographic variation in the acoustic signals of *C. orni*, at least for widely separated populations, a result which does not appear to conform with the Paterson prediction that the SMRSs would be largely invariant within species (Paterson, 1985; Quartau *et al.*, 1999, and Pinto-Juma *et al.*, 2004; but see also Claridge, 1985b; Claridge *et al.*, 1988).

SPECIES OF *CICADA* IN EASTERN MEDITERRANEAN EUROPE

In eastern Mediterranean Europe, species of *Cicada* occur usually in high densities during summer time from mid-June until October. Their habitats consist usually of open woodlands and males sing in general on trees such as *Olea europaea, Pinus halepensis* and *Cupressus* sp. (Quartau and Simões, submitted) (Figure 17.9).

Six nominal *Cicada* species are known so far from this area: *Cicada cerisyi, C. cretensis, C. lodosi, C. mordoganensis, C. orni* and *C. permagna*. As mentioned above, very little taxonomic and biological data are available for *C. permagna* and *C cerisyi*. In respect to *C. permagna,* the type locality (Antalya: Alanya, Turkey) was investigated during the summer of 2003, but no specimens were found. As such, the only information available here was from a single specimen, in poor condition, lent by Prof. Michel Boulard (Muséum national d'Histoire Naturelle, Paris, France); Concerning *C. cerisyi,* only the data from the literature were available. As such and as stated in the introduction, these two species were not considered in the present study.

As in the western pair of species, these four are very similar in morphology (Figure 17.10 to Figure 17.13), with the exception of *C. lodosi,* which is bigger than the others and the general colour is more greyish. On the other hand, *C. mordoganensis* is brownish, slightly bigger than *C. orni,* smaller than *C. lodosi* and also with the four conspicuous wing spots, yet not so well defined as in *C. orni*. Moreover, *C. cretensis* is very similar to *C. mordoganensis* in body colour and size, and with four spots as in *C. orni* (Quartau and Simões, submitted). With respect to the tymbal, there are apparently no detectable differences within this group of species.

Concerning the male genitalia, *C. lodosi* is also the only species easily separated from the remainder by its bigger size and the small apical pygophore spine (Figure 17.6d). Conversely, this is not the case of the remaining species, *C. orni, C. cretensis* and *C. mordoganensis*, which show no clear differentiation (Figure 17.6b, c and e).

C. orni is the species with the widest distribution, stretching from Portugal in the west to Greece, Turkey and some countries in the Near East and around the Black Sea. *C. mordoganensis* is an eastern species found in Turkey and in several Greek islands along the Turkish coast. No overlap between this species and *C. orni* has yet been found. *C. cretensis* is the only *Cicada* species occurring on Crete. Earlier records referring *C. orni* (Nast, 1972) or *C. mordoganensis* (Dlabola, 1984) to this island are undoubtedly misidentifications. Dlabola (*op. cit.*) also restricted *C. barbara* to north Africa only and did not refer to *C. orni* in Portugal. In respect to *C. lodosi,* it is present in Turkey, on the oriental part of the Mediterranean area and in sympatry with *C. mordoganensis*.

As with *C. orni* and *C. barbara*, the calling song profiles are the easiest way to discriminate and identify this eastern group of species (Figure 17.7.2 to 17.7.4 and Figure 17.7.6). *C. lodosi*, like *C. barbara*, has a continuous call pattern, while *C. cretensis, C. mordoganensis* and *C. orni* have signals with discontinuous pattern made up of echemes with characteristic interecheme intervals.

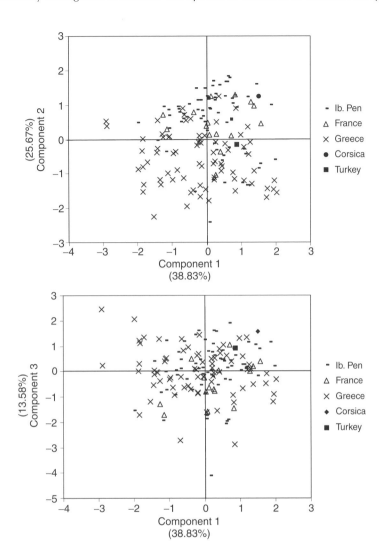

FIGURE 17.8 Bidimensional diagrams of relationships between specimens of *C. orni* (176 OTUs) of a principal component analysis based on a correlation matrix between 13 acoustic characters (after Pinto-Juma *et al.*, 2005).

FIGURE 17.9 The typical south-eastern Mediterranean habitat where the *Cicada* species occur. (Peloponnessus, Greece; Photo P.C. Simões. With permission.)

FIGURE 17.10 *Cicada cretensis.* (Crete, Greece; Photo P.C. Simões. With permission.)

The time and frequency-based analyses of the songs of these species showed that they differ in their temporal parameters, even at similar temperatures. Frequency characteristics for all are similar, discrimination being possible only through the duration of the echemes and interecheme intervals (Table 17.1).

Generally, echeme duration is shorter in *C. orni*, intermediate in *C. cretensis* and longer in *C. mordoganensis* (means of 0.07 s, 0.22 s and 0.32 s, respectively, and for the samples considered in Table 17.1). On the other hand, *C. cretensis* has the smallest interecheme interval (Quartau and Simões, submitted): mean of 0.10 s in relation to 0.19 s in *C. orni* and 0.13 s in *C. mordoganensis* and in respect also of the same samples considered in Table 17.1.

DISCUSSION AND GENERAL CONCLUSIONS

As typical in cicadas and other sound-producing insects, the calling song is the most common acoustic signal and is produced by males in order mainly to attract females which, in turn, are only attracted to the calls of conspecific males. Its structure is, therefore, an important component of the

FIGURE 17.11 *Cicada lodosi.* (Gundogmus, Turkey; Photo J.A. Quartau. With permission.)

FIGURE 17.12 *Cicada mordoganensis.* (Ikaria, Greece; Photo J.A. Quartau. With permission.)

SMRS, and is probably a major factor responsible for the premating reproductive isolation between closely related species of cicadas. As such, the analyses of the calling songs provide important taxonomic information at a specific level, namely, for deciding on the status of allopatric populations showing small morphological differences or in detecting sibling species. This is indeed so for species of *Cicada*, where the male calls are the best diagnostic characters for species delimitation and discrimination (*e.g.* Boulard, 1995; Quartau, 1995; Simões *et al.*, 2000; Quartau and Simões, submitted). On the other hand, the analysis of the song variables of these calls may also have strong microevolutionary interest by providing invaluable information on the speciation process of closely related sound-producing species such as the present cicadas. Alterations in the echeme duration and to a lesser extent also in the interecheme interval of the calling song might have been crucial in the speciation process among these species in the Mediterranean basin. It is clear that details cannot be known until relevant phylogeographic information is additionally acquired (Pinto-Juma and Quartau, in preparation). So we need to combine acoustic, morphological and biogeographic data with the new results from molecular phylogenetic investigations now in progress (Quartau *et al.*, in preparation). However, the extant acoustic and biogeographic knowledge already gathered allows one to garner some general perspectives on this issue: present evidence points to the fact that during the Pleistocene, and possibly earlier, some North African *Cicada* ancestor resembling the present-day existing species with a continuous signal, such as in the modern *C. barbara*, might have been split into several disjunct populations, thus providing opportunities for acoustic differentiation and speciation. This would have given rise to the closest living descendants with a continuous signal, which is assumed to be the plesiomorphic state, as well as to the present-day cluster of species with the derived discontinuous time pattern, where the

FIGURE 17.13 *Cicada orni.* (Athens, Greece; Photo P.C. Simões. With permission.)

TABLE 17.1

Descriptive Statistics of 13 Acoustic Parameters in the Samples of *Cicada cretensis*, *C. orni*, and *C. mordoganensis* Investigated (Original Data)

	N	Mean ± SD	Range
***C. cretensis* (Crete, Greece)**			
No. echemes/s	20	3.25 ± 0.58	2.15 to 4.45
Echeme duration	20	0.22 ± 0.04	0.16 to 0.28
Interecheme interval	20	0.10 ± 0.04	0.05 to 0.19
Echeme period	20	0.32 ± 0.06	0.23 to 0.46
Ratio echeme/interech. interval	20	2.85 ± 1.18	1.50 to 6.20
Peak frequency	20	5355.16 ± 335.93	5056.08 to 6475.40
Band width (−20 dB)	20	8256.63 ± 989.90	6969.85 to 10072.50
Quartile 25%	20	5246.04 ± 148.26	4915.00 to 5605.71
Quartile 50%	20	6116.68 ± 224.68	5595.00 to 6468.02
Quartile 75%	20	7683.55 ± 357.85	6949.40 to 8247.50
Quartile 75% to quartile 25%	20	2437.51 ± 285.57	1656.00 to 2900.91
Minimum frequency	20	2378.96 ± 180.15	1921.89 to 2758.57
Maximum frequency	20	10640.88 ± 1015.03	9280.46 to 12596.82
***C. orni* (Greece and Turkey)**			
No. echemes/s	103	4.15 ± 1.13	2.23 to 7.08
Echeme duration	103	0.07 ± 0.02	0.04 to 0.17
Interecheme interval	103	0.19 ± 0.07	0.06 to 0.35
Echeme period	103	0.26 ± 0.07	0.14 to 0.45
Ratio echeme/interech. interval	103	0.52 ± 0.37	0.21 to 2.48
Peak frequency	103	5015.36 ± 507.28	4061.71 to 6584.29
Band width (−20 dB)	103	7565.95 ± 1400.63	4391.51 to 11280.30
Quartile 25%	103	4817.92 ± 359.92	4047.52 to 5784.14
Quartile 50%	103	5687.44 ± 442.46	4812.96 to 6640.51
Quartile 75%	103	7211.54 ± 849.99	5597.78 to 9948.06
Quartile 75% to quartile 25%	103	2393.62 ± 669.89	1257.78 to 5205.60
Minimum frequency	103	2179.28 ± 289.94	1244.27 to 3235.48
Maximum frequency	103	9751.75 ± 1432.80	6966.84 to 13347.91
***C. mordoganensis* (Greece and Turkey)**			
No. echemes/s	84	2.26 ± 0.29	1.58 to 3.03
Echeme duration	84	0.32 ± 0.26	0.19 to 0.49
Interecheme interval	84	0.13 ± 0.03	0.08 to 0.27
Echeme period	84	0.45 ± 0.06	0.33 to 0.63
Ratio echeme/interech. interval	84	2.62 ± 0.82	1.20 to 5.59
Peak frequency	84	4442.31 ± 296.92	3716.35 to 5106.98
Band width (−20 dB)	84	6175.66 ± 1282.01	3132.40 to 10355.00
Quartile 25%	84	4450.83 ± 254.49	3781.06 to 4921.40
Quartile 50%	84	5070.65 ± 296.71	4376.00 to 5645.86
Quartile 75%	84	6472.28 ± 576.17	5360.59 to 7739.29
Quartile 75% to quartile 25%	84	2021.45 ± 488.77	960.33 to 3792.19
Minimum frequency	84	2238.04 ± 427.44	1441.29 to 3903.21
Maximum frequency	84	8418.66 ± 1163.82	6549.36 to 12392.86

Note: The time and frequency parameters are in seconds and in Hz, respectively. N = number of specimens analysed; SD = standard deviation.

duration of the echemes and the interecheme intervals might have diverged independently in small isolated populations. From studies on a wide variety of other organisms, it is generally agreed that southern Mediterranean refugia must have allowed small populations to survive and diverge during the Pleistocene glacial cycles and possibly also during late Pliocene (*e.g.* Avise and Walker, 1998; Hewitt, 2000; Paulo *et al.*, 2001). Two main areas were important in this scenario: the first was located in the western part of the Mediterranean area (Iberian Peninsula and north-western Africa), where the present-day *C. barbara* and *C. orni* coexist. The second encompasses a larger assemblage of present-day closely related species — *C. cretensis*, *C. lodosi*, *C. orni*, and *C. mordoganensis* — and was located on what are now the Balkans, the Aegean islands and the Turkish mainland in the eastern part of the Mediterranean basin.

ACKNOWLEDGEMENTS

For help in the field thanks are due to Genage André (in Portugal), to Sakis Drosopoulos, Penelope Tsakalou (in Greece) and to Irfan Kandemir (in Turkey). Moreover, both Gabriela Pinto-Juma and Sofia Seabra (Department of Animal Biology, Faculty of Science, University of Lisbon) are acknowledged for critical reading of the manuscript. Finally, we also thank Michel Boulard (Muséum National d'Histoire Naturelle, Paris, France) for loan of cicada material and Michael Wilson (National Museum of Wales, Cardiff, U.K.) for loan of literature.

18 Acoustic Communication, Mating Behaviour and Hybridisation in Closely Related, Pseudogamic and Parthenogenetic Species of Planthoppers

Sakis Drosopoulos

CONTENTS

INTRODUCTION

It would be a failure not to include here an account of the work on reproductive isolation in closely related species and their gynogenetically reproducing "species" of planthoppers (Hemiptera-Delphacidae), which started about 30 years ago and still continues. In my opinion, planthoppers provide one of the best insect groups for studies on systematics, ecology–ethology, physiology, cytogenetics, which forms a distinct branch of biology and "biosystematics", a term widely used by many workers since the 1940s. After so many years this term is now widely misunderstood and misinterpreted so that many biologists use it as synonymous with systematics. As a result much work published under this title has not been widely cited. I take the opportunity here to say that biosystematics is the sum of biological knowledge which can be used in separating species morphologically difficult to distinguish. In all examples where it has been used in the study of Hemiptera, it has resulted in fundamental discoveries. A recent example is a 15-year programme on the spittle bug, *Philaenus*, where a revolution has occurred in what we understand of the colour polymorphism (Drosopoulos, 2003). This was purely due to the fact that taxonomists thought that the genus comprised only two species world-wide, while now there are at least eight species recognised in the Mediterranean region alone and many more are waiting to be discovered, particularly by the application of molecular methods.

The same happened when I initiated field and laboratory work in 1972 at Wageningen, the Netherlands, by studying over six years as many biological differences as possible between two closely related species of *Muellerianella*. This genus was chosen because among 150 species sampled quantitatively at intervals of two weeks, one, *M. fairmairei*, exhibited clearly an abnormal sex ratio in favour of females. Further faunistic investigations in the Netherlands, and later in central Europe, showed that in these regions all populations had an excess of females (75 to 100%). However in Greece and some other Mediterranean countries, it displayed a normal sex ratio (1:1) as does the other species, *M. brevipennis*, in the Netherlands and elsewhere in Europe (Drosopoulos, 1977). During this period several biological studies on females showed clearly that the high proportion of females in *M. fairmairei* was due to the fact that two morphologically indistinguishable types of females coexisted (Drosopoulos, 1976). One of these types consisted of diploid females, producing females and males in equal proportion. The other type was triploid and pseudogamous (not parthenogenetic), reproducing repeatedly triploid pseudogamous females in succeeding generations, but only after copulation with males produced by the normal diploid females. Copulations with sterile hybrids were not effective (Drosopoulos, 1976, 1977, 1978). Sperm therefore is necessary to initiate embryogenesis, but it does not contribute hereditary material to the egg nucleus of pseudogamous females. This enigmatic and paradoxical phenomenon in biology is difficult to detect in invertebrates since pseudogamous females usually are not different morphologically from normal females with which they coexist. Only tedious caryological analysis of the egg nucleus reveals the presence of these organisms (Drosopoulos, 1976).

In other organisms pseudogamy (or gynogenesis) has also been found, both among invertebrates, for example, fresh water triclads (Benazzi and Lentati, 1966), the oligochaete worm, *Lumbricillus* (Christensen and O'Connor, 1958) a Ptinid beetle (Moore *et al.*, 1956), and in vertebrates, for example, fish (Schultz and Kallman, 1968; Schultz, 1969) and salamanders (MacGregor and Uzzell, 1964). These findings demonstrate that pseudogamy is not associated with a particular group of animals but is widely spread among various groups. In addition there are strong indications that in several cases of pseudogamy the phenomenon was at first wrongly interpreted as embryonic or larval male mortality or as an unbalanced sex determining mechanism, *etc.*

This scientifically unique material was naturally further studied by selected postgraduate students at Wageningen and to a lesser degree by the author in Greece (Drosopoulos, 1983). These studies reveal further that two more bisexual species existed in Europe, one of them taken out of its synonymy (*M. extrusa* Scott, 1871) by Booij (1981), morphologically very closely related to *M. fairmairei*, but using only a different grass host, *Molinia caerulea*, than the two then known species, *M. fairmairei* on *Holcus lanatus* and *M. brevipennis* on *Deschampsia caespitosa*. Simultaneously, Drosopoulos (1983) discovered another triploid pseudogamic female population coexisting with a bisexual species, almost identical morphologically to *M. extrusa*, at a locality in NW Greece, but feeding on a different food plant. We decided then to call it a race of *M. extrusa*, but two years later true *M. extrusa* consisting of both male and females in equal number were discovered also on *M. caerulea* (Drosopoulos, unpublished). Thus in a relatively small area, Epirus–Greece, four bisexual species (one of which still needs to be named) and another pseudogamic species existed. Such pseudogamic species were found both by Drosopoulos (unpublished) in northern Greece and in west Wales, U.K., *H. mollis* and *D. flexuosa*, by Booij and Guldemond (1984). These authors even found associations of pseudogamic with the biparentally reproducing species, *M. brevipennis*. It is now questionable whether these associations are related to the bisexual *M. brevipennis* or to another closely related species, as is the Russian species, *M. relicta*, and the undescribed Greek one. In addition, one species under the name *Delphacodes laminalis* (Van Duzee) was redescribed morphologically by Du Bose (1960) from

North America, but certainly this species belongs to the genus *Muellerianella*. Du Bose states that this species is widely distributed in many states of the U.S.A. and Canada, but gives no ecological data. However, these findings are very important since they prove that *Muellerianella* is widely distributed in the whole northern hemisphere between 30° and 65°N (Drosopoulos, 1977).

Finally, there is no clear association of pseudogamic triploid females with *M. extrusa*, which in the laboratory can maintain themselves by using four sperm donor species biparentally reproducing while in nature they use only one. I fully agree with Booij (1982a) who, after his and Drosopoulos' (1978) hybridisation studies, stated, "As far as I know such a complex has not been reported in nature before".

The above details are necessary for the reader of this book to follow the role of acoustic communication and mating behaviour in reproductive isolation between this species complex. Including similar studies made by collaborators and the author on two other planthopper species complexes of the genera, *Ribautodelphax* and *Delphacodes*, the author believes that a review such as that made here should have been done earlier.

ACOUSTIC COMMUNICATION IN *MUELLERIANELLA* SPECIES

Initially acoustic studies were made with primitive equipment by a student in Wageningen in 1975 (Houwink, unpublished report). At that time recordings were made only of the males and females of the two bisexual species, their male and female hybrids, and the then only known pseudogamic female associated with *M. fairmairei* (Drosopoulos, 1985). Further studies were made on the same species initially, and in addition on *M. extrusa* and the male calling songs of sympatric and allopatric populations of *M. fairmairei* and *M. brevipennis* (Booij, 1982b). Here I will refer briefly to the main findings of these studies. Details are in Booij (1982b). However, Booij did not report on the acoustic signals of sterile hybrids from various crossings between the bisexual species. I report here that male hybrids are produced especially readily in the laboratory, and are very active and copulate more easily with parental females, maybe due to their intermediate songs and genitalia (Drosopoulos, 1985). Considering here only the calling songs emitted by both sexes for pair-forming, there are clear differences among the three species (Figure 18.1). These differences are more pronounced in males than in females. Thus, the songs are more characteristic for each species and more readily distinguished than the male genitalia (Figure 18.1). Differences between populations were smaller and variable when measured at the same temperatures ($\approx 25°C$). In the field, temperature, light intensity and humidity were variable, and it is then very difficult to conclude that this variability is of great significance. It is clear however, that, as for example in genitalial characters, the calling songs of *M. brevipennis* are more differentiated than those of the other two bisexual species, *M. fairmairei* and *M. extrusa*, as will be shown below also for interspecific crosses.

INTRASPECIFIC BEHAVIOUR IN *MUELLERIANELLA* SPECIES

Detailed studies on reproductive isolating mechanisms are rare. Despite acoustic differences, our studies on postmating isolation were the most fascinating ones (Booij, 1982b; Drosopoulos, 1985). In the mid-1970s I spent several months, for more than three hours daily, observing all combinations of intraspecific and interspecific mating behaviour and crossings of the two bisexual species, the pseudogamic and the hybrids between them. Most were done in pair crossings and subsequently insemination embryogenesis, larval development, chromosome analysis of the egg nucleus and male spermatogenesis, and adult progeny were measured. Booij (1982a) also followed the same techniques, especially in crossing the third bisexual species, *M. extrusa*, with the other two. Very crucial and supportive for these studies were others related to the life history, ecology, physiology, geographic distribution and genetics of all species, both under field and under the same conditions of laboratory rearings for all species (Table 18.1).

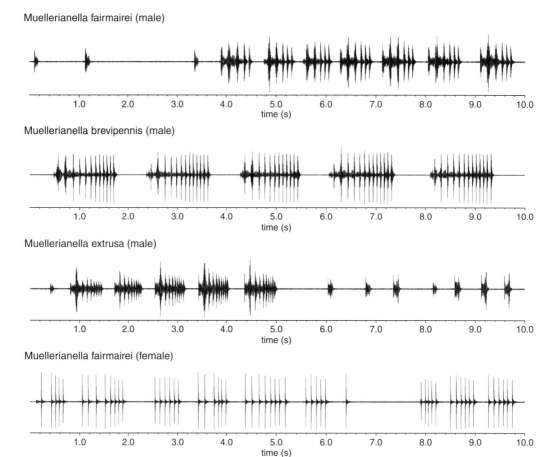

FIGURE 18.1 Male and female calls from *Muellerianella* species. (Oscillograms from recordings by C.J.H. Booij.)

Concerning mating behaviour, sexual maturity, precopulatory behaviour, copulation, frequency of copulation, duration of copulation and competition of males were accurately measured both in intra- and interspecific crosses. The main results were:

1. In intraspecific crosses, copulations started clearly earlier in male and females of *M. fairmairei* than in *M. brevipennis*.
2. Whenever females of *M. fairmairei* call for a male or respond to a calling male, copulation follows shortly, but in *M. brevipennis*, especially in virgin females, there is a threshold to overcome between readiness to be courted and to copulate despite continuous courtship. In males courtship before copulation (emission of courtship songs and rapid raising of the wings) was more intense in *M. fairmairei* than in *M. brevipennis*.
3. Copulation in both species occurred in a similar manner (intensification of courtship, female raising the abdomen and males turning about 180° with raised genital segment and parameters through the ovipositor lead the aedeagus into the female genital organ).
4. Frequency of copulation. Females of both species, after mating once, can produce fertile eggs during their entire oviposition period. Despite this, females readily copulate several times before, during and after their oviposition period. Males and females of *M. fairmairei* copulated significantly more frequently than those of *M. brevipennis*.

TABLE 18.1
Differences between Two *Muellerianella* Species

	M. fairmairei		*M. brevipennis*
Characteristics	Bisexual	Unisexual	Bisexual
Morphology	♂ different	—	♂ different
Chromosome number	$2n = 28$	$3n = 41$	$2n = 28$
Sex chromosome (♂)	trivalent	—	Heteromorphic bivalent
Range	European-wide	W. and C. Europe	Western Siberian
Food plants in field	*Holcus lanatus* and *H. mollis*		*Deschampsia caespitose*
Food plants in lab	Oligophagous		Monophagous
Oviposition Plants			
Long day	*H. lanatus*		*D. caespitosa*
Short day	*J. effusus*		*D. caespitosa*
Habitats	Common in wet natural meadows		Stenotopic in wooded places
Phenology			
In the Netherlands	Bivoltine	Bivoltine	Partly bivoltine
In Greece	Polyvoltine	—	Univoltine
Wing Dimorphism			
Long winged under long photoperiod)	± 25%	± 15%	± 60%
Sex ratio (♂:♀)	1:1	0:1	1:1
Reproduction			
Sexual maturity (♂)	Earlier	—	Later
Sexual maturity (♀)	Earlier		Later
Preoviposition period	Shorter		Longer
Fecundity	High	Higher	Lower
Reproduction rate	High	Higher	Lower
Longevity (♂-♀)	♂♂ > ♀♀	—	♂♂ ≈ ♀♀
Diapause			
Induction	Later		Earlier
Intensity	Ateleo-oligopause		Teleo-oligopause
Termination	Earlier		Later
Short-day rearing	Possible		Impossible
Acoustic communication (signals)	♂ different, ♀♀ the same		♂♂ different, ♀♀ little different
Mating Behaviour			
Frequency of copulation	Higher		Lower
Duration of copulation	Shorter		Longer
Interspecific crossings	♀♀ easier		♀♀ difficult

5. Duration of copulation. In both species females were found to experimentally control the duration of copulation ranging from a few minutes up to one and a half hours. Females of a Dutch colony mated by males of either a Greek or Dutch origin appeared to copulate significantly longer than females from Greek colonies. However, females of *M. brevipennis* copulated even longer than *M. fairmairei*. In both species there was a tendency to increasing time of copulation as the days between two matings increased (statistically significant correlation in *M. fairmairei* but not in *M. brevipennis*).

6. Post copulatory behaviour. It was found that males of *M. fairmairei* start mating behaviour with another female in a shorter time after copulation than *M. brevipennis*. Mated females of both species may start laying eggs only a few minutes after finding a suitable oviposition site.

7. Competition. When males of *M. fairmairei*, which had emerged on the same date, were placed in a tube with one female, competition was not only obvious but violent, even when one of them accidentally happened to copulate first. Such competition was stopped when in a short time the dominant male had twice copulated with one more female. No corresponding competition was observed in males of *M. brevipennis*. Females with one male, either of *M. fairmairei* or *M. brevipennis*, never competed for the male.

It may be concluded that *M. fairmairei* is more active than *M. brevipennis* in all of the above mating behaviour activities. This is fundamental for interspecific crosses as described below. I will now refer to the third species, *M. extrusa*, which was taken out of its synonymy with *M. fairmairei* by Booij (1981). This species appeared from morphological and bioacoustics differences to be more closely related to *M. fairmairei* than to *M. brevipennis*, and this was the reason that it was synonymised with *M. fairmairei*. This species was discovered independently at about the same time feeding on the same food plant (*Molinia caerulea*) in several places in northwestern Europe and Greece, and its sex ratio was normal. Booij (1982a, b) noted that *M. extrusa*, with regard to ecology, voltinism–diapause and oviposition plants, resembled most *M. brevipennis*. However, morphologically, acoustically and genetically it is more related to *M. fairmairei*. Unfortunately, Booij did not focus on hybridisation studies between *M. extrusa* and the other two bisexual species *M. fairmairei* and *M. brevipennis*. Instead of trying to insert a third genome into the crosses of *M. fairmairei* with *M. brevipennis* and all other combinations he repeated several crosses, the synthesis that was made twice of *M. fairmairei–brevipennis* (Drosopoulos, 1978). Thus, chronologically, Kees Booij in the autumn of 1978 found and separated *M. extrusa* as a distinct species. Drosopoulos in 1980 discovered a new species (presently unnamed) associated with another triploid pseudogamic female most probably of hybrid origin between the new species and *M. fairmairei*, which exists in the same area but only as bisexual species. Finally in the summer of 1983 *M. extrusa* was found also in that area of Greece on its host and oviposition plant.

INTERSPECIFIC MATING BEHAVIOUR IN *MUELLERIANELLA* SPECIES

As in intraspecific, so also in interspecific, cross mating behaviour was studied in test tubes so that female progeny could be tested: (a) by dissecting spermathecae, (b) by embryonic development during oviposition period and (c) by measuring the number of adult progeny. The mating behaviour of interspecific crosses was similar to intraspecific and females accepted courtship from all males. However, in these crosses when copulations occurred, they appeared to be more difficult and of shorter duration than in intraspecific ones. As in intraspecific crosses males and females of *M. fairmairei* were more active in copulating than *M. brevipennis*. Remarkably half the number of females of one species mated even when they were caged with both sexes of the other species, while all conspecific females and pseudogamic ones were mated. This is fundamental and strongly suggests that hybridisation may occur in mixed populations of the two species in the field. No differences were found when females of *M. brevipennis* were crossed either by sympatric (Dutch) or far away allopatric (Greek) males of *M. fairmairei*.

It must be mentioned here that crosses between the two species were started in 1973 and went on until 1977 (Drosopoulos, 1985). The most important methods used were described by Drosopoulos (1977, 1985). The tedious original experiments were made to test the frequency of interspecific inseminations (Figure 18.2). In this figure is demonstrated the process of egg fertility

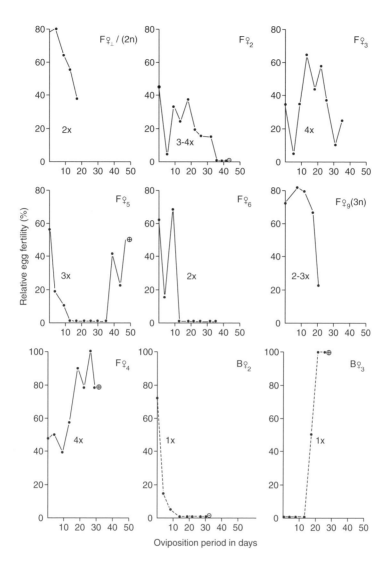

FIGURE 18.2 The fertility of eggs deposited by each female of *M. fairmairei* (solid line) and *M. brevipennis* (dashed line) during oviposition period, when each female was reared in a test tube together with a male of the other species. The symbols ($\oplus$, $\ominus$) indicate whether females had or had no sperm in their spermathecae, respectively. $\times$ indicates times of possibly successful copulations. (From Drosopoulos, S. *Meded. Landbouwhogeschool Wageningen*, 77, 1–133, 1977. With permission.)

during the oviposition period and the times of females they were inseminated. It must also be noted here that females of all *Muellerianella* species can oviposit even without prior insemination.

Back crosses of any F$_1$ hybrid females with males of the parental species were made more readily than interspecific crosses. Even hybrid males (all sterile) copulated freely with the females of the parental species. This suggests that the intermediate genitalia of the hybrids (as in acoustic signals) are more suitable for this. The fact that sterile males are so active in copulating without contributing to reproduction (hybrid vigour) is remarkable.

An important finding also in interspecific crosses was that females of one species that could not copulate with a certain male could do so with another. Worth mentioning here also is that hybrid progeny of all combinations of crosses between *M. fairmairei* and *M. brevipennis* appeared under the same conditions (20 to 25°C) at least two times longer than that of the parental species. The same happened also when crossing back the hybrid females with the parental males. The sex ratio of the hybrid progeny was variable. Some hybrid progeny originated by crossing only one female of *M. brevipennis* with one male of *M. fairmairei* and consisted of eight males and 136 females, while some other crosses produced only female hybrids.

As expected from morphological and bioacoustics data, hybridisation between *M. extrusa* and *M. fairmairei* was easy and even fertile males were produced. Some crossings of the hybrid males with females produced F_2 which appeared to be fertile despite low fecundity (Booij, 1982b). Crosses between *M. extrusa* and *M. brevipennis* are less easy and the progeny is small. Hybrid males produced from such crosses were all sterile. Finally, all three bisexual species can serve as sperm donors, but the triploid females associated in nature with *M. fairmairei*, when crossed in the laboratory, also gave the larger numbers of progeny.

ACOUSTIC COMMUNICATION IN *RIBAUTODELPHAX* SPECIES

Initially *Ribautodelphax* was suggested to me early in the 1970s as more problematic in morphology than *Muellerianella*. Rearings of this genus had started, but *Muellerianella* was more than enough for biosystematic studies over six years by me and another five years by Kees Booij. Later this genus was taken up by K. Den Bieman at Wageningen. After five years of intensive investigations on biosystematics in a similar manner as for *Muellerianella*, he finally discovered 10 bisexual European species and two, surprisingly, also pseudogamic triploid associations with only two bisexual ones (Den Bieman, 1987). Using the same methods of recording as in *Muellerianella* (De Vrijer, 1984), the calling songs of 11 species and the pseudogamic triploid females, it appeared that males and females could be separated by clear acoustic differences (Figure 18.3). Analysis of these revealed that one "out group species", *R. albostriatus*, differed from another species, *R. pallens*. Den Bieman defined all the other bisexual species as belonging to a super species complex, *R. collinus*, while the "out group" species mentioned above are composed of one species each.

As in morphological characters, also in bioacoustics ones, *R. pallens* resembled more the *R. collinus* species group than *R. albostriatus*. Additional studies using allozyme polymorphism data supported this relationship (Den Bieman and Eggers-Schumacher, 1987), but resulted also in some additional subgroups. One of two triploid females, despite some variation in their calling songs, also did not differ in this aspect from diploid females with which they coexisted in the field.

INTERSPECIFIC BEHAVIOUR IN *RIBAUTODELPHAX* SPECIES

If *Muellerianella* may be considered as an unusual complex of species consisting so far of four clear bisexual and two pseudogamic ones, the situation in *Ribautodelphax* is even more complicated. It was known earlier that taxonomic problems existed in clearing the species before biosystematic studies, including bioacoustics analyses, could be started (Den Bieman, 1986). Thus, he had to face a big problem, so much so that, in my opinion, it was impossible to make detailed interspecific behaviour studies on so many species in such a short time. However, the discovery of two pseudogamic species, exactly as in *Muellerianella*, was really unexpected since no excess of females or other morphological indicators had been detected in field populations. Den Bieman, using parallel methods to those used for studies on *Muellerianella*, on *Ribautodelphax* species across a wide European range, exposed a very complex situation that required more detailed studies. The dendrogram of electrophoretic data, based on Nei's D over five loci by Den Bieman and Eggers-Schumacher (1987), suggested that investigations should be focused on three new species groups: *R. imitans*, *R. colinus* and *R. pungens*, which, respectively,

R. collinus complex – male calling signals

Ribautodelphax pungens (male)

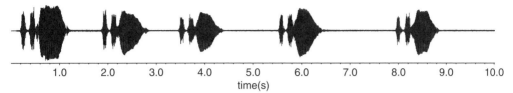

Ribautodelphax imitans (male)

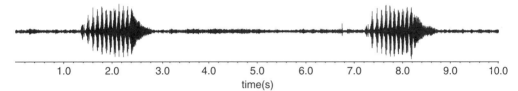

Ribautodelphax collinus (male)

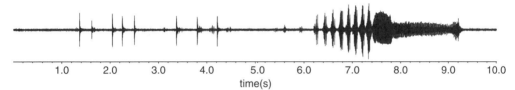

Ribautodelphax angulosus (male)

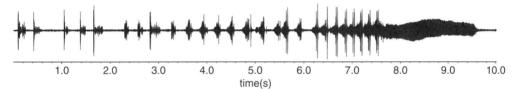

Ribautodelphax vinealis (male)

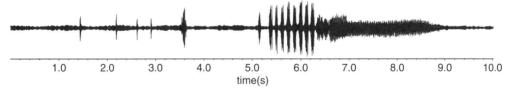

Ribautodelphax ventouxianus (male)

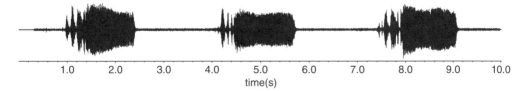

FIGURE 18.3 Male and female calls from *Ribautodelphax* species. (Oscillograms from recordings by C.F.M. Den Bieman.)

R. collinus complex – female calling signals

Ribautodelphax pungens (female 2n)

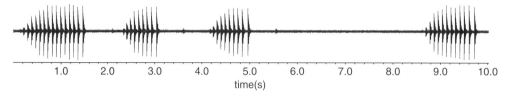

time(s)

Ribautodelphax pungens (female 3n)

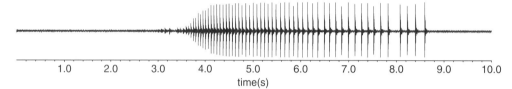

time(s)

Ribautodelphax imitans (female)

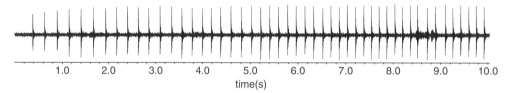

time(s)

Ribautodelphax collinus (female)

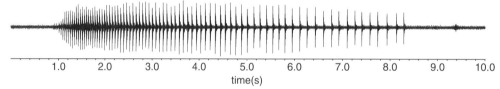

time(s)

Ribautodelphax angulosus (female)

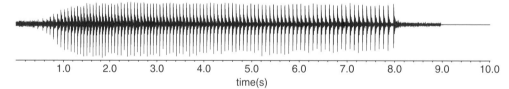

time(s)

Ribautodelphax vinealis (female)

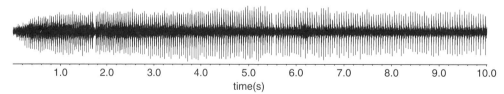

time(s)

FIGURE 18.3 Continued.

Other Ribautodelphax species

Ribautodelphax pallens (male)

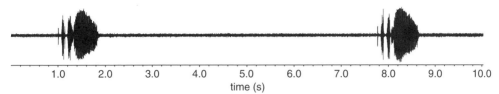

Ribautodelphax albostriatus (male)

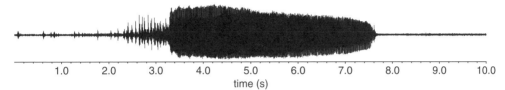

Ribautodelphax pallens (female)

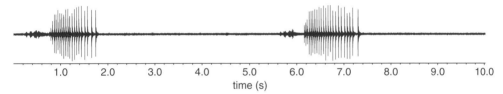

Ribautodelphax albostriatus (female)

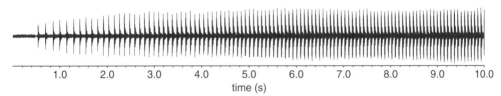

FIGURE 18.3 Continued.

include five species with one pseudogamic, four species and two species with one pseudogamic. The two remaining species, *R. albostriatus* and *R. pallens*, are well differentiated from the others more or less like *M. brevipennis* in the *Muellerianella* species complex, also lacking any pseudogamic females association. The situation in these species groups had to be investigated first by intraspecific pair crosses, then interspecific crosses of the cleared species of each species group and finally interspecific between species of the three species groups. Thus interspecific crosses made by Den Bieman (1986) of all species he defined in his extended studies on the complicated *R. collinus* species group, in my opinion, makes an analysis for any group data (even acoustic calling signals, electrophoresis and morphology) not sufficiently detailed to make clear conclusions. I cite here an example where Den Bieman (1986) reported that he found active mate discrimination by males against pseudogamous triploid females resulting in low insemination ratios of triploid females in field populations with high triploid frequencies. If this is true, these triploid females should be of hybrid origin and not a result of autopolyploidy, as he suggested elsewhere. In addition, the cultures of some of his species reared in the laboratory do not exclude the possibility that they were already hybrids in the field or mixed populations of clearly related species which hybridised in the rearing cages (see also Strübing and Drosopoulos,

Chapter 19). In my opinion Den Bieman supported autopolyploidy in many parts of his studies too strongly and with little evidence.

Conclusions here could be made that hybridisation, as in *Muellerianella*, can be done easily under laboratory conditions and there is evidence that this also happens in the field (Drosopoulos, unpublished data).

ACOUSTIC COMMUNICATION IN *DELPHACODES* SPECIES

Extensive field investigations in Greece on *Delphacodes* suggested that it also consisted of several related species and two all female populations, one in Florina province (NW region) and another north of Thessaloniki. In the first locality several samples taken exclusively on *Carex riparia* were all females, while in the second locality on another food plant one sample also revealed only females, despite intensive searching. My collaborators, Den Bieman and De Vrijer (1987), offered to investigate the first all female population by rearing it in the laboratory, initially on *Carex riparia* and later on *Avena sativa* as a substitute food plant. After several generations in the laboratory the insects appeared to be truly parthenogenetic. This was the first report of parthenogenesis in the family Delphacidae and only the second time in the whole of the Auchenorrhyncha. Cytological investigation showed that these females were also triploid. Most surprising was that this parthenogenetically reproducing species emitted calling songs and responded to the male songs of the bisexual species, *D. capnodes* (Scott), which occurs both in Greece and the Netherlands and which we had thought it most resembled in morphology. Recording the calling songs of males and females of *D. capnodes* and the parthenogenetic species showed them to be very similar. Also the two *Delphacodes* females responded equally to the calling songs of the male (Figure 18.4). Copulation between the parthenogenetic species and the males was performed after the exchange of a few acoustic signals and lasted from four to ten minutes. By checking the spermatheca of the copulated triploid females it appeared that successful insemination had occurred, but surprisingly the progeny of these females were again all females and triploids, as in pseudogamous females of *Muellerianella* and *Ribautodelphax*.

DISCUSSION AND PHILOSOPHICAL CONSIDERATIONS

Results from these investigations together with similar studies on other animal species (*e.g.* salamanders, fish, birds) provide me with sufficient data to discuss some general evolutionary topics. Earlier statements that pseudogamy is the necessary prerequisite for parthenogenesis is very much supported here as an important evolutionary process.

In extensive faunistic studies over the whole area of Greece, we reasonably concluded that more than 95% of the planthopper species and their food plants are now known (Drosopoulos *et al.*, 1985). Among them, *Delphacodes* comprises a complex of more than four bisexual species plus the parthenogenetic one. There is strong evidence that the other female population of the genus from Thessaloniki consists of an additional unknown species associated with pseudogamic females. These studies, initially on pseudogamy and later on parthenogenesis, are in agreement with the views of Schultz (1969, 1973) and Norman Davis (Connecticut, personal communication). These workers agreed that, concerning hybridisation, unisexuality and polyploidy, the planthopper material was clearer even than the teleost fish species complex, *Poeciliopsis*. The suggestion was made by Drosopoulos (1978) that pseudogamy is rather a recent event in the evolution of the species and that complete parthenogenesis might be the next evolutionary step after pseudogamy. However, although most scientists support the hybrid origin of pseudogamous females there are few that support autopolyploidy, despite the fact that in insects at least, this phenomenon appears always in species complexes containing very closely related species, indistinguishable in morphology.

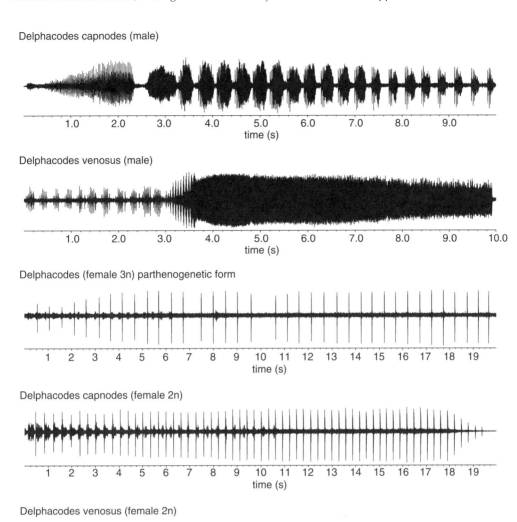

Delphacodes capnodes (male)

Delphacodes venosus (male)

Delphacodes (female 3n) parthenogenetic form

Delphacodes capnodes (female 2n)

Delphacodes venosus (female 2n)

FIGURE 18.4 Male and female calls from *Delphacodex* species. (Oscillograms from recordings by C.F.M. den Bieman and P.W.F. de Vrijer.)

After 30 years of experience in these studies, based on my own work and that of several students and postgraduates, I have acquired a fundamental understanding of what is reproductive isolation between closely related species and the nature of speciation. I am relying in particular on my knowledge of more than 2000 insect species and thousands of plant species, having also described several new species and even new genera in Auchenorrhyncha, being aware of the ability of migration of each species, a phenomenon related to "gene flow". Indeed, all combinations of wing polymorphism (brachypterous: unable to fly, semimacropterous and macropterous) in males and females can be found in the Greek Auchenorrhyncha. Thus, in species of Issidae ($\approx$70 species) there are only brachypterous males and females, in the Delphacidae ($\approx$120 species) all combinations occur and in the Cicadelidae (≈500 species) mostly macropters. Consequently, as also in many bushcrickets (Willemse, 1984), species of Issidae are highly endemic (Drosopoulos, 1990). I refer to this in order to show that speciation can be studied only by taking into account a plethora of different fields of research and this requires experienced workers.

That differentiation of a particular population which during time will result in a new species can originate either at one locality only, producing thus an endemic species, or at several localities that provide the same ecological requirements, is supported here with the following evidence. Differentiation at one locality only is known in the small fish, *Pungitius hellenicus*, Stephanides that survives only in a spring near Lamia-Central Greece (Economides, 1991). An example for the second case is provided by *Muellerianella fairmairei* that does not occur wherever its food plant (*Holcus lanatus*) is growing, but only at localities where this grass and the overwintering oviposition plant, *Juncus effusus*, grow together (see Table 18.1).

The most significant aspect of our studies, performed by intra- and interspecific pair crosses in the laboratory, was reproductive isolation. Through my own work, based on extensive faunistic and biosystematic studies and similar work of many colleagues, also in other species group complexes of planthoppers, leafhoppers, cicadas and recently froghoppers I came to acquire insight in the essential meaning of population and in extension of species and speciation (*e.g.* Gilham and De Vrijer, 1984; Claridge, 1985a; Claridge *et al.*, 1985b; De Vrijer, 1986; Drosopoulos and Loukas, 1988; Loukas and Drosopoulos, 1992a, 1992b; Wood and Tilmon, 1993; Drosopoulos, 2003; Hoch and Wessel, Chapter 13; Quartau and Simões, Chapter 17).

Thus, in my estimation, population is the most fundamental term in biology and can be defined here as a group of interbreeding individuals sharing a gene pool of *a well defined* species occurring, not necessarily in a particular area, but at a certain time, where genes differentiate gradually due to interactions with environmental conditions and food requirements. We have seen this definition of population to be involved in differentiation in various morphological characters including male genitalia, in acoustic and mating behaviour, food and oviposition preferences, polymorphism in planthoppers, leafhoppers and froghoppers. Such a differentiation occurs not because species require splitting, as is widely believed, but spontaneously and gradually and finally results in such a degree of differentiation that populations become reproductively isolated, demonstrated here by postmating experiments. Also in these studies there is evidence that even in populations of other congeneric species there are individuals (specimens) remaining less differentiated, due to some gene flow within a species, that are thus not fully reproductively isolated and can be crossed and even produce fertile females (*M. brevipennis* X *M. fairmairei*). Reduced gene flow naturally will be eliminated or absorbed in the gene pool of a newly differentiating species, thus forming various degrees of variation. In crossings of *M. extrusa* with *M. fairmairei*, on the contrary, many hybrid males and females were fully fertile. Pseudogamous and parthenogenetic triploid females carrying the genes not differentiated from their ancestors either need to copulate or be inseminated by males of species from which they most probably originated. This demonstrates that acoustic and reproductive organs, due to the recent origin of the parthenogenetic species, have not undergone degeneration, as shown, for example, in the blind cave dwelling planthoppers where the eyes degenerated gradually in each species since they were not necessary in the absolute darkness in caves (Hoch and Asche, 1993; Hoch and Wessel, Chapter 13).

Finally, there are many examples in the literature of various organisms where populations of a species appear, after careful investigations, to belong to another undescribed species. The reverse is also common in synonymised species.

Considering these data, we can call species: populations isolated from other populations of congeneric species, sharing common habitats, diets and other behaviour patterns, and in the process of various degrees of genetic differentiation not necessarily at a state of full reproduction isolation, which can be proved by sterile products of interspecific crosses. The examples reported here are cases of such speciation, *e.g. Muellerianella extrusa* isolated on its food plant *Molinia caerulea*, the new species (unnamed) on its food plant *Carex divulsa* and *M. fairmairei* on the grass *H. lanatus*.

Speciation may not necessarily result from geographic isolation, although this is the most common and obvious process. Such speciation has been demonstrated recently in the spittlebug, *Philaenus signatus*, differentiated on the same food plant, *Asphodelus microcarpus*, in the Balkan, Apennine and Iberian peninsulas of the Mediterranean region (Drosopoulos, 2003). Sympatric speciation initiated usually by host shift is more contentious, and I believe has already been demonstrated here and by several other authors (*e.g.* Wood and Tilmon, 1993; Bush, 1994). My own investigations suggest that sympatric speciation begins when some individuals of a population for various reasons (high densities of populations or shortage of food leading to migration, unsuitable climatic or environmental conditions, *etc.*) have to adapt to new resources for survival. Of course such a situation may occur several times during the differentiation of a population to species level and some contamination of the new formatting species by the parental populations may occur continuously. As a result hybrids can be found in nature or in our experiments as some individuals genetically resemble the ancestral species closely. This has been shown in interspecific hybridisations in this chapter. Under these circumstances I suspect that certain substances of the generally new food play a fundamental role by interaction in inducing changes in the genome of the newly differentiated population expressed in various polymorphic alleles we observe in any organism today. Adaptation by natural selection, often used to explain evolution, cannot be excluded.

EPILOGUE

Closing this brief but comprehensive discussion and in relation to the title of this chapter I would like to express some personal philosophical thoughts I have had in mind for several years concerning "assortative mating", a phenomenon quite common in many polymorphic animal species and in the human species expressed as racism, religionism, nationalism, eugenics, *etc.* Could it be that also acoustic (language) communication and differentiation are associated with this phenomenon? In my opinion many philosophers give little or no consideration to the fact that humans are, so far, one biological species.

ACKNOWLEDGEMENTS

I cordially thank Peter de Vrijer (Biosystematics group, Wageningen University) for providing the oscillograms of the signals presented in this chapter, all from original recordings.

19 Photoperiodism, Morphology and Acoustic Behaviour in the Leafhopper Genus, *Euscelis*

Hildegard Strübing and Sakis Drosopoulos

CONTENTS

INTRODUCTION

As late as the middle of the previous century leafhoppers were usually described according to external characteristics, *i.e.* according to morphological and colour criteria. Only when it was recognised that the male reproductive organ, the aedeagus, was an excellent characteristic for species diagnosis and was also reliably constant for many species, was the idea changed. Gradually one passed on to using this time-consuming way of species diagnosis which could be performed only in the laboratory by means of a binocular microscope.

However, the occurrence of day-length-induced seasonal variability in the genus *Euscelis* has been known for more than 50 years (Müller, 1954). In fact, Wagner had shown as early as 1939 that parallel to the development of pigmentation and size of *Euscelis incisus* and of *E. plebejus*, there was also a so-called eunomic line of transitional forms of the phallus form following a set pattern. Thus, even the aedeagus structure is not an infallible factor in the identification of all species. However, often a particular geographical distribution, temperature and climatic requirement, food plant preference and, when living material is available, behaviour and the acoustic signal, among others, are contributing factors in species identification. This chapter is meant to show that species identification comprises the use of as many factors as possible.

EUSCELIS INCISUS KIRSCHBAUM

Euscelis incisus, a central European species, is found practically everywhere in Germany. It is polyphagous and can be reared easily in the laboratory on *Vicia faba*. Outdoors, the small, darker *incisus* morphs occur only in spring and the bigger, lighter *plebejus* morphs occur only in summer, approximately from the beginning of June to sometimes late summer. The *plebejus* morphs used to be considered as separate species: *E. plebejus* Fall. The first stadia of a third generation may also occur. These hibernate in the mid-larval stage and may form as the beginning of the next spring generation. Eggs laid very late can survive the winter as diapause eggs, as later observations have shown.

The sensitive phase for the development into either long-day or short-day morphs lies in the period of the middle three out of a total of five larval stages, while the first and the fifth stages are to a large degree insensitive.

In the laboratory it is possible to obtain any morph desired, long-day or short-day, or also intermediate morphs under controlled day-length conditions, just as would develop under outdoor conditions (light/darkness in extremely long day 20/4 h, extremely short day 8/16 h, short day less than 14 h light). The shift in shape of the aedeagus is presented in Figure 19.2a.

The obvious question posed by Müller (1957) is whether there is a lock-and-key principle for the mating of animals of such different sizes. This was answered in the negative by Kunze (1959) in our institute, who studied the entire copulation process of *E. plebejus*. He found that females respond to acoustic signals of males without discrimination between long- and short-day morphs, and no matter whether they had been reared as *E. incisus* or *E. plebejus* males. In all cases males and females come together by acoustic signals and in all cases their offspring are fertile. This means that *E. incisus* and *E. plebejus* belong to the same species, and long-day females can breed with long-day as well as short-day males. Thus, by the international laws of nomenclature the name *Euscelis incisus* Kb. is valid for both the forms *incisus* and *plebejus*.

EUSCELIS LINEOLATUS BRULLÉ

Euscelis lineolatus, an Atlantic–Mediterranean species living on leguminous plants, occurs in Germany only west of the Rhine. Like *E. incisus* it is easy to rear in the laboratory on *V. faba*. These are both distinct species and may coexist in the same biotope without interbreeding. In the laboratory, however, crossings can be achieved (Müller, 1957).

There is a eunomic line for *E. lineolatus* as well, comparable to that of *E. incisus*. Besides the nominal morph *lineolatus* there are seven transitional morphs, which were all given names. Müller (1957) states: "These eunomic lines of phallus forms of plebejus and lineolatus can be seen as results of the varyingly strong influence of exogenous factors on the allometric growth of the phallus tip (compared with the body growth)".

Temperature as well as food quality may influence the strictness of the morphs, but cannot alter the photoperiodically induced seasonal morph; they just modify it.

The shift in shape of the aedeagus from short-day into long-day forms, as elaborated in our laboratory on material originating from the area of Barcelona, Spain, is presented in Figure 19.1b.

The *E. lineolatus* and *E. incisus* aedeagus lines are rather similar. In the short-day range they can be distinguished only with difficulty or not at all. However, in the long-day range the shape of the tip of the aedeagus is very specific.

The acoustic signals of both kinds, *E. lineolatus* and *E. incisus*, are rather similar, even if *E. lineolatus* displays a much greater variability in the "calling patterns" (Figure 19.3 and Figure 19.4). Figure 19.3a–e represent the various patterns of short and long calls of the male. Figure 19.4a–e represent the alternating calls of male (wide amplitude) and female (narrow amplitude). With increasing stimulation by the female, the males shorten their signals and do not produce stanzas any more, *i.e.* the long calls disappear. The female at first makes short bursts of rasping. In the course of the alternating calling, however, it no longer interrupts its rasping sounds, in turn becoming more like the *E. incisus* female. This may lead to errors under laboratory breeding, to interspecific mating and even to hybridisation. However, even if laboratory breeding acts in favour of "error", females of *E. incisus* and *E. lineolatus* frequently make errors. In other observations with very similar signal patterns, particularly with *Euscelis* species, they did answer but nevertheless rejected the male when it tried to copulate. There appears to be a great number of different stages of rejection between interspecific partners (Strübing, 1993).

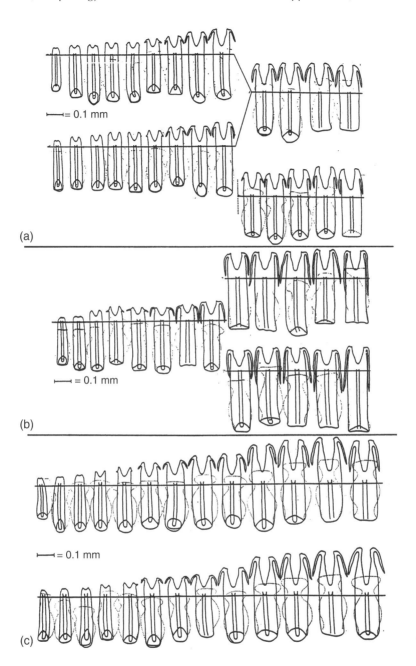

FIGURE 19.1 Shift in shape of aedeagus from short-day types (left) into long-day types (right): (a) *Euscelis ononidis* Remane, (b) *Euscelis lineolatus* Brullé, (c) *Euscelis remanei* Strübing. The close relation of *E. remanei* to *E. ononidis* mentioned in the text becomes obvious here. Only in the long-day form at the right end do the tips of the thorns reach the insertion depth or even exceed it slightly in the second case on the right side below. In the case of *E. remanei* the tip of the thorn reaches the insertion depth one time only in the upper line. When we look at the three species, *E. remanei* shows the deepest insertions by far at the tip of the aedeagus. In the case of *E. lineolatus* the lateral thorns are particularly long, they almost always exceed the insertion depth by far and often are situated close to the shaft of the aedeagus. Here the distinction of the three species becomes particularly clear, just as, on the other hand, their close relation to each other.

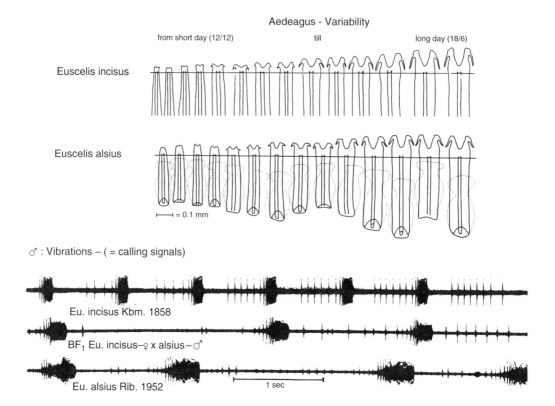

FIGURE 19.2 *Euscelis alsius* × *E. incisus*: (a) Aedeagus variability in *E. incisus*. (b) Aedeagus variability in *E. alsius*. (c) In the centre, a hybrid male BF_1 from *E. incisus* female × *E. alsius* male clearly intermediary call signals. Above that, calls of a *E. incisus* male, below calls of a *E. alsius* male. It can clearly be seen that the calls of *E. alsius* are of approximately double length compared with those of the *E. incisus* male.

In the case of *E. lineolatus-incisus* there was hybridisation in both directions, even crossbreeding *inter se*. Genetic and cytologic examination, generously performed by E. Wolf (Genetic Institute, Freie Universität Berlin), confirmed the close relationship of the two species. However, statements about passing on of calls cannot be made (Strübing, 1963).

EUSCELIS ALSIUS (RIBAUT)

For a long time there have been doubts whether the species *E. alsius* was really legitimate or whether it could just be a variant from the seasonal morph range of *E. plebejus* (*incisus*), perhaps an extreme southern morph. Remane (1967), on account of the extensive material he collected in Iraq and the Iberian Peninsula, was able to distinguish *E. alsius* from *E. plebejus* more clearly. He considered them to be two distinct species, possibly truly allopatric.

It was possible to rear *E. alsius* in the laboratory on *V. faba* and *Lolium perenne* from material obtained by courtesy of F. Heller (Ludwigsburg) in 1967 and R. Remane (Marburg) in 1972. Remane had collected the material in the area of Granada and Malaga, Spain on *L. rigidum*. Out of the material of Heller two generations were bred in the greenhouse under permanent long-day conditions. It was thus possible to obtain the long-day morphs and to distinguish them by the shape of the aedeagus from *E.* (*plebejus*) *incisus*. The acoustic signals were also recorded and compared. It was possible to get hybrids, and 82 larvae were obtained from three *Euscelis alsius* males and nine *E.* (*plebejus*) *incisus* females between 11/06/1967 and 12/18/1967. These developed into 43

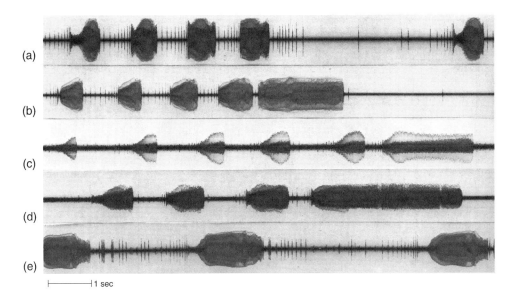

FIGURE 19.3 Variability in the sound patterns of the calls of a male of the species *Euscelis lineolatus*. Pulses are always short. The distance in between varies with the state of excitement of the male.

male and 30 female F_1 adults. However, in January and February more and more of the animals began to die as they needed short-day conditions for hibernation. The F_1 adult progeny proved to be better able to survive under the long-day winter breeding conditions than the original *E. alsius*. The females lived longer than the males, as is usual. The last hybrid female died on 07/01/1968.

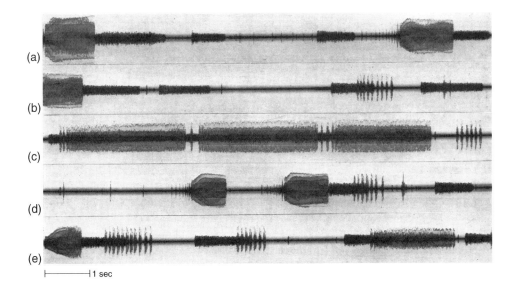

FIGURE 19.4 The same male in alternate dual singing with the female. The narrow amplitudes are those of the female answering the wide amplitudes of the male. Second line, towards the end: The single, short bursts of rasping are emitted by the male. Similar bursts of rasping may, however, also be emitted by the female, but not in this line. The sounds of the female in this case all have regular, low amplitudes. The female remains on the spot. The louder signals are all emitted by the male. Right at the end of (e), the signal of the male is slightly lower and almost completely overlaps that of the female which, however, answers again immediately.

In the meantime a new greenhouse became available with better temperature control ($+15°C$) and equipped with neon tubes. Therefore, it was possible to obtain short-day forms from the material provided by Remane. With falling temperatures and decreasing length of the day in autumn the larvae stored more and more pigments. Compared with the colourless long-day larvae they now became darker and finally totally black. The adults resulting later were smaller and darker, the females more or less dark grey, the males from dark grey to totally black. (Strübing, 1970, 1983; Strübing and Hasse, 1974).

Figure 19.2b presents the complete aedeagus line from short-day to long-day animals. Figure 19.2a refers to *E. incisus* for comparison and shows that here the lateral thorns begin lower and at least reach the insertion depth or even exceed it. In *E. alsius* they at most just reach the edge, but never exceed it. This is a significant characteristic for distinguishing these two species. Not only for these two, it also plays an important role for the distinction of the *E. lineolatus* group. In the case of short-day animals the aedeagus shaft sharpens towards the tip in *E. incisus;* in *E. alsius*, however, it is parallel sided.

Concerning aedeagus morphology and acoustic signals, *E. alsius* is even closer to *E. lineolatus* than to *E. incisus*. Crossings were made with *E. lineolatus* females as well as with *E. incisus* females and *E. alsius* males. With *E. lineolatus* more progeny was obtained than with *E. incisus*. However, the F_1 offspring turned out to be sterile.

Figure 19.2c shows the offspring between *E. incisus* females and *E. alsius* males to be intermediate with respect to the parent species. Furthermore it shows that the *E. alsius* calls are always approximately twice as long as those of *E. incisus*. In *E. alsius*, as in *E. incisus,* there are no other signals.

The morph cycle recorded for *E. alsius* is a seasonal dimorphism, even if outdoors no short-day forms have yet been discovered. This may be pure coincidence, however, because there is hardly anyone looking for them in March or April. Particularly the comparison of the 1967/1968 breeding results with those of 1972 to 1974 shows how much the species is bound to seasonal adaptation and how much its development depends on the photoperiod.

INTROGRESSION FORMS BETWEEN *EUSCELIS ALSIUS* RIBAUT AND *E. INCISUS* KIRSCHBAUM

E. alsius is a warmth-loving species, as the present observations have clearly shown. In many cases therefore there are also geographic reasons for not having both species at the same time. Populations of *E. alsius* are limited to regions of lower altitude, while *E. incisus* has retreated to the higher regions. At least this appears to be the case in the northern half of the Iberian Peninsula. Here *E. incisus* is distributed at approximately 1000 m altitude and, depending on temperature and humidity, even inhabits altitudes up to 1300 m or down to approximately 500 m (Remane, 1967). According to earlier reports, *E. incisus* in the north of the Iberian Peninsula and *E. alsius* in the south seem to be separated by a wide zone in the central Iberian region where neither was found.

After an excursion by Remane in 1972 it was important to make a thorough revision of the areas where *E. insisus* and *E. alsius* were known to occur, as it turned out that the empty intermediate zone no longer existed. Males were found in the central Iberian mountains which could not be assigned with certainty either to *E. incisus* or to *E. alsius*. These were clearly intermediate forms. After a wide-ranging collecting campaign and measuring these samples, Remane formulated a very laborious way of metric assessment of the aedeagi and statistical evaluation of the material of this *E. incisus-alsius* form range.

In his summary Remane wrote: "During determination of the material from the central Iberian comparison it turned out that besides animals belonging to *Euscelis incisus* there were also animals for which clear assignment was no longer possible due to the characteristics that usually enable a distinction of *Euscelis incisus* and *Euscelis alsius*".

I had received sufficient living material of these introgression forms from Remane and bred a further 25 generations. In the conclusions I wrote: "The wild populations of the introgression form collected in Spain and breeding of the same *inter se* over 25 generations in the Berlin breeding laboratory clearly confirmed Remane's hypothesis that this was a back-cross-breeding population". Remane assumed that *E. alsius* males flew into the region inhabited by *E. incisus* and possibly that resulted in hybridisation between the two species which were capable of reproduction only by back-crossing with one of their parents (in this case *E. incisus*). The outdoor population thus developing could keep its mixed character only by refreshening with the parental genotype. The alteration of the population character of the introgressive forms examined back towards the *E. incisus* parent in the course of the isolated laboratory breeding shows that the outdoor (wild) population is indeed a group which is not stable genetically. In the lab, there was no refreshening with *E. alsius* genes so that gradually they were swamped by the predominant *E. incisus* genotype. A similar situation is discussed for the *Ribautodelphax* species complex (Drosopoulos, Chapter 18).

EUSCELIS REMANEI STRÜBING

As Remane had pointed out in the *E. alsius* material collected on *Lolium rigidum* near Malaga, south Spain, there had been one female with dots on the forehead among them. The larvae bred from this female looked different from *E. alsius*. Fortunately they survived on *L. perenne* in the laboratory and grew into adults. Surprisingly, these looked like *E. lineolatus*, but the acoustic signals were absolutely different and unlike anything I had heard before. In addition, *E. lineolatus* does not occur so far south in Spain. There were enough offspring to retain a few animals in order to check the aedeagus. This, strangely enough, was also similar to *E. lineolatus* yet the "calling" was entirely different. Some of the material was sent for examination to R. Remane, who confirmed that this was indeed a new species: *Euscelis remanei* named after R. Remane to whom I owed the species (Strübing, 1980).

CONCLUSIONS

Figure 19.1 shows the relation between the three *Euscelis* species belonging to the *E. lineolatus* range. Apparently there is a particularly close relationship between *E. ononidis* and *E. lineolatus*, which occur sympatrically (Strübing, 1978). They seem to be separated only by their food plants. *E. ononidis* lives on *Ononis* bushes, which *E. lineolatus* not only avoids but rejects. Attempts to transfer *E. lineolatus* to young *Ononis* plants in the laboratory failed because after a few days all 25 *E. lineolatus* individuals had died. However, the two species can hybridise with each other as *E. ononidis* can be transferred to *Vicia faba* without difficulty. The resulting hybrids are fertile. The acoustic signals strongly suggest that the animals can communicate with each other as their signals are very similar. However, in nature this will probably never happen.

Amazingly, fertile hybrids also arose between *E. remanei* and *E. lineolatus* in the laboratory. The hybrids obtained by crossing both combinations of males and females of the two species all had an intermediary aedeagus structure. It was obvious that here, in the Berlin greenhouse, every species can hybridise successfully with every other species. What was going on here? *E. ononidis* was protected through its feeding plant and could live sympatrically with *E. lineolatus* outdoors. As in 1975 I had finished the examination of cross-breeding possibilities, I began diminishing the populations of the different crossing combinations so that nothing could get into a muddle. I was then able to begin a new culture when, in 1981, Remane gave me new material from a second region in Tarifa at the southern tip of Spain. This second approach did not alter anything regarding the results of the very extensive examination of the material from 1972 to 1975. In the meantime Remane had also made a correction: apparently *E. ononidis* can live only on *Ononis natrix* and lives in coastal areas. Samples he had found in 1963 and 1967 in Portugal, due to morphological similarity, had then been assigned to *E. ononides* and now had to be assigned to *E. remanei*. The samples had not been collected near the coast and not on *Ononis* bushes but in general herbage. Unfortunately in 1981 I had no *Ononis* plants available and, in spite of great efforts, I was not able to find any. If I had had *Ononis* plants, I could have tested whether *E. remanei* could live on them. *E. ononidis* could not survive permanently on *V. faba* plants either and had died in my laboratory in 1980. Only *E. lineolatus* had survived there.

The matter was clarified in 2002, when Remane travelled not only to south Spain, but also to Portugal and Morocco. During this expedition he found a great number of new populations of *E. remanei*. As early as in 1980, when I first described this species, he told me: south Spain, Portugal, probably Morocco, perhaps Algeria, now he added Tunisia as well and deleted the term "probably" for Morocco because there was now definite proof. Now it was clear: *E. remanei* clearly had to be distinguished from *E. lineolatus*, both species exist allopatrically and never meet in nature. *E. lineolatus* does not exist in the south of Spain. It is reported to exist as far south as Alicante (according to Remane), not further.

Some habitat observations are interesting. At home I still keep five different *Euscelis* species, one of them being *E. lineolatus*. Previously I had noticed that it was difficult to keep it over winter, as when the temperature fell it did not lay eggs. This situation has aggravated severely, as the temperatures are mostly as low as or below 20°C. Of a great number of animals only 12 have survived. The other four *Euscelis* species (*E. incisus*, *E. ormaderensis*, *E. marocisus* and *E. remanei*) survive well, with *E. remanei* being the most robust. H. J. Müller spent much time on *E. lineolatus*. He mentions that it loves warmth. According to R. Remane, apparently *E. remanei* is the species that has spread further to colder regions (*e.g.* the north of Morocco) and that it is much more robust. This seems to me another characteristic for distinguishing the two species.

This discussion shows that a species diagnosis requires as many features as possible about the insects. The last example showed that sometimes even the aedeagus structure, a factor ensuring a certain diagnosis in many cases, may lead astray. However, often particularly geographical distribution, behaviour, of course the acoustic signal, temperature and climatic requirements, among others, belong to the definite evaluation of a species diagnosis.

It remains unclear what is the degree of relationship among *E. lineolatus*, *E. ononidis* and *E. remanei*; at any rate they probably have to have belonged together once.

Worth mentioning here also is the photoperiodically determined dimorphic calling songs in the katydid, *Neoconocephalus triops* (Whitesell and Walker, 1978). In northern Florida the summer generation of *N. triops* sings simultaneously with *N. retusus*, whose song closely resembles that of the *N. triops* winter generation. However, it is not only in katydids that songs of two related species may overlap. This phenomenon may have very different causes and just as many different consequences. Unfortunately nothing further is known about the behaviour of these katydids. The postulation that the photoperiodically determined and different calling songs of *N. triops* are (or could be) analogous to the variation in the morphology of the aedeagus in the *Euscelis* species is insufficiently founded. The basis for this phenomenon in katydids is unknown. With regard to *Euscelis* we know that the insects develop under different ecological conditions and photoperiodism is not the only factor involved in aedeagus variability; temperature, food quality and quantity of the light are all effective together.

Finally, Jocqué (2002) refers to genitalic polymorphism in spiders, but not in relation to photoperiodism. He considers intraspecific polymorphism a phase in sympatric speciation and likely to occur much more generally than so far anticipated.

20 Mutual Eavesdropping Through Vibrations in a Host–Parasitoid Interaction: From Plant Biomechanics to Behavioural Ecology

Jérôme Casas and Christelle Magal

CONTENTS

INTRODUCTION

Insect host–parasitoid and prey–predator systems are in some sense a *parent pauvre* of acoustic communication, despite involving one the most diverse groups of insects, the parasitic Hymenoptera, with over 100,000 species. Reasons are difficult to give, but include the fact that loud sounds are usually not produced during these interactions. More insidious is probably the hidden assumption that these interactions can only be very rudimentary in terms of acoustic signal transfer as they are based on eavesdropping of unwanted signals. Furthermore, the emphasis is usually made on one or other side of the interaction, but not both. Here, we aim at a balanced view of vibration production and perception by both antagonists. This chapter is mainly a synthesis of the work done on a specific host–parasitoid system, the apple tentiform leafminer, *Phyllonorycter blancardella* (Lep. Gracillariidae), and one of its parasitoids, *Sympiesis sericeicornis* (Hym. Eulophidae). This is therefore not a review of vibratory information transfer in all known examples of host–parasitoids, since this topic has been reviewed recently (Meyhöfer and Casas, 1999). We refrain from using the term communication as the receiver normally benefits, which is not so here (Danchin *et al.*, 2004). We also present several unpublished and preliminary results on the leaf and mine as transmission channels. Indeed, one fundamental aspect in solid mechanics is that waves change form as they propagate. This behaviour implies very strong requirements on signal production to ensure signal integrity if individuals need to meet, as in sexual communication — a recurring theme in several chapters of the book. It also implies very strong requirements on signal perception to ensure proper recognition, whether partners want to be found (sexual communication) or not (predation). The final focus of this chapter is on the ecological context in terms of behavioural ecology.

LEAFMINERS AND THEIR PARASITOIDS AS MODEL SYSTEMS

Leafminers are a large ecological grouping of insects, mainly of the orders Lepidoptera, Coleoptera and Diptera. They spend a large part of their larval stages, if not the entire time, in a tunnel created by feeding on specific tissues within a plant leaf. In some, larvae pupate in their mine and only the adult instar is free living.

The apple tentiform leafminer is a secondary pest on apple and a rarer problem on other plants. Its mine has a peculiar and characteristic tentiform shape. This shape is the result of feeding activity of the *sap-feeding* instars, which delimit the extend of the mine by delaminating the two surfaces of the mine and spinning a dense carpet of silk on the lower surface, thereby forcing the upper surface to bend. *Tissue-feeding instars* eat the content of their mine. The upper surface of the mine has a spotted appearance, the origin of the other vernacular name, "spotted tentiform" leafminer. The whitish spots are in fact places where only the upper epidermis is left intact, while the green patches are areas with photosynthetically functioning plant cells. The order in which leaf tissues are eaten and the shape of the intact areas change during the course of larval development. Other species of the *Phyllonoryter blancardella* group produce similar tentiform mines on different plants.

This leafminer is attacked by a large array of mostly polyphagous parasitic wasps of the Chalcidoidea and Ichneumonoidea, in particular Eulophidae and Braconidae (Casas and Baumgärtner, 1990). The mortality from parasitism can easily reach 80%. This lifestyle is clearly not a good protection against attacks by parasitoids. *Sympiesis sericeicornis* is one of the most important parasitoids of this host and also one of the most agile while hunting on a mine. This is the species we have most studied and the one we have generally in mind when speaking of "parasitoids".

The interaction during an attack may be succinctly described as follows in qualitative terms. Shortly after landing, a female inserts its ovipositor into the mine, attempting to contact the host. However, the first trial is rarely successful, either because the host escaped or because of failure to estimate its position. This violent ovipositor insertion usually triggers immediately an equally violent reaction of the host, which wriggles and quickly moves to another location in the mine (if it is a larva). The following steps of the interaction may be very variable both in terms of behavioural success and of duration. Females then relocate their hosts, which themselves in turn avoid ovipositor insertions, and so on. This hide-and-seek game may last up to 20 min and up to 200 missed hits. The female either eventually hits its host or leaves the mine to hunt for other hosts. The interaction runs at a spectacular speed that hints at the use of signals of a physical nature such as vibrations. It also implies that this is a two-way system, both antagonists producing and perceiving vibrations.

VIBRATIONS PRODUCED BY HOSTS AND PERCEIVED BY PARASITOIDS

Hosts produce vibrations while moving inside mines and eating. The classification of these behaviour patterns on the basis of vibrations alone is difficult due to their variability and dependence on life stages. *Sap-feeding* larval instars produce signals that are usually lost in background noise. *Tissue-feeding* instars produce similar broadband signals when both moving and eating. Their highest peak frequencies are usually a few hundred Hz up to 10 kHz, with a corresponding intensity of some 10 to 20 dB (Meyhöfer *et al.*, 1994). The strongest signals are produced when a host is violently wriggling. Pupae wriggle too, producing the strongest signals with peak frequencies over 10 kHz and intensities over 30 dB. These values are rather high compared with other insects. One of the reasons may be that the signals were recorded very near to the insect, often on the mine itself. In general, signal acceleration is below 4 mm/sec^2 and displacement amplitude ranges from 0.025 to 6.5 mm, both ranges of values known to be perceived by insects.

Perception of host vibrations determines the time parasitoids will spend on a particular mine (Djemai *et al.*, 2004). Playing back host vibrations in otherwise empty mines led parasitoids to continue foraging for more than 5 min, while the majority of them had long abandoned the empty mines in the absence of vibrations. We studied in detail some of the possible mechanoreceptors located in different parts of the body which could be involved in vibration perception. We studied a conspicuous long mechanoreceptive hair that may act as a pick-up once the arolium at the end of the tarsus is fully evaginated (Meyhöfer *et al.*, 1997b).

VIBRATIONS PRODUCED BY PARASITOIDS AND PERCEIVED BY HOSTS

Parasitoids produce faint vibrations which are difficult to record when they walk on leaves and mines. They are broadband signals (DC to 10 kHz) with intensities often above 3 dB. By contrast, vibrations produced when inserting the ovipositor are strong, with peak frequencies around 2 to 10 kHz, intensities usually above 10 dB and velocities of *ca*. 1.5 mm/sec. These signals are thus largely above background noise (Bacher *et al.*, 1996). There is a puzzling lack of congruence between the short time scale at which strong vibrations are produced (10 msec) and the long behavioural scale at which the ovipositor is inserted, twisted or withdrawn (at least a fraction of a second) (Bacher *et al.*, 1996). We have no clear explanation for this, but it has been confirmed by other workers using new optical equipment (Djemai *et al.*, 2004). Sharp, impact-like transients are specific to this behaviour and were never observed in another context. Insertions away from the host often triggered behavioural responses of the leafminer while other foraging behaviour of the parasitoid failed to elicit similar changes in the host's behaviour (Meyhöfer *et al.*, 1997a). One can therefore safely conclude that vibrations produced during ovipositor insertions are the key vibrations picked up by hosts.

Hosts perceive vibrations produced from searching wasps sometimes very far from the host location, as the violent wriggling behaviour following insertion of the parasitoid ovipositor testifies. They do not react to other vibrations produced by parasitoids during foraging (Bacher *et al.*, 1997; Meyhöfer *et al.*, 1997a). Hosts react to a wide band of frequencies, the typical signature of such a signal being the transient, impact-like nature (Bacher *et al.*, 1997; Djemai *et al.*, 2001). Host reactions to strong stimuli include freezing, wriggling with no net displacement, freezing and evasive displacement (which may include wriggling) and finally no change of behaviour. The behavioural transition after perceiving a stimulus depends on the behaviour displayed before the stimulus (Meyhöfer *et al.*, 1997a). For example, a larva perceiving vibrations while feeding may display freezing reactions, while the same larvae will wriggle if vibrations are detected while moving inside the mine. Using biotests with synthetic signals mimicking real ones, we showed that the best frequencies, *i.e.* the frequencies at which the host reacts at the lowest signal amplitude, matches nicely the frequency range of signals produced during insertion of the ovipositor. The frequency range over which hosts react best is as broad as are the parasitoids signals (2 to 5 kHz).

Response curves may be considered on the basis of their shape as low-pass filters, thus filtering out frequency levels typical of background noise (see below). Larvae are more sensitive than pupae, maybe explaining why all larvae, but only 60% of pupae, reacted to synthetic stimuli. Another possible explanation is the remodelling of the nervous system in pupae, which may involve a temporary shutting down of the peripherical nervous system (Huber, 1964). This is the first demonstration of a good sensory matching between predator stimuli and prey behavioural responses in this exceptionally large class of interactions. It is too early to claim that this match implies coevolution, even though similar results obtained for moth–bat interactions are used to make such claims (Fullard, 1998).

THE OVERRIDING ROLE OF THE TRANSMISSION CHANNEL: WAVE PROPAGATION IN LEAVES

The interaction takes place on a leaf and on a mine, two very complex structures. The mine itself is made of leaf tissues as well as silk, adding another layer of complexity. The heterogeneity of plant tissues, their widely differing mechanical properties, the interwoven roles of geometry and material properties are simply mind boggling. Still, we believe that the influence of the transmission channel is a key to understanding vibratory information transfer between these two antagonists as well as for a whole cohort of insect–insect interactions using plants as substrate (see Barth, 2001, for a similar call). Despite its ecological importance and maybe because of the sheer complexity of the problem, wave propagation in plant tissues is a topic seldom studied. We tackled this problem from both ends, by empirical measurements on the one hand and simple modelling approximations on the other. We first present new results obtained on the statics of apple leaves using finite element modelling. We want to emphasise that these new results are preliminary and should only be considered as hints for further work. We believe that these results are nevertheless worth presenting, given the dearth of information on such topics.

THE INTERPLAY OF LEAF VEINS AND LAMINA IN DETERMINING THE MECHANICAL RESPONSE OF LEAVES

We used a finite element method (FEM) to simulate the deformation of a leaf by a "static" force of the order that one applied, for example, during the insertion of the ovipositor. The main goal was to characterise the relative role of veins and limb in the deformation of a leaf. Simulations were made with the Structural Dynamics Toolbox of Matlab. We constructed a three-dimensional mesh of a leaf with two types of finite elements, 452 triangles for the homogeneous regions (leaf lamina) and some parts of veins and 81 rectangles for veins. The three principal parts of a leaf (midvein, minor veins and the leaf lamina) were made of materials having their own properties. Young's modulus was set at 10 GPa for all veins and at 1 GPa elsewhere. The Poisson ratio was 0.3 and the density 1000 kg/m^3 for all leaf materials. We used a thickness of 1 mm for the midvein, of 0.5 mm for minor veins and of 0.1 mm elsewhere. All but the first two nodes could move in three directions.

Deformations strongly depend on the point of application of the force (Figure 20.1), but general rules about the interplay between leaf lamina and veins could be extracted comparing four approximations of real apple leaves:

  (i) A "real" leaf incorporating above characteristics
  (ii) A leaf considered as a plaque made of leaf lamina material only
  (iii) A skeleton of a leaf made of the veins only
  (iv) A leaf considered as a plaque made of midvein material only

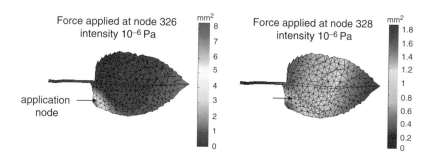

FIGURE 20.1 Influence of the force application on the leaf deformation. Only local deformations are observed when the force is applied at node 326, while deformations can be observed both in the region of force application and in the tip of the leaf when the force is applied at node 328.

All leaves had the same geometry except the skeleton. The force was applied at the same point in all models. Amplitudes of deformation values vary considerably from one approximation to another one (Figure 20.2). Compared with the values of a "real" leaf, which are of the order of 4×10^1 mm^2, amplitudes were much smaller for the skeleton made of vein materials only (10^{-9} mm^2). By contrast, amplitudes were much larger than in the "natural" leaf, of the order of 10^5 mm^2, for the approximated leaf made of limb material only. An approximated leaf made of vein material experiences deformations of the same order of magnitude as the "real" leaf. The comparison of these four models enabled us to understand the role of the different components of a leaf in its overall

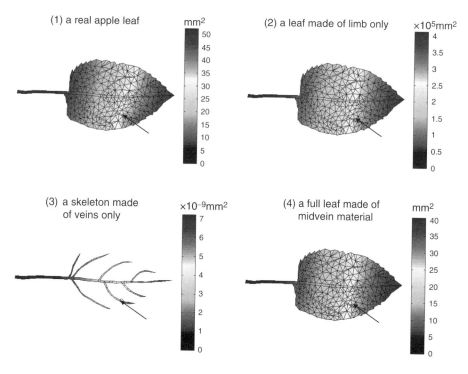

FIGURE 20.2 Four approximations of an apple leaf. The force is applied on a minor vein in all cases. While the overall pattern of deformation is similar, with the strongest deformations at the tip of the leaf, the absolute values of deformation amplitude vary greatly, as shown in the scale bars.

mechanical behaviour. An approximation of a real leaf must have the *geometry* of a plaque in order to ensure that deformations spread over the entire leaf. The *material properties* of this plaque must on the other hand be as stiff as the midvein material, this structural element determining the range of amplitudes of deformation.

WAVE PROPAGATION IN LEAVES

Turning now to wave propagation, its study is greatly eased if one can measure a transient wave simultaneously at two points in space. We therefore designed a prototype system made of two identical laser Doppler vibrometers to characterise the energy flow and frequency filtering of the leaf lamina and through veins of varying diameters (Magal *et al.*, 2000). As predicted by the FEM analysis, leaf lamina transmits vibrations better than veins. The loss in total energy is, however, not negligible even when remaining in the same sector of a leaf (a sector is defined as the leaf lamina between two adjacent minor veins). The energy loss through the midvein and the numerous minor veins is a function of vein diameter. The smallest minor veins near the apex of a leaf and at the leaf border attenuated signals to the same extent as the leaf lamina. By contrast, energy loss may be up to 80% for the midvein. A wavelets analysis showed that veins of both types act as low pass filters, the high frequencies being strongly damped.

The implication of our results for the leafminer–parasitoid system and other insect–insect relationships is that signals change over very short distance. Consequently, some parts of the signals, particularly the high frequencies, are quickly lost. This "handicap" in terms of signal recognition is offset by the broadband nature of the vibrations produced by both host and parasitoid. Sexual communication, however, often requires high signal integrity. Our results highlight the importance of the choice of exact positioning for both partners, in particular when high-frequency components are the key aspect of the signals.

TRANSMISSION OF LEAF VIBRATIONS TO AIR

It is usually assumed that vibration perception proceeds mainly through mechanoreceptors tuned to perceive substrate vibrations either in or on the legs. The possibility that substrate vibrations are transmitted through air and then picked up by air movement-sensitive receptors has seldom, if ever, been thoroughly studied in a behavioural and ecological context (but it has been somewhat studied in a neurophysiological framework). Why not? Maybe because substrate vibrations are of such small amplitudes that a transmission through air seems too far-fetched, but this has to be shown. We conducted a study combining laser Doppler vibrometry for measuring leaf vibrations and laser Doppler anemometry for measuring air velocities in the vicinity of a vibrating leaf (Casas *et al.*, 1998). The results are striking: leaf vibrations can be measured in the air up to 1 cm away from the vibrating leaf, and most likely further as we had records showing faint signals 2 cm away from the leaf boundary. The intensity of the air movements at 1 cm distance was, however, two orders of magnitude smaller than the intensity of leaf vibrations.

Despite this caveat, vibrations of biotic origins propagating within leaves must trigger air movements in the near surroundings of the vibrating substrate. Given the incredible performance of air flow sensing hairs in arthropods (*e.g.* Shimozawa *et al.*, 2003), these signals can be picked up well within a few millimetres from the substrate. The solid substrate, due to its geometry and material properties, acts as a frequency filter, a behaviour which also applies when transmitting energy to air. This leads to a rich and spatially distributed information pattern available to insects over a leaf. This area of research is a virgin field of investigations whose immensity is only matched by its importance. Amazing technological advances in optical flow measurement methods are good news to prospective students.

BIOTIC SIGNALS: LOST IN BACKGROUND NOISE?

While preceding results have, we hope, convincingly demonstrated the production and perception of vibrations for both antagonists, their use during attack and escape and the pattern of information flow in the "leaf" channel, one may still wonder about the relevance of all this in the field. What happens during rain and, even more importantly, under the presence of even the slightest breeze? Maybe the vibrations recorded with lasers on heavily damped tables in cosy laboratories are simply lost in the ambient natural environment.

Field observations give a clear-cut, but only partial, answer to some of these questions (Casas, 1989; Casas *et al.*, 1998). Parasitoids do not appear to forage under rainy conditions, nor in strong wind. We do not know how leafminers react under such conditions, but we observed occasionally violent wriggling when watering plants in the greenhouse. A false alarm is a small price to pay compared with a lack of reactivity when attacked. Wind of moderate speed may be a problem when parasitoids are foraging. We therefore studied leaf vibrations under moderate flow in the laboratory by exposing a leaf to wind of varying intensities. Leaves responded with steady oscillations of broad frequency bands with basic oscillations usually below 10 Hz and with an intensity decreasing with frequency (Casas *et al.*, 1998). Comparing these signals with those produced during an interaction, we can safely state that vibrations of abiotic origin do not interfere with vibrations of biotic origin for both protagonists. Indeed, biotic signals are, as described above, characterised by their transient nature, lasting less than a second with a peak frequency above a few hundred Hz up to a few kHz. As these two types of signals cannot be confounded in a time–frequency representation, biotic signals, irrespective of their origin, should be of importance to both leafminer and parasitoid as they can only be produced by another insect on the same leaf. For the leafminer, this is enough bad news to become alert and behave accordingly. For the parasitoid hunting on a leaf or mine, this is good news and encouragement to continue the hunt. What to do next is not obvious for either protagonist. Game theory is a useful framework to interpret the chosen behaviour patterns.

VIBRATIONS PERCEPTION AND BEHAVIOURAL ECOLOGY: THE PRINCESS–MONSTER GAME

Functioning sensory systems, attack weapons and defensive counterstrategies, all combined with ample information of a vibratory nature, are only parts of the success in escaping or getting the host. When to escape and in which direction? When to attack and where? Not acting may be fatal or postpone the successful strike for too long, acting will necessarily produce vibrations used by the other party. These are questions of a different nature, best tackled within the field of behavioural ecology.

Questions related to when and where to search for a target have been thoroughly studied in a military context under the "princess–monster" or "pursuit games" heading. Less offensive, if totally analogous, problems run also under the "rescue search" and the "rendezvous" problems (Stone, 1989, and references therein). The mathematics are invariably difficult and many assumptions are violated when applied to ecological relationships, but this approach is still the best so far to understand the host–parasitoid interaction (Meyhöfer *et al.*, 1997a; Djemai *et al.*, 2001). Students wishing to know the latest about such games may consult Alpern and Gal (2003) but must be warned: reference to the leafminer–host system is the only biological element in 300 pages full of equations.

Optimal Hunting and Escaping Theoretical Strategies Include Significant Amounts of Randomness

The optimal strategies to be followed by the monster and the princess in order to minimise (monster) or maximise (princess) the time until capture depend on the value of many parameters, such as the geometry of the arena, the costs of search and escape, the detection law (vibrations), *etc.* Assume here a homogeneous bounded two-dimensional arena without any particular place to hide, a monster moving at velocity 1 and a princess moving at velocity x. It is important to note that $x < 1$ for otherwise the game becomes trivial: the princess always escapes. Let Q1, Q2, … be identical independent distributed random variable of points uniformly distributed in the arena. The princess starts at Q1, stays there for time T, moves to the next point, Q2, at full speed (wriggling), stays there for time T, *etc.* Thus the points are chosen at random, leaving no information to the parasitoid. Such an algorithm may explain why the host shows much variability in its response to insertions of the ovipositor. It may be advantageous to react sometimes violently, sometimes not at all. If the princess is allowed some partial information in the form of knowledge about the monster's position at some time interval, then she could elude it indefinitely. This condition highlights one aspect related to the insertion of the ovipositor: inserting it at wrong places in the mine delivers information to the host. The monster strategy has therefore to be more complex. Lalley and Robbins (1988) defined sets of trajectories leading to a uniform distribution of sites visited. A high degree of randomness in the successive visited locations is another key condition for the monster as it is most uninformative to the princess about the threat of immediate attack. We discover here another aspect related to the insertion of the ovipositor: if a parasitoid missed its host, it should next time insert its ovipositor in a location that cannot be guessed by the host on the basis of previous locations of ovipositor insertions. That is now very interesting as it also builds randomness in the heart of the search process of the parasitoid.

APPLYING GAME THEORY TO THE HOST–LEAFMINER INTERACTION

How far does this approach truly apply to the leafminer which produces a tentiform mine enabling escape in the third dimension? What about the feeding pattern determining the locations in which parasitoids can insert their ovipositor and those where they cannot, making such areas refuges? This spatial heterogeneity introduces a spatial heterogeneity in preferred locations of hosts, valuable information for a parasitoid. We cannot answer this question fully, but Monte Carlo simulations of feeding patterns and larval positions gave a partial answer. Once a feeding pattern made of large feeding sites and large refuge zones is produced, hosts *must* take full advantage of them. Otherwise, their entire body can be under feeding zones and hence fully at risk of being hit. This does not occur when feeding at random because the small feeding windows are spread over the entire leaf and the host body is only partially at risk. Furthermore, there is no particular spot to hide or to avoid in a random pattern. Thus producing a nonrandom pattern may lead to higher risk. Indeed, a random body positioning on such a pattern becomes even more risky than a random positioning on a random feeding pattern (Djemai *et al.*, 2000).

The aim of the princess is to delay the fatal strike as late as possible, betting on the albeit unknown to her, giving up time of the parasitoid. This does work, as we have observed in the wild, parasitoids giving up after more than 20 min and over 200 insertions of the ovipositor (Casas, 1989). The conclusion we make from this game theoretical approach is therefore that optimal strategies for both protagonists necessarily involve a high degree of randomness. It is advantageous to have an exquisite sensory level able to pick up the slightest vibrations, but it is equally advantageous to avoid reacting invariably in a stereotyped way, whatever that may be. Thus, an optimal strategy consists of a delicate combination between acquiring the best possible information and displaying a fair amount of randomness as response to threat or within the attack manoeuvres.

IMPLICATIONS FOR OPTIMAL FORAGING

When a parasitoid lands on a mine, it quickly searches, proceeds with attacking the host inside it, moves on to a next mine, *etc.* It will eventually leave the searched leaf after parasitising only a few leafminers, between two and three, even if that leaf harboured many more hosts (Casas *et al.*, 1993).

Interestingly, it may leave a mine suddenly and proceed with hunting on a mine located at the other end of the same leaf. Was it attracted by stronger vibrations produced by the second miner? Alternatively, the randomness described in the princess–monster game applies when hunting also among mines. Only further experiments will tell. A stochastic Markovian model for the succession of behavioural decisions showed that parasitoids leave a leaf with many hosts for reasons not related to egg exhaustion or parasitising all hosts; the exact reasons, unknown, have to do with the difficulty to get hosts in their mines (Casas *et al.*, 1993). The probability to abandon a viable host for another one during hunting on its mine is large, amounting to 0.35. When this happens, the parasitoid may well leave the leaf.

CONCLUSIONS

We have come full circle. In this particular host–parasitoid system, vibration production and perception lead to a complex bi-directional interaction running at full speed. While neither protagonist can control the vibrations produced during their actions, they have developed strategies to decide when and where to act. Furthermore they possess sensory systems enabling the perception of the slightest vibrations produced by the other party being transmitted either directly through the substrate or through air. Both antagonists cope with the large changes of signals over a short distance by focusing on signal characteristics typical of biotic origin. Parasitoids will leave a host if it is too difficult to hit. This giving up time is set partly by the history of the parasitoid, but can also be increased by the host itself through vibrations produced during escape. The interplay of successful hits and failed pursuits sets the number of hosts eventually parasitised on a leaf.

We believe that bi-directional interactions of this type are frequent neither among host–parasitoid interactions nor among many insect prey–predator interactions. However, many aspects developed here have much wider applications than this specific system, including in remote fields such as operation research (see Alpern and Gal, 2003). It is our firm belief that through a comparative study of long-term case studies generalities about the role of vibrations in the insect world will emerge — studies conducted very much in the spirit of Barth (2001). This host–parasitoid system has been dissected within a "system analysis" approach over more than 25 man years of work. As so often, the level of comprehensiveness of this study — despite its numerous shortcomings — was only possible through decades long personal, personnel and financial investments.

ACKNOWLEDGEMENTS

This chapter is the synthesis of the Ph.D. thesis of Rainer Meyhöfer, Sven Bacher and Imen Djemai, as well as the work of students and collaborators. We thank them as well as Isabelle Coolen for discussions, Olivier Dangles for improving the manuscript and the editors for their support and patience.

Part II

Sounds in Various Taxa of Insects

21

Vibratory Signals Produced by Heteroptera — Pentatomorpha and Cimicomorpha

Matija Gogala

CONTENTS

INTRODUCTION

The existence of sound emission and sound producing mechanisms in land-dwelling bugs Heteroptera: Geocorisae was discovered at least three centuries ago. To my knowledge the oldest written remarks about this phenomenon were published by Ray (1710) and Poda (1761). Stridulatory structures of Reduviidae were investigated by Handlirsch (1900a, 1900b) and similar sound producing stridulatory structures of other bugs were described in series of papers, together with descriptions of the acoustic behaviour of these bugs, reviewed by Leston (1954, 1957) Leston and Pringle (1963) and Haskell (1957, 1961). Independently a German entomologist, K.H.C. Jordan (1958), published interesting results about low frequency signals in Pentatomidae and Acanthosomatidae. Haskell (1957) published the first sound recordings of the Cydnid bug, *Sehirus* (*Tritomegas*) *bicolor*. The first spectrographs (sonographs) were published by Moore in 1961. Other researchers published data about stridulatory mechanisms in other groups of Heteroptera, *e.g.* Coreidae (Štys, 1961), Lygaeidae (Ashlock and Lattin, 1963; Sweet 1976). More works will be cited in the systematic part of this chapter. It early became clear that stridulatory structures evolved many times independently in Heteroptera Geocorisae (and also water bugs) (Usinger, 1954; Ashlock and Lattin, 1963; Schaefer, 1980).

A few years later the first papers on Cydnid sound signals using different recording methods were published (Gogala, 1969, 1970). It was found that Cydnidae emit low frequency sounds in addition to stridulatory sounds. Since small insects are very poor emitters of low frequency sound, the question arose: what is the channel of communication in these bugs? This question became even more important when, in many other groups of land bugs, low frequency signals (usually in the range of 100 to 500 Hz) were discovered — generally in combination with song constituents of higher frequencies (Gogala, 1978b). Gogala *et al.* (1974) proved that in Cydnidae only the substrate-borne part of the vibratory signals is important for communication. In 1972 detailed studies of different aspects of vibratory communication through plants in *Nezara viridula* and some other Pentatomidae started with the description of airborne components of the emitted signals (Čokl *et al.*, 1972). Only narrow band low frequency signals have been recorded for species of this group (see further in the text) and sense organs are tuned to such low frequency signals (see Čokl *et al.*, Chapter 4). Michelsen *et al.* (1982) showed that bending waves are most important for the communication on plants by true bugs. Reviews on vibratory communication in small plant-dwelling insects have been published recently by Čokl and Virant-Doberlet (2003), Virant-Doberlet and Čokl (2004), and Virant-Doberlet *et al.* (Chapter 5). Čokl *et al.* studied the vibratory communication of Pentatomidae in detail (see a review in Čokl and Virant, 2003).

Vibration producing organs have been investigated further by scanning electron microscopy (Schuh and Slater, 1974; Drašlar and Gogala, 1976; Gogala, 1985b; Péricart, 1998).

SOUND OR VIBRATION

In literature authors describing acoustic emissions of bugs speak of sound, *i.e.* airborne vibrations. It is true that, with very sensitive microphones in sound-insulated chambers or in environment with very low level of background noise, one can make good recordings of these signals as sonic events in a strict sense (Gogala, 1985a). Nevertheless, as mentioned above, in intraspecific communication of Heteroptera only the channel of substrate vibrations is apparently relevant (Gogala, 1985b). There still exists a possibility that in interspecific communication high frequency stridulatory signals reach receiving animals as airborne sounds. But there is little doubt that Heteroptera use as a primary acoustic communication channel substrate vibration. However, one cannot exclude in close proximity, near field particle velocity communication, as in *Drosophila* and some other arthropods (Bennet-Clark, 1971).

METHODS USED TO RECORD VIBRATORY SIGNALS OF BUGS

Workers previously using only ears, simple hearing aids such as a stethoscope (Leston, 1954 and Figure 21.10a) or microphones to pick up acoustic signals of bugs, did not detect all sounds and missed especially the low frequency vibrations. Now there are some simple and some more sophisticated methods available for recording vibratory communication signals.

The best but most expensive method is the use of laser vibrometers. Such a device does not change the physical properties of a substrate or animals and one can measure vibrations exactly at an exactly defined point (Figure 21.3a and b). On the other hand, the signal-to-noise ratio of these devices is not very high. To get good reflections usually one has to use some reflective material (reflecting paints or sphaerulae) and one has to do all the experiments in a well-equipped laboratory. However, there are now also portable laser devices which can be used in the field.

Another possibility is to use piezoelectric accelerometers to detect substrate vibrations. The problem is that the mass of such transducers is not negligible and may change the physical properties of the structure and therefore also the vibration pattern. Some of the smallest devices of this kind can still be used for bioacoustic measurements of the vibratory songs of bugs, but these accelerometers are not sensitive enough for low frequencies and the whole measuring system may also be quite expensive.

One simple and cheap device is the magneto-inductive transducer system described by Strübing and Rollenhagen (1988). A small magnet (*ca.* 0.1 g) is fixed to a plant stem with bugs on it. A coil then has to be brought as close as possible to the magnet without touching it (Figure 21.8a). The connectors of the coil can be plugged directly into a low impedence microphone input of a cassette recorder, DAT or similar recording device. The problem is that the coil has to be fixed to a rigid and stable support to avoid additional vibrations of the system and the coil has to be shielded.

One can also use a gramophone cartridge attached to a plant with the insects (Claridge, 1985a). For some experimental purposes one can also glue the animal to the small rod attached to such a phono pick-up, which again can be plugged directly into a microphone input of any recording device.

Another inexpensive method available to anybody is the use of small electrodynamic loudspeakers or earphone speakers as contact microphones. Such transducers of a few centimetres diameter can be used to pick up vibratory signals of bugs. One can put the bugs on the diaphragm of such an earphone (Figure 21.10b) and cover it with a transparent cage made of plastic and suitable mesh. If the cage has a top made of thin glass one can observe or film the insects performing during their vibratory communication. Such a transducer may be again directly connected to a low impedance microphone input of a tape or cassette recorder, DAT or even audio input of a computer. A critical point in using such small speakers is sometimes a noise produced by locomotion and in some cases a physical difference to a specific substrate (plant stem, wooden twigs) used by single species for vibratory communication. Vibratory song characteristics may be adapted to the resonant properties of plants or plant organs where many bugs live and "talk" to each other.

Nowadays, there is a lot of software available to analyse digitised vibratory or sound signals with a computer, to produce oscillograms, spectrograms and frequency spectra or to play back signals as airborne sound to get an impression of the recorded songs. One can also play back previously recorded vibratory songs through a suitable transducer to stimulate other animals or to investigate acoustic behaviour of experimental animals.

COMMON PROPERTIES AND PECULIARITIES OF SINGLE SYSTEMATIC GROUPS

The vibration producing structures, characteristics of vibratory signals and corresponding behaviour is described here for most families of terrestrial Heteroptera.

PENTATOMOMORPHA

Cydnidae

Let us start a systematic overview and comparison of bioacoustic characteristics of single systematic groups with this family of bugs. There are many papers dealing with sound producing structures and acoustic emissions of the Cydnidae (Dupuis, 1953; Haskell, 1957, 1961; Leston and Pringle, 1963; Gogala, 1969, 1970, 1978b; Lawson and Chu, 1971; Gogala et al., 1974; Drašlar and Gogala, 1976; Schaefer, 1980; Gogala and Hočevar, 1990).

One of the characteristics of the acoustic behaviour of Cydnidae is a combination of two vibration producing mechanisms, stridulation and abdominal low frequency vibration. The main moving part of this system is a tergal plate, formed from the fused first and second tergum (Drašlar and Gogala, 1976). At the fusion line there is a fold with strong curvature at both sides (Figure 21.1e). To this structure, strong longitudinal TL1 and TL2 and lateral dorsoventral muscles are attached. The frontal edge of the tergal plate has on both sides parallel ridges forming a special type of a stridulatory plectrum, often called the lima (Figure 21.1c–e). The counterpart of this structure is a stridulatory vein, the stridulitrum on the postcubitus vein of the hind wings (Figure 21.1a and b). The anal field of the wings is folded anteriorly downwards and is therefore in most species lying between the lima and stridulitrum when the wings are in the resting position. Through the stridulatory movements of the tergal plate the lima is rubbed against the stridulatory teeth and thus makes perforations in the anal field of the wings (Drašlar and Gogala, 1976).

The function of the tergal plate as a low frequency vibration producing organ was clarified many years ago through some simple experiments (Gogala, 1970, 1985b). The stridulatory parts of the vibratory signals can be eliminated by cutting off the hind wings, and low frequency signals can be eliminated at least in males of *T. bicolor* by prevention of deformation of the tergal plate by putting small amount of wax–resin mixture into the curved part of the tergal fold (Figure 21.1e). The production of the low frequency vibration may occur simultaneously or separately to the stridulation, depending on species, sex and type of song. During singing, one can see also dorsoventral movements of the abdomen, possibly acting as a pendulum and thus amplifying the low frequency vibrations. Nevertheless, there are some observations justifying the assumption that, at least in the case of *T. bicolor*, each side of a tergum does not vibrate completely in synchrony. In airborne sound recordings one can see interference in low frequency parts of the calling song which disappear when one side of the tergal fold is immobilised. In view of these facts we may conclude that the vibration producing system in Cydnidae comprises the high frequency stridulatory mechanism and a low frequency system where the tergal plate functions in a similar way to the apparent tymbals of Auchenorrhyncha. The tracheal sacs below the tergal plate probably serve as an elastic support, but possibly also as a resonator (Gogala, 1985b).

Vibratory songs have been recorded and analysed in many species of Cydnidae, both in the subfamilies Sehirinae and Cydninae. Some characteristics are different for each subfamily and some shared by the whole family. In all species investigated vibratory songs are emitted by both males and females, and are sex- and species-specific. They all emit disturbance or distress stridulation calls if attacked or disturbed (Figure 21.2f). The males usually produce two types of songs, considered a calling or first courtship song and a courtship or second courtship song, respectively (Figure 21.2a and d). Two males at mating time often begin to alternate with a rivalry song (Figure 21.2e) (Sehirinae) or sing in a chorus (Cydninae).

The following species of Sehirinae have so far been investigated: *T. bicolor*, *T. sexmaculatus*, *T. rotundipennis* (Gogala, 1970; Gogala and Hočevar, 1990), *Sehirus luctuosus*, *Canthophorus impressus* (Gogala, 1978a), *C. dubius*, *C. melanopterus* (Gogala, 1970), *Legnotus limbosus* (unpublished data).

In the subfamily Cydninae songs are known in two species: *Cydnus aterrimus* and *Macroscytus brunneus* (Gogala, 1978a). Songs have been heard also in some other species.

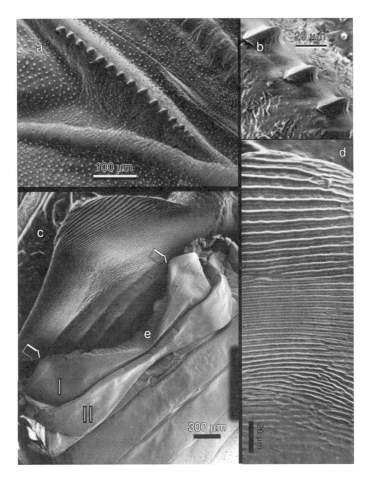

FIGURE 21.1 Vibration producing structures of the bug *Tritomegas bicolor* (Cydnidae). Scanning electron microscopic pictures of (a) the hind right wing stridulitrum, (b) three stridulatory teeth, (c) lima (plectrum) on the right front edge of the tergal plate, (d) its structure under higher magnification and (e) insert with a tergal plate representing fused terga I and II. Parts of the following abdominal segments are seen as well. Arrows are pointing to the left and right lima.

Thyreocorinae are considered by some specialists as a separate family and by others as a subfamily of Cydnidae. In this group the tergal plate is again an important part of the vibratory system, but instead of the lima on the laterofrontal margins of the first tergite a simple ridge is present in *Galgupha ovalis* (Lawson and Chu, 1971), and in *Thyreocoris scarabaeoides* only males have a simple ridge and females have a lima (Drašlar and Gogala, 1976). Signals of a stridulatory origin and low frequency non stridulatory ones were recorded in *T. scarabaeoides* (unpublished data).

Thaumastellidae, Parastrachiidae

Thaumastellidae is a family related to Cydnidae which has a similar, probably homologous but still different stridulatory apparatus (Štys, 1964; Schaefer, 1980). In contrast to Cydnidae, the lima is situated on the second tergum and not on the first one. The stridulitrum is on the postcubitus as in some other related families.

Another exotic group of bugs which were raised to the family level by Sweet and Schaefer (2002) are Parastrachiidae. Also this family has a similar stridulatory apparatus, but there are no reports on their bioacoustic repertoire.

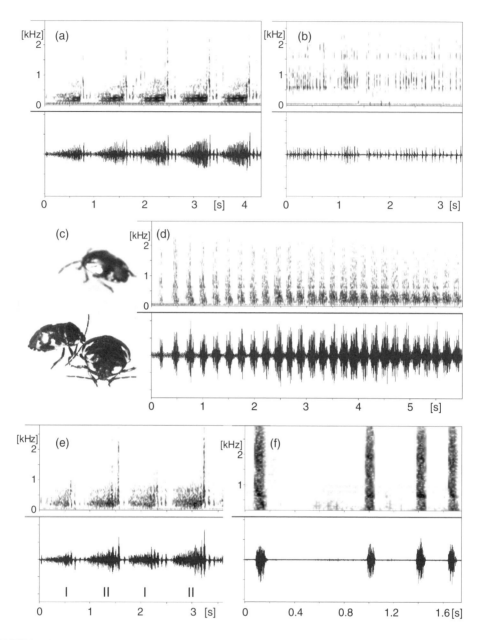

FIGURE 21.2 Songs of *Tritomegas bicolor*. Oscillograms (above) and spectrograms (below) of the (a) male calling song, (b) female acceptance song, (d) male courtship song, (e) rivalry song alternation between two males I–II–I–II and (f) male disturbance calls (different time scale). In (c) two courting males and a female (lower right) are shown.

Tessaratomidae

To our knowledge only a few morphological papers have been published on vibratory structures of this group, but they are similar to the Cydnidae (Muir, 1907; Dupuis, 1953; Leston, 1954). Recordings of stridulatory signals (probably disturbance signals) have been described only for *Tessaratoma javanica* (Puranik *et al.*, 1981), but without a clear biological context and there are no data on low frequency components in vibratory signals.

Pentatomidae, Pentatominae

Pentatominae is the largest subfamily of Pentatomidae and all are plant feeders (Panizzi *et al.*, 2000). Vibratory songs have been described for many species, but no stridulatory signals have been recorded. Bergroth (1905) and Leston (1954, 1957), however, mentioned the rugose areas on sterna II, III (in some species also IV), and tubercles on the hind femora as a possible stridulatory apparatus in another subfamily, the Australian Mecideinae (Halyinae). The function of vibratory songs during mating behaviour has been described in detail for the southern green stink bug, *N. viridula* (Mitchel and Mau, 1969; Fish and Alcock, 1973; Harris and Todd, 1980; Čokl and Bogataj, 1982; Borges *et al.*, 1987; Kon *et al.*, 1988; Čokl *et al.*, 1999, 2000). Mating behaviour includes communication at long and short range. A species-specific male emitted pheromone (Aldrich *et al.*, 1987; McBrien and Millar, 1999) attracts females to plants occupied by males (Borges *et al.*, 1987) and there, triggered by the pheromone or some other chemical signal, females start emitting the calling songs. Males respond with their calling song (Čokl *et al.*, 2000), increased pheromone emission (Miklas *et al.*, 2003) with directional movement to the calling female (Ota and Čokl, 1991; Čokl *et al.*, 1999). When in close proximity, other senses like vision, touch, smell and taste are probably also involved. In small species, like *Holcostethus strictus* (Pavlovčič and Čokl, 2001) and *Murgantia histrionica* (Čokl *et al.*, 2004), in which the calling phase of females is reduced, males start to sing first and the first female song is evoked by male calling.

The following species of Pentatominae have been investigated: *Acrosternum hilare* (Čokl *et al.*, 2001) *A. impicticorne*, *Eushistus heros* (Blassioli-Moraes *et al.*, in press), *E. conspersus* (McBrien and Millar, 2003), *H. strictus* (Pavlovčič and Čokl, 2001), *M. histrionica* (Čokl *et al.*, 2004), *Nezara antennata* (Kon *et al.*, 1988), *N. viridula* (Čokl *et al.*, 2000), *Palomena prasina*, *P. viridissima* (Čokl *et al.*, 1978), *Piezodorus lituratus* (Figure 21.6) (Gogala and Razpotnik, 1974), *P. guildinii* (Blassioli-Moraes *et al.*, in press), *Thyanta custator accerra*, *T. pallidovirens* (McBrien *et al.*, 2002), *T. perditor* (Blassioli-Moraes *et al.*, in press). Unpublished data also exist for *Eysarcoris aeneus*, *Graphosoma lineatum italicum*, *Aelia acuminata*, *Raphigaster nebulosa* (Figure 21.3 to Figure 21.5), *Sciocoris cursitans* and *Eurydema oleraceum*. In most species both males and females emit premating songs, and in some there is also rivalry alternation.

Vibratory signals are produced by vibration of the abdomen as a consequence of simultaneous contraction of a series of muscles connecting the fused first and second abdominal tergites ("tymbal") with other parts of the abdomen and with the thorax on the other side. Simultaneous recording of muscle potentials from different "tymbal" muscles revealed that they contract synchronously and in phase with the cycles of the emitted vibrations (Kuštor, 1989; Amon, 1990; Gogala unpublished: see below and Figures 21.3 to 21.5).

In order to clarify the physical mechanism of low frequency vibration production in this and some related groups, experiments with immobilisation of parts of the abdominal terga in *R. nebulosa* were made. In contrast to the Cydnid bug, *T. bicolor*, the filling of the suture in the tergal plate (fused terga I and II) with wax and resin mixture did not diminish the amplitude of the vibratory signals measured on the tergum by laser vibrometry (Figure 21.3b and Figure 21.4). However, when the front edge of this plate was sealed with wax to the metanotum the signal diminished substantially to about 20% amplitude. After wax removal the amplitude increased again to normal. Experiments with similar results were made by Numata *et al.* (1989) on *Riptortus clavatus* (Alydidae, see below).

If during vibration the abdomen behaves like a pendulum, one would expect that additional load would slow down the vibrations. Adding a 27 mg load, however, caused an overcompensation and increased the carrier frequency. Even a 84 mg load did not substantially change the frequency, but the amplitude dropped to about 25% as compared with the intact animal. The conclusion is that the vibrations of the abdomen produced by longitudinal and dorsoventral muscles are neurally

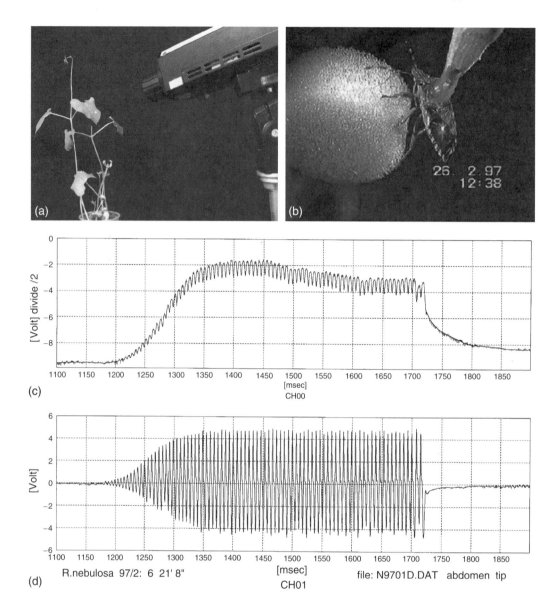

FIGURE 21.3 Measurements of vibratory signals with a laser vibrometer. (a) Setup for measurements of plant vibrations, (b) measurements of the body vibrations of the bug *Raphigaster nebulosa*. The animal is glued to a fixed support, wings are removed to allow vibrometry of the fourth tergite (bright spot) and with the legs it is holding a piece of plastic foam as support. (c) A trace representing a displacement and (d) velocity parameter of the vibratory signal.

controlled, and do not depend much on the physical properties of the abdominal dimensions and mass. As described for *N. viridula*, so in *R. nebulosa* there is a 1:1 relationship between muscular electric activity (EMG) and emitted vibrations (Figure 21.5).

According to behavioural context, songs of stink bugs may be divided into calling songs emitted at longer distances with the function to locate a mate (Ota and Čokl, 1991; Čokl *et al.*, 1999), courtship songs emitted mainly at close distance with the function of mate recognition and rivalry songs involved in male competition for females. Female repelling and copulatory songs were recorded in only a few species (Čokl and Virant-Doberlet, 2003).

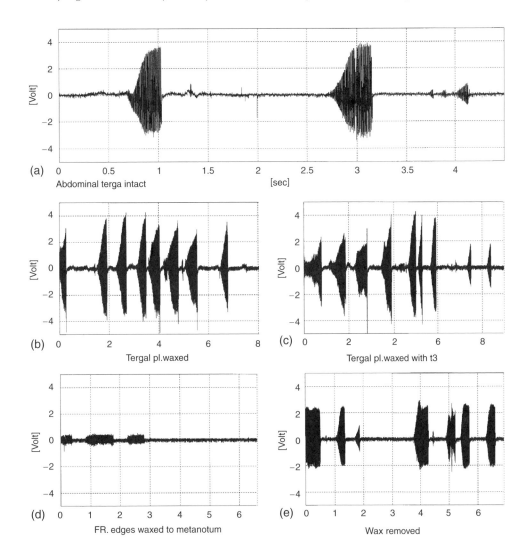

FIGURE 21.4 Vibratory signals of *Raphigaster nebulosa* during a tergal plate manipulation (wings removed). (a) Signals of the animal with the terga intact, vibrations measured on the surface of the fourth tergite (different time scale compared with b–e), (b) the same, tergal plate waxed, (c) the same as (b), tergal plate waxed and glued to the third tergite, (d) the same as (c), front edges of the immobilised tergal plate waxed to metanotum, (e) the same, wax removed from tergum. Note that only a rigid connection of the front edge of a tergal plate to metanotum substantially reduced the vibration signal amplitude.

Spectral characteristics of emitted signals were determined first for signals recorded as airborne sound (Čokl *et al.*, 1972; Čokl *et al.*, 1978; Kon *et al.*, 1988) or as vibrations of a nonresonating loudspeaker membrane used as the substrate for the singing animals. Spectral properties of vibratory signals recorded by these methods reflect vibrations of the body without the influence of mechanical properties of the substrate. Such recorded songs of all investigated stink bugs have low frequency and narrow band characteristics with the fundamental and dominant frequency peak ranging from 70 to 150 Hz, with higher harmonics not exceeding 600 Hz. Although broadband stridulatory signals are just an exception in the whole subfamily, spectra are to some extent expanded in the female calling song (FS1) of the southern green stink bug or male courtship song (MS2) of *P. prasina* and *P. viridissima* (Čokl *et al.*, 1978) combining spectrally different units within the same pulse train.

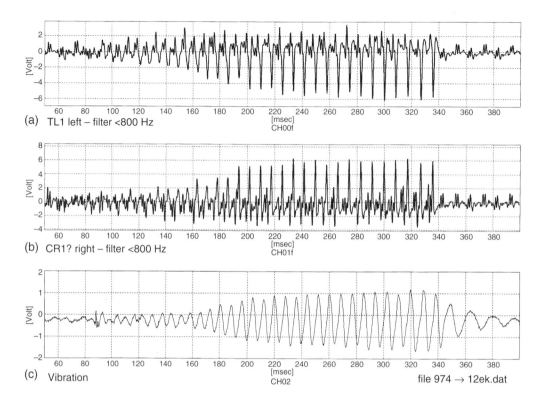

FIGURE 21.5 Vibratory signal of *Raphigaster nebulosa*. Simultaneous electromyography of the left tergal longitudinal muscles TL1 (a), right dorsoventral compressor CR1 (b) and the vibratory signal (c), recorded by the piezoelectric gramophone cartridge, to which the animal was fixed through a short wooden rod in a similar way as shown in Figure 21.3b.

In some species of Pentatominae songs with pronounced frequency modulation (FM) were recorded, as for example in *P. lituratus* (Gogala and Razpotnik, 1974) or *Thyanta* spp. (McBrien *et al.*, 2002). The importance of this song characteristic in acoustic communication has been studied with play-back experiments in *P. lituratus* (Figure 21.6) (Brvar and Gogala, unpublished data). Males of this species emit eight different songs. Males respond to the male FM rivalry signal (MS-R) with the same song. When exactly the same rivalry signal is played back from a tape in reversed direction, it elicits in other males either no or a different acoustic response (MS5, Figure 21.6b and c). Also changes of other parameters, like transposing the stimulation signal up or down (Figure 21.6g–i) by an octave or changing the duration of the signal by a factor of three (Figure 21.6d–f), are critical for responding males. A prolonged rivalry song phrase was more effective than the original song. However, the most interesting conclusion is that the "melody" or FM pattern is relevant for the song recognition.

Species specificity of vibratory songs of other species of Pentatominae is mainly expressed by the different temporal pattern of courtship songs which are emitted during courtship in close vicinity (Kon *et al.*, 1988; Čokl *et al.*, 2000; Čokl and Virant-Doberlet, 2003). In many species like *N. viridula*, a calling male increases the repetition rate of calling song pulses and finally fuses them into a complex courtship song pulse train. The stable pattern of courtship songs is characteristic during alternation between mates. The stable temporal structure of the courtship song is disrupted prior to copulation during intensive antennation and male butting. Females usually stop singing in this period but males of some species may continue singing for some time, even in copula. The function of the copulatory song is not clear. In *N. viridula* a female which is not ready to copulate sometimes emits the repelling song, which then silences a courting male (Čokl *et al.*, 2000).

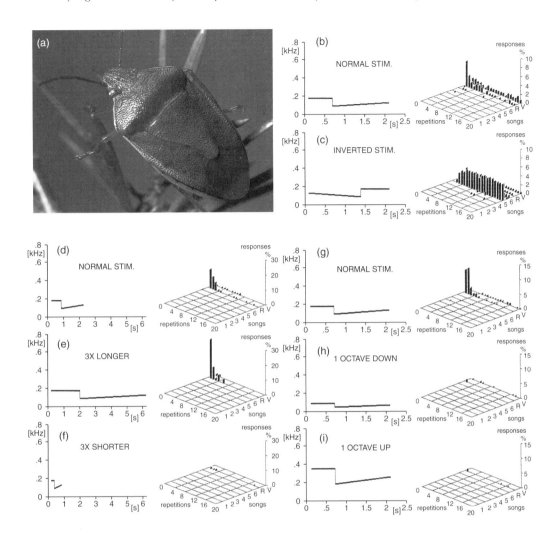

FIGURE 21.6 Results of the playback experiments with the synthetic signals to the males of *P. lituratus* (a). In graphs b–i on the left side a spectrogram of the stimulus is shown and on the right side the three dimensional graph showing type and number of responses to 20 successive stimuli. Normal stimuli with synthetic rivalry signals (b, d, g) elicit mainly rivalry responses (R) which decline exponentially and some disturbance responses (V). To the inverted signal (c) animals react mainly with the MS5 signals (5). A prolongation of the stimulus by a factor of 3 (e) are very effective but the time compressed stimuli (f) are not. Stimuli transposed one octave down (h) or up (i) are also not effective.

The male rivalry song is present in most species and shows similar temporal characteristics. Two (or more) males alternate in an a–b–a–b–a… fashion until one of them stops singing and the dominant one proceeds to calling and courting a female (Čokl and Virant-Doberlet, 2003).

Although pheromones and songs show species and sex specificity, they are probably, at least in this group of bugs, not the exclusive barrier to hybridisation. Interspecific mating was observed in the field between *N. viridula* and *N. antennata* in regions where they are sympatric (Kiritani *et al.*, 1963; Kon *et al.*, 1993, 1994). Furthermore, as with pheromones (Aldrich *et al.*, 1987; McBrien and Millar, 1999), statistically significant differences have been demonstrated in the temporal structure of different songs between individuals of the same population (Miklas *et al.*, 2003) as well as between geographically isolated populations (Čokl *et al.*, 2000). Nevertheless, members of different populations normally mate and copulate at least under laboratory conditions.

Pentatomidae, Asopinae

Recordings of vibratory songs and observations of mating in two species were made. The main peculiarities observed in the acoustic behaviour of *Picromerus bidens* (Figure 21.7a) were a kind of dancing movements observed in males (Gogala, 1987, unpublished data). During courtship, males orient towards the female and rhythmically rise and extend the front leg or legs in her direction, drop them to the substrate and pull them back after touching the substrate. These movements produce vibratory signals in a regular rhythm (Figure 21.7e) and are combined and occasionally interrupted by a low frequency courtship song (Figure 21.7f).

Similar observations made recently by Virant-Doberlet on *Podisus maculiventris*, another predatory species of this group. She observed in males a similar tapping behaviour with corresponding vibratory signals, normal low frequency song, and vibratory emissions during jerky shuttering lateral movements (Figures 21.7b–d) (Virant-Doberlet, unpublished data).

Scutelleridae

Stridulatory structures of this group of bugs were described in older papers (Leston, 1954; Leston and Pringle, 1963a, 1963b). Two songs were recorded in males of *Odontotarsus purpureolineatus*, different in frequency characteristics due to different vibration producing mechanisms (Gogala, 1984). One is the low frequency signal as in most related families and the second a true stridulatory signal, produced most probably by scraping the setous tubercles on the hind tibiae against rugose areas on abdominal sternites four to six during vigorous lateral movements of the abdomen. Schaefer (1980) and Au (1969) described completely different stridulatory structures in *Calliidea* and *Scutellera*. The plectrum described has longitudinal striations along the body axis in the medial fifth of the anterior border of the first tergite. In another 12 genera of this group, the same authors found a stridulitrum on the postcubital vein, but other details and songs are unknown.

Leston (1957) described an unusual probably pygophoral stridulatory mechanism in *Sphaerocoris* where conjunctival appendages function as plectra and six transverse rows at the tip of a genital segment act as a stridulitrum. No vibratory or sound signals were reported.

A new clade of Pentatomidae from S and W Australia (1 gen.n., 2 spp.n.) and Queensland (1 gen.n., 2 spp.n.) is characterized by the presence of diagonal stridulitra on abdominal ventrites 3 or 3 + 4, respectively, in adults (more developed in males) and plectrum formed by a row of denticles on the dorsal face of metafemur. The same kind of plectrum occurs also in larvae of 2nd–5th instar, while larval stridulitrum is not homologous to that of the adults, being formed by striate ridges distinctly present on abdominal ventral laterotergites 2 and 3 (and in a rudimentary form on more posterior ones) in both genera. Larval strigilatory organs present in other Heteroptera (Reduviidae, one species of Rhyparochromidae) are always homologous to those of the adult bugs (Pavel Štys, pers. comm.).

Plataspidae

The only species of this group investigated acoustically (Gogala, 1990) is *Coptosoma scutellatum*. Vibratory courtship song of the male peaks around 200 Hz (basic frequency) probably produced by abdominal vibrations (Figure 21.8).

Acanthosomatidae

The first data on emission of vibratory signals in the Acanthosomatidae were published by Jordan (1958) for *Elasmucha grisea*. According to this author males emit long lasting (1 to 3 sec) calls with a basic frequency of only 67 to 83 Hz during courtship.

Vibratory emissions of some other species (*e.g. Cyphostethus tristriatus*, Gogala, unpublished) have been recorded, but a thorough investigation of the acoustic behaviour of any species is still lacking.

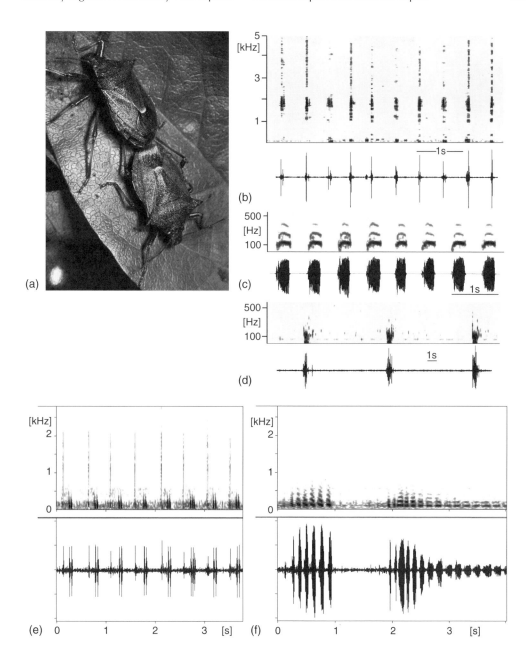

FIGURE 21.7 Dancing behaviour of Asopinae, *P. bidens* (a, e, f, Gogala unpublished) and *P. maculiventris* (b–d, Virant-Doberlet, unpublished). During courtship males of both species rhythmically rise and extend the front leg or legs in the direction of the female and drop them loudly to the substrate (tapping signal: b, e). They produce also low frequency signals by vibrations of the abdomen, as most other Pentatomidae (c, f). *P. maculiventris* shows also a third type of vibratory signal, produced during lateral shuttering of the body (tremulation: d). b–f: above spectrogram, below oscillogram.

Piesmatidae

A stridulatory mechanism found in *Piesma quadrata* represents a variation of the mechanism found in Cydnidae, Thaumastellidae, Tessaratomidae and related groups. Here, the plectrum (lima) is on a

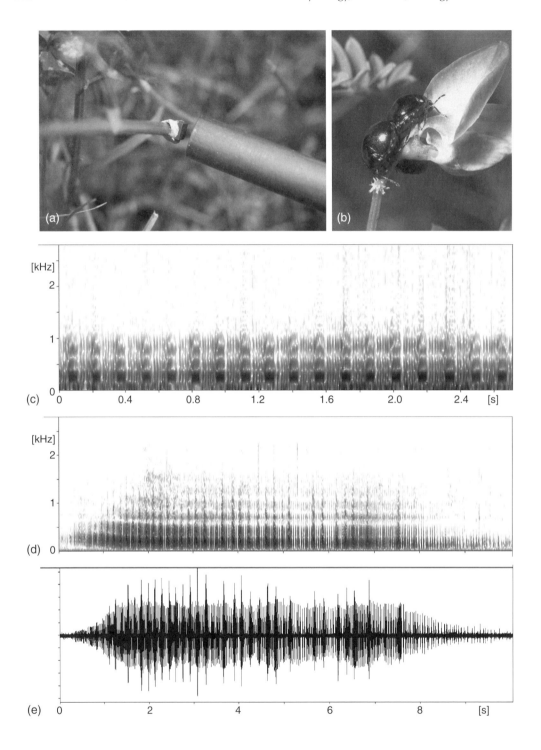

FIGURE 21.8 Vibratory signals of *Coptosoma scutellatum* (b), recorded by magnetoinductive system (a, d, e) or by the loudspeaker (c). (a) A small magnet is glued to the plant stem with the insects and the shielded coil is brought very close to the magnet. The spectrogram (d) and oscillogram (e) of such recording of the song is shown in comparison with a loudspeaker or earphone cartridge recording (c, extended time scale) (see also Figure 21.10b).

separate sclerite laterally in front of the tergal plate and the stridulitrum is on the cubital vein (Leston, 1957). Haskell (1961) recorded and published also stridulatory signals in this species, but reinvestigation of the vibratory communication signals is needed. Leston (1957) described deformation of the tergal plate during the action of longitudinal muscles, causing also the movement of the sclerite. Thus, one can expect not only high frequency stridulatory vibrations, but also low frequency vibrations as in most other Heteroptera.

Lygaeidae *sensu lato*

The oldest reports on sound or vibration producing structures in this group are in Leston (1957) and Leston and Pringle (1963a, 1963b) who reported for *Kleidocerys resedae*, a stridulatory apparatus with a stridulitrum on the hind wing (see Figure 21.9a and b) and as a supposed plectrum, the ridge on the postnotum. Ashlock and Lattin (1963) pointed out that the postonotal ridge is not mobile and cannot function as a plectrum, the moving part of this stridulatory mechanism is the fused terga I and II, which have striate areas on the anterior part ("lima", Figure 21.9c and d). Nevertheless, even in some recent textbooks authors repeat this old misinterpretation of the stridulatory structures based on Leston's description (*e.g.* Chapman, 1998). Ashlock and Lattin (1963) described in the same paper four different stridulatory mechanisms with the stridulitrum on the hindwings in *Kleidocerys*, on the prothorax as in *Pseudocnemodus*, on the costal and abdominal margin in *Xyonysius*, *Abpamphantus* or *Balboa* and on the abdomen in *Heteroblissus*. In *Plinthisus* spp. the stridulitrum is on the hind wings, ever in brachypterous animals and a plectrum (lima) on the first tergite (Sweet, 1967; Péricart, 1998). It is clear that these stridulatory structures have evolved independently many times within the Lygaeidae. At least in some of them sounds or vibratory signals have been heard or detected (see Figure 21.9e) (Leston, 1957; Thorpe and Harrington, 1981).

Another low frequency vibration producing mechanism and corresponding signals were discovered in *Platyplax salviae* (Gogala, 1984) and *Chilacys typhae* (Gogala, unpublished). The low frequency signals in Lygaeidae are probably produced by abdominal vibrations similar to the system in Pentatomidae (see above).

An extraordinary behaviour pattern has been observed during courtship in the Lygaeid bug, *Ryparochromus vulgaris*. The male shakes the body laterally during courtship, oriented with the head towards a female, and produces vibratory signals which could at least partly be produced by rubbing the legs against the body (Figure 21.10b and c). Such stridulatory device is mentioned for Cleradini and Myodochini in Péricart (1998).

Colobathristidae

In the American genera of Colobathristidae pedo-cephalic stridulatory structures have been described (Štys, 1966; Schuh and Slater, 1974). The stridulitrum is supposed to be the infraocular ridges on the head and the plectrum spines on the inner side of anterior femora and tibiae. No vibratory signals have been described.

Pyrrhocoridae

There are no reports of vibratory organs in this family. There are no reports either on vibratory signals, so common in most species of Heteroptera. This may be due to lack of bioacoustic investigations.

Largidae

Lattin (1958) described a stridulatory apparatus in *Arhaphe cicindeloides* and mentioned that also other species or genus have a similar structure, a strigil on the lateral edge of hemelytron and a plectrum in the form of a rugose area on the hind femur.

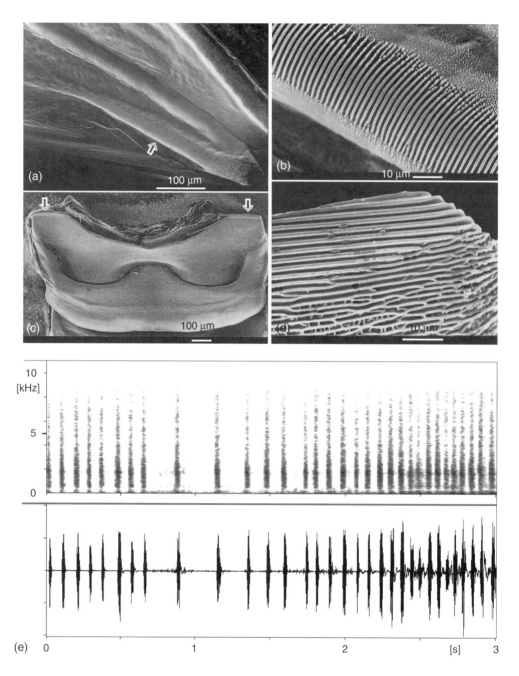

FIGURE 21.9 Stridulation of the lygaeid bug *Kleidocerys resedae*. SEM photographies of the stridulitrum (a, b) are shown as well as the tergal plate (c: fused terga I and II) and the lima (plectrum) at the front edge of tergum I (c: arrows, d). (e) Stridulatory signals of the same species (male), spectrogram above and oscillogram below.

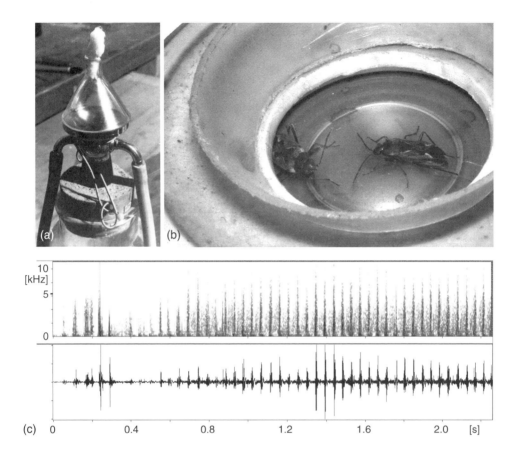

FIGURE 21.10 Simple recording method to detect vibratory songs of *Ryparochromus vulgaris* or other insects is the use of a loudspeaker or earphone cartridge as shown (b). In olden days people were using a stage stethoscope for similar purpose (a). *R. vulgaris* shows during courtship a kind of dance with rhythmic lateral body vibrations (shuttering) producing a vibratory (stridulatory?) song with a regular pattern of 45 pulses per second (c).

STENOCEPHALIDAE

To my knowledge no reports on acoustic or vibratory behaviour of these bugs have been published.

COREIDAE

Also this large group of Heteroptera has been neglected from a bioacoustic point of view. The main reason for this is probably a lack of stridulatory structures in most genera and species. Štys (1961) however described stridulatory structures on ventrolateral edges of the pronotum and the plectrum (lima) in the form of a modified process on the articular region of the forewing. This structure was described in *Centrocoris* and *Phyllomorpha* and is probably developed also in some other taxa. Sounds have been heard, but no recordings made. However, unusual high frequency vibratory emissions connected with locomotion were reported in *Enoplops scapha*, as well as in *Coreus marginatus* (Gogala, 1984). In addition, in both species also low frequency vibratory signals with frequency sweeps were recorded and analysed, similar to the vibratory songs in many other Pentatomomorpha (see also Gogala, 1990). Low frequency vibratory signals have been recorded also in some other Coreidae (*e.g. Gonocerus juniperi*) and also the recent nearctic intruder into Europe, *Leptoglossus occidentalis*, which is able to make similar low frequency signals (Gogala, unpublished data).

ALYDIDAE

Stridulatory structures have been described in many taxa of this group (Schaefer and Pupedis, 1981). A femoral alary strigilation system has been found in *Megalotomus junceus* and *Alydus calcaratus* (Gogala, unpublished). Songs were investigated in *R. clavatus* (Numata *et al.*, 1989). They found not only stridulatory signals, but also low frequency vibrations not produced by the same stridulatory mechanism. Immobilisation of various parts of abdomen showed that the low frequency vibration producing mechanism is not a tymbal formed by fused terga I and II, as postulated by analogy with similar structures in Cydnidae. Probably, the low frequency vibrations are produced by vibration of the whole abdomen, as in Pentatomidae.

Vibratory emissions of *A. calcaratus* gave similar results (Gogala, 1990). Also these bugs produce low frequency signals (200 to 800 Hz) very similar to those of Coreidae, and the stridulatory signals produced by rubbing the femora against the edges of hemelytra can occur simultaneously with low frequency vibrations in the higher frequency band (main emission between 1 and 3 kHz).

RHOPALIDAE

The only species which has been investigated is *Corizus hyoscyami*. The vibratory signals of males with highly damped transients are frequency modulated and have a main intensity peak at about 80 Hz (Gogala, 1990). Other species need to be investigated.

ARADIDAE

Usinger (1954) and Usinger and Matsuda (1959) described different stridulatory devices, mainly with the stridulitrum on abdominal ventral segments (second or third) and the plectrum or scraper on the hind femora. Nevertheless, in *Artabanus lativentris* series of teeth are situated on the hind tibiae, and on the second ventral abdominal segment is a ridge with many parallel striae completing the apparent stridulatory apparatus. No reports on acoustic behaviour or vibratory songs have been published.

CIMICOMORPHA

Tingidae

The sycamore bug, *Corythuca ciliata*, introduced some decades ago to Europe from North America, is to my knowledge the only species so far investigated (Gogala, 1985a). Males emit broadband ticking sounds during mating. The mechanism of a vibration production is not known. Nevertheless, it is clear that singing animals move the abdomen jerkily laterally from left to right in the rhythm of the clicking sounds (Gogala, 1984).

Miridae (Capsidae)

Some authors reported stridulatory structures with a stridulitrum on the edge of the embolium and the plectrum on the hind femora in some African Mirinae and Hyaliodinae (Schuh, 1974; Akingbohungbe, 1979). Nevertheless, no sound or vibration emissions have been reported. On the other hand, Groot (2000) described vibratory signals in males of *Lygocoris pabulinus* during courtship. The carrier frequency is about 200 Hz, 2 to 3 sec long phrases comprise pulses with a repetition rate of 5.2 ± 0.7 pulses per second. Strong *et al.* (1970) also noted vibratory signals in the species *Lygus hesperus*.

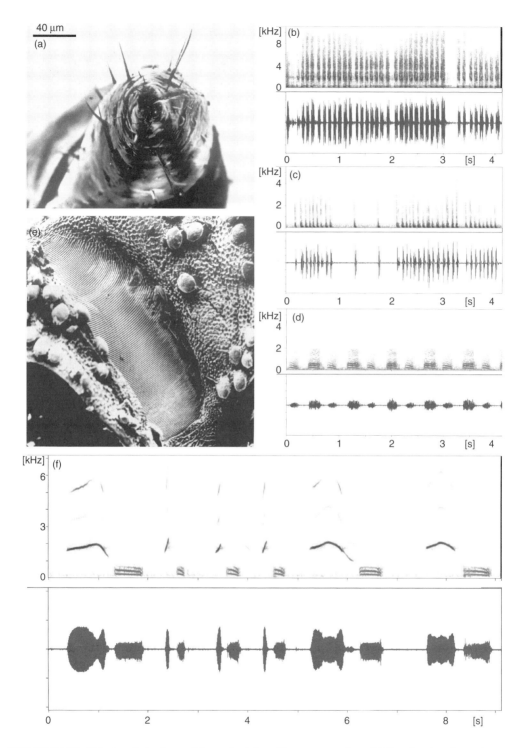

FIGURE 21.11 Vibratory behaviour of the ambush bug *Phymata crassipes*. (a) tip of proboscis as plectrum (above) and stridulitrum (below, SEM). (b) Stridulation signals, (c) locomotory signals and (d) alternation of two animals with the low frequency signals. (e) Human whistling sounds elicit low frequency alternation signals in a single ambush bug. Note that the duration of the response depends on the stimulus sound duration. (b)–(e) spectrograms above, oscillograms below.

It is possible that vibratory communication is also much more common in this group than reported.

Nabidae, Anthocoridae and Cimicidae

We have no data on vibratory communication in the families Nabidae, Anthocoridae and Cimicidae. Some authors described setae on the genital segment and hind tibiae of Nabidae as a stridulatory mechanism, but no sounds were ever heard (Leston, 1957). The behaviour of some Nabids during courtship, previously described as stridulatory movements, is apparently connected with pheromone dispersal from cuticular glands.

Reduviidae (Including Phymatinae)

As mentioned earlier, the vibration or sound producing structures were recognised first in this group of bugs (Ray, 1710; Poda, 1761). The strigil is a striated furrow between front coxae and the plectrum the tip of the rostrum. Most species of this family of bugs produce disturbance or alarm sounds with this device, which can be heard without any special apparatus by the human ear. These signals may be important, not only for intraspecific communication, but also for warning which may account for the high frequency component transmitted as airborne sound. Nevertheless, also vibratory signals of low frequency have been detected (Gogala and Čokl, 1983; Gogala, 1985b). The mechanism has not been studied in detail, but dorsoventral vibrations of the abdomen during low frequency emissions are obvious.

Some aspects of the acoustic behaviour of Reduviidae Triatominae are discussed by Lazzari *et al*. (Chapter 22).

Acoustic behaviour of *Phymata crassipes* (Reduviidae: Phymatinae) has been studied in more detail (Gogala and Čokl, 1983; Gogala *et al*., 1984). Males, females and larvae stridulate when disturbed (Figure 21.11a and b). Popping sounds are heard or detected as vibratory signals during locomotion (Figure 21.11c). Low frequency alternation signals are emitted by both sexes as a response to vibratory signals of another conspecific or to other sounds or vibratory stimuli from the environment (Figure 21.11d and e). This bioacoustic response is very interesting, since ambush bugs, *P. crassipes*, mimic the duration of sound stimuli (Figure 21.11e) in an complicated way — not linearly but in relation to the logarithm of the stimulus duration (Gogala *et al*., 1984). The most probable explanation for this mocking behaviour is in the attraction of potential prey by the imitation of their acoustic or vibratory signals.

CONCLUSIONS

Many authors describing different stridulatory mechanisms of the Heteroptera have mentioned the diverse mechanisms evolved in families like Lygaeidae, Aradidae and Scutelleridae independently many times. Despite the fact that in some other taxa the type of the vibratory organ is uniform and typical for the group (*e.g.* Cydnidae, Alydidae, Reduviidae) this diversity was surprising. Another surprising fact is that very similar stridulatory mechanisms are found in groups certainly not closely related, like in some Pentatomomorpha (*e.g.* Cydnidae, Tessaratomidae, see above) and Leptopodomorpha (Leptopodidae, Pericart and Polhemus, 1990) with tergal plectrum (lima) and stridulitrum on the hind wings. Another example is the Alydidae (Schaefer and Pupedis, 1981) and some Miridae (Schuh, 1974; Akingbohungbe, 1979) with the plectrum on the hind femora and stridulitrum on the lateral edge of the hemelytrae.

We know that the basic and widespread vibratory communication system in Heteroptera is a low frequency one, as in small Auchenorrhyncha (Claridge and de Vrijer, 1994). Apparently the physical mechanism is not exactly the same in all examples, but the integumental structures and muscles involved are homologous. The longitudinal and dorsoventral muscles attached to the fused terga I and II can produce a click (tymbal) mechanism as in Cydnidae, or the whole abdomen and body vibrates as in many other families (*e.g.* Pentatomidae, Alydidae, Miridae). With some exceptions bugs use this basic communication system and in addition stridulatory mechanisms evolved on many adjacent body parts. Stridulatory signals broaden the frequency range toward higher frequencies and enrich the diversity of species- and sex-specific signals.

Experiments have shown that for intraspecific communication bugs use substrate-borne signals (Gogala *et al.*, 1974; Michelsen *et al.*, 1982; Čokl and Virant-Doberlet, 2003). This is not surprising since small sources are bad emitters of low frequency sound, but can be efficient sources of substrate vibrations. Nevertheless, the stridulatory signals of bugs in higher frequency ranges can be detected by the human ear and also by other animals. Thus, such stridulatory disturbance or alarm sounds can be effective also as interspecific airborne signals. In addition, the case of the mocking behaviour of the ambush bug, *P. crassipes*, shows also that airborne sounds may elicit acoustic reactions in Heteroptera, probably indirectly through excitation as substrate vibrations.

ACKNOWLEDGEMENTS

I am sincerely grateful to Andrej Čokl and Meta Virant-Doberlet for contributing material for subchapters on Pentatomidae and Pavel Štys for many useful suggestions and remarks. I appreciate also the work and patience of both editors of this book.

22 Vibratory Communication in Triatominae (Heteroptera)

Claudio R. Lazzari, Gabriel Manrique and Pablo E. Schilman

CONTENTS

INTRODUCTION

Triatomines are haematophagous bugs (Heteroptera, Reduviidae, Triatominae), vectors of the flagellated parasite, *Trypanosoma cruz*, which causes Chagas disease. The American trypanosomiasis constitutes one of the most serious sanitary problems in Latin America, and generates important social and economic impacts (Dias and Schofield, 1999). Currently, about 16 to 18 million people are infected with this disease and approximately 120 million more are at risk of becoming infected. That is, about 25% of the population of Latin America is in danger (WHO, 2002). Triatomines comprise more than 130 species that inhabit a number of different habitats, such as nests, burrows, hollow trees as well as human dwellings and peridomestic structures where they live in association with birds, mammals, reptiles and human beings. Their ability to adapt to the human habitat, which offers abundant food (*e.g.* blood of humans, domestic animals, associated rodents, *etc.*) and many resting places that are easy to colonise (*e.g.* cracks and crevices in walls made of dried mud and thatched roofs) define, among other features, the importance of each species as a vector of Chagas disease.

A thorough understanding of the biology of triatomines is imperative because of their potential importance as a health hazard to humans, including a study of their communication systems. The ability of Triatomines to stridulate was recognised many years ago (Readio, 1927; Hase, 1933; Leston, 1957; Schofield, 1977). Some species are able to produce sounds that are audible to man, but the biological significance of stridulation has remained elusive principally because early studies ascribed a role in acoustic communication to the associated sound. However, these bugs lack obvious hearing organs and there is no behavioural evidence that they could respond to airborne sounds (Schofield, 1977). The subject remained neglected until research was focused on the vibratory component of the signal (Manrique and Lazzari, 1994). Almost no studies on sensitivity to substrate-borne signals had been made, except for that of Autrum and Schneider (1948). More recently, the signals produced in different behavioural contexts have been studied and their probable implications have been analysed.

Here we describe the morphology of the stridulatory organ, as well as the structure of the signals that the bugs produce. Based on evidence from different species, we suggest a biological significance for signals produced in different behavioural contexts.

SIGNAL PRODUCTION

Triatomines produce vibrations using a stridulatory organ located on the ventral part of the prothorax in both larvae and adults. It consists of a longitudinally oriented groove, located between the insertions of the forelegs and lined by transverse ridges (Figure 22.1). Stridulation occurs when the insect rubs the tip of its rostrum against the prosternal groove, scraping across the ridges with alternate backwards and forwards movements (Moore, 1961; Schofield, 1977; Di Luciano, 1981). About 90 ridges, separated by 6 to 9 μm, form the stridulatory groove of adult *Rhodnius prolixus* (Figure 22.1, also Manrique and Schilman, 2000). During long chirps, the number of pulses recorded was approximately 90. If there is a 1:1 ratio between the number of pulses recorded and the number of ridges rubbed during disturbance stridulation, *R. prolixus* would use the whole groove to produce these 90 pulses. However, analysis of vibratory signals and frame-by-frame analysis of simultaneous video films (Schilman *et al.*, 2001) suggested that they rub around a third to a half of the central portion of the groove during stridulation, making a 1:1 ratio between the number of pulses recorded and the number of rubbed ridges impossible. It is not yet clear whether each ridge is impacted only once by the tip of the rostrum in every cycle or, alternatively, whether more than one structure plays a role in the production of vibrations.

Early reports were typically based on the observation of stridulation when bugs were grasped with forceps (Moore, 1961; Schofield, 1977), and when bugs sometimes produced sounds that were audible to humans. Even though adults of both sexes, as well as larvae of all instars, are able to stridulate, the spontaneous occurrence of stridulation of untethered bugs has rarely been observed. However, about ten years ago, two of us (Manrique and Lazzari, 1994) observed the occurrence of spontaneous stridulation during mating in *T. infestans*. Thus, for the first time, stridulatory behaviour was described in another context than disturbance.

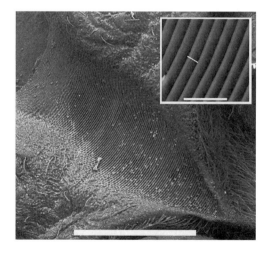

FIGURE 22.1 Prosternal stridulatory organ of a R. *prolixus* female viewed by environmental scanning electron microscope (scale bar = 350 μm). Inset: magnification of the transverse ridges rubbed by the tip of the insect's proboscis (scale bar = 20 μm; short diagonal line indicates distance between ridges = 5 μm). (Modified from Manrique, G. and Schilman, P. E., Two different vibratory signals in *Rhodnius prolixus* (Hemiptera: Reduviidae), *Acta Trop.*, 77, 271–278, 2000. With permission from Elsevier.)

MALE-DETERRING SIGNALS

Although some interspecific differences exist, mating in Triatomines is essentially similar across species. It is characterised by a series of behavioural steps, performed mainly by the male. Briefly, the male approaches the female and either suddenly jumps on to her or mounts her slowly. Then, the male grasps the female with its legs and attempts to copulate. The male's copulatory attempts are not always successful since female receptivity varies, being a main factor affecting success of copulation, at least in species such as *T. infestans* (Manrique and Lazzari, 1994) and *Panstrongylus megistus* (Pires *et al.*, 2004). Nonreceptive females usually display different kinds of rejection behaviour, including stridulation (Manrique and Lazzari, 1994). Female stridulation occurs once the male has mounted and in response the male stops his attempt to copulate. The male then remains motionless, irrespective of the mating step he was performing at the time of the stridulation. Then, when the female stops stridulating, the male resumes the mating sequence. If the female stridulates for a second time, the male stops once more. If the female produces several rejections, the male definitively abandons her. Males rarely further persist in copulation attempts, regardless of female stridulation (Manrique and Lazzari, 1994). Once a female is in direct contact with a male, these male-deterring stridulations are fully effective since the male ceases efforts to copulate. Playback experiments showed a *T. infestans* female artificially vibrated to be similarly effective in rejecting males' copulatory attempts (unpublished data). Therefore, as in other insects, such as the harvester ants *Pogonomirmex* (Markl *et al.*, 1977), stridulation behaviour by females during mating seems to play an important role in communication (Manrique and Lazzari, 1994; Roces and Manrique, 1996; Manrique and Schilman, 2000). No evidence is currently available about spontaneous stridulation of males in a sexual context.

The use of vibratory signals in behavioural contexts other than mating has been analysed in *T. infestans* and *R. prolixus* (Roces and Manrique, 1996; Manrique and Schilman, 2000). In both, under unrestrained conditions, stridulations by females were recorded as substrate-borne signals and were carried as waves propagating through the substrate or through the bodies of interacting individuals.

Male-deterring stridulation of female *T. infestans* consists of long series of repetitive syllables, each composed by two chirps (Figure 22.2), while those of *R. prolixus* are characterised by syllables consisting of a single chirp (Figure 22.3). Larvae, males and females of both species also emit disturbance stridulation when they are clasped with forceps. These differ in their frequency spectra. The maximum energy of disturbance stridulation in *T. infestans* occurs at 1500 Hz, a value well above that for male-deterring stridulation (700 to 800 Hz) (Figure 22.4). For *R. prolixus*, the main carrier frequencies of male-deterring and disturbance stridulation, respectively, reach 1500 and 2200 Hz. Clearly, both types of stridulation produced by *R. prolixus* are at a higher frequency than the comparable ones by *T. infestans*. Moreover, clear differences in the temporal pattern of both could also be seen. In both species disturbance stridulation was repeated at a lower rate: about 24 syllable/s in male deterring signals and 8 syllables/s in disturbance for *T. infestans*, and 36 and 6 for *R. prolixus*.

Like nonreceptive females of *T. infestans* and *R. prolixus*, nonreceptive females of *P. megistus* stridulate to reject male copulatory attempts (Pires *et al.*, 2004). Although there are some differences in the mating behaviour of different species of triatomines, stridulation of nonreceptive females during sexual behaviour seems to be a common feature.

DISTURBANCE SIGNALS

As in other insects, vibratory signals produced in different behavioural contexts should match the reception capacities of a receiver and require different degrees of specificity according to their biological significance. As originally proposed for Reduviidae (Leston, 1957) and for wasps

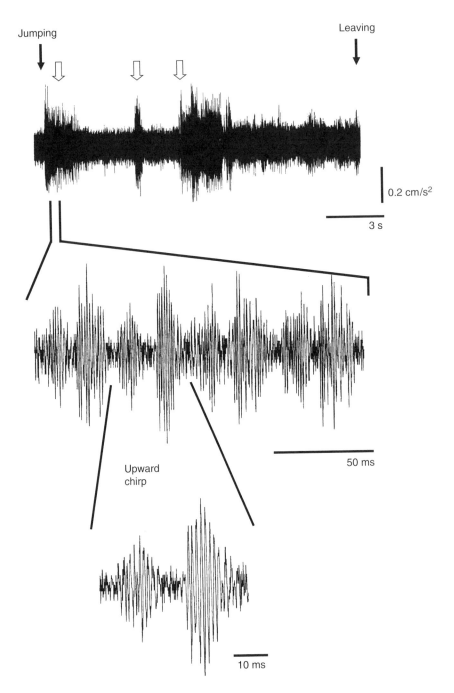

FIGURE 22.2 Male-deterring stridulations produced by a *T. infestans* female recorded as substrate-borne vibrations. Vertical calibration bar (acceleration in cm/s^2) applies to all three records. White arrows indicate copulatory attempts by the male; black arrows show when the male jumped on the female and when he retreated. (Reprinted from Roces, F. and Manrique, G., *J. Insect. Physiol.*, 42, 231–238, 1996. With permission from Elsevier.)

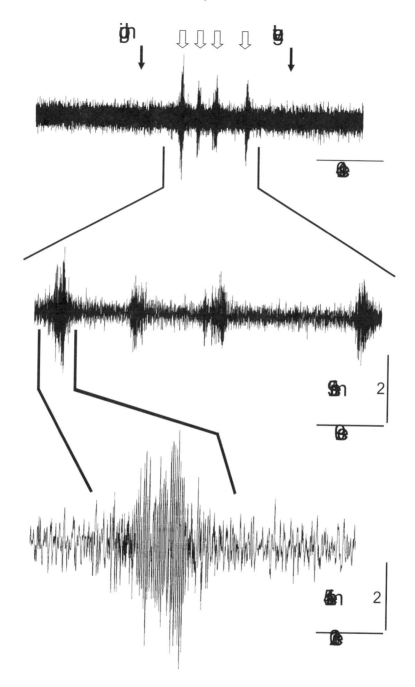

FIGURE 22.3 Male-deterring stridulations produced by a *R. prolixus* female. White arrows and black arrows are as in Figure 22.2. (Reprinted from Manrique, G. and Schilman, P. E., *Acta Trop.*, 77, 271–278, 2000. With permission from Elsevier.)

(Masters, 1979) and beetles (Bauer, 1976), stridulation by insects in response to disturbance would be defensive responses to predators. As a consequence, we should expect this signal to be similar among species, with a generalised common structure. To test this hypothesis, Schilman *et al.* (2001) analysed disturbance calls emitted by *T. infestans, T. guasayana, T. sordida,*

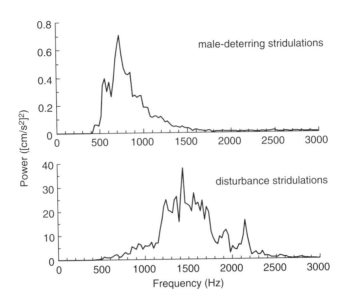

FIGURE 22.4 Power spectra of male-deterring (average of eight records) and disturbance stridulations (average of three records). Time windows of 100 ms were used for the fast Fourier transform analysis. Signals were amplified 104 times and filtered between 400 and 5000 Hz. FFT size: 2048. Bin width: 24.4 Hz. (Modified from Roces, F. and Manrique, G., *J. Insect. Physiol.*, 42, 231–238, 1996. With permission from Elsevier.)

R. prolixus and *Dipetalogaster maxima*. In all, stridulation consisted of long series of repetitive syllables, each one composed of two chirps, a short and a long one (Figure 22.5). The exception was *R. prolixus* which produced a series of short chirps and then a long one (Figure 22.5). The vibratory signals of all species analysed show similar frequency spectra and repetition rates (Table 22.1), even though the insects differ in size and their stridulatory grooves have different interridge distances. The analysis of the frequency spectra showed that the main carrier frequency of the vibratory signal reaches a value of *ca.* 2000 Hz. The temporal structure for distress signals of all five species showed repetition rates of around 8 syllables/s (Table 22.1).

The hypothesis that these calls are nonspecific disturbance responses receives further support from the similar power spectra and repetition rates produced by disturbed larvae and adults of *D. maxima* (unpublished results). Since larvae lack sexual behaviour, larval stridulation is probably elicited only in response to disturbance. To produce these general vibratory signals with similar frequencies and temporal modulation, insects of different species or larval stages should compensate for differences in body size, *e.g.* by employing a higher scraping velocity in species with longer interridge distances.

Disturbance stridulation is produced by many other insect species. For example, workers of the leaf-cutting ant, *Atta cephalotes*, stridulate when they are prevented from moving. This signal is repeated at a low rate, about seven syllables/s (Markl, 1968), a value near to the repetition rate of disturbance stridulation in Triatomines (Table 22.1, and also Schilman *et al.*, 2001). However, like *R. prolixus* and *T. infestans*, ants stridulate at higher rates in other contexts such as during leaf cutting (Roces *et al.*, 1993).

Finally, it is worth commenting that when disturbed, many adults (*e.g. R. prolixus, T. infestans, P. megistus*) release pungent odours (Schofield, 1979). This chemical signal may function to repel predators (Schofield, 1979) and also at low concentrations it seems to play a role as an alarm pheromone for conspecifics (Ward, 1981). The simultaneous use of both

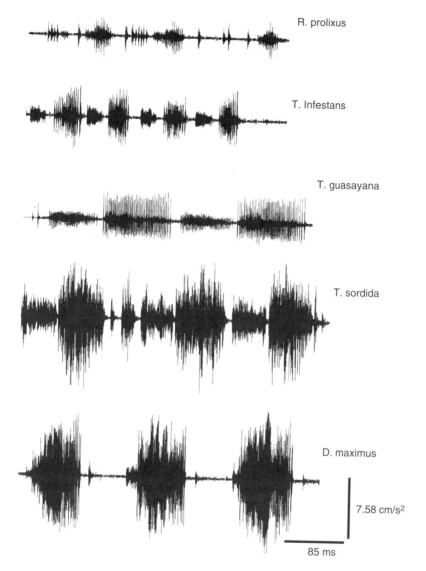

FIGURE 22.5 Disturbance stridulations produced by five different triatomine species. Insects were clasped with forceps by one leg and the signal was recorded directly as acceleration of the dorsal cuticula. Calibration bars (time in ms and acceleration in cm/s^2) apply to all five records. (From Schilman, P. E., Lazzari, C. R., and Manrique, G., *Acta Tropica*, 79, 171, 2001. With permission from Elsevier.)

chemical and mechanical signals during disturbance is known in several insect groups, as for example in some ants (Markl, 1968). The beetle *Elaphrus riparius* uses both signalling systems simultaneously when disturbed, *i.e.* stridulating and releasing the content of pygidial glands (Bauer, 1976). Although in Triatomines disturbance stridulation has not yet been proved to discourage predation, the available information supports the idea (Schilman *et al.*, 2001). At this point, two possible mechanisms of action can be proposed: either stridulations simply frighten or confuse predators, or the vibratory signal may serve as an indicator of bad taste or toxicity, *i.e.* aposematism. The emission of pungent secretions accompanying adult stridulation seems to support the second hypothesis.

TABLE 22.1
Repetition Rate and Frequency Spectra of Disturbance Stridulations and Body Sizes of Adults of Five Triatomine Species

Triatomine Species	Total Body Length (mm)	Repetition Rate (syllable/s)	Frequency Spectra (Hz)
Rhodnius prolixus	17.2 ± 0.2	6.75 ± 0.6	1900–2900
Triatoma infestans	24.2 ± 0.4	12.29 ± 0.24	1600–3200
T. sordida	19.7 ± 0.4	5.22 ± 0.09	1800–2000
T. guasayana	16.6 ± 0.3	9.64 ± 0.33	1800–2400
Dipetalogaster maxima	35.5 ± 0.8	5.78 ± 0.13	1800–2600

Values are mean ± SE.
Source: From Schilman, P. E., Lazzari, C. R., and Manrique, G., *Acta Trop.*, 79, 171–178, 2001.

CONCLUSIONS

To combat the physical problem that high frequency signals become steeply attenuated during propagation through the air, triatomine bugs, like many other insects, use the substrate or their bodies to propagate vibratory signals. For example, when females stridulate to reject mating, the signals are propagated through direct body–body contact with the male. It is interesting to note that the disturbance signal and the male-deterrent signal within one species, *e.g.* either *T. infestans* or *R. prolixus*, differ more than do signals emitted in the same circumstance by multiple species, *i.e.* disturbance stridulation. Triatomine bugs are capable of producing different vibratory signals in accordance with the behavioural context. This suggests that the respective signals should be matched in terms of both frequency and repetition rates to the sensory discrimination capacities of either males in the case of male-deterrent signals, or those of predators in the case of disturbance signals. This may explain the divergence between mating-discouraging signals in *T. infestans* and *R. prolixus*. On the other hand, the lack of divergence between species in disturbance stridulation supports a role as a generalised response to attack by predators.

Finally, the question arises of how Triatominae are able to produce signals differing in frequency and temporal modulation from a single stridulatory organ? Also, the interridge distances of the prosternal stridulatory organ of, for example, *R. prolixus*, are similar along the whole length (Figure 22.1, and Table 22.1 from Manrique and Schilman, 2000). It thus seems likely that the differences found in the repetition rate and in the carrier frequency of the two kinds of signals are produced by scraping the stridulatory organ at different velocities or with different parts of the rostrum.

ACKNOWLEDGEMENTS

The authors are deeply indebted to K. French and R. Barrozo for English corrections and valuable comments on the manuscript, A. Marin-Burgin for the preparation of figures and F. Roces for helping to digitalise and analyse the data. The support from the University of Buenos Aires and CONICET (Argentina) and the German Service for Academic Exchange (DAAD) is acknowledged. We also thank Fabian Gabelli for his help in the preparation of multimedia files and Elsevier for permission to reproduce published data. The authors contributed equally to this chapter and are listed in alphabetical order.

23 Vibratory Communication in Treehoppers (Hemiptera: Membracidae)

Reginald B. Cocroft and Gabriel D. McNett

CONTENTS

INTRODUCTION

Few insect families are as diverse in morphology, behaviour and ecology as the membracid treehoppers (Hemiptera: Membracidae). Studies of communication in membracids are uncovering a correspondingly rich variety of signals, transmitted in the form of substrate-borne vibrations. In this chapter we summarise what we have learned about vibratory communication in treehoppers, drawing on the small but growing literature and on our own recordings and observations of temperate and tropical species. We highlight aspects of their biology that contribute to an impressive diversity in communication signals. Membracids offer promising opportunities for studying the use of signals in cooperation and competition within social groups, the importance of signal divergence in the process of speciation and the evolution of communication systems.

The membracids are a clade of some 3200 species (Wallace and Deitz, 2004), with highest species diversity in the tropics and especially in neotropical lowland forests (Olmstead and Wood, 1990; Wood, 1993b). They are apparently derived from within the leafhoppers (Cicadellidae) (Dietrich *et al.*, 2001). Membracids are characterised by an expanded pronotum, which in some species forms a simple projection over the abdomen and in others takes on some of the strangest shapes ever sculpted from insect cuticle (Figure 23.1a and b).

Treehoppers are unusual not only in their morphology, but also in the diversity of their social behaviour (Figure 23.1c and d; Wood, 1993b, 1979; Lin *et al.*, 2004). Some treehoppers live

FIGURE 23.1 Membracid treehoppers. (a) *Cladonota biclavata* from Panama; (b) *Acutalis* sp. from Ecuador; (c) *Umbonia crassicornis* female and nymphs from Florida, U.S.A.; (d) *Calloconophora pinguis* nymphs from Panama, feeding on a new, expanding leaf of *Piper reticulatum* (attending female not shown). (Photos in (a) and (b) by C. P. Lin. With permission.)

essentially solitary lives, associating with others only for purposes of mating. In this regard they resemble most species in related groups such as the cicadas and leafhoppers (but see Dietrich and McKamey, 1990). At the other end of the spectrum are species in which individuals spend their lives in social groups. Some social groups are composed of related individuals; for example, many species have some form of maternal care, which ranges from guarding eggs (Lin *et al.*, 2004) to defending offspring from predators in response to specialised vibratory signals (Wood, 1984; Cocroft, 1999, 2002). Other social groups contain a mix of related and unrelated individuals that form aggregations during their nymphal development or throughout their lives (Wood, 1984). Immatures of some neotropical treehoppers even form aggregations consisting of individuals of several species (Wood, 1984). Each of these forms of social behaviour has consequences for the forms of social signalling that may occur among prereproductive individuals, and for mating systems and mate-searching strategies in adults.

Group living in membracids is often related to their mutualisms with Hymenoptera, especially ants. Ant mutualism is especially common among tropical species, and may have evolved in response to the threat of ant predation (Wood, 1993b). Although ant mutualism can occur in solitary species (Wood, 1984), it is more common in group-living species (Wood, 1993b). Indeed, there is a clear relationship between group size and the benefits of ant mutualism: ants reduce predation on the treehoppers they tend, and larger groups of treehoppers usually attract more ants and maintain more consistent ant attendance (McEvoy, 1979; but see Morales, 2000). Living in larger groups may also allow individuals to reduce their individual share of the costs of mutualism in the form of providing nutrients to attending ants (Axen and Pierce, 1998). Although ants are the most common mutualists, some stingless bees or vespid wasps also tend treehoppers (Wood, 1984). Treehoppers probably communicate with their mutualists, as do ant-attended lycaenid and riodinid caterpillars (deVries, 1991; Travassos and Pierce, 2000), but this has not yet been documented.

SIGNALS AND MATING BEHAVIOUR

DISCOVERY OF MATING SIGNALS IN MEMBRACIDS

Although most research on communication in membracids has come in the last 15 years, evidence that membracids produced some form of mechanical signal was provided more than 50 years ago by the pioneering work of Ossiannilsson (1946, 1949). Ossiannilsson (1949) described structures similar to the tymbals of cicadas in many species of Auchennorryncha, including the European membracid *Centrotus cornutus*. Ossiannilsson (1949) recorded and described low-amplitude airborne songs from many of the same species, but he suggested that the signals were probably transmitted to other insects as vibrations travelling through the substrate rather than as airborne sound. Evans (1946, 1957) described tymbal structures in two species in the Aetalionidae, a group closely allied to the Membracidae (Dietrich *et al.*, 2001). The first sonogram of a membracid signal was published by Moore (1961). After placing several individuals of both sexes of *Anisostylus elongatus* in a small glass vial along with some pine needles, Moore (1961) detected a short, broadband signal that he speculated was produced by the wings. The first clear evidence of substrate-borne signals in membracids was not published for another 30 years, when Strübing (1992) and Strübing and Rollenbach (1992) described a complex vibratory signal produced by males of the North American membracid *Stictocephala bisonia*. Shortly thereafter, Hunt (1993) conclusively demonstrated the function of these signals in mating behaviour. Hunt (1993) found that males of the North American treehopper *Spissistilus festinus* produced complex frequency-modulated signals during mate-searching behaviour and courtship, and that individuals engaged in male–female duets prior to mating. Using methods similar to those of Ichikawa and Ishii (1974), Hunt (1993) found that vibratory duetting between a male and a female on different plants occurred only when the plants were placed in direct contact, providing a continuous vibration-transmitting pathway. The plant-borne vibratory signals of treehoppers are transmitted as bending waves (Cocroft *et al.*, 2000; also see Michelsen *et al.*, 1982) and are detected by vibration receptors in the legs (Kalmring, 1985). Most vibratory signalling interactions occur within a range of 2 m or less, between insects on the same plant or on neighbouring plants in contact through leaves, stems, or roots (Čokl and Virant-Doberlet, 2003).

DIVERSITY IN MATE ADVERTISEMENT SIGNALS

Vibratory signalling is probably universal among the membracids. It is widespread among related taxa (Claridge, 1985b) and has been found in all membracids so far examined (we have recorded mating signals of approximately 75 species from North America and from the New and Old World tropics). With mating signals recorded from approximately 2% of the 3200 described species of membracids, it is already clear that there is a diversity of signals and signal production mechanisms. We can only guess at the diversity of signals in the other 98%, most of which occur in the tropical forest canopy. Among the species we have sampled, signals range from pure tones or harmonic series, which are usually frequency modulated (see Figure 23.2a and c), to trains of broadband clicks or other noisy elements (see Figure 23.2b). Many species produce complex signals incorporating both tonal and broadband elements, either simultaneously or in alternation (Figure 23.2d), and in combinations that can be startling when played back as airborne sound. In many species, males produce a series of low-amplitude percussive or rattling signals before each bout of advertisement signals (Hunt, 1994; Figure 23.2d).

The degree of complexity in membracid mating signals may be correlated with signalling rates. Species that produce simple signals typically repeat them relatively rapidly in bouts of 2 to 12 signals, separated from other bouts by silent intervals (Figure 23.2c). In some cases when multiple males are present, males may produce a continuous series of signals (*e.g. Notocera bituberculata*, some *Enchenopa binotata*; unpublished data). These groups of stationary, continuously signalling males are analogous to the choruses described for some leafhoppers (Ossiannilsson, 1949;

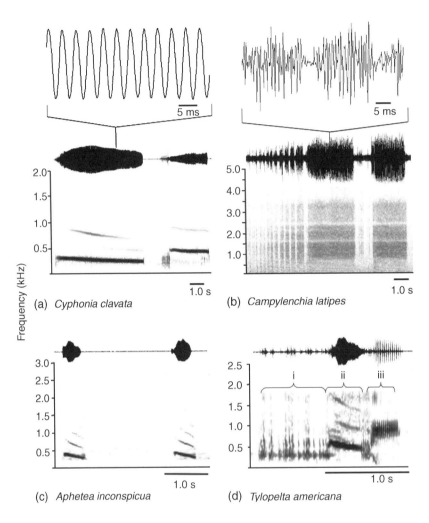

FIGURE 23.2 Male mating signals from four membracid species, illustrating pure tone and noisy elements as well as simple versus complex signals. The signal of the Panamanian *Cyphonia clavata* (a) largely consists of two pure tones that differ in frequency, while the signal of the North American *Campylenchia latipes* (b) contains broadband noise bursts. Male *Aphetea inconspicua* (c) from Panama produce simple signals, consisting of a short, repeated tone (two signals shown). In contrast, male *Tylopelta americana* (d) from North America produce a signal that begins with a series of clicks (i), continues with a tone that drops in frequency with simultaneous, low-amplitude clicks (ii), and ends with a series of pulses that rise in frequency (iii). The transducers used to make the recordings are as follows: (a) and (c) phonograph cartridge (PC); (b) accelerometer (ACC); and (d) laser vibrometer (LDV).

Hunt and Morton, 2001) and for many species that signal using airborne sounds (Gerhardt and Huber, 2002). In contrast to species producing simple signals, species that produce the most complex signals often produce them at more widely separated intervals. For example, *Potnia brevicornis* from Panama produces a characteristic alternation of broadband pulses and frequency sweeps terminating in a high-amplitude harmonic series that rises then falls in frequency; males (at least when recorded singly) then pause for tens of seconds before producing another similar signal. Similar signalling patterns were recorded in individually recorded males of *Oxyrachis tarandus* from India and *Campylocentrus brunneus* and *Cladonota biclavata* (see Figure 23.1a) from Panama. The apparent pattern of short, simple signals produced in series and long, complex signals produced singly represents the endpoints of a continuum, and there are exceptions.

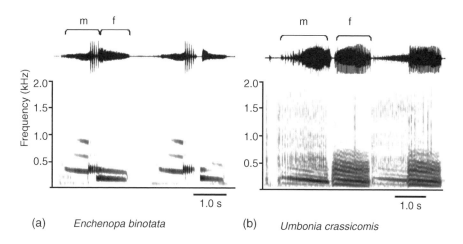

(a)   *Enchenopa binotata*   (b)   *Umbonia crassicornis*

FIGURE 23.3 Male–female duets in two membracid species. (a) Duet from a member of the North American *Enchenopa binotata* complex occurring on the host plant *Viburnum prunifolium* and (b) a duet from the neotropical species *Umbonia crassicornis*. (a) and (b) both recorded using LDV.

For example, male *Tylopelta americana* produce complex signals (Figure 23.2d) but in bouts, with each about proceeded by a series of broadband clicks. In any case, establishing the relationship between signalling rates and signal complexity will require the use of a rigorous comparative phylogenetic approach.

During the vibratory duets that occur during courtship (Hunt, 1993, 1994), males may either produce typical advertisement signals or switch to different signals used only in that context (Cocroft, personal observation). The relatively few recorded examples of female response signals suggest that they are simpler than male signals and relatively similar across species (Figure 23.3). It is unclear to what extent male duetting responses may be influenced by variation in female response time or by differences in other signal features. This broad-scale pattern of diversity in the male trait and conservation in female response is suggestive of the actions of sexual selection in the evolutionary history of these signals.

After a vibratory duet during which the male locates the female, the male may then climb onto the female's pronotum and continue to signal. The signals used by males in this close-range courtship are often more complex and lower in amplitude than their longer-range advertisement signals (Cocroft, 2003). This form of signalling via direct contact may be free of some of the sources of selection on longer-range signals (*e.g.* selection for effective long-range transmission); it would be worthwhile to examine their rate of evolutionary change relative to advertisement signals.

Males in some species also use different signals when interacting with other males, as is common in other related groups (Claridge, 1985b; Gogala, 1985a; Hunt and Morton, 2001; Čokl and Virant-Doberlet, 2003). In *Vanduzea arquata*, it is possible that these male–male signals represent a form of mate guarding and are produced by males associated with aggregations of females (Cocroft, 2003). In *U. crassicornis* the function of male–male signals is unclear as they are given when two mate-searching males meet, even in the absence of females (Cocroft, personal observation). Males do not defend territories or home ranges, and the production of signals that may or may not be as attractive to females as advertisement signals would seem to incur opportunity costs for mate-searching males.

SIGNAL RECEPTION AND RECEIVER PREFERENCES IN MEMBRACIDS

There are no published studies of the morphology or physiology of vibration sensing in membracids. Based on study of other insects (Kalmring, 1985), especially within the Hemiptera (Čokl, 1983), it is likely that membracids detect signals using subgenual organs located in the tibia.

In addition to the frequency tuning of the vibration receptors, vibration perception is also influenced by the mechanical response of the body to substrate vibration, as shown in a study of *Umbonia crassicornis* (Cocroft *et al.*, 2000). A treehopper's body resting on its legs is analogous to a mass on a set of springs, with resonant properties like other mass-and-spring systems (Cocroft *et al.*, 2000). The resonance of the body can change when the insect alters its position (unpublished data), and such changes may be important in vibration perception. In a fiddler crab, where the body likewise behaves as a mass-and-spring system, individuals alter their posture after detecting a substrate-borne signal and thereby sharpen the frequency tuning around the resonant peak (Aicher *et al.*, 1983). The mechanical response of the body to substrate vibration also provides a source of directional cues, but it is not known whether the insects use these cues (Cocroft *et al.*, 2000).

Although membracid vibration sensing has not been characterised physiologically, the signals produced by females in response to male signals provide a behavioural assay of the response to vibration. In treehoppers in the *Enchenopa binotata* species complex, vibratory playbacks to responsive females have shown that females have a narrowly tuned frequency preference; male signals are also narrowly tuned around the same frequency (Cocroft and Rodríguez, 2005). Whether female responses to variation in the frequency of male signals reflect perceived changes in amplitude or true frequency discrimination has not been investigated.

DIVERSITY IN MATING SYSTEMS

Mating systems have not been widely studied in treehoppers, but what is known suggests that their mating systems are diverse. This diversity will in turn influence the evolution of communication systems. Important determinants of mating systems include female mating frequency, the degree of spatial clumping of females and the degree of synchrony in receptivity among females (Schuster and Wade, 2004). All of these features vary widely among different treehopper species. For example, while females in a number of species typically mate only once (Wood, 1974), females in other species mate multiply (Wood *et al.*, 1984; Eberhard, 1986). In some species females are highly clumped, as in the ant-attended species *Vanduzea arquata* where there may be 100 or more individuals on a single host plant (Cocroft, 2003), while in others females are dispersed (Funkhouser, 1917). Finally, in some temperate species with one generation per year, female receptivity within the population is relatively synchronous (Wood, 1980), while in tropical species with several generations per year, there are likely to be receptive females present throughout the year.

In the *Enchenopa binotata* species complex (described in more detail below) the mating system appears to be a "cursorial polygyny" in the classification of Schuster and Wade (2004). This system is characterised by singly-mated, semelparous females that aggregate around resources, and roving males with alternative mating strategies including searching and possibly mate guarding and usurpation of other males. Within the species complex, variation in population density, synchrony of female receptivity and tendency for multiple females to be found on the same branch tip will influence sexual selection on male traits. For example, active mate searching and long-range advertisement signals may be more important in species occurring at lower densities, while mate guarding and male–male signals may be more important in species occurring at higher densities.

Alternative mate-searching strategies are widespread in membracids. When individuals are dispersed, males of many species engage in call-fly behaviour, as described for some leafhoppers (Hunt and Nault, 1991) and cicadas (Gwynne, 1987). In this form of mate searching, a male signals on a series of plants. After arriving at a new location, the male signals and then waits for a few seconds. If the male perceives a female response signal, he will begin to search locally while continuing to signal; if not, he will fly to another location and signal again. If individuals are aggregated, a male may instead remain near a group containing females and engage in close-range courtship. Finally, males may stay in one location and produce a long series of signals, often in alternation with other males (see Hunt and Morton, 2001, for similar behaviour in a leafhopper). In the aggregating treehopper *Vanduzea arquata*, males engage in both call-fly/walk searching and courtship signalling on silent females that they encounter while searching (Cocroft, 2003). In the *Enchenopa binotata* complex, males of at least some species can engage in call/fly searching, extended courtship in contact with silent females and extended chorusing, which involves remaining stationary near other males and signalling in alternation (L.E. Sullivan, personal communication). These different mate-searching strategies are thus not fixed, species-typical behaviour patterns, but rather an aspect of phenotypic plasticity that may be adaptive for individuals which encounter a wide range of social environments. A corresponding plasticity in signal may be equally important.

Treehopper species vary not only in mating systems, but also in their level of inbreeding and outbreeding. Masters (1997) studied two closely related species in Costa Rica (*Umbonia ataliba* and *U. crassicornis*), both of which have extended maternal care of aggregated offspring. *Umbonia crassicornis* is an outbreeding species with high population densities, with mating taking place after dispersal from the natal aggregation (Masters, 1997; see also Wood and Dowell, 1985). In contrast, *U. ataliba* is an inbreeding species with low population densities, with mating taking place between siblings before dispersal from the natal aggregation (Masters *et al.*, 1994). Masters (1997) found that, while *U. crassicornis* experiences inbreeding depression when siblings mate, *U. ataliba* experiences outbreeding depression when nonsiblings mate. She concluded that inbreeding in *U. ataliba* was a form of mating assurance in light of the high costs of mate searching associated with outbreeding in this relatively rare species. Furthermore, sex ratios within broods of *U. ataliba* are female-biased, probably reflecting an adaptation to local mate competition (Masters *et al.*, 1994). These breeding systems will have important consequences for sexual selection and communication, with active mate searching by roving males in *U. crassicornis*, but highly localised competition among siblings in *U. ataliba*. Different traits and signalling behaviour are likely to be important for male mating success in the two species.

MATING SIGNALS AND SPECIATION IN THE *ENCHENOPA BINOTATA* COMPLEX

The *Enchenopa binotata* species complex is a clade of nine closely related species, each occurring on a different species of host plant. The species have not yet been given formal names, and here we refer to them by reference to their host plants. This group has been a model system for the study of sympatric speciation resulting from shifts to novel host plants (Wood and Guttman, 1983; Wood *et al.*, 1990, 1999; Wood, 1993a). The combination of divergent selection and assortative mating facilitates sympatric speciation (Schluter, 2000; Coyne and Orr, 2004). Host shifts result in divergent selection: survivorship of *E. binotata* can drop dramatically when individuals are transferred to a nonnatal host (Wood, 1993a). Changes in host use also lead to assortative mating because life history timing in the *E. binotata* complex is dependent on the phenology of the host plant (Wood *et al.*, 1990). Use of different host plants can thus cause an allochronic shift in the timing of mating, which, in combination with high host fidelity, can reduce mating between populations on different hosts (Wood *et al.*, 1990). However, some interbreeding is still possible due to partial overlap of mating periods and occasional dispersal, especially of mate-searching males. This possibility for gene flow between *E. binotata* populations on different hosts highlights the potential importance of mating signals, which, because of their role in assortative mating, have often been implicated in the process of speciation (West Eberhard, 1983).

Hunt (1994) described the mating signals of one species in the *E. binotata* complex. Males produce complex frequency- and amplitude-modulated signals, and receptive females respond with a simpler signal of their own. Rodríguez *et al.* (2004) used vibratory playback of signals recorded from several species in the complex to ask whether variation among the signals of different species in the complex was important for assortative mating. The authors showed for one species in the complex (*E. binotata* from *Viburnum lentago*) that females responded to signals of conspecific males but discriminated strongly against those of males of the most closely related species in the complex. Given that signals can contribute to assortative mating among extant species, studies are now underway to examine variation in male signals and female preference curves within and between species (unpublished data). These studies will reveal which signal traits are most important in assortative mating, and set the stage for further studies to examine how host shifts influence the evolution of those traits.

Research on *E. binotata* also suggests that the communication system can contribute to assortative mating in the early stages of a host shift before divergence in signals or preferences has occurred. For example, the communication system might contribute to assortative mating if host fidelity is reflected in male mate-searching behaviour, such that males invest less in signalling on nonnatal hosts. Sattman and Cocroft (2003) found that signalling behaviour is indeed influenced by plant identity: male *E. binotata* from *Ptelea trifoliata* produced fewer, shorter signals when on a nonhost plant. This host fidelity in advertisement signalling should have the consequence of reducing the likelihood of mating between host-shifted populations because females prefer males that produce more signals per bout (unpublished data). Mate-searching behaviour is also biased towards host rather than nonhost plants in *U. crassicornis* (Masters, 1997), although males do not alter their signals on nonhosts (unpublished data). Other aspects of phenotypic plasticity could also influence gene flow between host-shifted populations. Because mating periods of populations on different hosts are allochronic (Wood, 1980), interbreeding among host-shifted populations will only be likely between older males from the early population and receptive females from the late population. A decrease in the attractiveness of the signals of older males would reduce this probability. However, Sattman and Cocroft (2003) found no influence of male age on signal variation.

SIGNALS AND GROUP LIVING

FUNCTIONS OF SOCIAL SIGNALLING

For many group-living insect herbivores, vibratory communication may be important for solving the challenges of life on a plant, including avoiding predators and finding feeding sites (Cocroft, 2001). The first evidence that vibratory signals played an important role in treehopper social behaviour came from studies of group-living nymphs of the thornbug treehopper, *Umbonia crassicornis* (Figure 23.1c; Brach, 1975; Cocroft, 1996). Nymphs in this species develop to adulthood in dense, cylindrical aggregations on their host plant stem. Aggregations are attended by their mother, who typically remains stationary at the base of the group. In their exposed locations at the growing tip of a host plant stem, nymphal aggregations are vulnerable to invertebrate predators including syrphid fly larvae, predatory Hemiptera and vespid and sphecid wasps; and because aggregations are stationary, predators can either remain near the aggregation or make repeated visits (Cocroft, 2002). For most predators, the nymphs' principal or only protection is the mother's active defence, which involves wingbuzzing, approaching the predator and kicking the predator with specialised hind legs (Wood, 1983). Mothers travel rather slowly, however, and thus early information on the presence of a predator is critical. Coordinating signalling among the nymphs provides this early information.

When a predator approaches an aggregation of *U. crassicornis* nymphs, the first nymphs to perceive the predator each produce a short vibratory signal. This elicits additional signals from siblings and signalling rapidly spreads across the group, resulting in a coordinated group display. Coordinated displays, to which each individual contributes only one signal, are produced every 1 to 2 sec (Figure 23.4a). In response to a series of group displays from her offspring, a female leaves her usual position at the base of the aggregation and walks into the group attempting to locate and drive away the predator. The offspring continue to produce coordinated signals as long as the predator is present, suggesting that they continue to provide information to the female throughout a predator encounter. The potential fitness benefits of rapid offspring–parent communication are high: a field study in Costa Rica showed that when a female disappeared from an

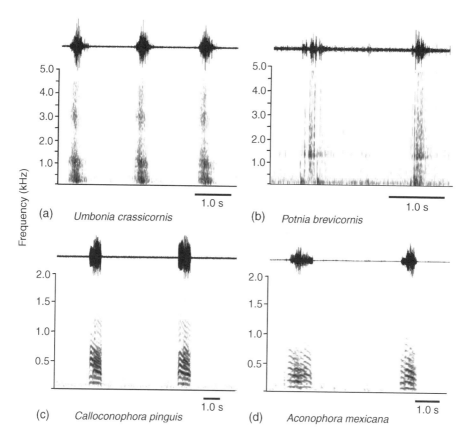

FIGURE 23.4 Signals of membracid nymphs showing conservation of signal form in two pairs of closely related species. *Umbonia crassicornis* (a) and *Potnia brevicornis* (b) are members of the tribe Hoplophorioni; in response to the approach of a predator, aggregated nymphs of both species produce brief, noisy pulses that are synchronised into group displays. *Calloconophora pinguis* (c) and *Aconophora mexicana* (d) are close relatives within the tribe Aconophorini; nymphs in both species produce relatively long, harmonically rich signals that function to recruit siblings to a feeding site. (a–c) recorded using ACC; (d) recorded using PC.

aggregation, attacks by predatory wasps were always successful; while in contrast, when a female was present, she was able to drive away wasps in about 75% of the encounters (Cocroft, 2002).

Although parent–offspring groups in *U. crassicornis* communicate largely or exclusively in relation to predators, the context of communication is very different in family groups of another neotropical treehopper, *Calloconophora pinguis*. Nymphs of this species likewise develop to adulthood in aggregations of siblings (Figure 23.1d), which are usually attended by the mother. Aggregations feed at the base of young, expanding leaves and, as one leaf matures and becomes unsuitable, the group must find another. The process of locating and moving to a new feeding site involves vibratory communication: individuals begin leaving the group and when one locates a suitable feeding site it produces a series of signals (Cocroft, 2005). Other individuals respond to these signals by approaching, and once at the site begin signalling in synchrony with the individuals already there (Figure 23.4c).

It is notable that in both *U. crassicornis* and *C. pinguis*, individuals produce coordinated group displays. One likely causal factor is that in both cases a group of individuals is producing signals that function to influence the behaviour of receivers outside the group. Synchronised signalling is not optimal for within-group communication because other potential receivers within the group would be producing signals at the same time. However, the superposition of multiple signals may increase overall display amplitude and enable receivers outside the group to assess the amount of signalling taking place. Such assessments might be important if female *U. crassicornis* need to assess the degree of threat or if dispersing *C. pinguis* nymphs are faced with deciding among more than one advertised feeding site.

DIVERSITY IN SOCIAL SIGNALS

Social signalling is widespread in membracids and may occur in all group-living species. Where social signals have been recorded from immatures, the signals and signalling behaviour of closely related species are similar, suggesting that social signals are more evolutionarily conservative than male advertisement signals (Figure 23.4). For example, *Umbonia crassicornis*, in which immatures produce group displays in an antipredator context, is a member of a clade with similar forms of offspring aggregations and maternal care (the Hoplophorionini; McKamey and Deitz, 1996). Signals very similar to those of *U. crassicornis* have been recorded not only in congeners including *U. spinosa* and *U. ataliba*, but also in other related genera including *Alchisme* and *Potnia* (Figure 23.4b). All were produced by maternally-defended immatures in response to disturbance (unpublished data). The signals of *Calloconophora pinguis* nymphs (Figure 23.4c; Cocroft, 2005) are very different from those of *U. crassicornis*. However, they are very similar to those of species in the closely related genera *Guayaquila* and *Aconophora* (Figure 23.4d), which are also produced by immatures aggregating at a feeding site (unpublished data). The consistent association of signal form and function is striking — the antipredator signals of *Umbonia* and its close relatives are brief, high-pitched, noisy pulses, while the recruitment signals of *Calloconophora* and its relatives are longer, lower-pitched harmonic series (Figure 23.4). Whether these differences are adaptive given the time scales and distances involved in each communication context remains to be tested.

Social signals are produced not only by immatures, but also by adults. In *U. crassicornis*, as mentioned above, females defending their offspring produce a series of short, percussive "clucks" after the predator has left (Cocroft, 1999). Similar signals are produced by adults in predispersal aggregations (Cocroft, personal observation). Unlike the synchronised signalling by nymphs attempting to elicit their mother's defence, signallers in these adult aggregations actively avoid signal overlap, which suggests that the intended receivers are other group members.

In addition to vibratory signals, group-living membracids also can use chemical cues to assess the presence of a predator (Nault *et al.*, 1974). These cues are released when an individual is injured. It is not known whether the chemical cues involved represent an evolved signal, favoured by selection because of its effect on the behaviour of receivers, or whether receivers have simply evolved the ability to detect incidentally-produced cues from injured conspecifics. It is likely that in social species such as *Umbonia crassicornis*, there is an interaction between vibratory signalling and chemical cues. For example, perception of cues from an injured sibling may trigger vibratory signalling.

SIGNAL PRODUCTION MECHANISMS

While signal production in Membracidae has received little rigorous attention, it is clear that multiple mechanisms underlie the diversity of signals. It is characteristic of the vibratory modality that multiple signal-producing mechanisms can easily be incorporated into an individual signal. Vibratory signals are known to travel with little attenuation along woody stems (Michelsen *et al.*, 1982; Čokl and Virant-Doberlet, 2003). As a consequence, nearly any movement of a body part can result in a signal or cue being propagated a reasonable distance and, hence, detected by other individuals. This is much less likely in airborne signals, where there exist inherent constraints in coupling a mechanical disturbance to air (Bennet-Clark, 1998a). This should leave airborne signallers with fewer signalling options and make it less likely for them to incorporate multiple mechanisms into an individual signal since most movements do not translate into a signal or cue that is loud enough to be biologically useful. Thus, complexity in the airborne signals of insects may be more restricted to variations in temporal pattern (see Bailey, Chapter 8), while complexity in vibratory signals may also involve variation in spectral features where different signal components are produced by different means.

Most of what is known about signal-producing mechanisms in membracids is based on the work of Ossiannilsson (1949), who described a signal-producing system including striated tymbals in the first abdominal segment with associated muscles and muscle attachment points. The tymbals and associated muscles described by Evans (1946, 1957) for two aetalionids closely match those of Ossiannilsson. This tymbal system appears to be homologous with that of cicadas, perhaps representing the ancestral condition from which the dramatic tymbal structures of cicadas evolved.

The tymbal mechanism described by Ossiannilsson (1949) and Evans (1946, 1957) presumably is responsible for click-like portions of vibratory signals. Yet as we described above, many membracids also incorporate pure tones or harmonic series which are likely produced by a different mechanism. Tonal signal elements are typically accompanied by the dorso-ventral abdominal tremulation (Virant-Doberlet and Čokl, 2004), which may be powered by the wing muscles. Additional signal production mechanisms in Membracidae are also evident from observations of signalling males (Figure 23.5). In *Vanduzea arquata*, for example, the signal consists of an initial series of taps produced by the male rocking forward and backward and apparently striking its head on the stem, followed by a harmonic series that is accompanied by abdominal tremulation (Figure 23.5a; Cocroft, 2003). In *Atymna querci* the signal contains percussive elements that appear to result from the male striking the substrate with his abdomen (Figure 23.5b). Another example can be found in *Tropidaspis affinis*, where signal production is accompanied by a rapid vibration of the wings (Figure 23.5c). It is clear from examination of the advertisement signals of some species that more than one sound production mechanism is used simultaneously. For example, male *Umbonia crassicornis* produce a frequency-modulated tonal component lasting about a second while simultaneously producing a series of higher-pitched clicks (Figure 23.3b).

Whether variation in pronotal shape has any relationship to variation in the properties of communication signals remains untested (see Montealegre and Morris, 2004, for a relevant study in Tettonigoniidae). A direct relationship between pronotal shape and signal variation seems unlikely given the lack of an airborne sound signal for which the hollow pronotum could serve as a cavity resonator. Furthermore, qualitatively similar signals are produced by species in related groups that lack expanded pronota.

Nymphal treehoppers likely have a diversity of vibration-producing structures given the range of signals they produce. The tymbals of the treehopper *Aetalion reticulatum* (Evans, 1957) are present not only in adults but also in every nymphal instar except the first. Evans (1957) inferred the production of social signals that might function in maintaining group cohesiveness; nymphs and adults in this species do indeed produce rather similar high-intensity, broadband vibratory signals (unpublished data) but their function has not been studied. A different mechanism must underlie the harmonic series produced by nymphs of *Calloconophora pinguis* and relatives. It would be interesting to investigate whether these nymphs use novel means to produce their signals, or whether they are simply using a form of the adult structure at an earlier ontogenetic stage as in *Aetalion*. It can safely be stated that the field of vibratory signalling in general would benefit from more detailed investigations of signal-producing mechanisms.

VIBRATORY COMMUNICATION IN THE FIELD

Most research on vibratory communication in insects has been conducted in the laboratory (Claridge, 1985b; Cocroft and Rodríguez, 2005), and studies of membracids are no exception. However, field research is necessary for understanding the social and ecological context in which communication takes place. Although solitary membracids might be difficult to study in the field,

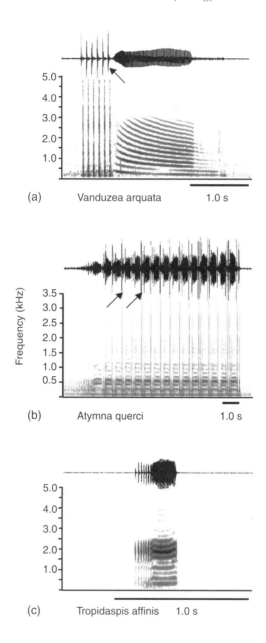

FIGURE 23.5 Mate advertisement signals of male membracids that illustrate a diversity of signal production mechanisms. In *Vanduzea arquata* (a), a male first produces a series of taps (arrow), apparently by striking its head on the substrate, then a harmonic series accompanied by abdominal tremulation. In *Atymna querci* (b), male signals incorporate percussive elements (arrows) that apparently result from the male striking the substrate with his abdomen. In *Tropidaspis affinis* (c), signal production is accompanied by rapid vibration of the wings. (a and b) recorded using ACC; (c) recorded using PC.

the high density and site fidelity of social species make them excellent subjects for field study. Research on mating behaviour and dispersal in the aggregating *Umbonia crassicornis* (Wood and Dowell, 1985) shows the potential for field study of membracids, as does the career-long series of studies by T. K. Wood on the biology of the *Enchenopa binotata* species complex (see Wood, 1993a). However, to date only one study has examined mate-signalling behaviour in a field study of treehoppers (Cocroft, 2003). This study found a high degree of spatial and temporal variation in the social context experienced by both males and females of the aggregating, ant-tended membracid *Vanduzea arquata*. This variation may be important in influencing the mating decisions of females and the mate-searching strategies of males.

The prospects for field study of antipredator or recruitment signalling are even more promising than for mating behaviour, especially among the immatures where dispersal is limited and groups can persist in the same location throughout nymphal development. Field studies of social behaviour and ant mutualisms have been conducted in several species, but so far only one study has examined social signalling in the field. Cocroft (1999) examined the signalling behaviour of offspring and mothers of *Umbonia crassicornis* during attacks by predatory wasps. This study revealed that, as predicted by laboratory studies, *U. crassicornis* nymphs produce group displays as soon as a predator arrives and continue to display until after the predator leaves. It also revealed behaviour not noticed in laboratory studies, such as signalling by the defending female after the predator had left, or sporadic signalling by mother and offspring at a low rate throughout the day. Furthermore, study of the details of predator attacks revealed which individuals in the group were at highest risk, allowing predictions of how signalling behaviour should vary within groups. Such fieldwork then provides additional predictions that can be experimentally studied in the lab, and the potential for this interplay of laboratory and field studies is one of the attractive features of research on membracids.

ACKNOWLEDGEMENTS

We thank Mike Claridge and Rafa Rodríguez for comments on the manuscript. We also thank Lew Deitz, Chris Dietrich, Chung Ping Lin, Stuart McKamey, Mark Rothschild and Rob Snyder for help and companionship in the field and identification of specimens, and Chung Ping Lin for the photos used in Figure 23.1. Financial support for GDM was provided by a Life Sciences Graduate Fellowship from the University of Missouri-Columbia, and for RBC by NSF IBN 0318326. This contribution is dedicated to the memory of Tom Wood.

24 Acoustic Characters in the Classification of Higher Taxa of Auchenorrhyncha (Hemiptera)

D. Yu. Tishechkin

CONTENTS

INTRODUCTION

The diagnostic value of songs in insects was first recognised in the middle of the 19th century. At about this time distinct differences in the songs of closely related species of Orthoptera were discovered (see Ragge and Reynolds, 1998). Recently most taxonomic papers on European Orthoptera and singing cicadas (Cicadidae) have included oscillograms of calling signals of the species under investigation. Recording the vibratory signals of small Auchenorrhyncha is a more difficult task because it requires complex and specialised equipment. However, investigation of signals of these usually small insects are often the simplest and most reliable method of solving taxonomic problems.

It is widely accepted that taxonomic utility of acoustic signals analysis is primarily at the species and subspecies levels. Striking demonstrations of this are to be found among the Orthoptera. Species that appear to be closely related by all other criteria often produce quite different signals (*e.g.* grasshoppers of the *Chorthippus biguttulus* group). On the other hand, it is widely known that certain higher taxa (subfamilies, families) may differ from each other in acoustic characters. For example, repetitive continuous signals with narrow-band carrier frequencies are characteristic of crickets (Gryllidae). Most Tettigoniidae (Conocephalinae, Tettigoniinae) generally produce signals with similar temporal patterns, but with broader frequency spectra reaching up to the ultrasonic range. Rather short echemes, *i.e.* signals with a determinable beginning and end, and with wide-band noise spectra, usually belong to Gomphocerine grasshoppers (Acrididae). Therefore, most higher taxa of sound-producing insects have their own unique "acoustic appearance" (Riede, 1996). Auchenorrhyncha are no exception.

Classification of higher taxa is one of the most difficult problems in systematics of Auchenorrhyncha. Certain large and heterogeneous families such as Cicadellinae are subdivided into numerous subfamilies and tribes. In spite of many works in this field, no generally accepted opinion concerning the limits, relationships and status of these taxa exists at present. Different classifications contradict each other. Moreover, sometimes the points of view of the same author expressed in different papers and in different years vary considerably. Most classifications are based only on morphological characters. A few attempts to involve caryological, biochemical and some other data have been made recently. The inclusion of acoustic characters in systematic analysis would seem to be useful.

Unfortunately, our knowledge of adaptive characters and trends in the evolution of vibratory communication in Auchenorrhyncha is very slight. For this reason, the judgement and interpretation of similarity in acoustic characters is a difficult task. Consequently, classifications based on these may be questionable. New characters for taxonomic analysis are needed and may be useful, even with such provisos. Data on the structure of communication systems allow us to obtain additional independent corroboration of alternative viewpoints and finally will encourage the stabilisation of classification.

Recently, I have published a series of papers on vibratory communication of more than 250 species of Auchenorrhyncha from Russia and adjacent countries, with notes on the classification of higher taxa (Tishechkin, 2000, 2001, 2003, 2004). Acoustic characters useful in the systematics of tribes, subfamilies and families are considered here, based on these published examples. Full data on collecting sites, temperature during recording, *etc.* are in the original papers.

FREQUENCY SPECTRA

Vibratory signals of most Auchenorrhyncha have low-frequency wideband spectra (Figure 24.1, item 13). Signals with pure-tone components are rare. Only two groups in which most species produce such signals are known at present. These are the Typhlocybinae (Figure 24.1, items *1* to *3*, *10*) and the Cicadellinae *sensu lato* (Figure 24.1, items *4* to *9*, *11* and *12*) of the Cicadellinae. Their signals sometimes sound like howling or whistling to the human ear. On high-speed oscillograms these signals appear as almost pure sine waves (Figure 24.1, items *3*, *6*, *9*). Sometimes the signal consists of a continuous succession of sine waves, in which any regular amplitude modulation (*i.e.* subdivision into pulses) is absent (Figure 24.1, items *7* to *9*).

The Typhlocybinae is a specialised group of leafhoppers differing from all others in a number of anatomical characters. Most Cicadellinae are xylem- or phloem-feeders, whereas almost all Typhlocybinae live on leaves and feed from mesophyll cells. Various opinions have been expressed concerning the position of this group among other subfamilies. More detailed discussions of the problem are cited by Tishechkin (2001). Here I simply note that the diversity of leafhopper subfamilies (more than 50 in certain classifications) is usually clubbed in three or four major groups. In different classifications Typhlocybinae are placed into different of these groups. Some authors create a separate group for Typhlocybinae since the origin of the group is obscure and it appears neither to be derived from, nor ancestral to, any other present-day group of leafhoppers.

If we take frequency characteristics of signals into consideration, then the opinion expressed earlier that Typhlocybinae is either the sister-group of Cicadellinae *sensu lato* or the derivative of one of their subdivisions is further supported.

ACOUSTIC BEHAVIOUR DURING MATING AND ACOUSTIC REPERTOIRE (THE SET OF FUNCTIONAL TYPES OF SIGNALS)

General acoustic characters provide useful distinctions between subfamilies and tribes of Cicadellinae.

The Athysanini is one of the largest tribes in the Deltocephalinae and many common and abundant Palaearctic leafhoppers belong here. The limits of the tribe are uncertain. Several taxa are

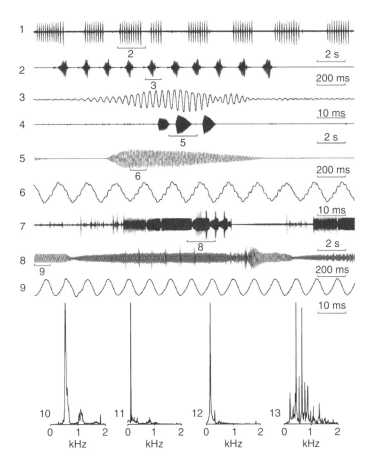

FIGURE 24.1 Oscillograms and frequency spectra of calling signals of Cicadellidae. *1* to *3*, *Alebra albostriella* (Typhlocybinae), oscillogram of calling signal; *4* to *6*, same, *Kolla atramentaria* (Cicadellinae, Cicadellini); *7* to *9*, same, *Evacanthus interruptus* (Cicadellinae, Evacanthini); *10*, *A. albostriella*, frequency spectrum of the part of signal presented on oscillogram *3*; *11*, *K. atramentaria*, frequency spectrum of the part of signal presented on oscillogram *6*; *12*, *E. interruptus*, frequency spectrum of the part of signal presented on oscillogram *9*; *13*, *Platymetopius gr. undatus* (Deltocephalinae, Platymetopiini), frequency spectrum of the part of calling signal. Faster oscillograms of the parts of signals indicated as *2–3*, *5–6* and *8–9* are given under the same numbers.

either regarded as separate tribes or included into Athysanini. Acoustic behaviour provides useful characters for solving such problems.

One of the groups that differs somewhat from Athysanini *sensu stricto* is the Cicadulini. In all representatives of Athysanini studied, the calling signal has a constant temporal pattern. Males produce signals of the same type when isolated and during all stages of courtship behaviour (Figure 24.2, items *1* to *2*). In Cicadulini both reduced and full forms of the calling signal are present (Figure 24.2, items *3* to *4*). The latter is more elaborate with additional components. Single males can produce signals of both types alternately, but produce only full calls during courtship. Therefore, in the acoustic repertoire of Cicadulini, calling and courtship signals are present, but these two types overlap considerably. Thus, based on acoustic characters the Cicadulini is a separate taxon, differing distinctly from the Athysanini.

The Platymetopiini and Fieberiellini are two other problematic groups closely related to Athysanini. Males of *Platymetopius* and *Fieberiella* in isolation produce only one type of call,

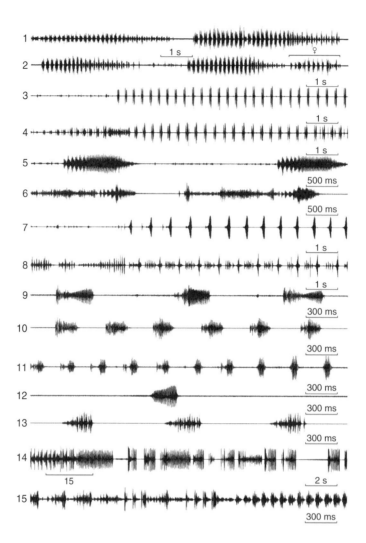

FIGURE 24.2 Oscillograms of vibratory signals of Cicadellidae. *1*, *Graphocraerus ventralis* (Deltocephalinae, Athysanini) male calling signal; *2*, same, male calling and female reply; *3*, *Elymana kozhevnikovi* (Deltocephalinae, Cicadulini) reduced form of male calling signal; *4*, same, full form; *5*, *Platymetopius gr. undatus* (Deltocephalinae, Platymetopiini), calling signal; *6*, same, courtship signal; *7*, Fieberiella septentrionalis (Deltocephalinae, Fieberiellini), calling signal; *8*, same, courtship signal; *9* to *12*, calling signals of *Aphrophoridae* from the tribe Philaenini (*9*, *Neophilaenus campestris*; *10*, *N. exclamationis*; *11*, *Philaenus spumarius*; *12*, *Aphilaenus ikumae*); *13*, same, *Aphrophora alni* (Aphrophoridae, Aphrophorini); *14*, same, *Lepyronia coleoptrata* (Aphrophoridae, Lepyroniini). Faster oscillogram of the part of signal indicated as *15* is given under the same number.

as in the Athysanini *sensu stricto* (Figure 24.2, items *5* to *7*). On the other hand, when courting a female, males produce signals that consist of the basic calling signal with some additional elements (Figure 24.2, items *6* to *8*). In this manner both taxa differ from Athysanini and resemble each other. Consequently, acoustic data support the opinion of Hamilton (1975), who united these two genera in the tribe with the valid name Platymetopiini and excluded it from the Athysanini.

Sometimes males of leafhoppers show peculiar courtship rituals which also can serve as taxonomic characters. For example, courting males of Alebrini and Empoascini (both from the

Typhlocybinae) move along the leaf with small paces, at the same time trembling the body and producing courtship signals, which are hard to distinguish by a human observer because vibrations resulting from male movements drown the signals. Courting males move slower than during normal walking. After walking two or three centimetres, males stop, call and start "dancing" again. In other tribes of Typhlocybinae, the courtship pattern is distinctly different and does not include any unusual movements. Apparently, the two above-mentioned tribes are more closely related to each other than to the remaining Typhlocybinae. Some authors came to the same conclusion based only on morphological characters (Wagner, 1951).

Distinctive behaviour of another kind was described by Ossiannilsson (1949) in the Megophthalminae and observed by me also in the Ulopinae (Tishechkin, 2003). A courting male mounts a female and performs rapid movements of the body, first running a little forward, then back and then sideways left or right on her back. This behaviour lasts for several seconds. The male then stops briefly, produces a succession of pulses and then resumes "dancing" again (the term "dancing" was introduced by Ossiannilsson, 1949). The behaviour described is unique for these two subfamilies. It may thus be used as evidence of their close relationship. There are different opinions concerning the position of Ulopinae among other Cicadellid subfamilies. The data obtained provide additional evidence for this discussion.

Similar courtship patterns were observed in two species of treehoppers, Membracidae, namely, *Centrotus cornutus* (L.) and *Gargara genistae* (F.) (Tishechkin, 2003). In both species, males perform rapid, abrupt, jerky movements on the female's back during courtship. This may be evidence of relationships between these three groups, *viz*. Membracidae, Ulopinae and Megophthalminae. The Membracidae are often regarded as the sister-group of Cicadellinae (Dietrich and Deitz, 1993). Quite different opinions concerning the position of this family were expressed recently by Rakitov (1998) and Dietrich *et al*. (2001). Based on molecular and some other characters, they proposed that the Cicadellinae are paraphyletic with respect to Membracidae + Aetalionidae, *i.e.* treehoppers are a specialised line, derived from within Cicadellinae, and are thus not their sister-group. Bioacoustic data support the phylogenetic reconstructions proposed by these recent authors and do not support the sister-group relationship of the Membracidae and Cicadellinae.

TEMPORAL PATTERNS OF CALLING SIGNALS

Temporal patterns may vary greatly even between closely related species. For this very reason it may serve as a good character for discriminating between sibling species or for establishing the status of local populations showing small morphological differences. Nevertheless, in several cases it can also be used in the classification of higher taxa. It should be noted that calling is generally easy to record because single males usually produces such signals quite readily when in isolation.

Identical patterns of calling signals are characteristic of Philaenine froghoppers of the Aphrophoridae. Signals in these species consist of single or repeated elements, including a short succession of pulses and also more prolonged monotonous elements (Figure 24.2, items *9* to *12*). Occasionally, the succession of pulses may be partly or entirely reduced (Figure 24.2, items *11* to *12*). Temporal pattern of calling in *Aphrophora* (Figure 24. 2, item *13*) and *Lepyronia coleoptrata* L. (Figure 24.2, items *14* to *15*) (Aphrophorini and Lepyroniini, respectively) is quite different. Thus, in froghoppers, analyses of calling signals provide good taxonomic characters at the tribal level.

Among *Cicadellinae*, examples of a similar kind may also be found. The tribes, Eupelicini, Dorycephalini, Hecalini and Chiasmini (=Doraturini) are similar to each other in many characters. This similarity is usually regarded as superficial. As a result, in certain classifications all or most of these taxa are included in the Deltocephalinae, at most as tribes. In other classifications the Eupelicinae Dorycephalinae and Hecalinae at least are regarded as separate subfamilies, while

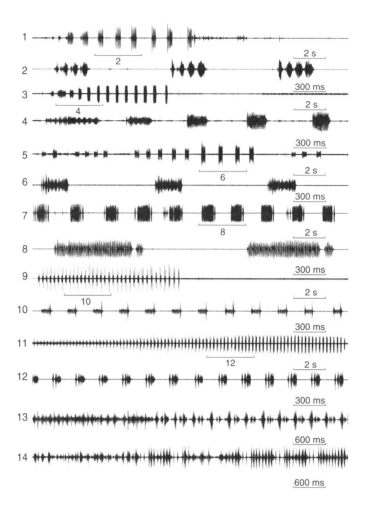

FIGURE 24.3 Oscillograms of calling signals of Cicadellidae. *1* to *2*, *Hecalus glaucescens* (Hecalini); *3* to *4*, same, *Dorycephalus hunnorum* (Dorycephalini); *5* to *6*, same, *Eupelix cuspidata* (Eupelicini); *7* to *12*, species from the tribe *Chiasmini* (=Doraturini); (*7* to *8*, *Doratura homophyla*; *9* to *10*, *Chiasmus conspurcatus*; *11* to *12*, *Aconurella diplachnis*); *13*, *Aphrodes bicinctus* (Aphrodinae sensu stricto); *14*, *Deltocephalus pulicaris* (Deltocephalinae). Faster oscillograms of the parts of signals indicated as *2*, *4*, *6*, *8*, *10* and *12* are given under the same numbers.

the Chiasmini are retained in the Deltocephalinae. Calling signals in representatives of all four taxa are similar in temporal pattern. Usually, these consist of rather short, simple elements repeated regularly (Figure 24.3, items *1* to *12*). If the durations of these elements increase, the breaks between them become more prolonged and less constant in duration. Therefore, sometimes a signal consists of one or irregularly repeated syllables. However, all the transition variants between regularly repeated and single signals may be found even in the same genus (*e.g.* in some Doraturini). As a result, in this case acoustic data provide additional evidence of close relationships between the taxa under consideration.

The Deltocephalinae is a large and heterogeneous subfamily of leafhoppers. Hamilton (1975) revised the limits of the subfamily and the status of included taxa. One of the most notable taxonomic changes proposed by Hamilton was the joining of the Deltocephalinae to a small subfamily, the Aphrodinae, which includes only two to four Holarctic genera. The latter name appeared to be the oldest and therefore this taxonomic rearrangement resulted in the change of the name of this huge group (Aphrodinae instead of Deltocephalinae). This point of view was not generally accepted and became a matter of considerable dissent (Emelyanov and Kirillova, 1989). Nonetheless, the temporal pattern of calling signals in several Aphrodinae *sensu stricto* and some Deltocephalini and Paralimnini (both Deltocephalinae) is quite similar. The signal is a prolonged phrase. Its initial part is a nonspecies-specific succession of pulses with very variable shape and repetition period. In contrast with it, the main part is a succession of elements with constant and species-specific temporal patterning (Figure 24.3, items *13* to *14*). The parallel origin of such a complex signal in nonrelated taxa seems unlikely to me. It is more likely that Deltocephalini and Paralimnini are closely related to the Aphrodinae *sensu stricto*. As a result, the opinion expressed by Hamilton (1975) is supported by the bioacoustic data.

A CASE STUDY — ISSIDAE AND CALISCELLIDAE

Traditionally, the Fulgoroid family, Caliscelidae has been included as a subfamily of Issidae. Hamilton (1981) was the first to separate these two taxa as separate families. Recently, several authors have supported this opinion based on morphological, molecular and bioacoustic characters. Differences in the communication system of these two taxa are briefly discussed here.

In most species of Issidae *sensu stricto* only male calling signals have been recorded (Figure 24.4, items *1* to *14*). Males produce signals of this type in isolation and in the presence of other individuals of either sex. Signals consist of short discrete elements, *i.e. pulses*, in all species studied. In contrast to Cicadellinae and Delphacidae, pulses in signals of Issidae *sensu stricto* are rather uniform and are quite similar in temporal pattern even in the songs of species from different genera. Evidently, the variation observed (Figure 24.4, items *4, 7* to *8*) is a result of differences in physical characteristics of the substrate, either plant stems or twigs, or in the relative position of the singing insect and the vibrotransducer.

The general structure of signals in different species is also rather similar. In *Mycterodus intricatus, Scorlupella discolor* and *Alloscelis vittifrons*, calling signals are successions of pulses of approximately 2 sec duration (Figure 24.4, items *1* to *8*). In *Scorlupaster asiaticum* signals are much shorter and average 200 to 300 msec duration (Figure 24.4, items *9* to *11*), whereas in *Agalmatium bilobum* calling signals generally last for 15 to 20 sec, with more elaborate temporal pattern (Figure 24.4, items *12* to *13*). Occasionally, a single male sings for several minutes, producing signals with more or less constant intervals. On the other hand, in less active individuals pauses between signals in certain cases can exceed 5 to 10 min.

Several males sitting on the same stem sometimes start singing more actively. In this case they alternate signals, but the structure of the song for the most part remains the same as in a single individual. Signals produced in such situation are much shorter in *A. bilobum* (Figure 24.4, item *14*). Rivalry behaviour in any form was not observed in Issidae *sensu stricto*.

In a single stationary male of *M. intricatus*, prolonged trills of pulses were recorded (Figure 24.4, items *15* to *16*). The function of this signal is obscure.

In *Caliscelidae*, temporal patterns of calling signals are quite different (Figure 24.5). Signals of *Aphelonema* species and *Ommatidiotus dissimilis* consist of more or less uniform groups of pulses, so-called *syllables*, repeated with constant or gradually changing period (Figure 24.5, items *5* to *8, 2* to *15*). In *Caliscelis affinis* and *O. inconspicuus*, a continuous, monotonous fragment follows the succession of syllables (Figure 24.5, items *1* to *4, 9* to *11*). In the former species, it consists of two parts with different waveforms and frequency spectra (Figure 24.5, item *4*). Calling signals in Caliscelidae are much more elaborate than in Issidae *sensu stricto*.

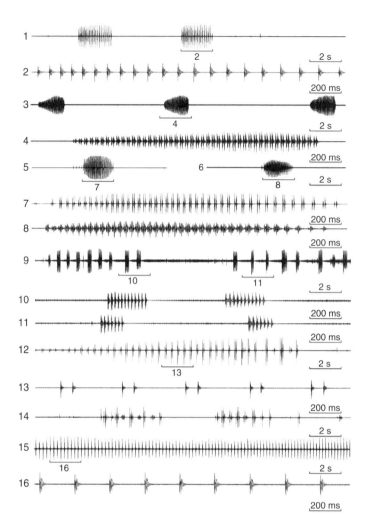

FIGURE 24.4 Oscillograms of vibratory signals of Issidae. *1* to *2*, *Mycterodus intricatus*, calling signal; *3* to *4*, same, *Scorlupella discolor*; *5* to *8*, same *Alloscelis vittifrons*; *9* to *11*, same, *Scorlupaster asiaticum*; *12* to *13*, same, *Agalmatium bilobum*; *14*, several males of *A. bilobum*, calling alternation; *15* to *16*, *M. intricatus*, trill of pulses. Faster oscillograms of the parts of signals indicated as *2*, *4*, *7–8*, *10–11*, *13* and *16* are given under the same numbers.

In *C. affinis* and *O. dissimilis*, signals of individuals from two localities remote from each other were studied. In both species, the temporal pattern of signals appeared to be quite constant and does not demonstrate any distinct geographical variation (Figure 24.5, items *12* to *15*).

The acoustic repertoire of Caliscelidae also differs from Issidae. In males of *A. eoa* and *Ommatidiotus* species territorial signals were recorded (Figure 24.6, items *1* to *6*). Insects can produce signals of this type in isolation but they sing much more readily in the presence of others. Two males sitting on the same stem generally alternate territorial signals. Sometimes in such situations, breaks between signals became shorter and shorter until one of them leaves the plant. Occasionally, more active individuals launch direct attacks, attempting to push the adversary from the stem. In *Ommatidiotus*, females also are capable of producing territorial signals, but their temporal pattern differs from that of males (Figure 24.6, items *7* to *10*). Evidently, territorial signals play substantial roles in the regulation of distance between individuals on a plant.

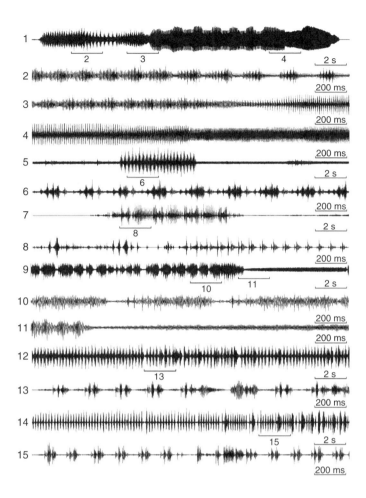

FIGURE 24.5 Oscillograms of calling signals of Caliscelidae. *1* to *4*, *Caliscelis affinis*; *5* to *6*, *Aphelonema punctifrons*; *7* to *8*, *A. eoa*; *9* to *11*, *Ommatidiotus inconspicuus*; *12* to *13*, *O. dissimilis*, male from Moscow Area; *14* to *15*, same, male from Rostov Area. Faster oscillograms of the parts of signals indicated as *2–4*, *6*, *8*, *10–11*, *13* and *15* are given under the same numbers.

O. inconspicuus at a high density in a cage made aggressive signals, differing in structure from territorial ones (Figure 24.6, items *11* to *12*). These signals are produced in reply to other calling individuals, interrupting their songs. Signals of this type were never recorded from single males or from groups of males if the individuals had the opportunity easily to leave the plant.

Acoustic communication during mating in *O. dissimilis* was observed. As in most of other small Auchenorrhyncha, males sing spontaneously whereas receptive females normally produce signals only in reply to a calling male. Female reply-signals have much simpler structure compared with male ones and consist of short successions of pulses (Figure 24.6, items *13* to *15*). After detecting female calls, a male starts singing more actively and runs in different directions searching along the plant. Occasionally, he stops and calls briefly. As a result, his signals become distinctly shorter. One can say that these are not complete calling signals, but just fragments. In contrast to males, females remain stationary producing reply-signals from time to time. A few seconds after meeting, the partners join their genitalia.

Comparison of temporal patterns of calling signals of Issidae with other Fulgoroidea reveals no major differences except for the Delphacidae and Caliscelidae. In most studied representatives of

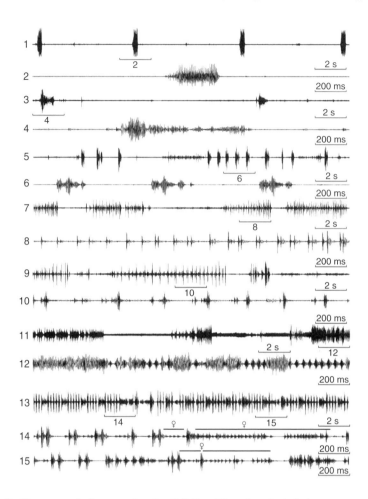

FIGURE 24.6 Oscillograms of vibratory signals of Caliscelidae. *1* to *2*, *Aphelonema eoa*, male territorial signals; *3* to *4*, same, *Ommatidiotus inconspicuus*; *5* to *6*, same, *O. dissimilis*; *7* to *8*, *O. inconspicuus*, female territorial signals; *9* to *10*, same, *O. dissimilis*; *11* to *12*, *O. inconspicuus*, male calling, and aggressive signals; *13* to *15*, *O. dissimilis*, male calling and female reply-signals. Faster oscillograms of the parts of signals indicated as *2*, *4*, *6*, *8*, *10*, *12*, and *14–15* are given under the same numbers.

Cixiidae, Derbidae, Dictyopharidae (both Dictyopharinae and Orgeriinae) and Tropiduchidae, calling signals have rather simple temporal patterns (Figure 24.7). For the most part these are syllables consisting of more or less uniform short pulses (Hoch and Howarth, 1993; Tishechkin, 2004). On the contrary, calling signals of Caliscelidae have very elaborate temporal structure and sometimes consist of several different parts (Figure 24.5). Some Delphacidae also produce calling signals with complicated temporal patterns (Tishechkin, 2004). Moreover, well-developed territorial and aggressive behaviour has been described in representatives of this family (Ichikawa, 1982).

Thus, Issidae *sensu stricto* on a basis of acoustic characters resemble most other Fulgoroid families than the Caliscelidae. My data support the opinion of Hamilton (1981) and other authors that the latter group, which deserves the rank of a separate family, must certainly be excluded from the *Issidae*. Also, it should be noted that in certain works (*e.g.* Nast, 1972), *A. vittifrons* is treated erroneously as a member of Caliscelidae (or Caliscelinae). Nonetheless, in morphological and in bioacoustic characters, this is a typical Issid.

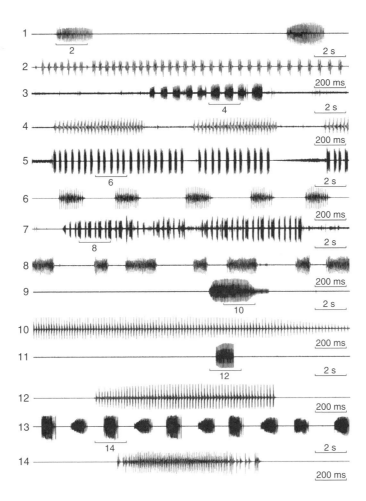

FIGURE 24.7 Oscillograms of calling signals of Fulgoroidea. *1* to *2*, *Cixius nervosus*; *3* to *4*, *Reptalus arcbogdulus*; *5* to *6*, *Hyalestes obsoletus* (Cixiidae); *7* to *8*, *Malenia sarmatica* (*Derbidae*); *9* to *10*, *Dictyophara pannonica* (Dictyopharidae, Dictyopharinae); *11* to *12*, *Mesorgerius tschujensis* (Dictyopharidae, Orgeriinae); *13* to *14*, *Trypetimorpha occidentalis* (Tropiduchidae). Faster oscillograms of the parts of signals indicated as *2*, *4*, *6*, *8*, *10*, *12*, and *14* are given under the same numbers.

CONCLUSIONS

It is generally accepted that behaviour evolves much faster than morphological structures and, consequently, behavioural characters as a rule cannot be used in taxonomy for discrimination between higher taxa. However, most higher taxa of Auchenorrhyncha differ distinctly from each other in bioacoustic characters. Over recent years the procedures of acoustic signal recording and analysis have become simpler, so the body of information on acoustic communication is increasing rapidly. The use of signal analysis in taxonomy at the species level is well established, but possibly this can now be extended also to the systematics of higher taxa.

25 Acoustic Signals, Diversity and Behaviour of Cicadas (Cicadidae, Hemiptera)

Michel Boulard

A proprement parler, les insectes ne chantent pas.
Les Cigales, à le bien prendre, font de la musique instrumentale

J. Grolous, 1880

CONTENTS

INTRODUCTION

Acoustic communication is widely known in the animal kingdom as an important factor in reproduction. Various groups of animals possess often quite elaborate organs for the emission and reception of sounds. True cicadas (Cicadidae) belong to this interesting category. Contrary to other acoustic insects they are often referred to as "singers" or "musicians" which use added attributes in different parts of the body like legs, wings, thoracic or abdominal parts whose first function is not sound production. Cicadas have very elaborate devices exclusively for the production and reception of sound, the first being present only in males, but the auditory complex is present and is almost identical in both sexes.

Acoustic communication is a rapid process and very reliable in that it leaves no material traces. It consists of the following phases:

- The behaviour of the individual emitting signals
- The sound emission or "acoustic signal"
- The transmission of this signal through a medium
- The perception of the signal by the targeted individual
- The reaction of the receiver to the signal
- The possible perception of the signal by nontarget individuals, conspecific, friendly or not (predators), and their possible reactions

The present work concerning acoustic communication in cicadas from all over the world will focus mainly on the first two phases while underlining the importance of the last three in respect of communication.

Although cicadas are unmistakably noisy, research on their sound production and ethology has been rare and fragmentary for a very long time (Carlet, 1877a, 1877b; Distant, 1897). Myers (1929) recognised the specificity of their signals and stated that "*every cicada with which we are familiar may be recognised with certainty by its song*".

Up to the last quarter of the 20th century, the only way of analysing sounds was with the aid of heavy and very expensive devices, oscilloscopes and vibralysers (*e.g.* Kay Sona-Graph), which allowed only very partial transcriptions, not always clear, of the sounds studied. Nevertheless, substantial progress was made (Pringle, 1953, 1954, 1955; Alexander, 1957a; Popov, 1975, 1981, 1997; Young, 1975; Boulard, 1977; Dunning *et al.*, 1979; Leroy, 1979).

The enormous development of computers and the digitisation of insect sounds made it relatively easy to translate sounds into images and show emissions, while at the same time allowing an in-depth analysis of the acoustic signals and rhythms of action which structure them.

Thus, every sonogram obtained proved to be typical of the species considered, a sort of ID card, completing and sometimes overruling the specific status elaborated using more usual tools of systematics like general and genital morphologies, not to mention electrophoresis and genetics. It has thus been possible to draw up distinct Cards for Identification by Acoustics (CIA) and Cards of Ethology by Acoustics (CEA). These are precious documents for the analysis of sounds and their diversity, for the creation of sonotypes and of etho-sonotypes as well as for the preservation of their images. In fact, the digitisation and analysis of acoustic recordings, laborious but very reliable, have allowed for the definitive separation of very close species, sibling or cryptic species previously often not recognised or confused (Boulard, 1995, 2004c; Gogala and Trilar, 2004; Prešern *et al.*, 2004).

In what follows, my own recordings have been used from cicadas which were certainly correctly identified. They derive from all over the world and are grouped arbitrarily in this essay.

Signals were mostly sampled at 22,050 or 44,100 Hz, but sometimes only at 11,025 Hz for species emitting at relatively low pitch, were computerised (on an Apple Macintosh computer) and then were shown by means of two software packages for the analysis of sounds, firstly "Signalyze" developed by Professor Eric Keller (University of Lausanne) and secondly the

"MacSpeech Lab" finalised by the U.S. firm "G. W. Instrument". The terminology adopted will be that of Eric Keller (1994).

The objective of this chapter is to give an outline of the taxonomy of cicada sound signals as well as their enormous diversity, while demonstrating their behavioural significance. Some new or little known data on the behaviour of males during sound emission are included.

ACOUSTIC SIGNALS IN CICADAS

SOUND PRODUCTION, THE TYMBAL AND ITS SOUND

Anatomy of the Tymbal

With respect to behaviour it is of particular significance that the fully developed tymbal system is only found in males. Detailed descriptions are given elsewhere (notably Carlet, 1877a, 1877b; Vogel, 1923; Pringle, 1954, 1957; Young, 1972; Michel, 1975; Bennet-Clark, 1999b). It consists of two symmetrical membranes, the drums or cymbals of Casserius (1600) and Réaumur (1740), usually known as tymbals. They are found dorso-laterally on the first abdominal segment (Figure 25.1). In some species, females have small rudimentary tymbals, which are, however, totally sclerotised and ineffective. This is the case notably in the large *Tosena albata* Distant. On the other hand, there are also species in which males lack a tymbal. Apart from the archaic Tettigarctidae, this is so in some rare species in a few genera of Tibicinidae, but otherwise phylogenetically only distantly related, mainly from equatorial Africa and North America (Boulard, 1976a, 1986, 1990a, 1993b, 2002b).

From outside, the tymbal appears as a slightly convex plate of cuticle, strengthened with sclerotised ribs and disks framed in a roundish rim of cuticule. In some species the tymbals are (partly) hidden under the protruding tergite of the second abdominal segment. Each tymbal is connected, eccentrically, to a very powerful muscle by a short triangular tendon. The tymbals buckle inwards by muscular contractions and then pop back with the slackening of the muscles.

These actions, in very rapid repetition, produce a train of pulses, often very loud, as the ribs are alternately or simultaneously compressed and relaxed in a rhythm varying from species to species (Pringle, 1954; Boulard, 1990a; Bennet-Clark, 1999b). The sound produced is amplified by what is truly an abdominal resonant chamber formed by a huge bag in the tracheal respiratory system, occupying the largest part of the male abdomen and pushing back the digestive and procreative viscera towards the end of the body. This resonant chamber lightens the body of the males. So, when these are seen against the light, their abdomen, astonishingly, seems to be empty (see *Pomponia cyanea* Fraser, in Boulard, *op. cit.*).

A somewhat similar organ is found in males of several groups of moths, notably Arctiidae and Nodolidae (Fullard, 1977, 1992; Conner, 1999; Skals and Surlykke, 1999).

Such studies need also to take account of tymbal-like structures found widely in other related auchenorhynchous families (*e.g.* Ossiannilsson, 1949; Strübing and Schwartz-Mittelstaedt, 1986), and also of the very strange "acymbalic (or atymbalic) cicadas".

The Sound of the Tymbal, Fundamental Acoustic Parameters

The comparative study of cicada acoustic signals is based on three essential and measurable variables, which define a sound:

- Frequency
- Amplitude
- Duration of call units

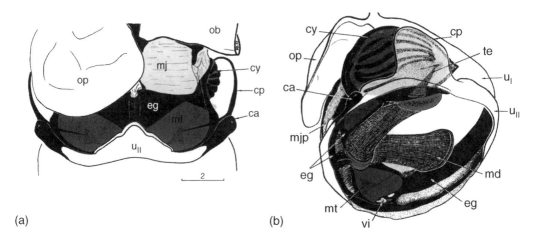

(a)   (b)

FIGURE 25.1 Sound and auditory acoustic system of a male cicada: (a) Base of abdomen seen from below, very stretched, with the left opercula turned upwards "to open" the ventral acoustic chambers and to show the components of the acoustic system. (b) Acoustic system (isolated abdominal segments I and II) seen for 3/4 from behind and left. ca = left auditive capsule; cp = tymbal cover; cy = tymbal; eg = entogaster (internal armature); md = right tymbal muscle; mjd = yellow membrane (distended, the organ in function); mjp = yellow membrane (wrinkled, at rest); mt = tympanum; op = operculum (latero-ventral cuticular disc); you = tendon connecting the tymbal muscle with the internal face of the tymbal (seen here as through a transparent medium); UI, UII = the first two abdominal segments (Acousticalia) carrying the acoustic organs; vi = thin memebrane within the right auditive capsule. (From Boulard, M. and Mondon, B., "Vies et Mémoires de Cigales" Éditions de L'équinoxe, Barbentane, 1995. With permission.)

In addition to these three variables is the timbre, which often allows the human ear to distinguish between two sounds having the same basic or fundamental frequency and amplitude. The timbre depends on the existence and number of harmonics, the respective frequencies of which are multiples of the basic or fundamental sound frequency.

The acoustic signals of animals and especially of insects differ from pure sounds and casual noises by their structure and the regular repetition or rhythm of emission. Insect sounds are consisted of basic units or echemes that are repeated. The echemes vary in duration and speed of attack and decay, and may be repeated, unitary or grouped in simple or complex motives or phrases. Thus there is an almost infinite range of possibilities leading through simply temporal characteristics to the establishment of a very high degree of specificity. Of particular importance, in fact determining the sound produced, are not only the form, number and spatial distribution of the constituents of the emitting device (specific dimensions, morphology and topography of the tymbals), but also its dynamics, the time of the year, the hour and the impact of the environment, in particular temperature (See Sanborn, Chapter 7).

Visualisation of Acoustic Signals: CIA and CEA

Thanks to the development of microcomputers and of software for the study of audio data, it has now become rather simple to transform digitised sounds into images and thus to allow them to be analysed in a frequency, time or amplitude domain and archived. In the first applications of this new technology, cicada specialists have been able to establish CIA and CEA which can be transformed into sonotypes and etho-sonotypes. Programs allowing the multiwindow stacking of sonograms have revealed unique graphic representations of the sounds of the species under study. Other authors use different but similar presentation of sound characteristics.

For the sake of authenticity, recordings should be made wherever possible in nature as calls in captivity are often atypical. However, recording of the smallest species is rarely possible in the field.

Once computerised, multiwindow sonograms containing several different, but complementary graphs are constructed. It is customary to include:

(a) The temporal oscillogram reproducing an ethological unity, generally the phrase or a regularly repeated sequence of a signal, or a relatively substantial portion of one, expressed in time.
(b) The curve of the average spectrum (analysis of relative intensity over frequency) or the curve of relative energies in time called "amplitude envelope".
(c) A portion of the temporal oscillogram (a) stretched in arbitrarily chosen space–time revealing the spatiotemporal structure of the emission (Figure 25.2).
(d) The spectrogram showing the temporal variation of emitted frequencies and intensities. Sometimes, it can be constructive to add a second sector, an additional card depicting the periodicity, often very narrow in cicadas, the pulses and their phonatomic composition providing the morphology and the average number of sound units transcribed in a reference span of time. This second section will contain usually also:
(e) A visualisation of the displayed signal over a time period of 1/10th of a second (sometimes more, sometimes less, according to species), a matrix for the following analysis.
(f) The structure of sound units shown over a reference span of time of about 12 msec, as defined by the computer tool.

Anatomy of the Auditory System

Cicadas of both sexes can perceive sounds thanks to some rather remarkable "ears" which they also have on their abdomen. These are two small capsules, called the auditory capsules, hemispherical or in the form of a "skullcap", implanted symmetrically on the sides of the second abdominal segment. Each auditory capsule encloses sound-sensitive organs, *scolopidia*, which receive and integrate the

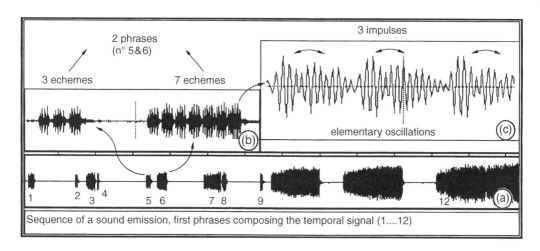

FIGURE 25.2 Schematic graphic transcription of the beginning of a sound emission (a) and structural analysis after more and more stretching in time scale and spaces (b), (c), until it is possible to visualise individual elementary oscillations, original groupings of which constitute impulses (phonatomes). (From Boulard, M. and Mondon, B., "Vies et Mémoires de Cigales" Éditions de L'équinoxe, Barbentane, 1995. With permission.)

vibrations perceived by huge, rather thin, tympanic membranes (Figure 25.3) tightly embedded in the first intersegmental base of the abdomen. Often iridescent and shining, the tympana can also be completely transparent or more or less opaque according to species. Réaumur (1740) and Vogel (1923), examining the apparatus of the big European cicada, *Lyristes plebejus*, called them mirrors. The *scolopidia* with their long axons pass on to the axial nervous system the information about sound vibrations that have activated the tympana. Rather curiously, the tympana are always larger in males than in females.

FUNDAMENTAL ROLE AND ENVIRONMENTAL IMPACT OF ACOUSTIC SIGNALS

When males call, it is above all to reveal their presence to the females of their species. It is a specific message from males to females, an invitation to mate, that the latter can hear and decode. A receptive female in the neighbourhood will join a male that is calling and copulate with it. This is the fundamental etho-physiological result of the cicadas' call. In other words, males use sound to guide the females to them and, by what might be called a "male-led control system", assures the perpetuity of their genes and the survival of their species.

In the second place, calling may be interpreted by other males as a signal for aggregation, very common in gregarious species, or to exploit a sort of territorial claim as the acoustic behaviour of some diurnal species suggests. For example, *Aola bindusara* and *Pomponia scitula*, living side by side in the same tropical Asian habitats, call from time to time without there being any signs of courtship. In these cases the signal often lacks the urgent vigour and rhythm characteristic of the mating call.

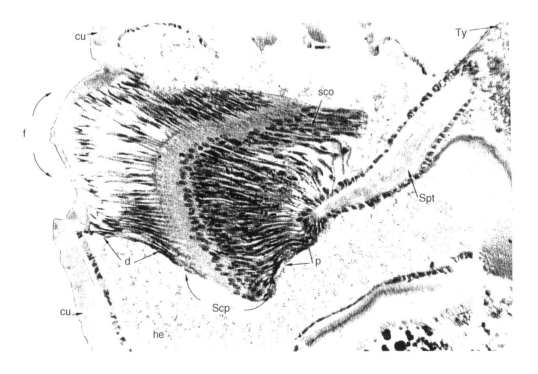

FIGURE 25.3 Histological section in the left auditory capsule of *Cicada orni* L. ca = sclerotic wall of the capsule; f = dimple of external insertion of the scolopophore scp; sco = scolopidies; sp = spatula of internal insertion for Scolopophore; Ty = left prescolopidial part of the tympanum. (From Boulard, 2000c.)

Whatever the ethological meaning of cicadas' calling, there is also an inherent ecological disadvantage as calling can also give away the location of an individual and thus attract the attention of potential predators such as birds, lizards and small arboreal mammals.

Nevertheless, many species of cicadas have developed numerous defences against these fatal threats. Some are aposematic, showing conspicuous warning colours such as *Huechys sanguinea* with its black and red livery warning for its emetic properties which make it distasteful (Boulard, 2004a). Many are cryptic in colouration and resemble their normal resting place and are thus difficult to localise (Boulard, 1985) while others, often weak fliers, call mainly if not exclusively at dusk in conditions unfavourable to predators that hunt by sight. This is notably the case of *Ayuthia spectabile*, the "ghost cicada", an almost "nonchalant" species which makes the most of two defensive strategies together. Also, the two types of defences can be supported by callings in chorus, which impede the detection of individual insects and may also scare away predators. This is the case with many gregarious species such as the European *Cicada orni* L. and the neo-tropical *Fidicina mannifera* (Fabricius). In some diurnal and particularly noisy species, such as the large *Tosena*, males callings only stay for a very short time in the same place and so escape fatal detection.

ACOUSTIC BEHAVIOUR OF CICADAS

POSTURES OF SOUND EMISSION (FIGURE 25.4A TO FIGURE 25.7)

Calling males of each species adopt a specific calling site (trunk, branch, stem or even a stem of grass). Many species do not adopt a particular posture when calling except for a slight extension of the legs and an almost imperceptible trembling of the abdomen, as shown by the European *Cicada orni* L. (Figure 25.4a), *Tettigetta argentata* and *T. pygmea* (Olivier) and the large *Malagasia inflata* Distant from Madagascar. Others may also show more or less elongation of the abdomen, as in the neo-tropical *Fidicina* and *Fidicinoides* (Boulard, 1990a; Cocroft and Pogue, 1996) and the Asian *Aola* (Boulard, 2003c).

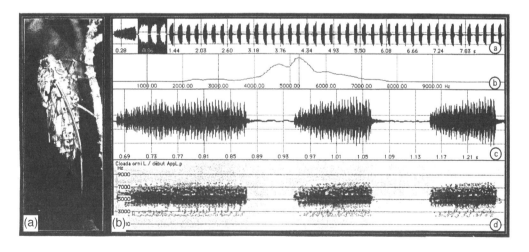

FIGURE 25.4 *Cicada orni* Linné. A, Male calling, Provence, France. B, ID Card based on the beginning of the calling signal. (a) Temporal signal transcribing in real time, the first 10 sec of the call; (b) Average spectrum; (c) Oscillogram obtained in an arbitrary space-time centered on three modules of the temporal signal (in colour inverted in a); (d) Spectrogram corresponding to the previous oscillogram. (From Boulard, M. and Mondon, B., "Vies et Mémoires de Cigales" Éditions de L'équinoxe, Barbentane, 1995. With permission.)

Many other cicadas do adopt very characteristic postures during sound emission, head raised or lowered, males may disengage their anterior wings from the meso-notal grooves and, abandoning the resting position, lower them laterally. The abdomen so freed may then be:

- Stretched and maintained static, as in the Malagasy, *Yanga pulverea* (Distant) (Figure 25.5a DVD), in the Neo-tropical, *Pompanonia buziensis* Boulard and in the Australian, *Chrysocicada franceaustralae* Blrd (Figure 25.5g DVD); or vibrated by short longitudinal accordion-like movements, as *Chremistica numida* Distant.
- Arched, the tip raised as much as possible and maintained so during calling, as in the Polynesian, *Raiateana oulietea* Blrd (Figure 25.5c DVD) the Mediterranean, *Cicadatra* (Boulard, 1992) (Figure 25.5h DVD) and the Malagasy, *Abroma inaudibilis* Blrd, whose head is also lowered (Boulard, 1999).
- Arched, but with the tip of the abdomen pointing downwards. Many species take up this posture, including the African, *Brevisiana brevis* (Walker), *Ioba limbaticollis* (Stål), *Soudaniella melania* (Distant), *Karscheliana parva* Blrd and *Oxypleura spoerryae* Blrd, the Malagasy, *Antankaria signoreti* Metcalf, the Asian, *Platypleura mira* Distant, *P. nobilis* (Germar) and *Purana jdmoorei* Blrd as well as the New Caledonian, *Poviliana sarasini* (Distant) (Boulard, 1975, 1990a, 1995, 2001a) (Figure 25.5d–f DVD).
- Vibrated in fast vertical oscillations of cyclical amplitude, so modulating the signals. This very spectacular method is best known in the Mediterranean species of *Lyristes* (Figure 25.6 DVD) and the SE Asian *Maua albigutta* which shows also other abdominal movements during its complex sound emission (Gogala *et al.*, 2004).

Others raise their wings to an almost horizontal position over their back without decoupling them from the mesonotum, like *Tibicina haematodes* (Scopoli) (Figure 25.7a DVD) and many Cicadettini, such as *Cicadetta fangoana* Blrd and *Tympanistalna gastrica* (Stål) (Figure 25.7b and e DVD). *Tibicina* species usually also show audible or silent wing-flicking just before calling (Figure 25.7c DVD) (Boulard and Mondon, 1995; Sueur and Aubin, 2004).

Similar or identical behaviour has been recorded also by others, notably Moulds (1990) in Australia, Villet (1988, 1992) in South Africa and Sueur *et al.* (2003) in western Europe.

DIVERSITY OF ACOUSTIC SIGNALS

There is an extreme diversity in the acoustic signals of cicadas. Each species calls in its characteristic way so that one might even say that each species has its own dialect.

Essentially, one can distinguish between the calling signal, demonstrated in almost all species, and the courtship sound that may precede mating, as in the Mediterranean, *Tibicina* (Boulard, 1995, 2000b; Sueur, 2002b) and in the Asian, *Purana tigrina* (Walker, unpublished) (Figure 25.8).

In the extensively studied *Cicada orni* L. (Figure 25.9a–c), sounds of supposed "anxiety" and "annoyance" have been identified. There may also be a signal of escape out of fear shared by all species, short drumbeats resulting in a hurried flight accompanied by a sudden ejection of watery excreta (Figure 25.9d).

Finally, there is a "cry of distress", emitted in frightened protest by male cicadas when caught. However, this signal of distress has no alarm value in the ecoethological sense (Boulard, 1994) (Figure 25.9e).

SOUND REGISTER AND BIOTOPE

Seemingly the environment in which every species normally lives has an influence on the type of sound emission. For example, forest species generally call in a frequency range totally different from that of grassland species. In fact the relative size of the various cicadas is the determining factor. Forest species are mostly large or larger than average. By contrast, species of very open areas are almost always small or smaller than average.

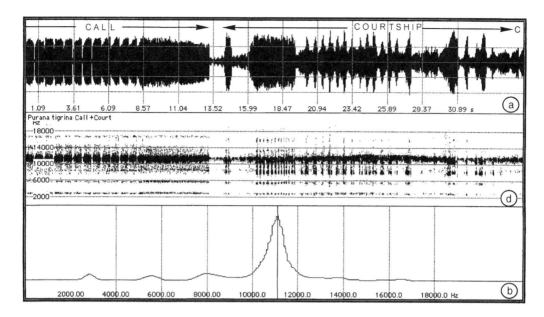

FIGURE 25.8 *Purana tigrina* (Walker). Acoustic ethological card based on the transition call/courtship signal: (a) Temporal transcription. Oscillogram reflecting the end of one calling appeal sequence (CALL) at once followed by the courtship sound, which the male stops as soon as it is coupling the female (c); (b) Average spectrum; (d) Spectrogram corresponding to the previous oscillogram. The harmonic III and IV turn out largely disrupted throughout the courtship.

For the most part, forest cicadas have a relatively low frequency range with a fundamental of 1500 to 3000 Hz, strengthened by harmonics only very rarely exceeding 12,000 Hz. In grassland species, on the contrary, the fundamental is high pitched, often exceeding 4000 Hz, and the harmonics may exceed 20,000 Hz. However, especially in this latter group, there are exceptions, notably *Spoerryana llevelyni* Blrd, *Paectira feminavirens* Blrd (Boulard, 1977), *Purana khuanae* Blrd (Boulard, 2002a) and *Pomponia fuscacuminis* Blrd (Boulard, 2004b).

Other aspects of sound quality are related to solitariness and gregariousness.

DISTINCTIVE SIGNALS OF SIBLING SPECIES

It may seem surprising that the tymbal, constructed according to the same plan in all cicadas, can allow for such a range of specific sounds. This organ has proved, in fact, to be so sensitive that either the least morphological variation or the slightest change in the rhythms activating the two tymbals, either synchronous or alternately (Pringle, 1957; Fonseca and Bennet-Clark, 1998) is enough to alter the sounds produced. So much so that it has been possible, thanks to acoustic analysis, to discriminate between twin species, that is to say, species morphologically identical or very close but of elusive, unconfirmed, or even unsuspected specific identity. In this respect, the acoustic recognition of various N. American *Magicicada* species was pioneered by Alexander and Moore (1962).

Amongst the most sensational applications of the comparative analysis of sound production and the creation of Acoustic Identity cards, here are some examples, either little known or as yet unpublished:

- Sound detection of *Lyristes gemellus* Boulard, occurring sympatrically, syntopically and on the same food plant with its sibling species *L. plebejus* (Scopoli) in Asia Minor (Boulard, 1988) (Figure 25.10 and Figure 25.11 DVD). A third species, *Lyristes isodol* Blrd, its call very accelerated as compared with the former two species, was discovered

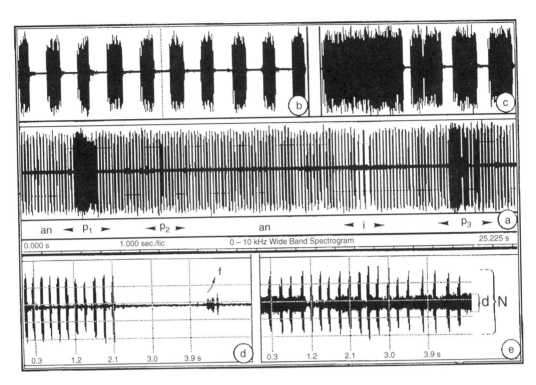

FIGURE 25.9 *Cicada orni* Linné. Pluriwindowed Acoustic Ethological Card (AEC) showing various acoustic signals according to its etho-physiological conditions: (a) Visualisation of 25 sec of a nuptial call recorded in nature and ecological adventures of which were carefully noted. This temporal signal, transcribed in real time, includes: Five ranges of the normal calling (an), which is a serial access of speech in a continuation of "closer and regular sound modules", as already shown in the previous card, but here amplified in the window. (b) Two ranges translating more or less violent protests (p1 and p2) caused by an ant having nibbled the wing, the rostrum, then touched one of the legs of the cicada; A demonstration of anxiety (i) due to the fast passage of a bird in the field of vision of the insect and being translated by one brief slowing down tymbalic. A reaction of opposition (p3), protest activated by the pressing approach of a second male. This signal of opposition, constituted by numerous and deep tymbals bangings moved closer and crumpled up as indicated in the window (c), had the effect to stop homosexual attempt. (c) Irregular aspect of modules of the cymbalisation, portion taken in (a) and 10 times amplified; (d) Image of a temporal signal of 4.30 sec showing a calling signal abruptly interrupted by sudden danger and swiftly followed by a hasty departure, accompanied with a brief "signal of flight" f. (e) Concomitant recordings, for 4 sec, of two calling males at short distance from each other; the first, having been caught, emits muddled "signals of distress" (small irregular central range) d, whereas the second continues to call regularly (wide range composed of normally structured signals) N. This last document gives clear evidence of a direct observation verified many times: the signals of distress of the captured male do not affect the behaviour of the free male. In cicadas the distress call has no ecological value of alarm. (From Boulard and Mondon, 1995.)

in Turkey sympatrically with *L. plebejus* and *L. gemellus*, but not syntopically and on a different food plant (Boulard, 1988).

- The replacement of *Cicada orni* L. by *C. mordoganensis* Blrd (Boulard and Lodos, 1987) in the islands of Samos and Ikaria in the East Mediterranean, confirmed by acoustic analysis (Simões *et al.*, 2000).
- *Tosena albata*, occurring sympatrically, syntopically and on the same food plant with its sibling species *T. melanoptera* of which it had been considered a variant for more than a century, has recently been described as a valid species (Boulard, 2004d, 2005) (Figure 25.12 and Figure 25.13 DVD).

- The discovery of *Cicadetta cerdaniensis*, replacing its sibling species *Cicadetta brevipennis* (Scopoli) in the mountainous South of France (Puissant and Boulard, 2000). This species has recently been reported also from other parts of Europe (Gogala and Trilar, 2004; Trilar and Holzinger, 2004).

- The detection in Slovenia of three *Cicadetta* species, *C. montana* (Scopoli), *C. cerdaniensis* (Puissant and Boulard, 2000) and *C. brevipennis* (Fieber). *C. brevipennis* is more widely distributed in France where it had been confused with *C. montana*. In Macedonia, there exists another species of this species complex *Cicadetta macedonica* Schedl discovered by its unique signal pattern (Gogala and Trilar, 2004).

- The acoustic distinction in a rain forest of the north of the Thailand between *Pomponia dolosa* Blrd and *P gemella* Blrd that are almost identical in general and genital morphology, but occur allopatrically (Boulard, 2001c, Figure 25.14 and Figure 25.15 DVD).

- Alternate presence of two sibling species in Ranong Province in the south of Thailand; two large *Pomponia* have been found at different times in the same forest. As like as identical twins (the same size and appearance, the same emerald green eyes), they are impossible to differentiate through examination of museum specimens. It is only in their natural environment that they can be separated on the basis of their very different calls. The call of the first species, *Megapomponia pendelburyi* (Blrd), prevailing in January and February sounds like a trumpet signal (Figure 25.16 DVD); the second, *Mp. clamorigravis* Blrd, sounds more like a tuba and is mainly found in November and December (Figure 25.17 DVD).

- Maybe even more spectacular, museum material labelled "*Pomponia fusca*" has been shown to contain three other species, morphologically almost identical but differentiated by their calls, namely, *P. fuscoides* Blrd *P. dolosa* Blrd, and *P. gemella* Blrd (Boulard, 2004b).

By contrast, entities considered as different species have been returned to subspecific rank. This is notably the case for two French allopatric *Tibicina* today named as *T. corsica corsica* (Rambur) and *T. corsica fairmairei* (Blrd) (Boulard, 2000a). Another example is supplied by *Salvazana imperialis* Distant from Indo-China, which was put in synonymy with *S. mirabilis* Distant (Chou Yao *et al.*, 1997) although morphologically it is very different by virtue of the contrasting colours of their wings; acoustics confirmed that it was indeed a case of two morphs of the same species (Boulard, 2002).

FUNCTIONS OF ACOUSTIC SIGNALS

Calling Signal and Acoustic Specificity

Cicadas' secondary sexual dimorphism is centred on the tymbal complex, as indicated above. Thus, males exclusively produce sound and calling as an invitation from males to females. These, when receptive, are attracted by the males' invitation and react phonotactically, *i.e.* they move towards a calling male (Boulard, 1965; Fonseca, 1993). In fact, in the cicadan environment as in other *Cicadomorpha*, acoustic signals are above all, if not the only means of intraspecific sexual recognition and constitute a premating isolating mechanism (Claridge, 1990).

In cicadas, sight does not play an important role in specificity of contact and courtship. Males are unable to distinguish by sight between species or sex. They cannot even distinguish a dead from a living cicada near them. For example, in an assemblage of *Dundubia*, a male of *D. feae* which is calling will attract a female of its own species, not one of *D. nagarasingna* that lives on the same food plant. But, if a male of *D. feae* by chance approaches a female of *D. nagarasingna*, he will attempt to mate. Attempts at interspecific mating, usually brief, have been observed and filmed in nature (Boulard, 1973a). In gregarious species it is quite common to see males approach each other and attempt to mate (Figure 25.50 DVD). This will cause the courted male to protest vehemently (Boulard, 1995, and here Figure 25.9a).

Courtship

The sexual encounters of a species are dominated by the distinctiveness of their signals. In strictly gregarious cicadas, the calling signal alone is sufficient, the individuals already being close and easily visible to one another. In solitary species, however, the female that has been attracted does not necessarily reach the calling male straight away and may settle some distance from it. On responding to the female, the male changes his call to a courtship signal which represents an urgent invitation. This courtship signal, sometimes referred to as "the whispered call", generally leads to the meeting and physical contact of male and female. This is clearly shown by *Cicadatra atra* (Olivier), *Pagiphora yanni* Blrd, *Angamiana floridula* Distant (see Boulard, 1992, 1993a, 2001a) and also by *Meimuna durga* Distant (unpublished) (Figure 25.17 DVD and Figure 25.18). Calling signals have been shown also to induce substrate transmitted vibratory signals

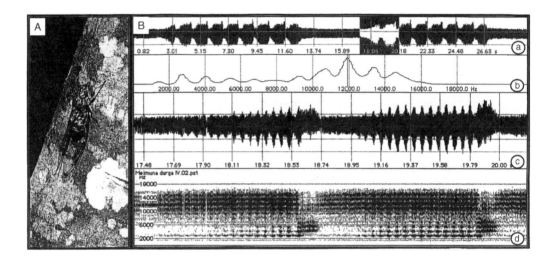

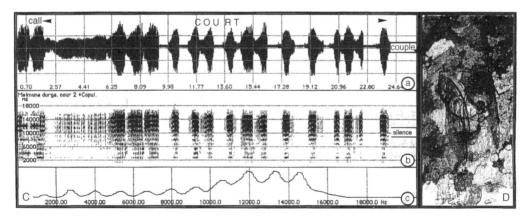

FIGURE 25.18 Acoustic behaviour and function of an heliophilous species, *Meimuna durga* (Distant). A = photographic ID of a calling male; B = calling signal; C = courtship signal; D = male and female coupling. Thailand.

(Stölting *et al.*, 2002). It is therefore possible that vibratory signals also play a role in short-distance localising and recognition of the partner.

As a general rule, courtship signals stop once contact has been made. I know of only one case where the male continued his serenade (in this case even while in the act of mating) and that was *Pomponia dolosa*, a medium-sized crepuscular, almost nocturnal species given the name of "Hand-bell cicada" because it sounds very much like a vigorously shaken bell (Figure 25.19 DVD).

SECONDARY WING FLICKING AND ITS MEANING

In some species, as soon as a receptive female arrives nearby a calling male (range of 1 to 1.5 m), the latter changes its posture, the register and the tempo of its signals and at the same time it moves its wings down on its body.

While modifying the call, the male turns in the direction of the female, contracts its abdomen and returns its wings to the resting position, the *clavus* of the *tegmina* fixed in the corresponding grooves on the thorax, and begins moving rhythmically the biggest part of the *tegminal* area, the *remigium*, bringing the latter to strike more or less strongly the underlying surface of the body (Figure 25.20a–d DVD). The abdomen, threequarters empty as has been pointed out, amplifies the sound of successful strikes so that these are perfectly audible and transcribable, especially in *Cicadatra atra* because they alternate with tymbal signals (Boulard, 1992) (Figure 25.21 DVD). Wing-flicking usually does not involve the total wing, but, allowed by the brachial fold along vein A1, it is rather only the *remigium* that strikes the abdomen hard.

Let me emphasise once again that audible wing-flicking does not result from a release switch (click mechanism) of the *tegmen* from the *scuto-claval* joint, as suggested by Ewing (1989a, 1989b), apparently based on observations by Popov (1981) on the small Russo-Asian cicada, *Cicadetta sinuatipennis*. However, this mechanism of clicking seems to me technically improbable given the actual speed of this exhausting movement (Figure 25.21 DVD) and the type of coupling system implied, which calls for a double action "engaging/disengaging" of the mesothoracic and hemelytral structures in swift repetition. Moreover, Figure 25.20d (DVD) presented here, an abstract from an accelerated photographic sequence, sufficiently demonstrates that the *clavus* remains in a groove of the meso-thorax during this wing-flicking movement, at least in *Cicadatra atra* and also some others. Gogala and Trilar (2003), using high-speed video recording, confirm the fixation of the *clavus* during wing-flicking in *C. atra* and *C. persica*, but claim that the click is produced just before the forewing returns to its resting position and suggest that only folding of the wing along the A1 vein is involved.

In *Tibicinini*, by the same procedure, audible wing-flicking occurs in males preceding the courtship signal and also in some females such as *Tibicina haematodes* (Scopoli) before copulation in duet with the male. Similar wing clicks produced by females during courtship were reported also by Cooley and Marshall (2001) in *Magicicada*. In addition males and females of *T. tomentosa* produce silent wing-flicks and so, fanning their wings, probably (proof lacking) facilitate pheromone diffusion (Sueur and Aubin, 2004).

In *Spoerryana llevelyni*, a small brachypterous African species, on the other hand, the total *tegmina* seem to be driven in the receptive female (Boulard, 1974) (Figure 25.49 DVD).

Audible wing-flicking has also been reported in some other species such as the New Zealand, *Amphipsalta cingulata* (F.) and *A. strepitans* (Kirkaldy) (Myers, 1928; Fleming, 1971) and *Clidophleps beameri* Davis, a Californian species (Davis, 1943).

STRUCTURE OF CALLING SIGNALS

Depending on the species, the calling signal may be simple, more or less compound or particularly complex. Typical examples of a simple call are given by *Cicada orni* L. (Figure 25.4b), *Guyalna nigra* (Boulard, 1999), *Attenuella tigrina* (P. de Beauvois), *Purana atroclunes* Blrd (Boulard, 1999,

2003a, 2003b, 2003c) and *Baeturia uveaiensis* Blrd (Figure 25.22 DVD), with each in its own environment relentlessly repeating the same short unit in its specific register and rhythm. Rather simple is also the call of *Tettigetta pygmea* (Olivier) (Boulard, 1995; Popov *et al.*, 1997) (Figure 25.23 DVD).

As examples of the second type of calling signal, mention may be made of the lively and continuous clacking of *Chrysocicada franceaustralae* Blrd (Figure 25.24 DVD), that "blown" in regular jerks of the eagle cicada, *Cryptotympana aquila* (Walker) (Figure 25.25 DVD), the surprising composition of *Ueana tintinnabula* Blrd from New Caledonia that interrupts its intonations by sudden bending of the abdomen (Boulard, 1991) and that of *Poviliana sarasini* (Distant) from New Caledonia, every sequence of which begins with a train of signals resulting in a sort of rolling cracking ever accelerating until an optimal level of speed and intensity is achieved and sustained for 2.5 to 2.7 sec and then abruptly coming to a stop (Boulard, 1997) (Figure 25.26 DVD), somewhat like the call of the south Iberian, *Tettigetta josei* (Boulard, 1982a).

There are very many examples of complex calling signals, each more fantastic than the rest, notably in African, Malagasy, Guyano-Amazonian species and those, maybe even more compelling, of southeast Asia. All proceed according to rhythms and subrhythms, which are always very fast and/or fractionated. The following may be mentioned:

- The long double-clicking sequences of *Dundubia feae* Distant (Figure 25.27 DVD), differently but regularly structured in low frequencies while having the same amplitude as *Ayuthia spectabile* Distant, the ghost cicada (Boulard, 2002)
- The calling in two speeds of *Karscheliana parva* Boulard (Figure 25.28 DVD) and *Strumosella strumosa* (F.) (Boulard, 1999)
- The calling in a rhythmical metallic timbre in a glissando of *Agamiana floridula* Distant (Figure 25.29 DVD)
- The crackings interrupted with bird-like sounds of *Orientopsaltria cantavis* Blrd (Figure 25.30 DVD) and *Meimuna tavoyana* Distant (Figure 25.47 DVD) (Gogala, 1995)
- The rolling call of *Spoerryana llewellyni* Blrd (Figure 25.31 DVD) and *Sadaka radiata* (Karsch), that of *Ugada limbalis* (Karsch) evoking an electric circular saw (Boulard, 1996b) and that of the large *Pomponia linearis* (Walker) strongly grating and jerky (Figure 25.32 DVD)
- The tenuous whistlings (Figure 25.33 DVD) of *Antankaria signoreti* (Metcalf) or those piercing in different registers from *Chremistica bimaculata* (Olivier) and *Chr. numida* (Distant) (Boulard, 2002, 2003c), those modulated over a relatively long time by *Ioba limbaticollis* (Stål) (Figure 25.34 DVD) and *Soudaniella seraphina* (Distant) (Boulard, 1999) and the hyper-sharp whistle by *Leptopsaltria draluobi* (Boulard, 2003c)
- The performances of the trumpeters *Megapomponia intermedia* (Distant) and *Mp. pendleburyi* (Blrd), at first strong "soundings" then stretching, decrescendo, into long fragmented tails to be strongly boosted once more (Boulard, 2001b) (Figure 25.15 DVD)

To end this subjective enumeration, let us recall the extraordinary calls in several ranges of tonality by the large *Tosena melanoptera* White and *T. albata* Distant already quoted (Figure 25.13 and Figure 25.14 DVD), supreme examples of the apparently unlimited potentialities of the tymbalic organ.

Some species such as *Salvazana mirabilis* and *Chremistica numida* emit sounds which are almost pure and at relatively low frequencies (Figure 25.35 DVD). *Pycna madagascariensis* (Distant) emits a piercing whistle, its fundamental strengthened with three or four powerful harmonics (Figure 25.36 DVD) while other species, particularly those smaller ones, may produce signals rich in harmonics ascending towards ultrasounds; *Tanna ventriroseus* and *Pomponia littldollae*, recently discovered in Thailand thanks to their acoustic signals, provide perfect examples. Their calls no doubt go beyond the technical limits of the most portable sound recorders available for use today (Figure 25.37 DVD). For detection and recording of songs of smaller high-pitched cicadas, the use of ultrasonic (bat) detectors is very useful, especially in combination with the parabola (Popov *et al.*, 1997). When used with a detached microphone and parabola, one gets a highly directional ultrasound detector and can easily localise the singing insects.

To detect acoustic signals of even smaller species in nature is not easy, most operating in high frequencies ranges only permitted by the narrow width of the tymbal and its speed of action. Not only are our ears rather unsuccessful, but also our normal recording equipment is hardly more reliable. To record, for example, the surprising Malagasy, *Abroma inaudibilis* Blrd, which calls perched upside down on a stem of grass with the acoustic chambers clearly freed from the *opercula* by means of a sharp bending of the abdomen, or *Huechys sanguinea* (Degeer), the small aposematic Asian cicada, it was necessary to place the head of the microphone at less than 5 cm from the males which, luckily, did not frighten them (Figure 25.38 and Figure 25.39 DVD).

CALLING OF SOLITARY AND GREGARIOUS SPECIES

Many species of cicadas lead a solitary life, including the large *Tosena* and *Platypleura*, the small Asian *Purana*, the European *Lyristes plebejus* (Scopoli), the Guyano-Amazonian *Dorisiana bicolor* (Olivier) and *Fidicinoides pseudethelae* Blrd and Martinelli (Figure 25.40 and Figure 25.41 DVD). Many others, however, such as *Cicada orni* L., *Fidicina mannifera* (F.), *Quesada gigas* (Olivier), *Ugada limbalis* (Karsch), *Trismarcha excludens* (Walker) and *Tosena splendida* Distant are gregarious, living in colonies of varying densities, some such as *Tibicina haematodes* (Scopoli), *Dundubia nagarasingna* Distant, or *Balinta tenebriscosa* Distant rather loosely (Boulard, 1995, 2001a, 2002a; Sueur and Aubin, 2002). However, *L. plebejus* may appear in certain years in the Provence in such great numbers that, from fundamentally solitary, it becomes gregarious and calls in deafening chorus, making it impossible to single out an individual call. This may also be the case with the chorus *impromptu* by *Chremistica moultoni* Blrd from Thailand, which more often is solitary.

Males of gregarious species often seem to call in chorus; in fact, the synchronisation is rarely perfect, but the slight interval between even the closest males is lost, at least to our ear, as it merges seamlessly into the whole orchestra. This is the case, very impressively, in the Provençal olive groves with *Cicada orni* where, during beautiful warm sunny days, one has the fantastic sensation of a single powerful call produced on all sides by a thousand tymbals!

ACOUSTIC SIGNALS AND ASSEMBLAGES OF CICADAS

Concomitant Usage of the Sound Space

Several species may very well evolve together, side by side, in the same environment. Cicada species, living in the same biotope at the same time, can be heard:
- Separately, at different moments and intervals
- Together, but in different frequency ranges and action rhythms

A typical example is the North American species pair, *Magicicada cassini* and *M. septendecim*, where males of the former synchronise their calling while those of the latter do not (Alexander, 1957a; Alexander and Moore, 1958; Huber *et al.*, 1990).

The following example concerns three species of heliophilous diurnal cicadas belonging to different genera, *Ugada taiensis* (Blrd), *Sadaka radiata* Karsch and *Trismarcha excludens* (Walker), recorded in 1986 in a beautiful forest in the south of Ivory Coast. The first two, (Ut) and (Sr), solitary species, were calling at the same time, but without common measure in intensity, rhythm or frequency. After one (Ut) then the other (Sr) had stopped, the males of the third (Te), a gregarious species, would call in chorus in another range of frequencies and a very different rhythm as compared with the former two species on the same tree (Figure 25.42).

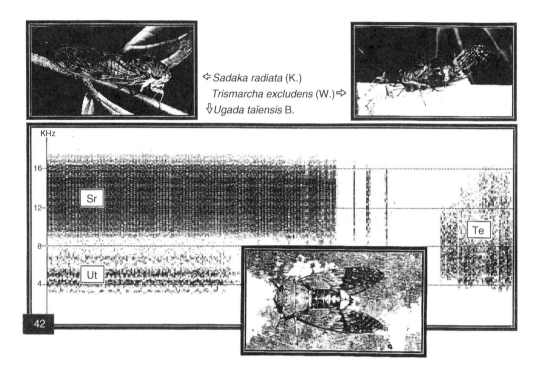

⇐ *Sadaka radiata* (K.)

Trismarcha excludens (W.)⇒

⇓ *Ugada taïensis* B.

FIGURE 25.42 Concomitant, relative or successive usage of the cicadan sonic space in a rainforest (Ivory Coast). Ut = *Ugada taiensis* (Blrd), Sr = *Sadaka radiata* Karsch and Te = *Trismarcha excludens* (Walker). Explanations in the text. (From Boulard, 2000c.)

Yet another example concerns four diurnal species from Doi Mon Kia, North Thailand, *Tosena splendida*, *Platypleura nobilis*, *Cryptotympana aquila* and *Meimuna tavoyana*. The first two are equally stationary to the extent that they can be found all day on the same tree, *T. splendida* more densely aggregated and *P. nobilis* sparser. They are active at the same time but call in different and nonoverlapping frequency ranges (Figure 25.43 DVD). The last two, compulsive travellers, appear only now and then. *C. aquila* calls from the top of the canopy in low pitch first, between 4800 and 6200 Hz, then higher, between 9000 and 10,200 Hz (Figure 25.25 DVD). At other times the high-pitched (6000 to 19,000 Hz) crackling call of *M. tavoyana* can be heard, interrupted by sounds in low pitch which are astonishingly reminiscent of bird songs (Figure 25.44 DVD).

In view of the great complexity of the signals emitted by cicadas living in assemblages in the same geographical area, it appears that this complexity may be a function of the number of species evolving at the same time. In support of this, mention can be made of the banality of the acoustic signals of the rare Seychellian, *Chremistica*, and of that, poorer still, of *Raiateana oulietea* Blrd, the only and very beautiful cicada of French Polynesia (Figure 25.4C and Figure 25.45 DVD).

Territorial Claims: Calling in Competition and Cooperation

Species with relatively sparse populations can be observed to take part in a sort of acoustic joust. This is quite common in *Talainga binghami* Distant (Figure 25.46 DVD) and *Dundubia terpsichore* when two males are in relatively close proximity (5 to 10 m apart). Often it is as if each one tries to call down the other. This seeming possibility of calls being territorial claims needs to be investigated in depth, as on other occasions the supposed rivalry rather resembles a dialogue, even a cooperation in attracting the most distant females. This was observed in the large, diurnal forest species, *Ugada giovanninae* Blrd., from central Africa. Two males, about 5 m apart, were recorded in what seemed a "conversation", both males calling alternately at regular intervals (Boulard, 1996b), (Figure 25.47 and Figure 25.48 DVD).

By contrast, in the Kra isthmus (south Thailand) at dusk, the large crepuscular *Megapomponia pendleburyi* starts calling in deafening cacophony, their signals overlapping each other, making it difficult for the human ear to hear individual calls. However, when two males are fairly close to each other (some 12 m or less apart) without other congeners nearby, they will engage in frenetic competition for acoustic prevalence. Belligerent by necessity, the two males remain in place, the second, even before finishing his first normal sequence, striving to call an ever more impressive verse as soon as the first cicada starts up again. As a result, each male may call many sequences (a good 20 or so) from the same calling site, as long as no female arrives, before one of the two (often the second) abandons his place. Thus, the whole environment seems to be in dispute. As the combat takes place in the dark, females when sound-attracted with increased fervour should be able to catch up with males more easily. This, however, remains to be confirmed. Comparable observations were recently reported by Sueur (2000, 2002b) and Sueur and Puissant (2003).

INVERSION OF ROLES OF THE SEXES

There are many exceptions, some still unexplained, to the rule that calling males invariably remain stationary awaiting females. For example, females of *Spoerryana llewelini*, a small species from the vast prairies at the foot of Mt Kenya, are brachypterous and do not fly, but the males are normally winged and more or less capable of flying in spite of their abdomen being inflated by a large air sac. And it is they who, while emitting a curious croaking call, move in search of the females by fluttering jumps from 3 to 4 m. Here and there, they drop on a tuft of grass and cling to it while strengthening and embellishing their signals. If a receptive female is in the tuft, she clambers up to join the male, flipping its small wings against the sides of the abdomen, and mating may occur. Otherwise the male flies away after a short time to search somewhere else (Boulard, 1974) (Figure 25.49 DVD).

DIURNAL AND SEASONAL RHYTHMS IN ACOUSTIC ACTIVITY— UNACCOUNTABLE BEHAVIOUR

There are numerous diurnal species active in sunny periods during the day, like all European, numerous Asian, African and Madagascan and certain New Caledonian species. Others call in the period just before dusk, as the Thai *Dundubia somraji*. Many are essentially crepuscular, as in the discreet Brazilian, *Pompanonia buziensis* (Boulard, 1982b), the ombrophilous, Wallisian *Baeturia uveiensis*, the central African, *Musoda flavida* Karsch and *Nablistes heterochroma* from the Ivory Coast (Boulard, 1996b), the New Caledonian, *Kanakia annulata* (Distant) and the large *Pomponia* with their impressive thundering calls (often nicknamed "six o'clock cicadas") and some others of the *fusca* group such as *P. fuscoides*, *P. dolosa* and *P. gemella*.

Others are also active at dawn and start calling as soon as day is breaking, as in the Guyano-Amazonian, *Quesada gigas* (Olivier) (Figure 25.51 DVD) and the Asian halophilous, *Dundubia andamansidensis* (Boulard). Other "alarm-clock cicadas", each in its own habitat, are *Soudaniella seraphina* (Distant), *Platypleura mira* Distant, *Terpnosia nonusaprilis* Blrd, *Dundubia oopaga* Distant and *D. feae* Distant. In tropical Asia, this seasonal and diurnal variation has been used pragmatically by ethnic groups of bygone days as a socialo-periodic means of marking the rhythm of days and seasons (Terradas, 1999). However, they all stop calling as soon as the sun rises above the horizon and, as far as I know, no proper calling signals are concerned. Moreover, at least the last four species mentioned may call also throughout the day with breaks to rest and feed.

Some species are also active during the first hours of warm, moonlit nights such as *Cicada orni*, *Lyristes plebejus*, *Ugada limbimacula* (Karsch) and *Megapomponia intermedia* (Distant), or even during the whole night such as *Kanakia typica* Distant, a New Caledonian species which is also active during the day. I have recorded only one truly nocturnal species of Cicadidae, *Kanakia gigas* Blrd (Figure 25.52a DVD), but it is doubtless not the only nocturnal cicada; there is another record of a "Midnight Cicada" from Malaysia, possibly some large *Pomponia* (Gogala and Riede, 1995). *K. gigas* is active a good hour after dark. In view of the morphology of its eyes and antennae it is not like a nocturnal insect. Apparently it can use the weak lunar or stellar light. After calling its rather stiffening signal, a brief "mooing" repeated two or three times (Figure 25.52b DVD), it flies off to settle at some distance and call again. This is behaviour not uncommon in large Asian cicadas such as various crepuscular *Pomponia*, diurnal, spectacular black-winged *Tosena* and *Formotosena montivaga* that cling to a tree trunk when calling (Figure 25.53a and b DVD). Although some cicada females have been shown to locate a calling male during flight (Doolan and Young, 1989; Moore *et al.*, 1993; Daws *et al.*, 1997), it is difficult to understand how they do this, especially in species that occur rather sparsely, like *K. gigas*. I have never been able to observe at first hand any "foreplay to mating", only couples already engaged in the act as, shown here with the "hand bell cicada", *Pomponia dolosa* (Figure 25.19b DVD).

More enigmatic is the acoustic behaviour of the crepuscular New Caledonian, *Ueana rosacea* (Distant), the "twirling rattle". First the male emits a plaintive call lasting about 40 sec from the branch where it rests. Then, abruptly leaving this support, it produces a powerful rattling noise while flying quickly and in an apparently uncoordinated way, whirling around the tree tops then dropping quickly halfway down or even to ground level. This goes on for 2 to 3 min whereupon it settles on some other support and falls silent.

After a short rest, this bustle is repeated again and again, accompanied by its rattling noise, until nightfall. In Figure 25.54 (DVD) the CEA is presented of four turns made at varying distances from the microphone, hence the spindle shape. I have never come across a mating couple of this species and so far fail to understand this very strange behaviour.

STRIDULATORY SIGNALS

Certain cicadas possess potential stridulatory devices. They show considerable differences between them and it is very remarkable that, with one exception, the apparatus discovered by Arnold Jacobi (1907) in the genus *Tettigades* (Tettigadesini) and relatively widespread, each is characteristic of only a single genus or even a single species. The mechanisms that allow stridulation always consist of two main elements, the *pars stridens* or strigil rubbed by the *plectrum* or bow. There is also a possible third component, the "mirror", the particular area of thin and tense cuticle, amplifying by its resonating power the stridulation arising from the action of the first two. A brief outline follows of the main actually or potentially stridulatory devices that have so far been encountered in cicadas.

The Tettigadesian Device

Of the "mesonoto-tegminal" type, this device includes two streaked areas embedded in the fore angles of the *mesonotum*, which rub the thickened and sclerotised anal base of the corresponding *tegmina*. Such a system, present on each side of the body and identical in the two sexes, characterises not only the family Tettigadesini, in which it was discovered, but is also perfectly developed in Zammarini, Plautillini and in various Tibicininae, all far distant from the Tettigadesini and from each other (Varley, 1939; Boulard, 1976b, 1978, 1990a).

The Ydiellian Device

Identical in the two sexes, this is of "wing-tegminal" type strikingly elaborate (Boulard, 1973b). The mirror is very striking as it covers the better part of the tegmina, largely modifying their vein topography. On every costal edge of the hind wings, a stridulating crest consisting of about 50 conical teeth, all similar, regularly implanted, constitutes the plectrum, whereas the third anal nerve of the *tegmina*, covered with many spines, represents the *pars stridens*. This type of stridulatory device, exceptional for cicadas, characterises the genus *Maroboduus* Distant, 1920 (= *Ydiella* Boulard, 1973), including two species that have no tymbal for which it is the only differentiated sound-producing system (Boulard, 1973b, 1986).

The Moanian Device

Revealed in *Moana expansa* Myers from the Samoan islands, this stridulatory device is "tegmino-scutellar" and also found on each side of the body. The *pars stridens* occupies the *scutellar* ribs of the *meso-thorax*, whereas the *plectrums* and mirrors are on the *tegmina*. These are characterised by the unusual dimension of the radial cell, the total hyalinity, the peculiar tension of the *clavus* and the robustness of the *cubito-anal* nerves (Boulard 1976b; Duffels, 1988).

So far, only the tettigadesian mechanism has been recognised as functional (Van Duzee, 1915; Varley, 1939). Precise observations as to the ethological significance of other structures are lacking.

DRUMMING SIGNALS — WING BANGING

There is a last method to send acoustic signals (to make noise, to mark one's presence?) adopted by some small groups of cicadas. It consists in banging the costal edge of the *tegmina* on the support holding the insect. This is a sort of drumming of the wings, resulting in a soft crepitating sound.

This behaviour I noted myself, in October 1979, in Maranhão State in Brazil, in a small population of the tiny *Taphura debruni* Blrd. Males and female drummed in unison on twigs and leaves on or under which they were resting by means of fast lowering and raising of the *tegmina* which were kept parallel to the body. At the point where it touches the support the *costa* is very thick and curved. The males of these small and very agile cicadas, of which I collected only a few specimens, were shown to possess small, but perfectly functional tymbals (Boulard, 1990).

As suspected since the 1920s (W. Knauss, in Davis, 1943) and confirmed since then by various authors, wing banging is the only means allowing the North American species of *Platypedia* to make sound because they totally lack any other obvious mechanism. It is perhaps also the means used by the other cicadas that lack a tymbal, such as *Karenia ravida* Distant and *K. caelatata* Distant from China, *Lamotialna condamini* Blrd and *L. couturieri* Blrd from the Low Ivory Coast (Thailand) and *Bafutalna mirei* Blrd from Cameroon (Boulard, 1993b). In all species of the *Platypedia* group the costal edge of the forewings is strong and more or less curved. We have no ethological data about these peculiar cicadas which, *Platypedia* and *Taphura* excepted, have not yet been observed in nature.

ACKNOWLEDGEMENTS

Very cordial thanks are due here to my close coworkers, Khuankanok Chueata, Hélène LeRuyet-Tan and Gilbert Hodebert, as well as to my esteemed colleagues Sakis Drosopoulos and Michael Claridge, who have invited me to draw up this hasty summary. In addition, I am greatly indebted to John D. Moore, my friend and linguist, thanks to whom I have not tortured Shakespeare's language in the course of this essay.

26 Vibratory Communication and Mating Behaviour in the European Lantern Fly, *Dictyophara europea* (Dictyopharidae, Hemiptera)

Hildegard Strübing

CONTENTS

INTRODUCTION

Dictyophara europea, the European lantern fly, is the only representative of the mostly tropical Dictyopharidae in Germany, where it occurs locally, and also in Berlin and the surrounding area. It may be found in relatively dry grassland, but always in the direct vicinity of open water.

It hibernates in the egg stage and adults appear only as late as July or August up to the beginning of September. Several times I have seen the animals really sunning themselves in the late afternoon. When the sun went down they moved to the tops of grass stems and always held their bodies towards the sun, as is in their tropical nature.

Attempts over several years to rear *Dictyophara* in the laboratory were not successful. In winter it is very difficult under laboratory conditions to regulate the humidity appropriately. A further complication may be that females usually oviposit on soil, as described by Müller (1942). They very rarely lay eggs in captivity. Only once did I succeed in keeping a few larvae in the spring from rearing attempts. The larvae did not reach the adult stage. Thus, fresh outdoor material had to be collected every year.

Dictyophara is not monophagous. It can probably exist on various legumes. In its original habitat I found *Medicago lupulina*, on which captive animals survived for more than 6 weeks. In the laboratory it was usually kept on *Vicia faba* on which it could live equally well.

HISTORICAL OVERVIEW

Ossiannilsson (1949) was the first to notice and describe the generation of acoustic signals in Auchenorrhyncha, a fact that had been assumed but never proved up to then. As there were no steel tape recorders yet at that time, he tried to describe the sounds in different ways (*e.g.* on music paper according to tuning fork analyses). He could make the animals' signals audible to himself by putting one or several animals into a little glass tube with some plant parts and holding the open tube end to his ear. He did not rear the animals but used only outdoor materials freshly caught, never knowing whether the females had already been copulated or not. His rather small cages presumably did not provide sufficient space for normal copulation behaviour, particularly concerning the foreplay. Therefore his experiments to prove the role of those signals were not conclusive, although he could discern duet-calling between males and females of *Doratura stylata* (Boh.), as well as search actions by the male up to attempted copulation (Ossiannilsson, 1953). He did, however, prove an ability of producing acoustic signals in Auchenorrhyncha, which at that time was entirely unknown.

The significance of acoustic communication between male and female for finding the proper partner for copulation could be proved for the Delphacid *Struebingianella* (*Calligypona*) *lugubrina* Boh. (Strübing, 1958, 1959). When a few virgin males and females that had been separated in the last larval stage were introduced into a test cylinder of approximately 20 cm height and 10 cm diameter put over a flowerpot planted with *Glyceria aquatica*, mutual responses could be observed after a very short time. The males started running around agitatedly on the *Glyceria* stalks, conspicuously moving towards the females. Females immediately began to vibrate, not drumming on to the plant, which they did not touch with their abdomen, only their feet in contact with the plant. They set their whole bodies in vibration, as I noticed much later. The females remained stationary while vibrating. The males usually reached the female in little time and took a characteristic position behind or at the side of the female. Sometimes copulation started very quickly or in some cases the male additionally performed a regular courtship song, clapping his wings and dancing from one side to the other and back again behind the female. Thanks to my knowledge of Ossiannilsson's observations, I immediately recognised all this behaviour as acoustic communication. However, I did not hear anything during all these observations. When not just a few but 25 males were put into the glass cylinder closed with a moistened pig bladder instead of gauze, I could hear the calling clearly. Samples of male and female acoustic signals, including the courtship song, recorded with an S-VHS camrecorder, are given on the DVD with this book.

At that time our institute did not possess a steel tape recorder and the first recording was made by courtesy of my student, Johannes Kinzer and his friend Fred Methner, who possessed a gigantic tape recorder with a recording speed of 76 cm/sec and a recording frequency of 40 to 16,000 Hz, an amplifier and a special underwater microphone, which was particularly sensitive (Figure 26.1).

Wagner (1962) revised the Delphacidae of central Europe and split the genus *Calligypona*. A new genus name *Struebingianella* gen. *novo* W. Wag. was erected in appreciation of my examinations, particularly the acoustic communication of Delphacidae.

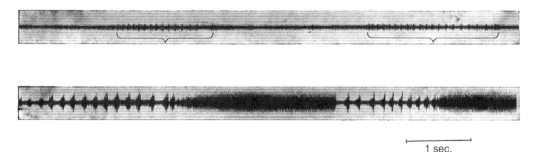

1 sec.

FIGURE 26.1 The first recording in 1957 of the alternating calls between male and female of the Delphacid *Struebingianella* (*Calligypona*) *lugubrina*. The 50 Hz deflection from the baseline is caused by the amplifier of the band recorder. Above: the rhythmic drumming of the female (indicated by the two braces); below: the male call (in this case the male is very excited by the female's drumming, oscillogram continued from above). (From Strübing, H. *Zool. Beitr. (N.F.)*, 4, 15–21, 1958. With permission.)

MATERIAL AND METHODS

All test subjects came from the botanic garden in Berlin.

The magneto-dynamic (MD) system, an inductive transducer working without direct mechanical contact with the vibrating substrate, used to record the vibratory signals of males and females (Strübing and Rollenhagen, 1988) is briefly described here.

The set-up of the MD is simple. A stem of a suitable plant is fixed in a 2.7-cm wide acrylic glass cylinder covered on top with gauze, allowing the test subjects to move freely. A small, light weight (approximately 0.1 g) magnet is attached to the stem opposite a coil, mounted to a sturdy support. The magnet should be as close to the coil as possible without ever touching it. It should be light enough not to interfere with the vibrations of the stem, yet powerful enough to induce a measurable voltage in the coil by its movements. The induced voltage is in general proportional to the displacement velocity of the vibrating substrate and can be recorded on tape without additional preamplifier (input for low impedance microphones) (Figure 26.2). It is a great improvement over

(a)

FIGURE 26.2 The recording system called MD recorder. Above, overall picture of the set up; below, in more detail (a) broad bean plant; (b) test subject; (c) magnet; (d) coil; (e) coil support; (f) glass cylinder; (g) foam plastic stopper around the coil support; (h) gauze cover; (i) base. (From Strübing, H. and Rollenhagen, T., Ein neues Aufnehmersystem für Vibrationssignale und seine Anwendung auf Beispiele aus der Familie Delphacidae (Homoptera-Cicadina) — A new recording system for vibratory signals and its application to different species of the family Delphacidae (Homoptera-Cicadina), *Zoologische Jahrbücher, Abteilung Allgemeine Zoologie und Physiologie der Tiere*, Gustav Fischer Verlag Jena, Vol. 92, pp. 245–268, 1988. With permission.)

the ultrasound microphones (1 to 3 kHz) and accelerometers used before (Strübing, 1967), which were not as sensitive. Furthermore, the damping of the accelerators was too high for certain tests and they were also very expensive.

MATING BEHAVIOUR

As in other Auchenorrhyncha, *Dictyophara* communicate by vibratory signals transmitted through the substrate, and males and females come together in this way (Traue, 1978a, 1978b, 1980; Michelsen *et al.*, 1982).

The assumption that the medium of signal transmission is air and that in the direct vicinity of the animals a medium flow is generated that is registered as particle velocity by hair sensilla at the abdomen end of the animals cannot be maintained. Examination of these hair sensilla by electron microscope showed them to be purely mechanical receptors (Schönrock, 1976).

Amputation of the antennae which are equipped with very many sense organs, does not interfere with mating as long as the animals are on their nutrient plant. However, when they jump off, they are no longer able to find their way back. So it seems that the sense of smell keeps the animals together in a certain area of their nutrient plant. This area may be quite large, much larger than the range of the acoustic signals. Thus it seems that the sense of smell forms the foundation for the local distribution of the species and keeps them together in their biotope, irrespective of the acoustic communication of the individuals with each other.

The females react and mate only once in their lives in answer to the males' calling, apart from reactions of very old females after the oviposition period. While males are ready for further copulation immediately, females, once copulated, refuse and push the males away with their hind legs, shake their abdomens or run away. Females that are not vibrating are seemingly not even recognised by the males as possible partners. Probably this behaviour is without biological significance.

In all observations the female remained stationary, waiting for the searching male attracted by the vibratory signals of the female. The intensity of the vibratory signals depends not only on the distance between sender and receiver, but also on the extent of stimulation and excitement of both partners (Figure 26.3d). These parameters seem to be important for orientation, especially for the searching male (Strübing, 1977a, 1977b).

No rivalry or special courtship songs (Figure 26.3a and b) were observed, but the males seemed to show a sense of direction in finding the females.

The male has two vibratory signals:

• The call, which is answered by the female, is a "tooting" sound (Figure 26.3a). When the male is highly stimulated by the reactions of the female it produces a rhythmically "honking" sound, sometimes of long duration (Figure 26.3b).

• The female's "honking" (Figure 26.3c), very similar to that of the male, is emitted by the female when it is ready to copulate and it answers the male's "tooting", but when the male reacts with "honking", the female answers this too, so at last overlapping of both signals occurs (see Figure 26.3d and Figure 26.3e). Here the situation shortly before copulation is shown. In this record (a–e) one can follow the searching male, who continuously lessens the distance to the female, but energetical emission of signals depends also on the extent of stimulation. Both lines (d and e) belong together without interruption. For better understanding they are marked at some places with the symbols for male and female. In the second part of Figure 26.3e, male and female are emitting the "honking" sounds with the same intensity. They have met shortly before mating and the signals overlap (see also Figure 26.3c and soundtrack on the accompanying DVD).

Dictyophara produces a very soft sound during copulation, which is emitted at more or less regular intervals. The human ear is able to filter this sound out of background noises, but it was not possible to display it in the males' calling presented here, the "tooting" sound. Possibly it has the

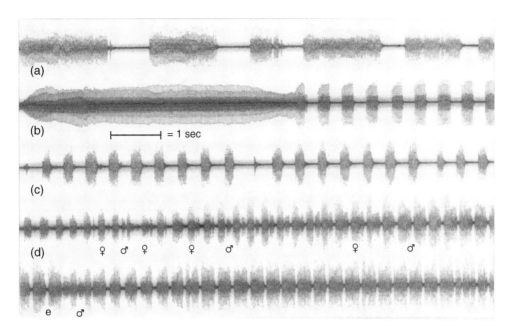

FIGURE 26.3 The vibratory signals of *Dictyophara europea*. (a) The call of the male, which is here answered by the female, is a "tooting" sound. (b) The male answers with its long "tooting", but stimulated by the reactions of the female it presents a rhythmic "honking" of sometimes long duration. (c) The female's "honking", very similar to that of the male; when it is ready to copulate it answers the male's "honking". (d) continued in (e) alternate "honking" of male and female until at last both are emitting with the same intensity. After they have met, signals overlap shortly before mating. (From Strübing, H. *Zool. Beitr. (N.F.)*, 23, 323–332, 1977a. With permission.) (See also soundtrack on accompanying DVD.)

following biological significance. Copulation takes longer than one hour and the low sound may possibly be interpreted as serving the further stimulation of the female for such a long time.

All vibratory signals have very low frequencies here; they are distinctly lower than 50 Hz (see Figure 26.3a–c).

Film recordings made with a normal video camera will be deposited in the *Museum für Naturkunde* (Natural History Museum) in Berlin.

27 Vibratory Communication in Psylloidea (Hemiptera)

D. Yu. Tishechkin

Sound production by Psyllids was first reported by Ossiannilsson (1950). He described very faint, short buzzing signals produced by two species of *Trioza* and by *Livia juncorum*. For studying these sounds he placed insects in a glass tube and the open end of the tube was put to his ear. Individuals of both sexes appeared to produce sounds by wing vibration. Amplitude of the wing movements in insects studied did not exceed 10 to 15 °.

Using Ossiannilsson's technique, Heslop-Harrison (1960) heard sounds of more than 20 British species. Also, he found teeth on the second anal vein of the forewings, corresponding with well-developed metapostnotal epiphyses in several species. He suggested that they were associated with sound production.

Almost at the same time, Taylor (1962) studied both sexes of some species of *Hyalinaspis*. He illustrated "corrugations" on axillary cords of the meso- and metascutellum of the thorax and a hard "flange" at the base of the second anal veins of both wings. He suggested that these could be drawn over the axillary cords during stridulatory movements. Later Taylor (1985) made more detailed investigations of structures on the wings and scutellum using the scanning electron microscope. He presented SEM photographs and drawings of rows of teeth on the axillary cords of the meso- and metascutellum and corresponding rows of teeth under the second anal vein of both pairs of wings in two species of *Scheidotrioza* (*Triozidae*). Also, probable wing movements during stridulation were described. Nevertheless, no sounds were recorded and the structures described were referred to as "possible stridulatory organs".

Campbell (1964) reported sounds associated with rapid vibration of the wings when held at rest in males of two *Cardiaspina* species.

Peculiar behaviour, possibly associated with vibratory communication, was described in a number of Australian psyllids from four families (Carver, 1987). Adults of both sexes usually walk and stand on basal tarsomeres carrying the apical ones aloft. When stationary, both in isolation and with other individuals, the insects vibrate their apical tarsomeres of all legs. The function of this behaviour is unknown.

In a detailed study of taxonomy and reproductive isolation of two *Paurocephala* species (Carsidaridae) from Taiwan, Yang *et al.* (1986) made the first attempt for recording and analysis of Psyllid signals. A small cage enclosing the insects was placed close to the membrane of a microphone. Signals were recorded on magnetic tape and analysed by means of a KAY Digital Sonograph. Both sonograms and oscillograms at different speeds were published. Signals of *Paurocephala* consist of some ten chirps each and have a total duration of 1 to 2 sec. Each chirp is composed of separate pulses, visible on high-speed oscillograms. Production of chirps always coincided with a series of wings vibrations. It is likely that the sound-producing mechanism here is the same as that described by Taylor (1965, 1985) for Australian species.

It is now widely recognised that representatives of many insects orders use low-amplitude vibratory signals for communication rather than airborne waves.

There are several techniques available for the recording of low-amplitude vibratory signals in insects. Using a piezo-electric crystal gramophone cartridge and matching amplifier is the simplest one. This is adequate for simple song description and for the investigation of temporal patterns for the purposes of taxonomy and has been used here.

I have published oscillograms of the vibratory signals of eight species of Psyllids from the Moscow region of Russia (1989). In two species, communication between male and female during courtship behaviour was studied. Signals of 14 species including eight from the paper mentioned above are discussed in the present chapter. In *Craspedolepta nebulosa* (Zett.), a cartridge and a sensitive microphone (model MV-201, RFT, Germany) connected to the sound level meter RFT 00-017 were used. Oscillograms of sounds and vibratory signals appeared to be almost identical for both techniques (Figure 27.1, *9–11* and *12–14*). The list of species with data on collection sites

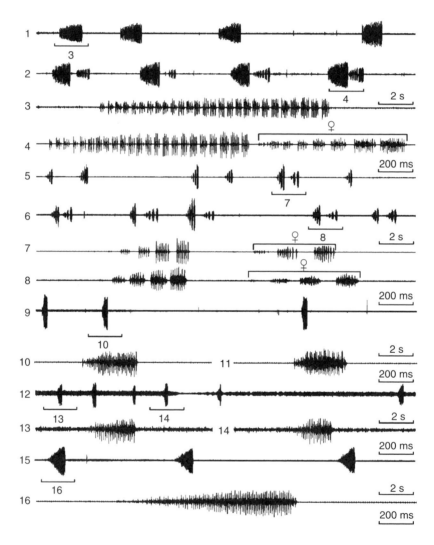

FIGURE 27.1 Oscillograms of acoustic signals of Psylloidea. *1, 3 Craspedolepta nervosa*, male calling signal, *2, 4* same, male calling and female reply, *5–8 C. flavipennis*, male calling signals and female reply, *9–11 C. nebulosa*, calling signals, vibratory component registered by piezocrystal cartridge, *12–14* same, airborne component registered by microphone, *15–16 Neocraspedolepta subpunctata*, calling signals. Faster oscillograms of the parts of signals indicated as "*3–4*", "*7–8*", "*10*", "*13–14*" and "*16*" are given under the same numbers.

and temperature during recording is given in the Appendix on the accompanying CD. The classification of families and subfamilies of Psylloidea is that of Ossiannilsson (1992).

Signals of all species consisted of distinct pulses, *i.e.* successions of sine waves, of increasing and following decreasing amplitude (Figure 27.2, *1*). Pulses were grouped into syllables. In some species signals consist of groups of syllables of more or less constant structure and with clear starts and endings. Such groups are referred to here as phrases (Figure 27.2, *1*).

Observing singing psyllids and hearing their signals at the same time established a clear association between wing vibrations and sounds produced. A singing insect holds its wings as in the

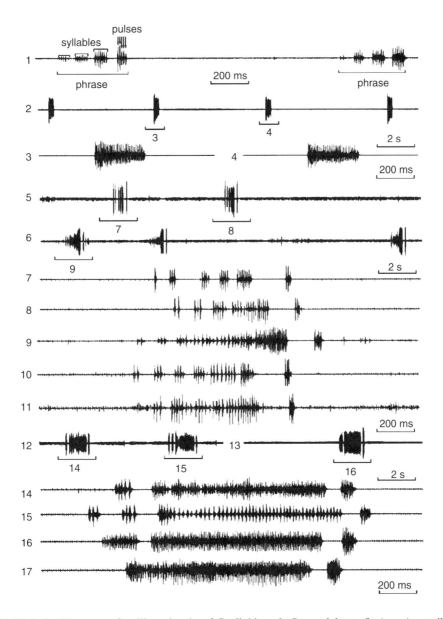

FIGURE 27.2 Oscillograms of calling signals of Psylloidea. *1 Craspedolepta flavipennis*, oscillogram showing the terminology used in this chapter, *2–4 Livia juncorum, 5–11 Aphalara affinis, 12–17 A. polygoni.* Faster oscillograms of the parts of signals indicated as "*3–4*", "*7–9*" and "*14–16*" are given under the same numbers.

resting position, but moves them up and down rapidly in an almost vertical plane. Undoubtedly, structures on the anal veins and dorsal part of thorax, as described by Taylor (1962, 1985), are the stridulatory organs.

Two distinctly different functional types of signals were recorded in Psyllids. Males produce calling signals both singly and in the presence of others of either sex (Figure 27.2, *2–17*, 2, *1–13*, 3, *1–16*, 4, *1–13, 15–17*). Occasionally, several males sitting on the same stem sing alternately or simultaneously. As in other insects, singing of one male stimulates acoustic activity in others nearby. Rivalry or aggressive behaviour and signals of corresponding types were never recorded even at a high density of insects on a plant.

Receptive females produce signals only in reply to a calling male. Calling and reply signals of the same species appeared to have more or less similar temporal patterns in all species studied (Figure 27.1, *1–4, 5–8*, 4, *9–14*). After detecting reply signals, a male starts singing more regularly and walks in different directions along the plant searching for a female. The latter remains stationary responding to the calling male. The break between calling and reply sometimes does not exceed 40 to 50 msec, so signals of the two types may almost overlap (Figure 27.1, *4*).

Usually calling males do not move directly towards receptive females, but they search the twig or leaf apparently at random so that meeting of partners seems to be an accidental event. Similar behaviour has been described in small Auchenorrhyncha that use vibratory signals for communication (Tishechkin, 2000).

In contrast to airborne sounds, there is no simple relation between the amplitude of vibratory signals and the distance to the insect singing on the plant (Michelsen *et al.*, 1982). Because of reflections of vibratory waves at the ends of a stem (or stems, if the plant has several branches) and some other factors, the resulting pattern of waves is often very complicated. Moreover, it is likely to be different in every particular case, depending on the physical properties of the plant and also on the exact position of the singing animal. Consequently, insects using vibratory communication cannot find a potential singing mate merely by moving along a gradient of amplitude. Thus, it is likely that Psyllids use vibratory, but not airborne, signals for intraspecific communication.

Single syllables following each other with irregular intervals were recorded in some species (Figure 27.3, *14–15*). Usually, insects produce such signals when disturbed by other individuals walking along the plant.

In certain cases insects of both sexes kept in a cage together mated actively, but produced no signals. It seems that acoustic communication is not always necessary for successful pair forming in Psyllids. Many species as a rule live in dense aggregations, so a similar phenomenon may occur also in nature.

Temporal pattern of calling signals in some species is variable. For example, in *Aphalara affinis* and *A. polygoni* (Figure 27.2, *5–11* and *12–17*) the number and duration of syllables in a phrase vary greatly. Occasionally, syllables are joined partially or completely, making a united succession of pulses (Figure 27.2, *9, 11*). In *Craspedolepta campestrella* the number of syllables in a phrase varied from four to seven in our recordings (Figure 27.3, *10–13*). In *Livilla ulicis* the structure of syllables sometimes, but not always, changed gradually towards the end of the song (Figure 27.4, *5–7*). This type of variability may be a result of changes in the functioning of the sound-producing apparatus, possibly in the manner of wing movements (Table 27.1).

The shape of pulses sometimes also varies in the signals of the same species (Figure 27.1 *7–8*, and Figure 27.4, *9–11*). In some signals pulses are separated by distinct gaps (*e.g.* Figure 27.1, *7*), whereas in others they are partially joined with each other so that gaps between them are lacking (Figure 27.1, *8*). This type of variability may be a result of different transmission properties of various parts of twigs or stems, and of differences in relative positions of singing insect and the vibrotransducer detecting the signals. As shown by Michelsen *et al.* (1982), these factors affect the spectral composition of a signal and, consequently, the waveform of single pulses.

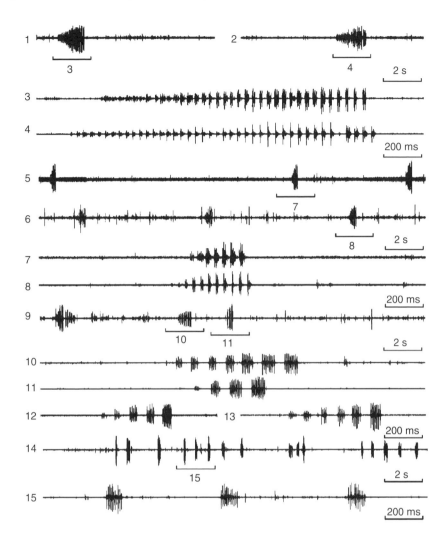

FIGURE 27.3 Oscillograms of vibratory signals of Psylloidea. *1–4 Craspedolepta omissa*, calling signals, *5–8 C. gloriosa*, calling signals, *9–13 C. campestrella*, calling signals, *14–15* same, single syllables. Faster oscillograms of the parts of signals indicated as "*3–4*", "*7–8*", "*10–11*" and "*15*" are given under the same numbers.

Comparative analyses of temporal patterns of acoustic signals have been widely used in taxonomy for discrimination between closely related species and for elucidation of the status of dubious forms. Comparison of oscillograms of vibratory signals of different species of Psyllids shows that their temporal pattern for the most part is species-specific. On the other hand, even among the 14 species discussed here, one can see at least two examples of similarity of signal structure in different species. Syllables in signals of *Craspedolepta nebulosa* and *Livilla ulicis* are almost identical in some examples (Figure 27.1, *9–11* and Figure 27.4, *5–8*). In *L. ulicis*, singing males usually produce syllables with more regular intervals, but sometimes the songs of these two species are indistinguishable. In two *Craspedolepta* species, *C. campestrella* (Figure 27.3, *9–13)* and *C. flavipennis* (Figure 27.1, *5–8*), calling signals sometimes are also impossible to tell apart. Several more examples of this kind were revealed in recent studies of the signals of *Craspedolepta* species from southern Siberia (Tishechkin, unpublished).

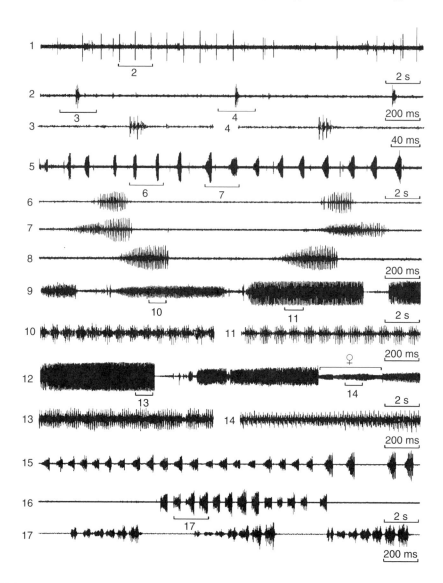

FIGURE 27.4 Oscillograms of vibratory signals of Psylloidea. *1–4 Strophingia ericae*, calling signals, *5–8* same, *Livilla ulicis*, *9–11, 13 Trioza urticae*, calling signals, *12* same, male calling and female reply, *14* same, female reply, *15–17 T. galii*, calling signals. Faster oscillograms of the parts of signals indicated as "*2–4*", "*6–7*", "*10–11*", "*13–14*" and "*17*" are given under the same numbers.

Undoubtedly, the use of acoustic signal analysis in the systematics of Psylloidea will be fruitful. Distinct differences in the temporal pattern of signals in closely related forms in most examples will indicate their specific status. Nonetheless, the similarity of signal structure itself is not conclusive evidence of the conspecificity of the taxa under investigation.

In sympatric species that use acoustic communication, calling signals as a rule differ from each other at least in one physical variable. In this way different species avoid interference of acoustic communication channels or so-called "acoustic niches". This phenomenon was described for vertebrates (Hodl, 1977) and invertebrates (see Bukhvalova, Chapter 14). However these studies were all made on airborne sound signals. In insects producing vibratory signals transmitted through the substrate, the position may be somewhat different. Such insects can hear the signals of each

TABLE 27.1
Data for Recordings of Signals of Psyllids Used for Oscillograms Reproduced in the Chapter

Species	Localities (all in Moscow Area) and Dates of Recording (DD.MM.YY)	Air Temperature, °C
Family Psyllidae		
Subfamily *Liviinae*		
Livia juncorum (Latr.)	Env. Pirogovo, at the SE boundary of Moscow, 13.VI.1988	22
Subfamily *Aphalarinae*		
Aphalara affinis (Zett.)	Same locality, 11.V.1988	21
A. polygoni Frst.	Same locality, 11–12.V.1988	20–22
Craspedolepta omissa Wagn.	Serpukhov Distr., env. Pushchino-na-Oke, from *Artemisia vulgaris* L., 11.VII.1993	22–23
C. gloriosa Log.	Same locality, from *A. abrotanum* L., 11.VII.1993	22–23
C. campestrella Oss.	Moscow, "Sokol'niki" Park, from *A. campestris* L., 15.VII.1993	22
C. nervosa (Frst.)	Env. Pirogovo, at the SE boundary of Moscow, from *Achillea millefolium* L. 20.VI.1988	23
C. flavipennis (Frst.)	Serpukhov Distr., env. Luzhki village, 5.VI.2002	20–21
C. nebulosa (Zett.)	Env. Pirogovo, at the SE boundary of Moscow, from *Chamaenerion angustifolium* (L.) Scop., 25.V.1988	24–25
Neocraspedolepta subpunctata (Frst.)	Same locality, from *Ch. angustifolium*, 24.V.1988	24–25
Subfamily *Strophingiinae*		
Strophingia ericae Hodkinson	Serpukhov Distr., env. Luzhki village, from *Calluna vulgaris* (L.) Hull in pine forest, 7.VI.2002	20–21
Subfamily *Arytaininae*		
Livilla ulicis Curtis	Same locality, from *Genista tinctoria* L., 12.VI.2002	23–24
Family *Triozidae*		
Trioza urticae (L.)	Env. Pirogovo, at the SE boundary of Moscow, from *Urtica dioica* L., 9.V.1988	18
T. galii Frst.	Same locality, from *Galium mollugo* L., 29.V.1988	24–25

other only if the substrate, such as a plant, on which the singer and the recipient occur are in physical, and therefore acoustic, contact. Various species of *psyllids*, as well as most small Auchenorrhyncha, as a rule have strong host preferences and usually cannot be found off their host plants. Possibly this is the reason why two sympatric species living on different plants which normally grow together, such as *L. ulicis* on *Genista tinctoria* L. and *C. nebulosa* on *Chamenerion angustifolium* (L.) Scop.), use signals with similar temporal patterns for intraspecific interaction.

ACKNOWLEDGEMENT

The work was partially supported by a grant of the Russian Foundation for Basic Research (No. 05-04-48586).

28 Mating Behaviour and Vibratory Signals in Whiteflies (Hemiptera: Aleyrodidae)

K. Kanmiya

CONTENTS

INTRODUCTION

Whiteflies (Aleyrodidae) include many pests of agricultural plants and vectors of plant viruses. The current wide distribution of pest whiteflies is mostly the result of accidental introduction. The pests have been introduced with their host plants, most noticeably over the past 20 years. Because of the relative uniformity of adult morphology, taxonomic diagnoses of the whiteflies are based usually on characters of fourth-instar nymphs (Mound and Halsey, 1978). However, it is very difficult to identify whitefly species based on characters of pupal cases when faced with such biological and genetic variation as host range, host-plant adaptability and plant virus transmission capabilities (Costa and Brown, 1991; Brown *et al.*, 1995a, 1995b).

Acoustic signals produced during the process of mating and mate finding are significant features of the specific mate-recognition system (MRS) (Paterson, 1980) in these insects. Therefore, if we can recognise any differences in the acoustic signals among different populations, the degree of dissimilarity in acoustic properties may be expected to reflect the degree of genetic differentiation from a common gene pool since acoustic properties are likely to be under genetic control (Hoy, 1974; Carson, 1985).

Over the past eight years, I have studied the mating behaviour of 36 species belonging to 20 genera of Japanese fauna. As a result, males of 32 species and 18 genera in the tribes Dialeurodini, Trialeurodini, Aleyrodini, Aleurolobini and Aleurocanthini have been shown to produce vibratory sounds. Thus, the evidence suggests that the family is characterised by such sounds in their mating behaviour, as are many families of Auchenorrhyncha (Claridge, 1985a, 1985b).

RECORDING OF THE MATING SOUNDS

Whitefly mating sounds are produced by contractions of the thoracic muscles so as to make vertical oscillations of the abdomen without wing movements (Kanmiya, 1996a). Such very low intensity

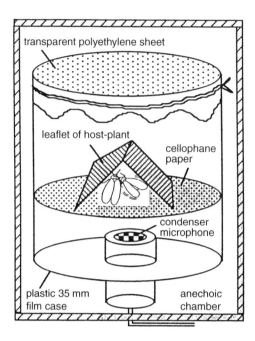

FIGURE 28.1 Apparatus used to detect substrate transmitted signals in courtship of whiteflies.

sounds are difficult to hear for the human observer and also difficult to record by vibration measuring devices (*e.g.* accelerometers, strain-gauge sensor, piezoelectric pick-up). A sheet of cellophane paper was used to pick up the substrate-borne sounds produced by whiteflies through contact with a piece of host-plant leaf on which the whiteflies were located (Figure 28.1). The top of a plastic film case was sealed with polyethylene sheet after a sample pair of insects were released. The sound pressure caused by the whitefly vibration produced standing waves in the cellophane, and the vibration of the cellophane resulted in greatly amplified oscillation of the diaphragm of a condenser microphone. This method detects both air- and substrate-borne vibration by small insects as effectively as a particle velocity microphone described by Bennet-Clark (1984a, 1984b). The best sounds were recorded when the male was located on the polyethylene roof of the cage, though whitefly mating usually occurred under the leaflet on the cellophane.

PROPERTIES OF MALE VIBRATORY SOUNDS

Kanmiya (1996a) was first to publish accounts of sound production and the roles of such signals in whitefly mating behaviour. Adult males of some whiteflies exhibit various movements during premating behaviour (Weber, 1931; Las, 1979; Ahman and Ekbom, 1981; Li and Maschwitz, 1985; Kanmiya, 1996a) (Figure 28.2). Li *et al.* (1989) described such behaviour in the sweetpotato whitefly, *Bemisia tabaci*. Perring *et al.* (1993) compared courtship behaviour between "biotypes" of the cotton strain (*B. tabaci*) and the poinsettia strain (*B. argentifolii*), but did not include analyses of vibratory signals transmitted via the host plant.

Kanmiya (1996a) showed that sexually mature males of the greenhouse whitefly, *Trialeurodes veporariorum*, during courtship frequently produced vibrations by rapidly and rhythmically oscillating the abdomen up and down. These abdominal vibrations were transmitted as substrate-borne vibrations to the female. In some species, males showed antennal drumming, body pushing movement, periodical wing-flicks and continuous wing-flicks before mounting the female (Kanmiya, 1998). Thus a complex MRS works effectively for female choice. Other whiteflies of several different genera were successively shown similarly to produce abdominal vibration for communication (Kanmiya, 1996b, 1998). Kanmiya and Sonobe (2002) described temporal features

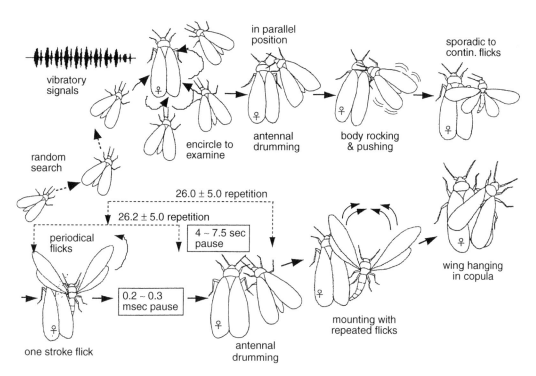

FIGURE 28.2 A sequence of stereotyped courtship behaviour in the greenhouse whitefly, *Trialeurodes veporariorum*. A temporal sequence of the male courting sounds and wing-flicking rhythm were shown in Kanmiya (1996a, 1998).

of the male vibratory sounds that varied with the behavioural context for the woolly whitefly, *Aleurothrixus floccosus,* and defined as the calling sound, courtship sound and premating sound (Figure 28.3).

MALE SPECIES-SPECIFIC MATING SOUNDS

Male vibratory sounds occurred irregularly as a single burst, as a chirp with a train of several pulses, as a series of pulses with a strict time interval or a mixture of such pulses and bursts. In terms of time duration, male sounds may be divided into two main categories:

(1) A sound unit consisting of bursts or pulses of more or less equal duration
(2) A sound unit consisting of a complex of bursts and pulses of different duration.

For time interval, vibratory sounds also may be divided into two:

(1) An independent burst or pulse with rather long interpulse intervals
(2) A series of pulses or short bursts with regular intervals, making a regularly placed chirp

Consequently, male vibratory sounds of 32 Japanese species may be divided into eight groups by defining the unitary sound and interval on the basis of acoustic properties of the time domain:

1. The sound unit consists of a single pulse or a set of very short pulses. Pulses are placed with intervals much longer than the duration of the pulse or a set of pulses (Figure 28.4). This type includes species of the genus *Pealius*, and exhibits a pulse duration of

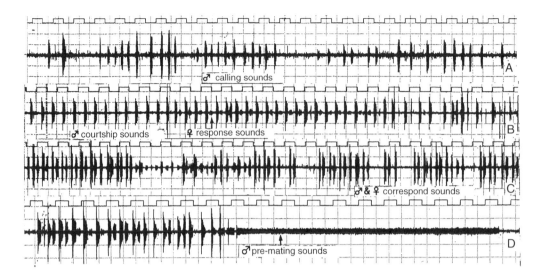

FIGURE 28.3 Oscillograms of a series of vibratory sounds observed in the entire courtship process in the woolly whitefly, *Aleurothrixus floccosus*. Upper square waves indicate a 1 Hz oscillation. (After Kanmiya, K. and Sonobe, R. *Appl. Entomol. Zool.*, 37, 487–495, 2002. With permission.)

25 to 60 msec. *Aleurocanthus spiniferus* is also tentatively included here, with a series of short pulses, the first of which is of 160 to 200 msec duration. In *P. polygoni* and *P. azaleae*, each pulse is bimodal or trimodal (preceded by one or two weak pulses).

2. The sound unit consists of a burst separated by long irregular intervals. I include here *Bemisia tabaci* and *B. argentifolii* for which independently occurring bursts with long intervals (more than 2 sec) are shown (Figure 28.5), with high-speed oscillograms and power spectral density plots (Figure 28.6). In these two species, courting males always produce wing-flicks before mounting.

3. The sound unit consists of a chirp of between 2 and 10 or more shorter bursts, with time interval between bursts longer than the duration of bursts and chirp interval of more than 2 sec or much longer and irregularly placed (Figure 28.7). This sound type includes *Bemisiella artemisiae*, *Siphoninus phillyreae*, *Aleurotrachelus ishigakiensis*, *Aleurolobus taonabae* and *Bemisia puerariae*. The interval between bursts is equal or slightly longer than the duration of bursts. The chirp interval is more than 2 sec and placed irregularly. Specific discrimination is based on the number of bursts and the peak of the fundamental frequency.

4. The sound unit consists of a chirp comprising less than nine short pulses, with interpulse interval usually a little shorter than pulse duration (Figure 28.8). This type includes *Asterobemisia yanagicola*, *Bemisia shinanoensis*, *Aleurotuberculatus aucubae* and *Odontaleurodes rhododendri*. Specific discrimination is possible based on the fundamental frequency of the chirp and chirp interval. Female responding vibration was observed in *B. shinanoensis* and *A. aucubae*.

5. The sound unit consists of a chirp, a series of more than 12 short pulses, with the pulse duration nearly equal to the pulse interval (Figure 28.9). This type includes *Aleyrodes japonica*, *Dialeurodes citri*, *Bemisia* sp. on *Lespedeza*, *Neopealius rubi* and *Aleurolobus marlatti*. The chirp interval is either irregularly placed with a long interval or regularly placed and of more than 2 sec. Specific discrimination is possible based on the number of pulses as well as the power spectral density plots.

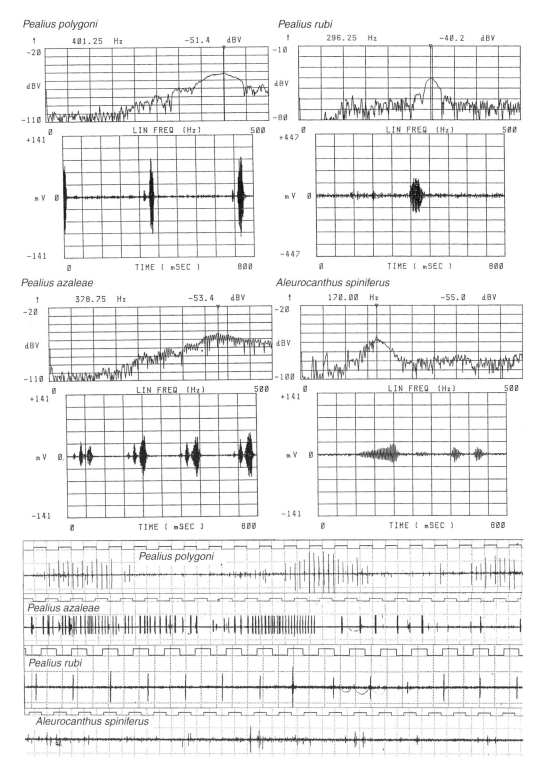

FIGURE 28.4 Acoustic properties of the male vibratory sounds in group 1: a sound unit consists of a single pulse or a set of very short pulses. Both upper 500 Hz instantaneous power spectrum and lower 800 msec oscillogram make a match for each species. The square wave on each oscillogram in the lower column indicates 1 Hz oscillation, respectively.

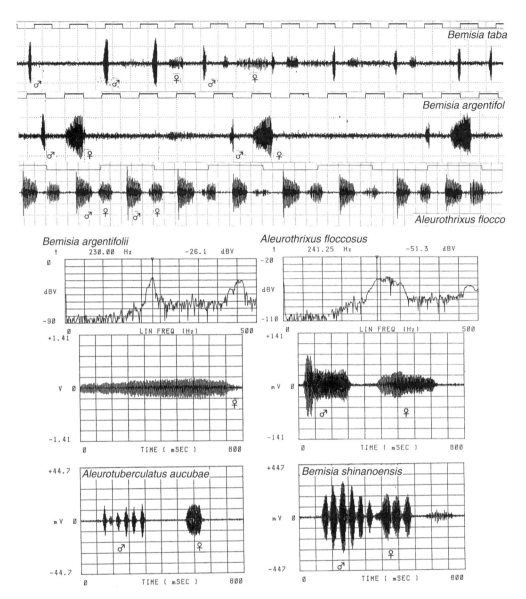

FIGURE 28.5 Oscillograms and instantaneous power spectra on the male courtship and female response sounds. The upper square wave on each oscillogram in the upper column indicates 1 Hz oscillation.

6. The sound unit consists of a chirp comprising a series of short bursts. The burst interval is more or less equal to, or much shorter than, the burst duration (Figure 28.10). Exceptionally, *Acanthobemisia distylii* has as a chirp duration of 5.5 to 8.1 sec consisting first of a long burst (732.4 ± 7.2 msec) and a shorter terminal pulse series (6.48 ± 0.4 sec duration) with 28.3 ± 3 pulses. Pulse intervals are sporadic in *Taiwanaleyrodes meliosmae* and more periodic with a long interval of more than 3.5 sec in *Acanthobemisia distylii*, *Neomaskellia bergii* and *Aleuroplatus daitoensis*. Specific discrimination is easily possible based on the number of pulses.

7. The sound unit consists of a set of the preliminary burst and the following intermittent and compact pulse series (Figure 28.11). This type includes three *Aleyrodes* species and *Pealius* sp. on *Clerodendron trichotomum*. The pulse interval is variable, the least

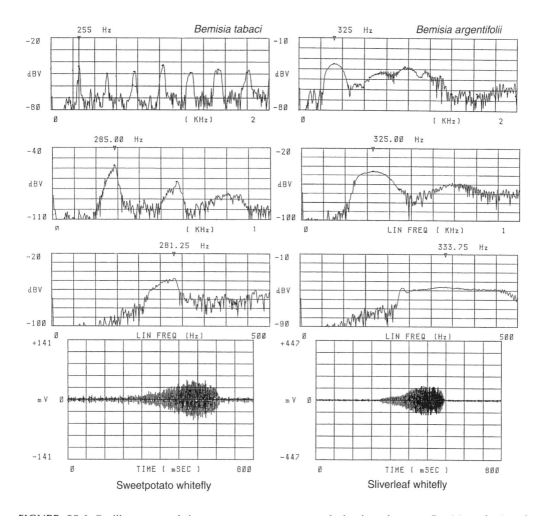

FIGURE 28.6 Oscillograms and instantaneous power spectral density plots on *Bemisia tabaci* and *B. argentifolii*. Small triangular marks in the spectra indicate a peak or a mean point for the fundamental frequency.

so being *A. sorinii*. Chirps are separated by long, often irregular intervals. The fundamental frequency is highest in *Pealius* sp. on *Clerodendron trichotomum* with a range of 400 to 435 Hz at the end of the pulse series. The preliminary long burst has a fundamental frequency of about 225 Hz.

8. The sound unit consists of a complex of long and short bursts of irregular duration and interval (Figure 28.12). This type includes *Trialeurodes vaporariorum*, *Aleurothrixus floccosus* and *Aleurotrachelus* sp. on *Pseucedanum japonicum*. According to Kanmiya (1996a), the vibratory sounds of *Trialeurodes vaporariorum* are composed of a sequence of discrete bouts of vibration, a "chirp". A chirp comprises 10 to 20 pulses, varying in number as courtship progresses. The pulses are defined as varying in a bimodal phase, alternating short (16.4 ± 1.8 msec) and long waves (29.8 ± 2.4 msec). The sounds are produced continuously at quite irregular intervals of 2.5 to 14.4 sec. On the other hand, the calls of *Aleurothrixus floccosus* and *Aleurotrachelus* sp. are composed of bursts and pulses of varying duration and interval. In *Aleurotrachelus*, duration of sounds varies from 43 to 276 msec with intervals of 17 to 184 msec with a peak frequency between 268 and 462 Hz.

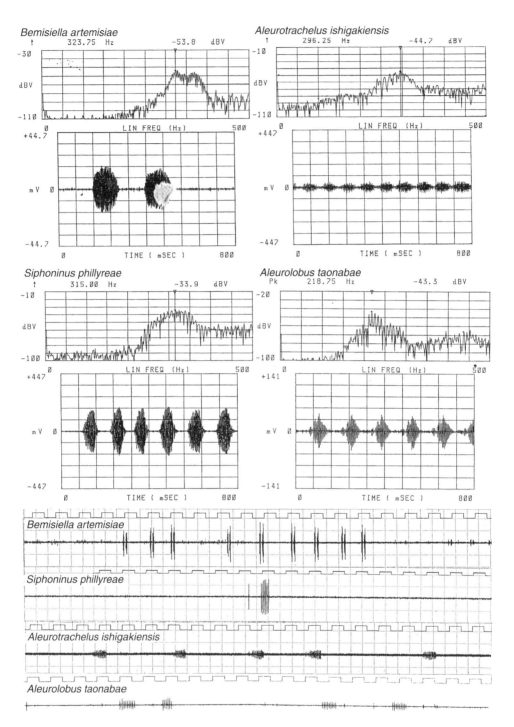

FIGURE 28.7 Acoustic properties of the male vibratory sounds in group 3: a sound unit consists of a chirp of 2 to more than 10 short bursts.

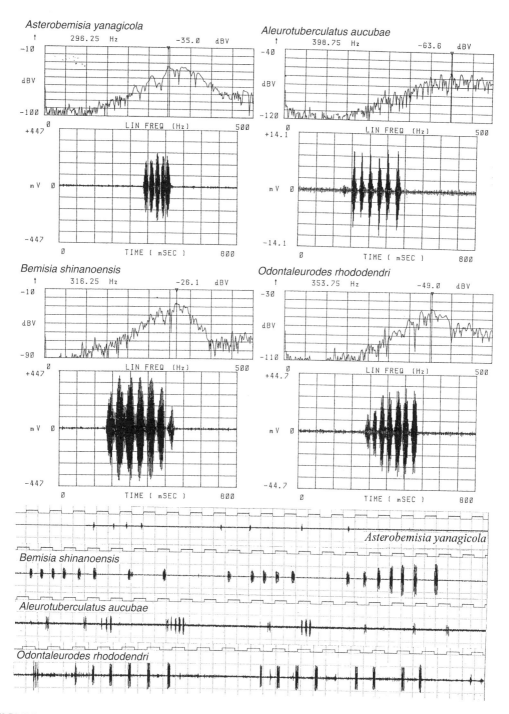

FIGURE 28.8 Acoustic properties of the male vibratory sounds in group 4: a sound unit consists of a chirp comprising less than nine short pulses. The interval is usually a little shorter than the pulse duration.

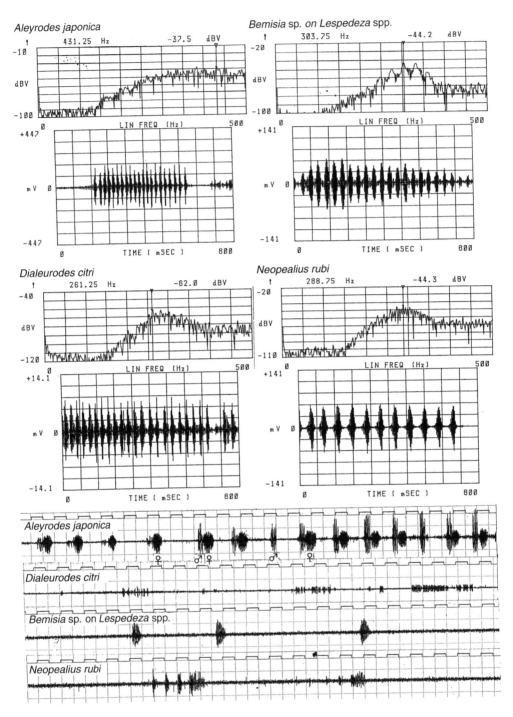

FIGURE 28.9 Acoustic properties of the male vibratory sounds in group 5: a sound unit consists of a chirp including a series of more than 12 short pulses. The pulse duration is nearly equal to the pulse interval.

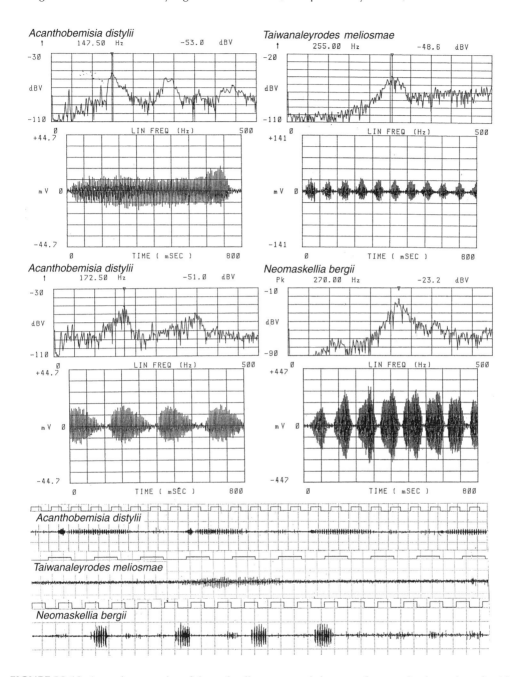

FIGURE 28.10 Acoustic properties of the male vibratory sounds in group 6: a sound unit consists of a chirp comprising a series of short bursts. The burst interval is equal to or much shorter than the burst duration.

FEMALE RESPONSE SOUNDS

Female vibratory sounds for response to the male calling and courtship sounds were first recorded in *Aleurothrixus floccosus* (Kanmiya and Sonobe, 2002). Female response sounds are here additionally recognised for five other species: *Trialeurodes vaporariorum*, *Bemisia tabaci*, *B. argentifolii*, *B. shinanoensis* and *Aleurotuberculatus aucubae* (Figure 28.5). In these species,

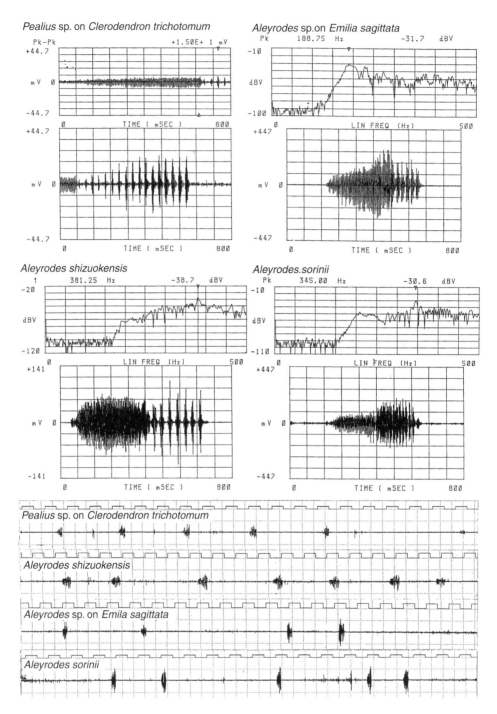

FIGURE 28.11 Acoustic properties of the male vibratory sounds in group 7: a sound unit consists of a set of the preliminary burst and the following intermittent and compact pulse series.

male and female exchange signals throughout the sequence of mate finding and mating. A receptive virgin female produces short bursts of response sounds when a male orients to her, and then the pair continues producing reciprocal sounds up to copulation. The rate of the female response sounds gradually increases towards copulation (Kanmiya and Sonobe, 2002) in the woolly

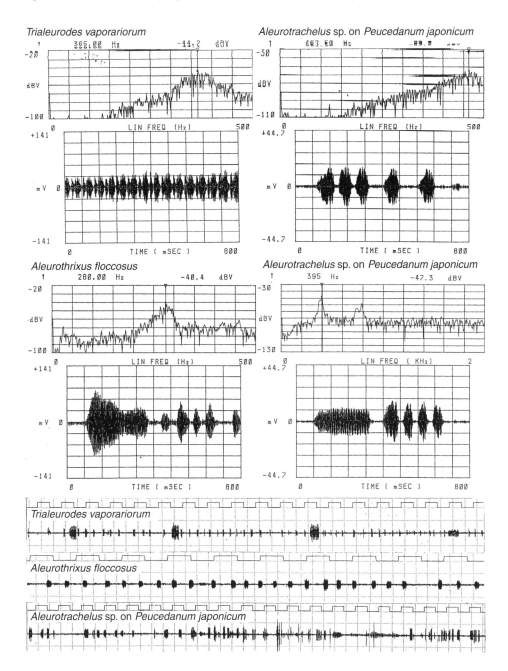

FIGURE 28.12 Acoustic properties of the male vibratory sounds in group 8: a sound unit consists of a complex of long and short bursts of irregular duration and interval.

whitefly. Female response calls are simple bursts for these species, except *B. shinanoensis* where they comprise several intermittent pulses. Fundamental frequencies of the female response sounds are always lower than male sounds of the same species. The duration of female response sounds is usually longer than male calling sounds. Most other known species of whiteflies do not produce response calls.

ACOUSTIC DIFFERENCES BETWEEN *B. TABACI* AND *B. ARGENTIFOLII*

The poinsettia strain, or so-called B-biotype, of the cotton (or sweetpotato) whitefly, *B. tabaci*, has become a serious pest and vector of plant viruses on vegetable crops in modern agricultural ecosystems. This exotic strain, the silverleaf whitefly, was established as a new species, *Bemisia argentifolii*, on the basis of RAPD-PCR methods (Perring *et al.*, 1993) and more detailed biosystematic research (Bellow *et al.*, 1994). However, more recent molecular and biochemical studies on *B. tabaci* have provided evidence for a substantial degree of polymorphism between populations from distinct geographic regions and host plants (Bartlett and Gawel, 1993; Gawel and Bartlett, 1993; Brown *et al.*, 1995a, 1995b, 2000; Frohlich *et al.*, 1999; Kirk *et al.*, 2000). The recently recognised differences among populations of *B. tabaci* have led to the suggestion that *B. tabaci* and *B. argentifolii* represent either different "biotypes" of *B. tabaci* or a species complex. In this case, it is a great practical problem that we cannot yet discriminate pest *Bemisia* species without genomic analysis!

Two high-speed oscillograms, with 800 msec frame, explain the differences in bursts of sound in *B. tabaci* and *B. argentifolii* (Figure 28.6). The duration in *B. tabaci* is always longer than that of *B. argentifolii*, exceeding 300 msec and usually about 450 msec. In *B. argentifolii* the duration is usually 230 msec very rarely exceeding 300 msec. The presence of remarkable frequency modulation on the latter half of the burst in *B. argentifolii* is a clear distinction from that of *B. tabaci* where frequency modulation is restricted only on the latter one third. The power spectral density plot on a burst of *B. argentifolii* shows the mean fundamental frequency of about 330 Hz with a trapezoid peak between 270 and 450 Hz, within 500 Hz frame, while that of *B. tabaci* shows a mean value of 245 ± 25 Hz with a narrow variance because of its weak frequency modulation. Both the 1 and 2 kHz spectral ranges also distinctly separate the two species by the number of harmonics (Figure 28.6). The harmonics in *B. tabaci* in the 2 kHz frequency range shows seven peaks, clearly contrasting with *B. argentifolii*. Differences in the individual values of burst duration and the peak or the mean values of the fundamental frequency showed clear differentiation between the two species in the scatter diagram (Figure 28.13). The samples of *B. argentifolii* are derived from nine different populations from five different host plants, and those of *B. tabaci* from six different populations on three wild host plants. As seen in the populations of *B. tabaci*, the duration of burst shows a wide deviation from 330 to 650 msec, whereas in *B. argentifolii* the mean value of the fundamental frequency shows a wide deviation from 230 to 405 Hz. Therefore, if we plot the difference of the minimum and maximum values of the fundamental frequency of the male burst

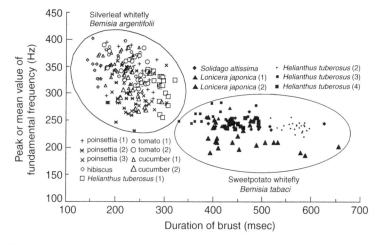

FIGURE 28.13 Scatter diagram between burst duration and peak or mean value of the fundamental frequency in the male courtship sounds for *Bemisia tabaci* and *B. argentifolii*.

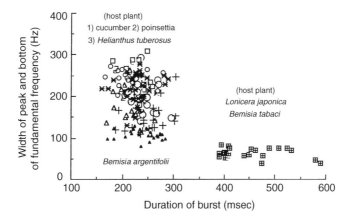

FIGURE 28.14 Scatter diagram between burst duration and difference of peak and bottom values of the fundamental frequency in the male courtship sounds for *Bemisia tabaci* and *B. argentifolii*.

against the burst duration, the discrimination between the two species is very significant (Figure 28.14). This clear difference between the two confirms the different biological species status of the two.

CONCLUSION

Since the initial discovery of vibratory signals in the Aleyrodidae (Kanmiya, 1996a), additional knowledge has rapidly accumulated. A total of 32 species belonging to 18 genera have been shown to produce substrate-borne vibratory signals in the course of mating behaviour. The sounds are all species-specific, showing characteristic rhythm and frequency spectra.

Acoustically, the vibratory signals are variously modified. They vary from a single burst of irregular intervals, a regularly placed chirp comprising a train of several pulses, a series of pulses with strict time intervals or a mixture of such pulses and bursts. The female vibratory response has so far been recognised for only six species. Males show different sound repertoires: calling, mating and mounting sounds. In addition to vibratory sounds, males of some species show a specific behaviour sequence of antennal drumming, body pushing, weak wing-flicking and strong continuous wing-flicking during mounting. Thus, a complex innate MRS works effectively during male courtship and female choice.

Here, I have described the species-specific acoustic characteristics in time- and frequency-domains. In addition, I have distinguished two very closely related species of economic importance on the basis of their acoustic signals.

ACKNOWLEDGEMENTS

I would like to express my gratitude to Michael F. Claridge for improvements to a draft of this manuscript and Rikio Sonobe for cooperative work with materials in data recording. I also would like to thank Chiun Cheng Ko and Yorio Miyatake for providing invaluable assistance in taxonomics and literature. The present work was supported in part by Grant-in-Aid for Scientific Research (Project No. 09660054) from the Ministry of Education, Science and Culture, Japan.

29 Communication by Vibratory Signals in Diptera

K. Kanmiya

CONTENTS

INTRODUCTION

Past bioacoustic studies on dipteran mating behaviour were restricted to a few families including some Culicidae, Chironomidae, Psychodidae, Tephritidae, Drosophilidae and Chloropidae. Adult flies have no cuticular structures that are obviously specialised either to produce clicking sounds, like the cicada tymbal, or stridulatory sounds, like beetles and crickets. They also lack ventral sclerotic plates to produce percussive sounds like stone flies by striking their abdomen against the substrate (see Stewart and Sandberg, Chapter 12). Diptera do not have obvious cuticular devices for making sounds. The question then arises of how they manipulate species-specific information to sustain their mating systems. Mating signals of flies have been reported as incidental sounds of free flight, as in mosquitoes and midges, or as intermittent sounds by controlling wing vibrations in Tephritids and Drosophilids. In addition to these airborne sounds, a unique sound production system of vibrating reed stems is known in the reed fly, *Lipara*. Thus, both airborne and substrate-borne sounds have been recorded from a very few dipteran groups.

    In this chapter I outline some case studies of flies that produce species-specific vibratory sounds while controlling wing movement. The acoustic data on the leaf-miner flies (Agromyzidae) are presented here for the first time. A unique mechanoreceptor to detect substrate-borne vibration is also illustrated.

ACOUSTIC SIGNALS IN DIPTERA

Sound signals reported in Diptera are generally divided into two categories:

- Type I: Airborne sounds produced by rapid vibration of wings or other appendages
- Type II: Substrate-borne sounds produced by forced vibration against the substrates such as a plant stems or leaves

Sounds of Type I are divided into three groups A, B, and C:

- Group A: Flight sounds caused by continuous free beating of wings while in flight, as in Culicidae (Roth, 1948; Kahn and Offenhauser, 1949) and Chironomidae (Ogawa, 1992).
- Group B: Controlled intermittent wing movements when on the ground, as in Drosophilidae (Shorey, 1962; Bennet-Clark and Ewing, 1967), Tephritidae (Monro, 1953; Fletcher, 1968; Webb *et al.*, 1983; Sivinsky *et al.*, 1984; Kanmiya *et al.*, 1987; Kanmiya, 1988), Psychodidae (Kanmiya, 1996b; Souza *et al.*, 2002), Sciaridae (Kanmiya, 1999), and Phoridae (Kanmiya, 1985) (see Figure 29.1).
- Group C: Rapid free vibration of appendages other than the wings, as in Syrphidae (Landois, 1867; Harris, 1903) and Glossinidae (Chowdhury and Parr, 1981). The controlled wing vibrations of group B are sounds produced by near field particle motion at less than 1 cm distance (Bennet-Clark and Ewing, 1967; Ewing, 1977, 1978). Sounds of Type II are mating signals produced by substrate-borne vibration, which had been regarded as uncommon in Diptera. This method has been known exclusively in the genus *Lipara* (Mook and Bruggemann, 1968, Chvala *et al.*, 1974;

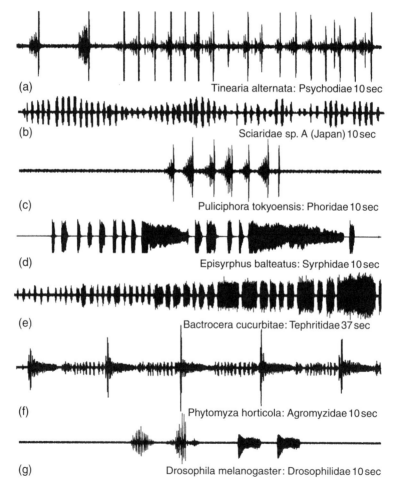

(a)   Tinearia alternata: Psychodiae 10 sec

(b)   Sciaridae sp. A (Japan) 10 sec

(c)   Puliciphora tokyoensis: Phoridae 10 sec

(d)   Episyrphus balteatus: Syrphidae 10 sec

(e)   Bactrocera cucurbitae: Tephritidae 37 sec

(f)   Phytomyza horticola: Agromyzidae 10 sec

(g)   Drosophila melanogaster: Drosophilidae 10 sec

FIGURE 29.1 Wing-generated, controlled airborne sounds by male dipteran flies. Time figures indicate full sound duration. Sound pressure level is relatively shown.

Kanmiya, 1981b, 1981c, and 1990). In addition, a Drosophilid, *Drosophila silvestris*, is exceptionally known to produce abdominal vibrations (Hoy *et al.*, 1988). This sound type is quite different from airborne sounds produced by controlled wing vibration for most *Drosophila*.

In neither the reviews of Kalmring (1985) nor Čokl and Virant-Doberlet (2003) on insect communication by acoustic and substrate-borne signals is mention made of the Diptera. Even in the most recent review of substrate-borne sounds (Virant-Doberlet and Čokl, 2004), only two examples from the Diptera (on *Lipara* and *Drosophila*) were cited compared with many studies on Coleoptera and Homoptera. A reason may be that the Drosophilid and Tephritid flies have long been known to produce courtship sounds by wing fanning.

PRODUCTION AND RECEPTION OF SUBSTRATE-BORNE VIBRATION

Courtship songs by airborne and substrate-borne vibration in Diptera are produced by moving both the indirect and direct flight muscles (Miller, 1965; Bennet-Clark and Ewing, 1968; Ewing, 1979a, 1979b). Flight, aerial manoeuvring, and sound production are known to be associated with both the functional power of the indirect myogenic flight muscle and the direct and axilliary neurogenic flight muscles (Figure 29.2). In making substrate-borne vibration, the vibratory power of indirect

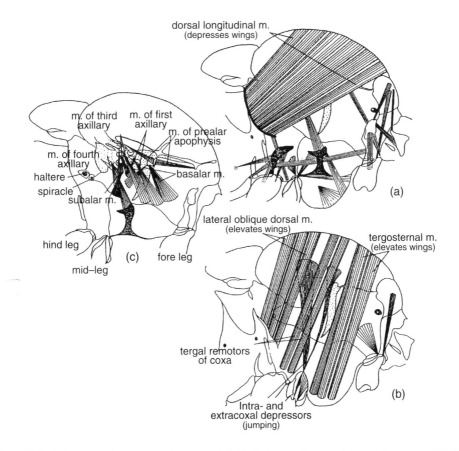

FIGURE 29.2 Diagrammatic cross sections of left half of thorax showing indirect depressors (a), indirect elevators (b) and direct flight muscles (c) in *Lipara japonica* (male). Terminology is after Miller (1965).

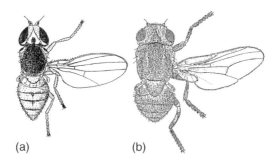

(a)   (b)

FIGURE 29.3 (a) *Dicraeus flavoscutellatus* (long-hovering species with body 1.5 mm, wing 1.66 mm in mean length), (b) *Lipara japonica* (short-straight flyer, body 5.4 mm; wing 5.6 mm). Both males are arranged artificially in the same body length.

flight muscle is transmitted through the middle legs to the substrate by antagonistic constriction of the direct flight muscles. Species of *Lipara* are characterised by their relatively large body size and short wings (Figure 29.3b), which enable them to produce such substrate-borne vibration in contrast to other smaller species of Chloropidae. The species figured as a typical Chloropid, *Dicraeus flavoscutellatus* (Figure 29.3a), is a hovering fly with a ratio of wing/mesonotum of 2.7 while that of *L. japonica* is 2.44. The enlarged thorax and short wings are not efficient for sustained flight performance, but are capable of containing large and powerful indirect flight muscles to create power for substrate-borne vibration.

There are several means whereby animals are potentially able to perceive vibrations through the substrate (Ewing, 1989a, 1989b; Čokl *et al.*, Chapter 4). In Nematoceran Diptera, airborne vibration is detected by delicate antennae with a well-developed Johnston's organ in the second segment. This organ has evolved primarily as a mechanoreceptor responsive to the tonic and phasic air movements that occur during active flight (Ewing, 1989a, 1989b). No subgenual organ or other type of vibration receptor has been reported in the Diptera. For response to substrate-borne vibration, cuticular stress detectors (CSDs) located at the joints of walking legs are known in other insects (Čokl *et al.*, Chapter 4). Kanmiya (1981b) recognised a possible mechanoreceptor associated with contact between two leg segments (coxa and trochanter). Adult male and female of *Lipara* have a protuberance on the middle coxa which detects substrate-borne vibrations through the legs. When a female *L. japonica* was placed on a reed stem, she often raised midlegs off the substrate in response to synthetic vibrations whenever high amplitude vibrations were applied to the reed. After closely examining the midlegs, I found a tapering protuberance on the coxa about 200 μm in length (Figure 29.4-2). The tip of the protuberance (Figure 29.4-1) always touches the dorsal surface of the trochanter when the fly is motionless and close to the substrate. The protuberance was assumed to be a mechanoreceptor to detect male vibratory signals through the substrate as the female did not give response signals back to the male's calling when the midlegs were cut between the coxa and trochanter (Kanmiya, 1981b). Both the males and females have a mid-coxal protuberance on the second coxa. All *Lipara* and both Calyptrate and Acalyptrate flies examined also have this protuberance. Thus, the mid-coxal protuberance appears to be a contact mechanoreceptor for detecting substrate vibration. Recently, such a protuberance was reported as a mid-coxal prong (MCP) by Frantsevich and Gorb (1998). A modified figure from Kanmiya (1981b) is given here (Figure 29.4) and also an additional example of a MCP in *Chrysomya pinguis* (Calliphoridae) (Figure 29.5). Generally, the coxae and femora of the Cyclorraphous flies are bent at the trochanter when motionless. In this position flies may be startled and escape danger quickly by detecting substrate-borne vibrations.

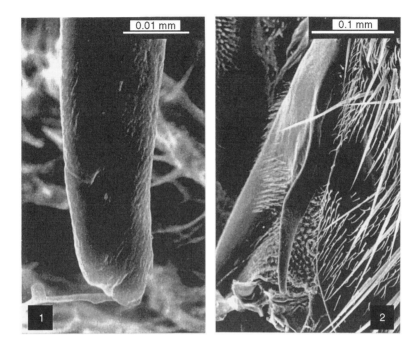

FIGURE 29.4 Mid-coxal prong of *Lipara japonica* (female), Chloropidae. The left is enlarged tip of the right prong.

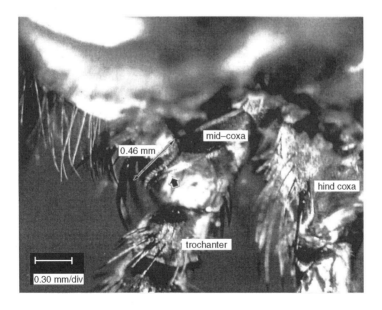

FIGURE 29.5 Mid-coxal prong of *Chrysomya pinguis* (male), Calliphoridae. The arrow indicates the mid-coxal prong.

VIBRATORY SIGNALS FOR TERRITORIALITY, AVOIDANCE AND DISTURBANCE

Many studies on male competitive courtship behaviour in insects have been reported (West-Eberhard, 1984). It would therefore be difficult for me to review all such studies,

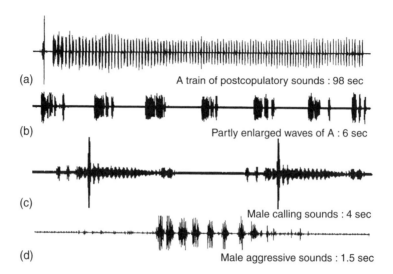

FIGURE 29.6 Oscillograms of the various vibratory signals in male. (a,b) postcopulatory sounds of *Phytomyza horticola*, (c) courtship sounds of the preceding species, (d) fighting sounds of *Formicosepsis hamata*.

even in Diptera. Observations have been reported on such social interactions between male flies in the process of mating, especially in Drosophilidae, Tephritidae, Sepsidae, Pyrgotidae, Cypselosomatidae and Ulidiidae, some species of which make patterned wing displays, head butting, boxing with the legs and other distinct behaviour patterns. Male–male interactions are known to influence mating success (Spieth, 1974; Sivinski and Webb, 1986; Sivinski, 1988; Poramarcom and Boake, 1991; West-Eberhard, 2003). However, there are few of offensive behaviour in Diptera by substrate-borne vibrations in relation to male. I have observed aggressive territorial interactions by males of *Formicosepsis hamata* (Cypselosomatidae). Substrate-borne vibrations were detected in this behaviour between males in close proximity. They produced short series of shock waves with closely placed pulses by body shaking when showing typical threatening behaviour by leg boxing or dashing as shown in Figure 29.6d. We have often observed such interruption signals in males of Acalyptrate flies where two males met or a prior calling was heard during sitting on the same substrate. In the mating behaviour of *Lipara*, when the preceding male starts to produce calling sounds for females, another male nearby often blocks the calling signals by producing short shock waves. Such aggressive signals by acute shock waves against rivals is shown for two Agromyzid species (Figure 29.7e,f). These are probably exchanges of information associated with territoriality.

Sexually immature or once-mated Acalyptrate female flies are often observed to produce short bursts of wing vibration or substrate-borne vibrations while head-butting against a courting male or the calling signals. Oscillograms (Figure 29.7a–d) shows a bout of rejection signals by females against male conspecific calling songs of Agromyzidae. Similar situations are also observed in many other Acalyptrates. It may be concluded that insects dwelling on a substrate suitable to be used as a vibratory medium can make the best use of substrate-borne vibration as an interactive function for mating, attacking and rejecting conspecific as well as for escape.

VIBRATORY SIGNALS IN MATING BEHAVIOUR

Male flies during sexual mounting keep their body balance by the use of their wings. At the moment of copulation the male usually freely vibrates his wings behind or on the back of the female. I have

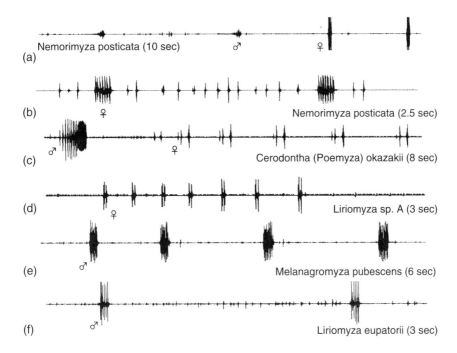

FIGURE 29.7 Oscillograms of the substrate-borne vibratory sounds during conspecific interactive actions between the same sex or other sex in Agromyzidae. Time figures also indicate full sound duration and so forth.

observed such long lasting wing sounds in Phoridae, Stratiomidae, Dolichopodidae, Sciomyzidae, Tephritidae, Chloropidae, Agromyzidae, Fanniidae and Muscidae.

Wing-fanning sounds in male Tephritids have various suggested functions, including calling songs (or sounds), courtship songs, approach songs, signalling songs, precopulatory songs, premating songs and copulatory songs (Webb *et al.*, 1976, 1983, 1984; Sivinski *et al.*, 1984; Sivinski and Webb, 1985; Kanmiya, 1987, 1989; Kanmiya *et al.*, 1987, 1991; Miyatake and Kanmiya, 2004). These song categories need to correspond with specific behavioural contexts concerning:

(1) Territorial signals
(2) Signals in response to conspecific males or females
(3) Sounds in continuous orientation to nearby females
(4) Sounds accompanying pheromone dispersal
(5) Sounds at the moment of mounting
(6) Sounds during in copulation

Copulatory sounds in Diptera occur at the moment of mounting, with continuous bursts of fluctuating amplitude, frequency and irregular duration. These sounds are different from other mating sounds in nature. Males of *Bactrocera pernigra* produce continuous wing vibrations at the moment of mounting, even though they do not have any repertoire of courtship songs. Thus, the copulatory sound is the most basic mating stimulus and its origin may be the incidental sounds which have been properly named as "mounting sounds". This continuous wing vibration is recognised in the behaviour of the *frigida*-group, which is the most primitive group of the genus *Lipara*. Some *Calamoncosis* and *Cryptoneura* also produce such mounting sounds.

In addition to wing vibrations for controlling posture, intermittent wing vibration in Acalyptrate flies are often observed while sitting on the back of the female during mounting after successful

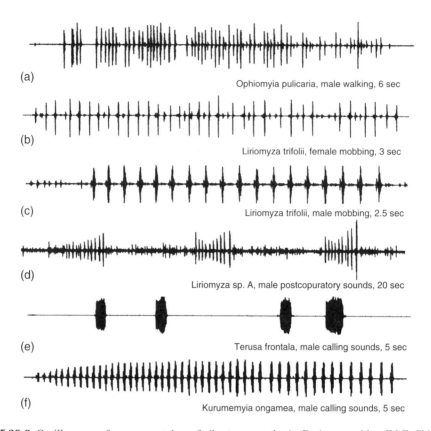

(a)

Ophiomyia pulicaria, male walking, 6 sec

(b)

Liriomyza trifolii, female mobbing, 3 sec

(c)

Liriomyza trifolii, male mobbing, 2.5 sec

(d)

Liriomyza sp. A, male postcopuratory sounds, 20 sec

(e)

Terusa frontala, male calling sounds, 5 sec

(f)

Kurumemyia ongamea, male calling sounds, 5 sec

FIGURE 29.8 Oscillograms of some repertoires of vibratory sounds. A–D: Agromyzidae, E&F: Chloropidae.

copulation. As already known in *Drosophila* species (Hoikkala *et al.*, 2000), this behaviour may be categorised as the copulatory courtship (Eberhard, 1991). In several species of Chloropid and Agromyzid males in copula, we can recognise continuous substrate-borne vibrations by shaking body with sporadic constriction of thoracic muscles. This appears to function to prevent movement of the female during injecting sperm or to continue insemination. Figure 29.6a and b shows a part of the postcopulatory sounds in the Agromyzid, *Phytomyza horticola*. After mounting, the male continues to produce vibrations which amount to a total of 526 pulses grouped in ten separated series with a total duration of 18.3 min. The interpulse interval was rather regular, 0.8 ± 0.03 sec. Figure 29.8d also shows male postcopulatory sounds which are produced on the back of the female. These postcopulatory sounds take more time than precopulatory ones in some Agromyzids. This appears to be an example of "cryptic female choice" (Thornhill, 1983).

VIBRATORY SIGNALS OF CHLOROPIDAE AND AGROMYZIDAE

In addition to *Lipara*, male mating signals for some other chloropid genera have been observed for species of *Calamoncosis* (Kanmiya, 1986), *Cryptonevra* and *Pseudeurina* (Kanmiya, 1999). I added here new records on genera of *Terura* (Figure 29.8e) and *Kurumemyia* (Figure 29.8f). These sounds are species-specific courtship signals for female receptiveness. It seems most probable that most of plant-dwelling Acalyptrates make substrate-borne vibrations for courtship signals or sexual interaction. Figure 29.9 shows bimodal (pulse and burst sounds) or trimodal waves of mating

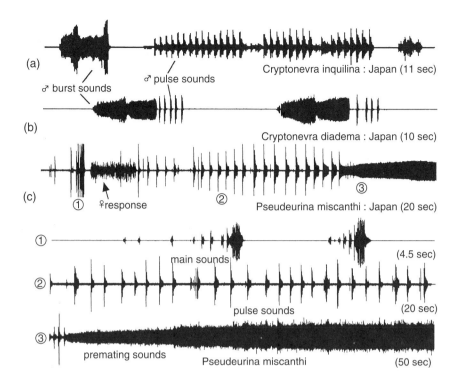

FIGURE 29.9 Oscillograms of some repertoires of the mating sounds in Chloropidae showing a long burst of sine waves or a train of pulse sounds.

sounds in three chloropid species. As in *Drosophila melanogaster* (Schilcher, 1976), females were subjected to pulse songs and sine wave songs as a prerequisite for total mating stimuli.

It seems that no acoustic analyses have been done on Agromyzidae in spite of their economic importance. Tschirnhaus (1971) recognised a haired file on the basal tergites and a ridged scraper on the inner part of the hind femur in *Agromyza* and *Liriomyza*. He regarded this as a stridulatory organ and discussed possible homology with the stridulatory organ (pectin) of the Tephritid, *Bactrocera*. This homology seems improbable, as the pectin is a sexual dimorphism and serves for pheromone dispersal, but the file is reportedly observed in both sexes. Kanmiya (1988) showed that the Tephritid pectin was not used for sound production. Adult males of *Agromyza* and *Liriomyza* produce sounds by rhythmical thoracic vibration accompanied by only very minute wing shivering. No contact was observed between the femur and file during sound production.

Agromyzid adults exhibit various substrate-borne vibrations other than courtship sounds, particularly those produced by normal movements on the plant surface (Figure 29.8a–c). We can distinguish different species by such walking sounds as well as courtship sounds by their acoustic properties. Courtship sounds of seven Japanese Agromyzids are shown here for the first time (Figure 29.10). We will soon have enough data for separating the injurious *Liriomyza* species.

REED FLIES, *LIPARA* SPECIES — A CASE STUDY

Adult *Lipara* are unable to fly actively and to find females by sight during flight. Thus, all *Lipara* species communicate by substrate-borne vibration of the reed stems on which they sit and exchange mutual courtship signals (Figure 29.11). Males make signals by setting a dried reed stem into vibration. Mature and virgin females quickly respond to male signals by vibration, as do males also in response to other conspecific male calls. If the male does not receive

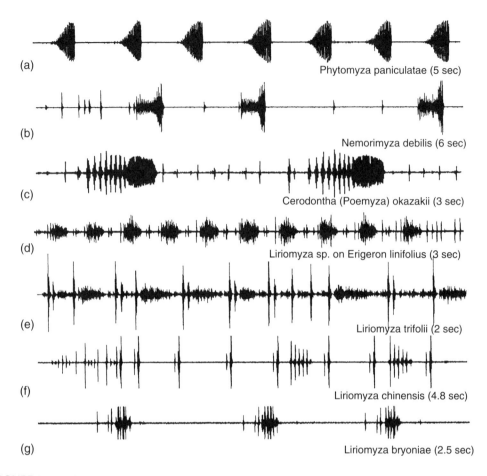

FIGURE 29.10 Oscillograms of substrate-borne courtship sounds occurred in the courtship of Agromyzidae.

responding signals from conspecific females, he flies to another stem. The females are always sedentary until copulation at the site of her gall. Thus, the female may take the initiative during mate choice.

I have examined mating signals in most described *Lipara* species in the world (Kanmiya, 1981c, 1986, 1990, 2001; Nartshuk and Kanmiya, 1996) and found that their signals are species-specific, being separated into four species groups (*pullitarsis*-group, *rufitarsis*-group, *similis*-group and *lucens*-group) by their acoustic properties (Kanmiya, 1986). I also examined mating signals of the genus *Calamoncosis*, which is phylogenetically placed near *Lipara* in the *Lipara* genus group (Kanmiya, 1983). Larvae of *Calamoncosis* do not make galls but feed gregariously in reed stems and cause rapid decay of shoot growing points. Vibratory signals of *Calamoncosis* (Figure 29.12) consist of several preliminary regular vibrations with the next main vibration as a single continuous burst. The *pullitarsis* group has a similar call pattern to *Calamoncosis*, but is distinguished by the preliminary vibration which has fewer and irregular intervals. In the *rufitarsis* group, type-*aino* of undescribed sp. has a preliminary vibration of long pulse series, then a main vibration of short pulse with close intervals follows. Signals of *L. rufitarsis* and *L. orientalis* also have irregularly spaced preliminary vibrations in fewer numbers and several pulses with long intervals follow. In the *similis* group, *L. similis* of England and *L. vallicola* from Japan are almost identical, without preliminary vibration. The main vibration is composed of long chirps with long intervals. In the *lucens* group, *L. lucens* from Europe has irregular preliminary

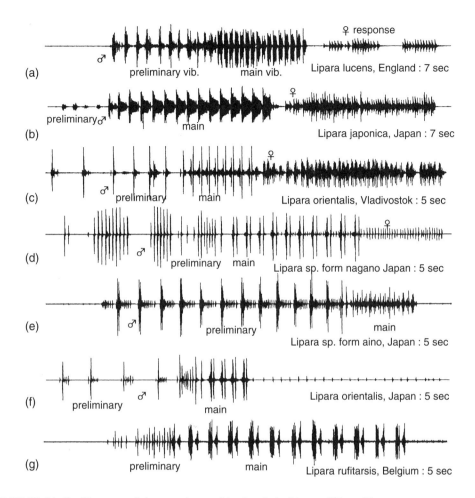

FIGURE 29.11 Oscillograms of the sexual courtship signals in *Lipara*, Chloropidae.

vibrations and the final main vibration is composed of many short, closely placed chirps. The pattern of *L. japonica* is quite different to that of *L. lucens*. Their chirps in the main vibration are longer and contain a weak accidental vibration between chirps. The pattern of *L. brevipilosa* and *L. salina* is also unique; the preliminary vibration is completely absent and very short chirps are continuously followed without rest. These two species can make high resonant vibration with the reed stem.

NATURAL HISTORY OF *LIPARA* AS IT AFFECTS VIBRATORY SIGNALS

The reed fly, *Lipara*, is well known as causing characteristic galls on common reeds, *Phragmites*. The gall is formed and occupied by only a single larva. The genus also includes the biggest species of the family (Figure 29.3). The shape of the gall is species-specific and easily distinguishable. Gall formation and gall shape are greatly influenced by the precise feeding behaviour of the larvae depending on whether they feed on the vegetative shoot apex or the reproductive shoot apex (Kanmiya, 1981a, 1986, 2001).

It is probable that some dynamic part of the acoustic properties of *Lipara* vibratory signals are related to their body size which is capable of moving reed stems as the muscle power is switched to

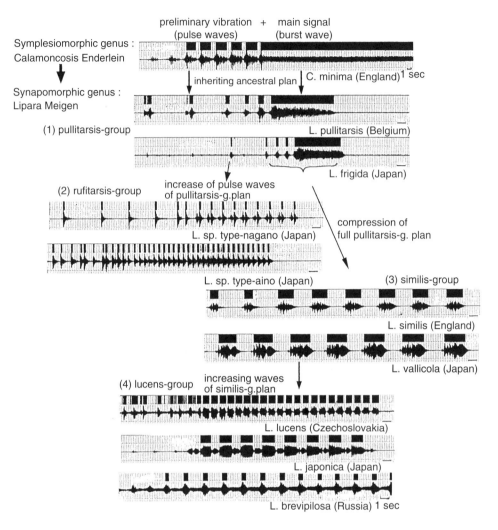

FIGURE 29.12 Inferred evolutionary progress of *Lipara*, based on acoustic development of the male vibratory signals, showing derivative from the ancestral *Calamoncosis*. (Redrawn after Kanmiya, K., Phylogeny and origin of the Reed-flies, In *The Natural History of Flies*, Shinonaga, S. and Shima, H., Eds., Tokai University Press, Tokyo, pp. 215–243, 2001, in Japanese. With permission.)

the substrate-forced vibration through the midlegs. The physical nature of flight power will be closely related to larval feeding activity and that will be closely related to the developmental stage of the reed shoot which is attacked. In early spring, if the larva slowly feeds on the shoot apical meristem, then it can obtain nutritious vegetative growth and achieve the biggest body size, as is the strategy of the *lucens* group. On the contrary, if the larva quickly feeds out the shoot apical meristem, it destroys the nutritious part rapidly, as in the *rufitarsis* group. If the larva hatches late in spring or early summer, it must feed on the host reproductive organs resulting in the smallest body size. Thus, body size is decided by larval feeding behaviour resulting from specific genetic backgrounds. Therefore, life-form and pattern of vibratory signal are closely correlated in relation to body size. The evolutionary progress is dependent on acoustic development of the male vibratory signals (Figure 29.12) and coincides with an evolutionary progress of larval life-form, reflecting adaptive species strategies (Figure 29.13).

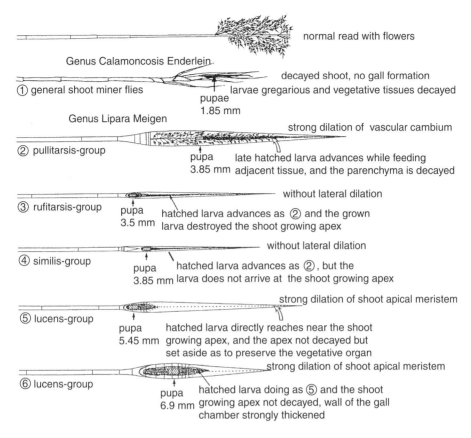

FIGURE 29.13 Cross section of the highest internodes of the reed galls induced by *Lipara*. Rectangles with oblique lines denote the point of pupation, and numerals under rectangles indicate a mean body-length in male. (Modified from Kanmiya, K., phylogeny and origin of the Reed-flies, In *The Natural History of Flies*, Shinonaga, S. and Shima, H., Eds., Tokai University Press, Tokyo, pp. 215–243, 2001, in Japanese. With permission.)

GEOGRAPHIC VARIATION OF MATING SIGNALS

I have examined vibratory signals produced by male and female *Lipara* from Japan, Korea, Vladivostok and seven European countries. Parameters of time- and frequency-domain properties among such disjunct local populations were measured with multivariate statistical analyses in order to elucidate the differentiation of acoustic signals across the range of distribution. Kanmiya (1986) demonstrated local variation in body size of *L. japonica*. Male body length was 83 to 88% shorter than that of the female. This size difference was constant at any locality and was found to be important for mating success. Mate success was decided not only by the nature of the vibratory signals, but also by body size. Mating failure was often caused by size differences of the sexes. For example, the smallest male from Hokkaido (northern Japan) could not copulate with the biggest female from Kyushu (southern Japan). For Japanese and Korean *japonica* and European *lucens*, I found geographic variation among spatially isolated populations in time-domain characteristics of male vibratory signals (Figure 29.14 to Figure 29.16). The signal variation does not show clear clinal variation, excepting the plots in numerals of 9 to 13, 15 and Korea, which show close relatedness on the phenotypic character as well as acoustics properties (Figure 29.12). Completely isolated local populations might incidentally yield acoustic specificity as a byproduct of genetic

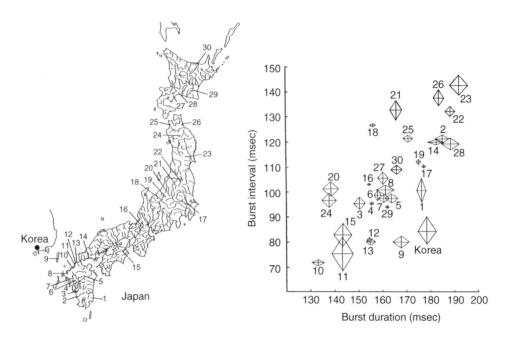

FIGURE 29.14 Geographic variation in vibratory courtship signals in *Lipara japonica*. Numerals in the map of Japan and Korea denote localities examined. Rhombi in the right diagram show the mean ± S.D. of the burst duration and burst interval which coincide with populations of the left localities.

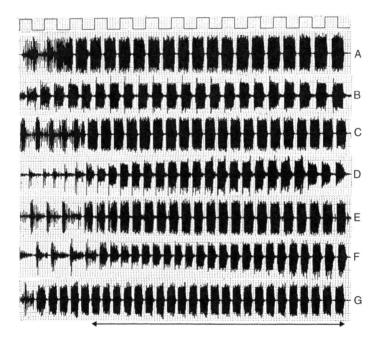

FIGURE 29.15 Representative oscillograms of a train of male calling signals from European *Lipara lucens*. A: Latvia (17), B: Germany (20), C: Belgium (22), D: Czechoslovakia (19), E: Hungary (20), F: Bulgaria (16) G: England (23), of which numerals in parentheses indicate the number of bursts within 2 sec duration shown by a both ends arrow. Upper rectangular waves indicate 10 Hz (100 msec) oscillation. (Modified from Kanmiya, K. *J. Ethol.*, 8, 105–120, 1990. With permission.)

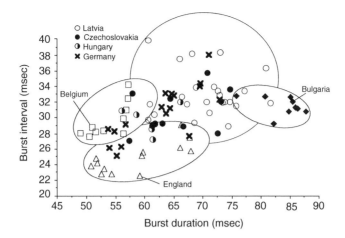

FIGURE 29.16 Scattered diagram between burst duration and burst interval in the male calling signals of seven localities in Europe. (Modified from Kanmiya, K., Phylogeny and origin of the Reed-flies, In *The Natural History of Flies*, Shinonaga, S. and Shima, H., Eds., Tokai University Press, Tokyo, pp. 215–243, 2001, in Japanese. With permission.)

divergence without different gene flow. Therefore, all *Lipara* species might have originated by allopatric speciation (Kanmiya, 1990). Figure 29.17 shows the vibratory signals of European *lucens* based on males and females.

CONCLUSION

I have presented here new data that show substrate-transmitted sounds are very significant in the communication of a wide diversity of Diptera. Many repertoires of substrate-borne signals in Agromyzidae and Chloropidae were recognised, and found that these flies have a delicate yet deliberate system of communication by way of substrate-borne vibration for mating and social interaction. I emphasised an idea that species-specific properties of vibratory signals in *Lipara* could be determined by dynamics of the muscle power being capable of moving the reed stem.

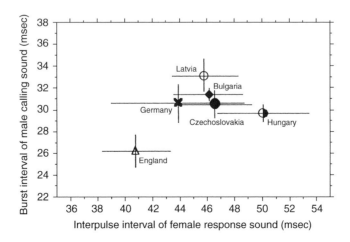

FIGURE 29.17 Diagram between interpulse interval of the female response signals and burst interval of the male calling signals in *Lipara lucens* from five different localities in Europe, indicating a mean ± S.D.

The body size and muscle power might be inevitably derived from adaptive process of innate larval feeding activities. All the acoustic data on Agromyzidae were newly discovered results. The common assumption that the life of flies is mainly one of the chemical senses may now have to be revised.

ACKNOWLEDGEMENTS

I would like to express my gratitude to Michael F. Claridge for revisions of earlier drafts of the manuscript which his appropriate comments improved. I also wish to deeply thank Mitsuhiro Sasakawa for kindly identifying Agromyzid specimens examined in this study.

30 Stridulation in the Coleoptera — An Overview

Andreas Wessel

CONTENTS

INTRODUCTION

The diversity of sound-producing organs in beetles is amazing and unmatched by any other order of insects. Beetles have achieved a number of fundamentally different (and in some respects evolutionarily independent) means of producing sounds by stridulation. Almost every part of the body technically capable of stridulation, *i.e.* adjoining parts moving in relation to each other, have been modified for this purpose in the adults and larvae of one or more groups of beetles (see Table 30.2). Surprising is the neglect of sound production in beetles in publications on insect acoustic behaviour and communication, even in standard texts, such as Pierce (1948), Lewis (1983), Ewing (1989) and Bailey (1991). Haskell (1961), Busnel (1963) and Tuxen (1964), however, provide substantial overviews.

HISTORICAL SKETCH

Sound production and the diversity of stridulatory organs in beetles were discovered rather late compared to those of well-known sound-producing insect groups such as orthopterans and cicadas, which were already studied by the Greek philosopher Aristotle (Balme, 1992; see also Meyer, 1855). The earliest mention of stridulation in beetles in a scientific study seems to be that of Swammerdam (1669) for Cerambycidae and Silphidae (see Landois, 1874). Frisch (1724) and Réaumur (1734 to 1742) were the first who accurately described the stridulatory apparatus of some beetles (Cerambycidae, Scarabaeidae and Chrysomelidae). Lesser (1738, cited from the 2nd ed., 1740) correctly described stridulation in two groups ("Holtz-Käfer", *i.e.* Cerambycidae and almost certainly *Polyphylla fullo*, Melolonthidae) and proposed hypotheses regarding the different biological functions of the sounds thus produced.

During the 19th century a couple of reviews on the occurrence of stridulation in beetles and the organs involved reflected the growing amount of descriptive knowledge (Goreau, 1837; Landois, 1874; Darwin, 1871, cited from the 2nd ed., 1877; Gahan, 1900; Arrow, 1904; and for larvae Schiödte, 1874). The work of these authors has rarely been surpassed since. In contrast to the well-studied orthopteran insects, practically nothing is known about the physiological mechanisms influencing sound production in beetles. A notable exception is the study on circadian rhythms of sound production in Cerambycidae by Tembrock (1960).

Considering sound reception, Reimarus (1798) early remarked on the necessary existence of hearing organs, which he considered undetectable by the technical means of his time. Hearing organs have been described only in the last two decades in two families: tympanal organs in Cicindelidae (Spangler, 1988; Yager and Spangler, 1995) and Scarabaeidae (Forrest *et al.*, 1997; Yager, 1999). Additionally, hearing by the use of Johnston's organ in Dytiscidae (Lehr, 1914) and an unusual sensitivity to substrate-borne vibrations in tibial spurs of the hind legs in Geotrupidae have been noted (Schneider, 1950).

SYSTEMATICS

Stridulatory organs are so far known from 30 beetle families (Table 30.1). This list also includes subfamily taxa classically considered as families. A restriction to recognised families is difficult, as all taxonomic ranks above the species level are to some degree arbitrary. The classification of beetles used here has been adapted from Beutel (2003).

From a systematic perspective, the distribution of stridulatory organs in beetles is challenging as both the occurrence of different types of sound-producing organs in one group, *e.g.* in the Curculionidae (Lyal and King, 1996), and the multiple convergent origin of the same type in several or even a single group probably blur the phylogenetic signal. Even in a comparatively well-known taxon such as the Cerambycidae, stridulatory organs have been interpreted as convergently derived (Michelsen, 1966a) or, alternatively, as a synapomorphy of the family (Breidbach, 1986, 1988). This uncertainty has prevented the frequent use of stridulatory organs in phylogenetic analyses. They have been used occasionally, for example by Kaszab (1936/37), Schmitt (1991) and Browne and Scholtz (1999). Sound-producing organs have been used as autapomorphies at supraspecific levels (Villatoro, 2002). The comprehensive study by Lyal and King (1996) on Curculionidae revealed a rather complex pattern of apomorphies pertaining to stridulating organs at different levels within the family.

Stridulatory organs may be of considerable value for species level taxonomy. Several authors have found species specificity both in distress and in courtship signals, but Schmitt and Traue (1990) provide a counter-example in the Chrysomelidae. The interspecific variation found may be correlated to many factors and need not necessarily indicate reproductive isolation seen in Geotrupidae (Carisio *et al.*, 2004); Passalidae (Palestrini *et al.*, 2003); Scarabaeidae (Kasper and Hirschberger, 2005); and Curculionidae (Riede and Stueben, 2000).

STRIDULATORY ORGANS

All stridulatory organs share the same basic structure consisting of two parts moving against each other: a usually elevated *pars stridens* with fine parallel ribs and a *plectrum*, which is essentially a sharply confined ridge moving across the *pars stridens*. In some cases this distinction is hard to make, however, as both parts may be ribbed and barely distinguishable. The terminology used to describe the various stridulatory organs is based upon the location of the part regarded as the *pars stridens*, which is morphologically more easily recognised than the *plectrum*. At least 14 types of stridulating organs have been described in adult beetles, three in larvae and one in pupae (Table 30.2, and comprehensive review by Dumortier, 1963a). In the last two decades new types of stridulatory organs have been described (for example, an alaro-abdominal type described quite recently by Hirschberger and Rohrseitz, 1995, in Aphodiinae). Also in taxa previously regarded as having only one type of stridulatory organ, additional types — though already described from other groups — have been found, *e.g.* in Bruchidae (Ribeiro-Costa, 1999). The counting of types of stridulatory organs in order to assess their variety is complicated by studies revealing that a single movement may produce sounds by using two different structures, *e.g.* Winking-Nikolay (1975) in Geotrupidae. A new development has been the increased recognition of substrate vibratory

TABLE 30.1
Distribution of Known Stridulating Organs and Behaviour in Beetles (See Text for Details)

Superorder/Series	Superfamily	Family	References
Adephaga		Amphizoidae	Arrow (1924)
		Hygrobiidae	Goureau (1837), Darwin (1877), Gahan (1900), Arrow (1942), Busnel (1963)
		Dytiscidae	Erichson (1837), Landois (1874), Darwin (1877), Gahan (1900), Arrow (1924), Busnel (1963), Smith (1973), Aiken (1985, Biström (1997)
		Carabidae	Marshall (1833), Goureau (1837), Landois (1874), Darwin (1877), Gahan (1900), Arrow (1924), Busnel (1963), Claridge (1974), Bauer (1976), Forsythe (1978, 1979, 1980)
		(Cicindelidae)	Landois (1874), Gahan (1900), Arrow (1942), Busnel (1963), Freitag and Lee (1972)
Polyphaga Staphyliniformia	Hydrophiloidea	Hydrophilidae	Gahan (1900), Buhk (1910), Marcu (1932a, 1932b), Busnel (1963), Van Tassell (1965), Maillard and Sellier (1970), Ryker (1975), Pirisinu *et al.* (1988), Watanabe (2000)
	Staphylinoidea	Silphidae	Burmeister (1832–1847), Goureau (1837), Landois (1874), Darwin (1877), Milne and Milne (1944), Busnel (1963), Lane and Rothschild (1965), Niemits (1972), Schumacher (1973a), Bredohl (1984), Huerta *et al.* (1992)
Scarabaeiformia	Scarabaeoidea	Lucanidae	Darwin (1877), Gahan (1900), Arrow (1904, 1942), Busnel (1963), Sprecher-Uebersax and Durrer (1998)
		Passalidae	Gahan (1900), Arrow (1904, 1924, 1942), Busnel (1963), Schuster (1975, 1983), Reyes-Castillo and Jarman (1983), Palestrini *et al.* (2003)
		Trogidae	Landois (1874), Darwin (1877), Arrow (1904), Arrow (1942), Busnel (1963), Alexander *et al.* (1963)
		Geotrupidae	Goureau (1837), Landois (1874), Darwin (1877), Gahan (1900), Verhoeff (1902), Fabre (1917), Arrow (1942), Busnel (1963), Winking-Nikolay (1975), Zunino and Ferrero (1989), Allen (1993), Carisio *et al.* (2004)
		Ochodaeidae	Horn (1876), Arrow (1904)
		Cerathocanthidae (= Acanthoceridae)	Hesse (1948), Alexander (1967), Crowson (1981)
		Scarabaeidae	Reaumur (1734–1742), Goureau (1837), Darwin (1877), Gahan (1900), Arrow (1904), Fabre (1917), Busnel (1963), Mini and Prabhu (1990)
		(Aphodiidae)	Hirschberger and Rohrseitz (1995), Hirschberger (2001), Kasper and Hirschberger (2005)
		(Cetoniidae)	Gahan (1900), Arrow (1904), Busnel (1963)
		(Dynastidae)	Landois (1874), Darwin (1877), Gahan (1900), Arrow (1904, 1942), Hinton (1946, 1955), Arrow (1951), Busnel (1963), Dechambre (1984)

Continued

TABLE 30.1

Continued

Superorder/Series	Superfamily	Family	References
		(Euchiridae)	Darwin (1877)
		(Melolonthidae)	Lesser (1840), Landois (1874), Gahan (1900), Arrow (1904), Baudrimont (1923a, 1925), Busnel (1963), Moron (1995)
		(Rutelidae)	Arrow (1899, 1904), Gahan (1900), Ohaus (1900), Busnel (1963)
		(Taurocerastidae)	Arrow (1904)
Elateriformia	Byrrhoidea	Heteroceridae	Darwin (1877), Gahan (1900), Busnel (1963)
Bostrichiformia	Bostrichoidea	Bostrichidae	Landois (1874), Gahan (1900), Arrow (1942), Busnel (1963)
		Dermestidae	Darwin (1877)
		Anobiidae	Gahan (1900), Busnel (1963)
		Ptinidae	Gahan (1900), Hesse (1936), Arrow (1942)
Cucujiformia	Cucujoidea	Nitidulidae	Gahan (1900), Arrow (1924), Busnel (1963)
		Languriidae	Gahan (1900), Arrow (1924), Busnel (1963)
		Erotylidae	Arrow (1924), Boyle (1956), Busnel (1963), Ohya (1996)
		Endomychidae	Gahan (1900), Arrow (1924, 1942), Busnel (1963)
	Tenebrionoidea	Tenebrionidae	Goureau (1837), Landois (1874), Darwin (1877), Gahan (1900), Dudich (1920), Remy (1935), Arrow (1942), Priesner (1949), Busnel (1963), Slobodchikoff and Spangler (1979), Pearson and Allen (1996)
		(Cistelidae)	Gahan (1900)
	Chrysomeloidea	Cerambycidae	Frisch (1724), Lesser (1840), Burmeister (1832–1847), Goureau (1837), Landois (1874), Darwin (1877), Gahan (1900), Alexander (1957a), Tembrock (1960), Busnel (1963), Michelsen (1966b), Breidbach (1986, 1988), Cheng (1991, 1993)
		(Prionidae)	LeConte (1878), Gahan (1900), Baudrimont (1923b, 1926), Marcu (1930a, 1932c)
		(Lamiidae)	Goureau (1837), Finn *et al.* (1972), Kaszab (1936/37)
		Bruchidae	Kingsolver *et al.* (1993), Ribeiro-Costa (1999)
		Chrysomelidae	Reaumur (1734–1942), Darwin (1877), Gahan (1900), Busnel (1963), Schmitt (1991)
		(Crioceridae)	Goureau (1837), Landois (1874), Darwin (1877), Gahan (1900), Dingler (1932a, 1932b), Schmitt and Traue (1990)
		(Hispidae)	Gahan (1900), Arrow (1904, 1924, 1942), Dudich (1920), Busnel (1963)
	Curculionoidea	Curculionidae	Goureau (1837), Landois (1874), Darwin (1877), Gahan (1900), Dudich (1920), Marcu (1930b, 1933), Arrow (1942), Busnel (1963), Claridge (1968), Harman and Harman (1984), Lyal and King (1996), Riede and Stueben (2000)

Continued

TABLE 30.1
Continued

Superorder/Series	Superfamily	Family	References
		(Scolytidae = Ipidae)	Gahan (1900), Arrow (1924, 1942), Kleine (1932), Busnel (1963), Barr (1969), Rudinsky and Michael (1973), Rudinsky and Ryker (1976), Lewis and Cane (1992), Lyal and King (1996)
		Platypodidae	Menier (1976), Lyal and King (1996), Ohya and Kinuura (2001)
		Brentidae	Kleine (1918)
		Nemonychidae	Kleine (1920)

TABLE 30.2
Types of Stridulating Organs (For Details See Text, for References Table 30.1)

Location of *Pars Stridens*		Families
Cephalic *pars stridens*	Maxillo-mandibular (Buccal method)	see Stridulation in larvae
	Cranio-Prothoracaic	
	Vertex	Nitidulidae, Languriidae, Endomychidae, Hispidae
	Gula	Anobiidae, Tenebrionidae, Chrysomelidae, Scolytidae
Thoracaic *pars stridens*	Prosterno-mesosternal	Melolonthidae
	Pronoto-femoral	Bostrychidae
	Mesonoto-pronotal	Cerambycidae
Abdominal *pars stridens*	Abdomino-femoral	Cetoniidae, Heteroceridae
	Abdomino-elytral	Carabidae, Hydrophilidae, Silphidae, Scarabaeidae, Tenebrionidae, Chrysomelidae, Curculionidae
	Abdomino-alary	Passalidae, Geotrupidae, Dynastidae
Pars stridens on the legs	Coxo-metasternal	Geotrupidae, Cerambycidae
	Femoral methods	Carabidae, Rutelidae
Pars stridens on elytra or hindwings	Elytro-abdominal	Hygrobiidae, Trogidae, Scarabaeidae, Curculionidae, Scolytidae
	Elytro-femoral	Carabidae, Cicindelidae, Lucanidae, Scarabaeidae, Tenebrionidae, Cerambycidae
	Alary-elytral	Dytiscidae, Erotylidae, Endomychidae
	Alary-abdominal	Scarabaeidae
Stridulation in larvae	Maxillo-mandibular	Cetoniidae, Dynastidae, Melolonthidae, Rutelidae
	Mesocoxa-hind leg	Passalidae, Geotrupidae
	Metatrochanto-mesocoxal	Lucanidae
Stridulation in pupae	Gin-traps of tergites 1 to 6	Dynastidae

communication in beetles, as opposed to the long-lasting and exclusive focus on air-borne sounds, *e.g.* Breidbach (1986) in Cerambycidae; Hirschberger and Rohrseitz (1995) in Scarabaeidae; and general discussion by Masters (1980).

BIOLOGICAL MEANING OF STRIDULATION

While a large amount of morphological knowledge on beetle stridulation exists, very little is known about its function. Since Darwin (1877) and even earlier authors (Lesser, 1740; Reimarus, 1798), sound production in beetles has been stressed as an important component of mating behaviour. Darwin consequently regarded it as a sexual character, potentially shaped by sexual selection. In contrast to this general assumption, empirical support is limited. This lack of observation may be partly an artefact due to the inadequate criterion used by Darwin and later authors to establish the role of stridulatory organs in mating: that is, the possession of these in both sexes and their frequent similarity have been interpreted as an argument against a function during mating as only males were supposed to use sound to attract females (Darwin, 1877; Crowson, 1981). In some groups the differences between species and sexes expected in accordance with that criterion were indeed found, *e.g.* in Tenebrionidae and Dynastidae (Darwin, 1877). Recent research has shown this assumption to be insufficient for several reasons as even the presence of the same structure in both sexes does not automatically mean identical use, as shown in Curculionidae (Wilson *et al.*, 1993), Scarabaeidae (Hirschberger, 2001), Platypodidae (Ohya and Kinuura, 2001), Scolytidae (Rudinsky and Ryker, 1976; Oester *et al.*, 1978) and Geotrupidae (Winking-Nikolay, 1975). However, in groups believed to provide clear evidence for the role of stridulation in mating, this assumption has been challenged and alternative hypotheses have been proposed, *e.g.* in Scolytidae (Pureswaran and Borden, 2003). Stridulatory signals may serve as an efficient species-specific mate recognition system (SMRS *sensu* Paterson, 1985) and thus act as an isolating mechanism and prevent interspecific matings, as shown for dung beetles by Kasper and Hirschberger (2005, and this volume). For some taxa, however, courtship stridulation is apparently inefficient as a premating ethological barrier, *e.g.* Lewis and Cane (1992) for Scolytidae. No clear picture is yet available.

Despite the early focus on a function during mating, much work indicates that stridulation also plays a role in various forms of defensive behaviour, for example, in Carabidae (Bauer, 1976), Chrysomelidae (Schmitt and Traue, 1990), Curculionidae (Riede and Stueben, 2000), and general studies (Dumortier, 1963c; Masters, 1979, 1980; Lewis and Cane, 1990). Some authors assume the existence of *acoustic mimicry* in some predators. For example, Crowcroft (1957) described a strongly repellent effect of *Cychrus* (Carabidae) stridulation on shrews and Lane and Rothschild (1965) suggested that the sounds produced by disturbed *Necrophorus* (Silphidae) resemble those of *Bombus* bees.

In several taxa it has been shown that one and the same structure may be used to produce different types of sound which may serve a variety of functions, including courtship, aggression, defense and aggregation, as in Cerambycidae (Alexander, 1957a), Scarabaeidae (Mini and Prabhu, 1990), Silphidae (Niemits, 1972), Hydrophilidae (Ryker, 1976), Scolytidae (Ryker and Rudinsky, 1976) and Geotrupidae (Winking-Nikolay, 1975).

For two taxa an involvement of stridulatory signals in social, or subsocial, behaviour (mostly larval–adult attraction) has been described in Silphidae (Niemits, 1972) and Passalidae (Schuster, 1983). For most groups the biological function of stridulation in larvae is unknown (Sprecher-Uebersax and Durrer, 1998).

Generally, it may be stated that many beetles produce a wide variety of sounds using a multitude of structures of the adult and larval body in an ethological context still widely unknown. Thus, biacoustic studies with both evolutionary and systematic focus and a behavioural–ecological context in the Coleoptera are much needed.

ACKNOWLEDGEMENTS

I am greatly indebted to my colleagues Jason Dunlop, Hannelore Hoch, Thomas von Rintelen (Museum für Naturkunde, Berlin) and Günter Tembrock (Institut für Biologie, Humboldt-Universität zu Berlin) for helpful comments and reviews on an earlier draft of the manuscript.

31 Vibratory Communication in Dung Beetles (Scarabaeidae, Coleoptera)

Julia Kasper and Petra Hirschberger

CONTENTS

INTRODUCTION

Aphodius dung beetles possess a stridulatory apparatus and produce substrate-borne vibrations, enabling communication within the dung (Hirschberger and Rohrseitz, 1995).

In the present study, numerous species of Aphodiinae, especially of *Aphodius,* were examined with regard to their sound production and the morphology of the stridulatory organ. The main questions were: how widespread the stridulatory organ is within the genus, how much morphology and sound differ among and within the species, and whether it is possible to distinguish potential sibling species or to use this character for phylogenetic analysis.

APHODIUS AND ITS LIFE HISTORY

With more than 1000 species the genus *Aphodius* is one of the most diverse groups of Scarabaeoidea. *Aphodius* are 2 to 9 mm long, mostly oval, elongate, moderately haired and yellowish, reddish or brownish to black in colour. The elytra often have dark spots. The pygidium is visible (Figure 31.1). *Aphodius* occur in the Palaearctic region, in sub- or boreo-alpine zones (Balthasar, 1964). They mainly live in and feed on the dung of herbivorous mammals (Hanski, 1991).

Dung is a nutrient-rich resource, which is spatially and temporally confined. Most dung beetles are generalists and do not specialise on dung of a particular mammal. However, preferences have been observed. Some species may avoid competition by differing in phenology, duration of use and

FIGURE 31.1 *Aphodius ater* on sheep dung. (Photo by K. Schütt. With permission.)

preferences for particular substrate conditions. Coexistence of several *Aphodius* species in the same dung pad often occurs. Inter- and intraspecific aggregation in dung pads can be observed. Physical proximity may be a prerequisite for communication via substrate-borne vibration (Hirschberger, 1998).

SOUND PRODUCTION

Arrow (1904, p. 709) seemed deeply impressed by the ability of lamellicorns (Scarabaeoidea) to produce sounds, stating that "The special importance of stridulation in that group is probably in part due to a mental development higher than that of most other beetles".

Among *Aphodius* sound production and the morphology of the stridulatory organ has only recently been described in detail for the European species *Aphodius ater* DeGeer (Hirschberger and Rohrseitz, 1995; Hirschberger, 2001) and for some North American species (Kasper and Hirschberger, 2005).

These studies revealed that beetles produce sounds when they are disturbed (for a description of disturbance sounds, see Masters, 1979). However, acoustic activity is more conspicuous during courtship and mating. Male beetles produce a complex song consisting of different temporal patterns when they meet a female and mate. Females have the same morphology of the stridulatory apparatus, but they appear to produce only disturbance sounds (Hirschberger, 2001).

As described in *A. ater* (Hirschberger and Rohrseitz, 1995), the *pars stridens* or file of the investigated species is a specialised area of a wing-vein (subcosta) with a large number of parallel ridges (Figure 31.2a). The *plectrum* or scraper on the first abdominal segment consists of a small number of parallel sclerotic lamellae (Figure 31.2b). As the abdomen extends and contracts the scraper is moved across the file on the underside of the hindwing. During stridulation, the oscillatory movement of the pygidium may be observed. Although suspected for a long time (Arrow, 1904), the larvae of *Aphodius* have no stridulatory organ (personal observation).

MORPHOLOGY OF THE STRIDULATORY ORGAN

The morphology of the stridulatory organs was investigated using scanning electron microscopy (SEM) (see Kasper and Hirschberger, 2005). Forty species of *Aphodius*, ten of other genera of Aphodiinae ($n = 2$ to 10) and additionally a genus of the sister group Aegialinae were examined. Morphometric analyses were made on the stridulatory apparatus in ten species ($n \geq 7$) (Figure 31.3).

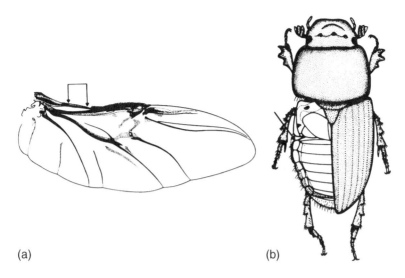

(a)   (b)

FIGURE 31.2 Stridulatory organ of *Aphodius* spec. (a) The file on the subcosta of the hindwing, (b) the scraper on the first sternite.

INTERSPECIFIC COMPARISON OF *APHODIUS* SPECIES

Our analyses provide evidence that the morphological configuration of the stridulatory organ is species-specific. The distance between the ridges, the width of ridges, and the intervals differ among species but are quite consistent within species (Figure 31.3).

The morphological differences in the stridulatory organ concern the following characters:

(1) Distance between costa and subcosta (Figure 31.4j and Figure 31.5a)
(2) The anterior end (Figure 31.4g, m and Figure 31.5a, d)
(3) Bristles on the costa (Figure 31.5d)
(4) Ridges on the costa (only *A. contaminatus*) (Figure 31.5b)
(5) The form of the ridges (Figure 31.4e, h, k, n and Figure 31.5e)
(6) The scraper (Figure 31.4f, i, l and Figure 31.5c, f).

A. luridus, *A. rufipes* and *A. depressus* are the only ones which have no ridges or lamellae, although the basic form of the stridulatory organ can be recognised (Figure 31.4a–c).

INTRASPECIFIC VARIABILITY OF *APHODIUS* SPECIES

Nine species of two to five populations were examined. The qualitative analysis yielded no significant differences nor the measurement of the stridulatory organ (Figure 31.6). Within-species variations in width of the ridges and the intervals are very small. For example, *A. erraticus* of three populations shows a deviation of 0.0 with regard to the interval, 0.4 with regard to the breadth of the ridges and 2.8 with regard to the total number of ridges. Only in *A. prodromus* could it be observed that the ridges of the three populations grow together in different dimensions.

A discriminant analysis of ten species was made for comparing individuals.

There is an assignment probability of 76.3%. The individuals of eight species were assigned correctly for 70 to 100%. *A. rufus* (78%), *A. convexus* (85%) and *A. ater* (71%) species of one subgenus are easy to distinguish. Even *A. convexus*, often discussed as the sibling species of *A. ater*, is obviously discriminated.

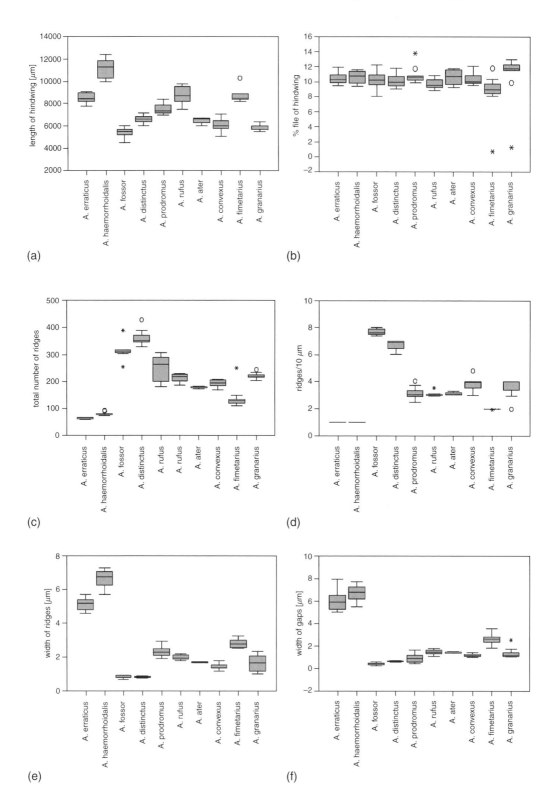

FIGURE 31.3 Morphometric analysis of the stridulatory organ of *Aphodius* species ($n \geq 7$) (boxplots).

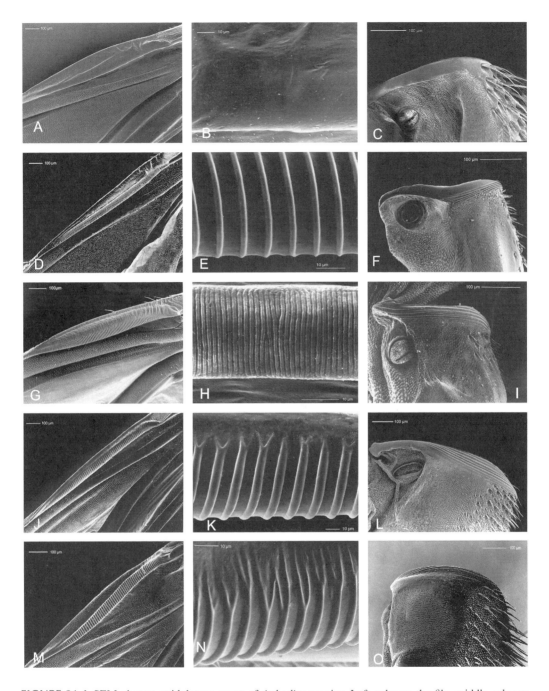

FIGURE 31.4 SEM photos: stridulatory organ of *Aphodius* species. Left column: the file, middle column: ridges on the file, right column: scraper; (A, B, C) *A. luridus*, (D, E, F) *A. coloradiensis*, (G, H) *A. pingius*, (I) *A. haemorrhoidalis*, (J, K, L) *A. fossor*, (M) *A. erraticus*, (N, O) *A. sigmoidaeus*.

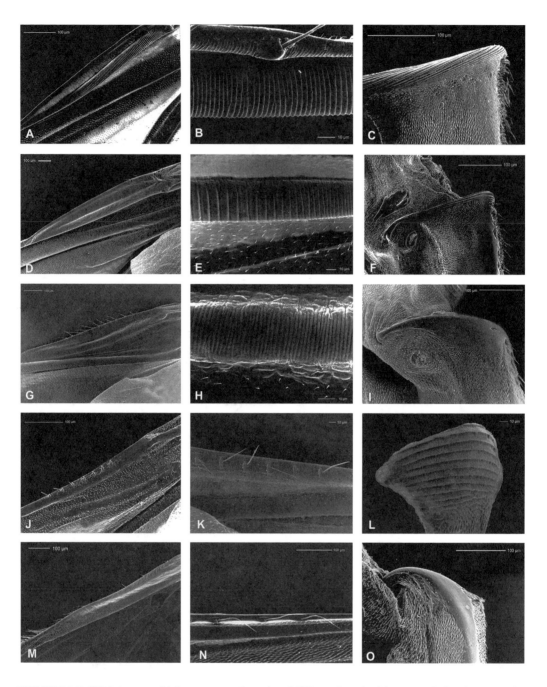

FIGURE 31.5 SEM photos: stridulatory organ of species of different Scarabaeide genera. Left column: the file or edge of the hind wing, middle column: ridges on the file or plates on the costa, right column: scraper or elevated structure on the first sternite; (A) *A. fimetarius*, (B) *A. contaminatus*, (C) *A. lapponum*, (D) *A. scrutator*, (E, F) *A. prodromus*, (G, H, I) *Heptaulacus carinatus*, (J) *Rhyssemus sindicus*, (K, L) *Psammodius laevipennis*, (M) *Aegiali blanchardi. erraticus*, (N, O) *Ataenius erratus*.

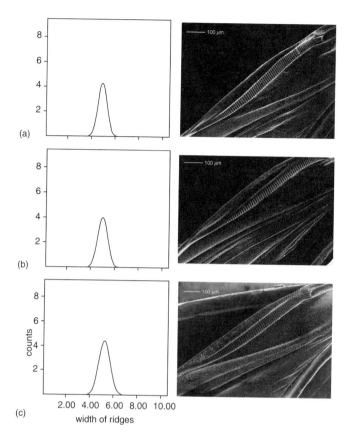

FIGURE 31.6 Left: histograms of width of ridges, right: SEM photos of files of three *A. erraticus* populations, (a) Erlangen, (b) Lethbridge, (c) Waterton.

Our investigation revealed no sexual dimorphism in the morphology of the stridulatory apparatus.

COMPARISON WITHIN AND BETWEEN APHODIINAE AND AEGIALINAE

The phylogenetic relationships among known Scarabaeidae have not yet been fully resolved (Figure 31.7). The question is whether it is possible to offer some explanations if we use the stridulatory organ as a character. Therefore we included two groups of Aphodiinae (Aphodiini and Psammodiini) and its sister group Aegialinae in our studies.

Within the taxon Aphodiini the stridulatory organ of *Heptaulacus carinatus* and *Oxyomus sylvestris* is very similar to that of *Aphodius*. (Figure 31.5g–i). One difference was that the bristles of the *costa* extend of small plates which could not be found in *Aphodius* (Figure 31.5i).

Ataenius also has long bristles on the edge of the hind wing. The subcosta and the first sternite have an elevated, but smooth structure without ridges or lamellae (Figure 31.5n, o).

Three species of Psammodiini are very different from *Aphodius*. *Pleurophoris caesus* has a file-like structure, but it is completely smooth without any ridges as also on the sternite.

Psammodius laevipenni and *Rhyssemus sindicus* have just one wing vein at the observed region (Figure 31.5j). On this vein there are small plates with thick and pointed bristles (Figure 31.5k), but a broad scraper with broad lamellae has also been found (Figure 31.5l).

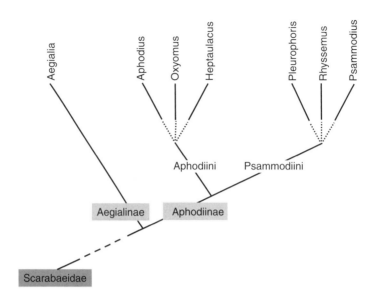

FIGURE 31.7 Phylogenetic relationships within the Aphodiinae with numerous polytomies. (After Krell, F. T., Fery, H., Lohse, G. A. and Lucht, W. H., Eds., Goecke and Evers, Krefeld, pp. 201–253, 1992. With permission.)

The species of the genus *Aegialia* (Aegialini) (Hatch, 1991) examined lacked the second wing vein and displayed a long, thin file on the edge of the hind wing (Figure 31.5m). The total number of about 450 ridges is high. This is interesting because *Aegialia* has reduced hind wings. The existing bristles on the file again extend from small plates, which are integrated between the ridges. The sternite also has lamellae in both species.

SOUND PRODUCTION IN *APHODIUS*

To examine the differences of the sound among and within the *Aphodius* species, nine European species were recorded and the sound analysed (for American species, see Kasper and Hirschberger, 2005). A detailed description of the method is given by Hirschberger (2001). The sound is divided into several elements, as is demonstrated in Figure 31.8. Every *verse* has many *syllables* divided into pulses. The gaps between verses have been termed *major-pause*.

The syllables are also divided into *pulses*. The syllables consist of two parts (the *prepulse* and the *main-pulse*) separated by a *minor-pause*. This pattern illustrates the back and forth movement of the abdomen during stridulation. The successive syllables are called *periods* which describe the time span from the beginning of one syllable to the beginning of the next.

Every pulse has many overlapping peaks, which were not considered in this investigation. This is caused by simultaneous scraping of many lamellae over many ridges of the file. As the frequency distribution of the sounds depends on the condition of the dung and the position of the beetle within the dung, absolute amplitudes were not considered (see Alexander, 1967).

INTERSPECIFIC COMPARISON

The species investigated produced sounds in two behavioural contexts: First, when the beetles were disturbed. Such disturbance sounds always consist of regularly emitted syllables with a high repetition rate. They are produced by males as well as by females and are relatively similar

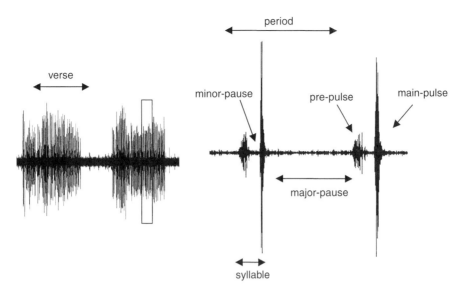

FIGURE 31.8 Overview of sound structures and their definitions.

in all species investigated. Second, sounds are produced when males and females meet within the dung in an undisturbed situation. Within this behavioural context the sounds are exclusively produced by males. In the few examples where beetles were observed while stridulating, only males sang during male–female interactions. In these cases copulation only took place during singing pauses. The males sing for a period of 10 to 20 min. Female responses or duetting were not observed.

Male songs are species-specific. The verses vary in number and duration. Frequency range of syllables was between 0.2 and 4 kHz with *A. piceus* the lowest and *A. fossor* the highest.

The long verses of *A. ater* and *A. convexus* show two different patterns (Figure 31.9a and b).

The verses of *A. distinctus* consist of variable number and length of syllables, producing a rattling sound (Figure 31.9c and d).

A. erraticus has variable pulses but consistent pauses (Figure 31.10). The syllables have a duration of 0.4 sec (Figure 31.10e). Both parts have similar intensity and length (Figure 31.10a and b).

The verses of *A. fimetarius* are clearly separated by long major-pauses (Figure 31.9e) and are comprised of a series of syllables in rapid succession with a consistent period (Figure 31.10f).

Males of *A. fossor* produce syllables that consist of three parts. The third part comprises a variable number of pulses (Figure 31.9f). The verses are long and the periods slow (Figure 31.10f). The pauses are variable (Figure 31.10d). *A. granarius* emits clearly separated verses (Figure 31.9g). The duration of syllables, major- and minor-pauses are consistent (Figure 31.10).

The sound of *A. piceus* consists of long verses with homogenous periods (Figure 31.9h). Both pulses have similar intensities.

Compared with *A. piceus*, *A. prodromus* has pulses of different length and intensity (Figure 31.9i).

Intraspecific Variability

To confirm interspecies-specificity, we analysed intraspecific variability. The sounds of the populations examined were very similar (Figure 31.11). The standard deviation of periods is low,

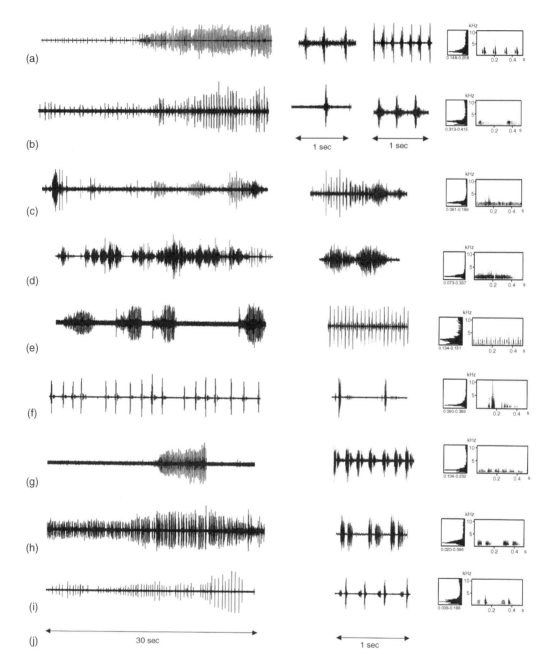

FIGURE 31.9 Oscillograms of *Aphodius* species, left column per 30 sec, middle column per 1 sec; sonograms per 0.5 sec. (a) *A. ater*, (b) *A. convexus*, (c) *A. distinctus*, (d) *A. erraticus*, (e) *A. fimetarius*, (f) *A. fossor*, (g) *A. granarius*, (h) *A. piceus* (i) *A. prodromus*.

which shows that period length may be a useful indicator for species-specificity. The populations of *A. ater*, for example, from northern Germany, the Netherlands and Denmark show nine, eight and seven syllables per second, respectively. The range of variation of pauses was comparatively higher. This parameter does not appear to be species-specific. The discriminant analysis confirms the species quality with an assignment probability of 80%.

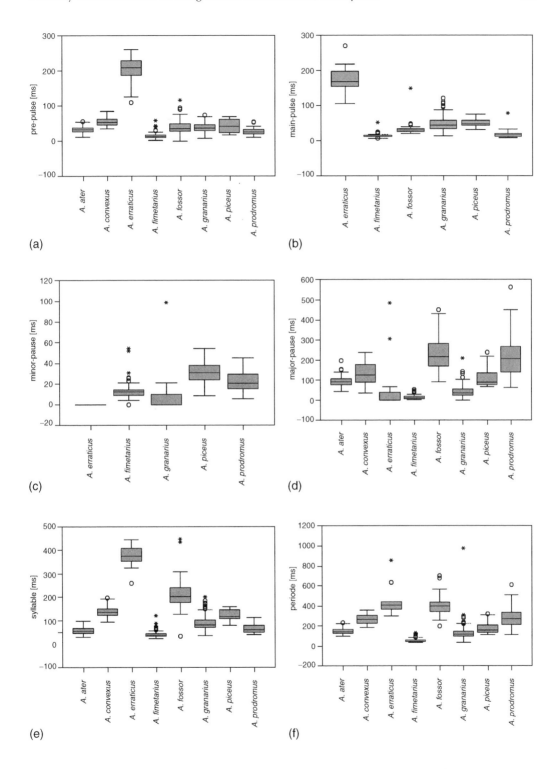

FIGURE 31.10 Analysis of various parameters of the sound in *Aphodius* species (boxplots).

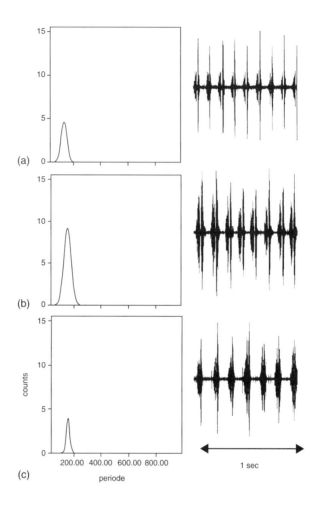

FIGURE 31.11 Left: histograms of period lengths, right: oscillograms of three *A. ater* populations: (a) Rodenäs, (b) Sjeilland, (c) Schiermonikoog.

ANALYSIS OF TWO SIBLING *APHODIUS* SPECIES

The investigation of further *Aphodius,* in addition to *A. ater*, has shown that songs are indeed species-specific. The question now is whether acoustic signals in *Aphodius* evolve more rapidly than morphological characters and if the sounds can be used to distinguish closely related species. In order to test this hypothesis, the stridulatory pattern of *A. ater* was compared with the supposedly sibling-species *Aphodius convexus* Erichson (1848). *A. convexus* has been treated variously as a synonym, subspecies, variation or race of *A. ater* for many years (see Pittino and Mariani, 1993, for detailed information). Recently, Pittino and Mariani (1993) have redescribed *A. convexus* as a valid species on the basis of more than 1500 specimens from a wide range of localities throughout its range in South Europe, from the Iberian Peninsula to Anatolia and the Caucasus. *A. ater* is widespread and common in Central and North Europe. According to Pittino and Mariani (1993), the two species are sympatric, with widely overlapping distributions throughout South Germany, Switzerland, Austria, North Italy and France. In their redescription they gave several morphological characters to distinguish between *A. convexus* and *A. ater*. They admit, however, that the variability in some of the characters is high so that several specimens of *A. convexus* strikingly resemble *A. ater*. The long controversy about the status of *A. convexus* make them ideal organisms in which to investigate the degree of distinction of sounds among sibling species and to assess the possible benefit of using sounds in systematics.

SEM revealed that the structure of the stridulatory organ in *A. ater* and *A. convexus* is nearly identical (Figure 31.3).

The present study describes the stridulatory patterns. Songs of males of four populations of *A. ater* from North Germany, Denmark and the Netherlands and of one population of *A. convexus* from South Germany were recorded in 1997 and 1998, respectively. A total of 28 recordings were available.

ANALYSIS OF SONGS

Males of both species stridulate for a considerable period of time when opposite sexes meet in the dung. The stridulation is a complex series comprised mainly of two alternating vibratory patterns that were termed "*basic*" and "*rhythmic*" in *A. ater* (Hirschberger, 2001). Syllable structure shows specific differences in both taxa. In *A. ater* the syllables of "*basic*" consist of two parts separated by a small pause (Figure 31.9). A sonogram shows that frequencies up to 4 kHz were recorded on the surface of the dung. All syllables have a frequency distribution with a maximum below 2 kHz. The "*rhythmic*" pattern consists of slow and loud syllables. In *A. convexus*, the syllables of "*basic*" do not have a characteristic pause. Instead, each syllable is followed by reverberation. The "*rhythmic*" pattern consists, as in *A. ater*, of alternating syllables with low and high amplitude. In *A. convexus* the syllables start with a buzzing sound of low frequencies, which was not observed in *A. ater*. A remarkable feature of the song of *A. ater* is the increasing velocity in "*rhythmic*" pattern. The syllables are emitted with increasing repetition rate (Hirschberger, 2001). This was not observed in any males of *A. convexus*.

The characteristic patterns of stridulation are shown in the acoustic ethograms (Figure 31.12). The two main patterns occur in both species. However, the stereotypic alternation between "*basic*" and "*rhythmic*" that has been described for *A. ater* was not observed in *A. convexus*. Figure 31.9a and b show oscillograms of a part of the song at which the beetles are switching from "*rhythmic*" to

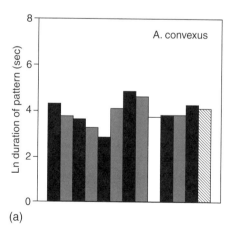
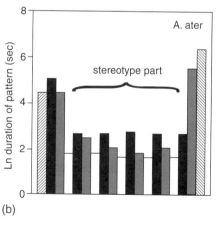

FIGURE 31.12 Acoustic ethograms of the song of males of *A. convexus* and *A. ater*. The consecutive order of the patterns is shown on the abscissa and the corresponding length of the patterns is shown on the ordinate. ▨ introductory phase with single chirps, ■ regular trains of syllables, defined as *basic*, ■ regular trains of syllables defined as *rhythmic*, □ phase of silence, ▧ final phase with single chirps. (a) Male of *A. convexus* without stereotypical alternation of patterns. (b) Male of *A. ater* showing a stereotypical alternation between the patterns (indicated by brace). Note that scale on the ordinate is logarithmic.

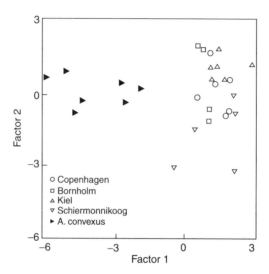

FIGURE 31.13 Scatterplot of the first two functions of the discriminant analysis of the measurements of period of song patterns and of song duration. Each data point represents a single individual, coded by population.

"*basic*". The pattern "*basic*" consists of syllables with more or less equal amplitude emitted quickly in regular intervals in both species. It is followed by the "*rhythmic*" pattern consisting of regularly alternating syllables with high and low amplitudes. This general pattern was observed in all populations of *A. ater* and in the population of *A. convexus*.

The analysis of temporal song patterns reveals a significant difference between the populations (ANOVA: "*basic*" pattern: $F = 4.554$, $P < 0.01$; "*rhythmic*" pattern: $F = 34.221$, $P < 0.001$; duration of song: $F = 3.600$, $P < 0.05$). The discriminant analysis of the temporal song patterns shows correct classification of 86% of the individuals in *A. convexus* compared to 46% in *A. ater*. Examination of the *F*-to-remove statistics reveals that the temporal structure of "*rhythmic*" has the highest relative influence on the separation of the groups.

To summarise, the general song pattern of *A. ater* and *A. convexus* are very similar, but specific differences in syllable structures and in the small-scale temporal structures of the song can be observed. Thus, compared with widely overlapping morphological features, the song phenotypes prove more reliable for distinction of both species (Figure 31.13).

CONCLUSIONS

In each of the *Aphodius* species examined, we found a stridulatory organ or at least its rudiments. The morphology of stridulatory organs in *Aphodius* is species-specific and we did not observe any sexual dimorphism in the morphology of the stridulatory organ.

Considering that several *Aphodius* species were found in the same dung pads at the same time at high densities, thus co-occurring spatially and temporally, it might be assumed that it would be most effective if their courtship songs were distinct. Indeed, we show that *Aphodius* species do produce species-specific songs, which play an important role during courtship behaviour.

In a comparison of two sibling species, sound shows better discrimination than morphology. The general song pattern of *Aphodius ater* and *A. concexus* is very similar, but species-specific differences in syllable structures and in the small-scale temporal structures of the song were observed. Thus, the sound may be a significant component of the specific mate-recognition system (Paterson, 1985).

If sound frequency is correlated with the density of ridges, we hypothesise that the morphological configuration of the stridulatory organ has limited phylogenetic value, as closely-related, sympatric species would be expected to differ significantly in this character. At present, morphology of the stridulation organ does not reflect the recent classification of Aphodiinae. Only one subgenus consists of species which all have a structure, apparently showing a reduced stridulatory organ (Figure 31.4a–c). However, it appears possible to reconstruct relationships among higher levels of taxa. Further investigation is needed to prove these hypotheses. Additional genetic analysis would be desirable.

ACKNOWLEDGEMENTS

We are grateful to Mikael Münster-Swendsen and Peter Holter (University of Copenhagen) and Frank-Thorsen Krell (Museum of Natural History London) for providing material and accommodation. We owe special thanks to John Kirly, John S. Edwards (Burke Museum, Washington), Jason Dunlop and Hannelore Hoch (Museum für Naturkunde Berlin) for comments on earlier versions of the manuscript and to the editors for inviting us to contribute to this volume.

32 Vibratory and Airborne-Sound Signals in Bee Communication (Hymenoptera)

Michael Hrncir, Friedrich G. Barth and Jürgen Tautz

CONTENTS

INTRODUCTION

When coming across the term "bee communication", almost automatically an image of the legendary "figure-of-eight movements" of the honey bee (*Apis mellifera*) crosses most people's minds. Expressions like "dance language" or "bee language" are associated with the name of the Austrian scientist Karl von Frisch (1886–1982), who observed that the body movements of foraging honeybees on their return to the nest from a valuable food source correlate with both its direction and its distance (von Frisch, 1965). Ever since this pioneering discovery, people have been fascinated by this concept of a "symbolic language" in which an abstract code (the waggle dance) is used to transmit information about remote objects (the food source). However, intensive investigations over the past decades, and steadily improving observation and recording methods, have revealed that the honeybees' communication is of an intriguing complexity which we are only beginning to understand in depth. The dance movements with their intrinsic complexity are but one chapter of the whole story of communication processes that coordinate several hundreds or thousands of individuals belonging to a single bee colony.

Since the beginning of the 17th century, it has been established that bees emit "sounds" audible to the human ear (Butler, 1609). It would, however, be misleading to denominate these "sounds" exclusively as "acoustic" in terms of human experience because it is the sensory world of the bees which has to be considered. What we humans perceive as "sound" (the pressure waves of airborne sound) is not necessarily the physical aspect of the signal perceived by the bees. This chapter is an attempt to outline the current knowledge of "sounds", and the other types of potential signals associated with these "sounds" emitted by social bees, honeybees (Apini), stingless bees (Meliponini) and bumble bees (Bombini). We ask the following questions: How are the signals produced by the bees? What is the true physical nature of the "signals" actually or potentially perceived by the bees? The important question about the *biological significance of the "sounds"* is not easily answered because the behavioural response of a bee to a certain signal much depends, not only on the behavioural context, but also on the recipient's motivation. Deciphering the messages and meanings of different "sounds" (*message*: information provided by the sender; *meaning*: influence on the behaviour of the receiver; Seeley and Tautz, 2001) is therefore essential for a comprehensive understanding of the bees' communication.

SIGNALS

The sense of "hearing" for a long time had been interpreted in human terms and restricted to the perception of airborne sound, specifically of the sound pressure waves it represents. Hence, it was attributed only to those animals equipped with sound pressure receivers, "eardrums" or similar membranes (von Buddenbrock, 1928, cited in Autrum, 1936). After suspicious membranous structures on bees' antennae, the antennal pore plate organs (*sensilla placodea*), had been demonstrated to be chemoreceptors instead of "eardrums" (von Frisch, 1921), bees were considered deaf to airborne sound for anatomical reasons. However, ever since the German zoologist and physicist Hansjochem Autrum (1907–2003) demonstrated the ability of insects to perceive minute substrate vibrations (Autrum, 1941; Autrum and Schneider, 1948) and that hair-like structures, such as some insect antennae, can function as sound velocity receivers (Autrum, 1936), it has been clear that "hearing" in arthropods is not at all confined to the perception of sound pressure waves. If we strive to understand the behavioural relevance of an animal's signal, the "sensory world" in which this animal lives has to be considered (Barth, 1998). To understand fully acoustic communication in bees, the physiological and ecological constraints on their communication have to be known — how behaviourally relevant signals are produced, transmitted and perceived (Figure 32.1).

SIGNAL PRODUCTION

A variety of different "sounds" have been described in bees (Armbruster, 1922; Hansson, 1945; Kirchner, 1993a, 1997). Most of them are characterised by a low fundamental frequency (300 to

Photo by MH

FIGURE 32.1 Acoustic communication in bees. Rhythmic thoracic oscillations produced by a bee (sender, S) are transmitted to potential receivers (R) as substrate vibrations (SV) and as air-particle oscillations (airborne sound, AS).

600 Hz, Table 32.1) and its harmonics (Woods, 1956; Simpson, 1963; Esch and Wilson, 1967; Michelsen *et al.*, 1986a, 1986b; Pratt *et al.*, 1996; Hrncir *et al.*, 2000, 2004a, 2004b; Seeley and Tautz, 2001; Aguilar and Briceño, 2002; Thom *et al.*, 2003) (Figure 32.2, also Figure 32.4 and Figure 32.5). Yet, in contrast to other insect taxa like cicadas and crickets, bees are not equipped with structures especially designed for the effective production of acoustic signals, a particular problem being their small size relative to the wavelength of the sound potentially emitted (Michelsen and Nocke, 1974). It has been speculated that bees produce sounds by a kind of "siren mechanism" (Woods, 1956), compressing air in the air sacs and releasing it through the abdominal spiracles (Woods, 1956; Lindauer and Kerr, 1958). Simpson (1963) refuted this theory of pneumatic signal production by demonstrating that the ventilation movements, expansions and contractions of the abdomen (Bailey, 1954) are not at all synchronised with the acoustic signals (piping) of honey-bee queens (*Apis mellifera*).

Thoracic Vibrations

Resonant sound-producing mechanisms occur in many insect groups (Bennet-Clark, 1999a). In bees, which do not possess a stridulation apparatus, a harp or a tymbal (Schneider, 1975), the "sounds" are generated by rhythmic thoracic oscillations (Figure 32.2). This is also true for the wing movements, which are generated with the help of an indirect wing mechanism (up- and down-strokes of the forewings are caused by rhythmic oscillations of the thorax, maintained by stretch activation of the antagonistic indirect flight muscles) (Pringle, 1957; Nachtigall, 2003). However, the oscillation frequency of the thorax is usually significantly higher during the presumed signal production than during flight (Armbruster, 1922; Hansson, 1945; Soltavalta, 1947; Esch and Wilson, 1967; King *et al.*, 1996). Experimentally, the wing beat frequency in bees during tethered flight can be increased by partially clipping or fully amputating the wings (Soltavalta, 1947, 1952; Lindauer and Kerr, 1958; Simpson, 1964; Esch and Wilson, 1967). This increase presumably is a result of a reduced damping of the oscillating system due to a reduced wing area (Esch, 1967a). "Signalling", however, is not necessarily equivalent to "flight with amputated wings". The motor pattern of the indirect flight muscles when producing "sounds" differs significantly from that during flight (Esch and Goller, 1991; King *et al.*, 1996). During signal production, an unequal shortening of the antagonistic indirect flight muscles and the resulting presumed closure of the scutal fissure are thought to alter the vibratory characteristics of the thorax in such a manner that the flight system vibrates at a higher frequency (King *et al.*, 1996).

A modulation of the fundamental frequency of the "sounds" has been described in honeybees (Esch, 1967a, 1967b; Esch and Wilson, 1967; Michelsen *et al.*, 1986a, 1986b; Pratt *et al.*, 1996; Seeley and Tautz, 2001; Thom *et al.*, 2003; see also Figure 32.5) and in stingless bees (Hrncir *et al.*, 2000; Figure 32.2). Increasing muscle contraction leads to an increasing stiffening of the thorax and thus to higher oscillating frequencies. A decrease of the damping of the oscillating system with decreasing wing area and wing-beat amplitude increases the frequency of the thoracic vibrations (Schneider, 1975). Hence, the highest frequency of the thoracic vibrations is seen when wings are completely folded (Esch and Wilson, 1967; Schneider, 1975). The fundamental frequency of a "sound" could thus be easily modulated by the bees through opening and closing the wings along with the signal (Seeley and Tautz, 2001).

Signal Transmission and Perception

To justify the terms "signal" and "communication", a crucial question has to be asked: Who understands these signals? The importance to identify potential recipients requires clarification of the physical nature of the signal and of the mechanisms of both transmission and perception of the signals.

TABLE 32.1

"Sounds" Produced by Social Bees: Honeybees (Apini), Stingless Bees (Meliponini), and Bumble Bees (Bombini)

Signal	Species	Main Frequency (Hz)	Signal Pattern	Sender	Possible Significance
Queen Signals					
Tooting	**Apini**				
	Apis mellifera [2,3,12,20,28,43]	350 to 500	P	Q[a]	Prevent hatching of further queens, trigger
	A. koschevnikovi [34]	2300 to 2900[c]	P		quacking
	A. cerana [34]	2000 to 3000[c]	P		
Quacking	A. mellifera [2,3,12,20,28,43]	300 to 350	P	Q[b]	Inform queen and workers about presence
	A. koschevnikovi [34]	330 to 3100[c]	P		and viability of confined queens
Defence Signals					
Hissing	**Bombini**				
	Bombus agrorum [11]	—	S		
	B. terrestris [13,22,36]	Broadband, up to 60,000	S	C	Aposematic warning signal directed against potential predators
Hissing (Shimmering)	**Apini**				
	A. florea [39,40]	Broadband, 90% of energy spectrum: 500 to 5000	S	C	Aposematic warning signal
	A. dorsata [24,39]	—	S		
	A. cerana [10,23,37]	Broadband, most energy: 300 to 3600	S		
Worker Piping	**Apini**				
	A. florea [40]	300 to 500	P	F	Triggers colony hissing
	A. mellifera [19,27,29,41]	300 to 550	S	T	Stop waggle dancers, facilitate recruitment of nectar receivers
	A. mellifera [38]	Frequency modulated: 100 to 2000	S	N	Stimulate swarm to prepare for lift-off
Recruitment Signals					
	Bombini				
	B. hypnorum [36]	—	—	F	Information about existence and quality of valuable food source
	B. terrestris [13,33]	—	P		

Bee species [reference of description]	Range of main frequency	Signal pattern	Sender	Information / possible significance
Meliponini				
Frieseomelitta sp. [18]	—	—		
Nannotrigona sp. [9,17,18,25,26]	—	P	F	Information about existence and quality of valuable food source; activation of additional foragers
Tetragonisca sp. [17,18,26]	300 to 350	—		
Cephalotrigona sp. [26]	—	—		
Scaptotrigona sp. [7,17,35]	200 to 500	P		
Plebeia sp. [26]	—	—		
Leurotrigona sp. [17,18]	—	—		
Meliponula sp. [9,17,18]	—	—		
Hypotrigona sp. [17,18]	—	—		
Melipona sp.	—	P	F	Activation of additional foragers, information about profitability of food source, information about food source location (?)
M. bicolor [32]	400 to 500	P		
M. costaricensis [1]	400 to 600	P		
M. mandacaia [32]	400 to 500	P		
M. panamica [30,31]	350 to 500	P		
M. quadrifasciata [7–9,14,24]	300 to 600	P		
M. rufiventris [Hrncir, unpublished]	250 to 600	P		
M. scutellaris [14]	350 to 550	P		
M. seminigra [7,9,15,16]	300 to 600	P		
Apini				
A. mellifera [4–6,27,42,44]	200 to 350	P	F	Increase attention of food receivers, information about profitability and location of food source
A. dorsata [21]	90 to 140	P		

For each type of potential signal, bee species [*reference of description*], range of *main frequency* (if described), and *signal pattern* (*S*, single pulse; *P*, pulse sequence), which bee emits the "sound" (*sender*: *C*, colony; *F*, forager; *N*, nest site scout; *T*, tremble dancer; *Q*, queen) and the possible significance of the signal are given.

[a] Hatched virgin queen.

[b] Virgin queens not hatched — confined in queen cells.

[c] The authors themselves (Otis et al., 1995) are in doubt about the frequencies measured.

[1] Aguilar and Briceño, 2002; [2] Armbruster, 1922; [3] Butler, 1609; [4] Esch, 1961a; [5] Esch, 1961b; [6] Esch, 1963; [7] Esch, 1967a; [8] Esch and Wilson, 1967; [9] Esch et al., 1965; [10] Fuchs and Koeniger, 1974; [11] Haas, 1961; [12] Hamson, 1945; [13] Heidelbach et al., 1998; [14] Hrncir et al., 2000; [15] Hrncir et al., 2004a; [16] Hrncir et al., 2004b; [17] Kerr, 1969; [18] Kerr and Esch, 1965; [19] Kirchner, 1993b; [20] Kirchner, 1997; [21] Kirchner and Dreller, 1993; [22] Kirchner and Röschard, 1999; [23] Koeniger and Fuchs, 1972; [24] Lindauer, 1956; [25] Lindauer and Kerr, 1958; [26] Lindauer and Kerr, 1960; [27] Michelsen et al., 1986a; [28] Michelsen et al., 1986b; [29] Nieh, 1993; [30] Nieh, 1998; [31] Nieh and Roubik, 1998; [32] Nieh et al., 2003; [33] Oeynhausen and Kirchner, 2001; [34] Otis et al., 1995; [35] Schmidt et al., 2004; [36] Schneider, 1972; [37] Schneider and Kloft, 1971; [38] Seeley and Tautz, 2001; [39] Seeley et al., 1982; [40] Sen Sarma et al., 2002; [41] Thom et al., 2003; [42] Waddington and Kirchner, 1992; [43] Wenner, 1962a; [44] Wenner, 1962b.

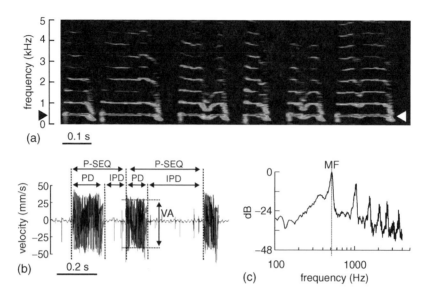

FIGURE 32.2 Characteristics of acoustic signals in bees demonstrated at an example of a recruitment signal by a stingless bee (*Melipona rufiventris*). Rhythmic thoracic oscillations (b) — measured directly on the thorax with a laser-vibrometer — show a low fundamental frequency (*e.g.* close to 500 Hz, arrowheads) and its harmonics up to several kHz (a, c). Bees emit either single pulses or a sequence of pulses as in the presented case (a, b). Measurable signal parameters include: duration of single pulses (PD), pulse amplitude (*e.g.* velocity amplitude, VA), main frequency (MF, 0 dB), duration of pauses between two subsequent pulses (IPD), pulse sequence (P-SEQ) and duty cycle (DC = PD/P-SEQ).

Substrate Vibrations

Transmission. The mechanism by which the thoracic vibrations of bees are transmitted to the substrate is most obvious during queen and worker piping in honeybees. Here, the signals are directly transmitted to the substrate through close contact between the vibrating thorax and the comb (Esch, 1964; Simpson, 1964; Michelsen *et al.*, 1986b; Pratt *et al.*, 1996; Thom *et al.*, 2003). However, even without such a close contact between the oscillating system and the substrate, signalling bees actually do generate measurable substrate vibrations (honeybees: Nieh and Tautz, 2000; stingless bees: Hrncir *et al.*, 2000; Nieh, 2000; bumble bees: Kirchner and Röschard, 1999), their legs acting as a mechanical link between thorax and substrate (Storm, 1998; Tautz *et al.*, 2001). In honeybees, the waggle movement was shown to improve the transmission of the signal from the vibrating thorax through the legs to the walls of the comb (Tautz *et al.*, 1996). Both the maximum application of force exerted by a bee on the rims of the cell walls and the production of the thoracic vibration occur when the bee is fully laterally displaced (Esch, 1961b; Storm, 1998). By increasing the effective mass and through a mechanical coupling to the substrate the bees optimise the conditions for injecting the energy of their thoracic vibrations into the comb at these turning points of the waggle movement (Tautz *et al.*, 2001).

Perception. Arthropods have a variety of mechanoreceptors capable of detecting substrate vibrations (Markl, 1973; Barth, 1986). In bees the vibration sensitivity has been predominantly attributed to the subgenual organ, a chordotonal organ in the proximal part of the tibia of each leg (Schön, 1911; Autrum and Schneider, 1948). This receptor responds to vibrations in the axial direction of the tibia (Figure 32.3a). When the leg accelerates due to substrate vibrations, the inertia causes the haemolymph, and the subgenual organ suspended therein, to lag behind the movement of the leg which mechanically stimulates the receptor cells (Kilpinen and Storm, 1997; Storm and

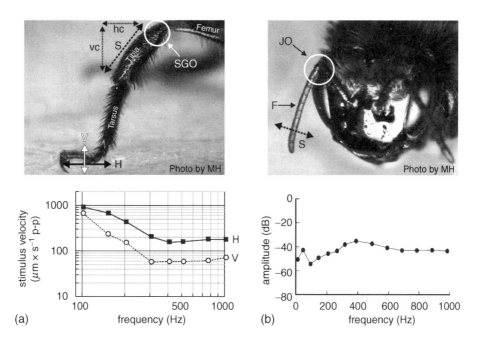

FIGURE 32.3 Some characteristics of potential receptors of acoustic signals. (a) Sensitivity for substrate vibrations have been predominantly attributed to the subgenual organ (SGO), which responds to vibrations in the axial direction of the tibia. This appropriate stimulus (S) consists of a horizontal component (hc) and a vertical component (vc). Legs of honeybees show a lower electrophysiological response threshold to vertical vibrations (V, open circles) than to horizontal vibrations (H, filled squares). Hence, the geometrical position of the leg (of the tibia) during signal reception is important for an individual's sensitivity to substrate vibrations. (After Rohrseitz, K. and Kilpinen, O., 1997 *Zoology*, 100, 80–84. With permission.) (b) Air particle oscillations induce vibrations of the antennal flagellum (F). These antennal vibrations (appropriate stimulus, S) stimulate the sensory cells of the Johnston's organ (JO). The antennal flagellum of honeybees shows a flat frequency response (vibration amplitude relative to stimulus amplitude, filled circles) without mechanical resonances. (After Kirchner, W. H., 1994 *J. Comp. Physiol. A*, 175, 261–265. With permission.)

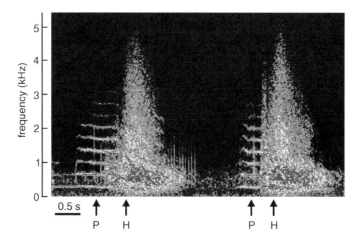

FIGURE 32.4 Piping and hissing in *Apis florea*. Sonogram of two consecutive alarm piping signals (P) by a forager after perceiving a potential danger in the vicinity of the colony (fundamental frequency of the piping about 300 Hz; note the harmonics up to several kHz). The forager's piping elicits a broadband hissing noise (H) by the colony members. (After Sen Sarma, M., Fuchs, S., Werber, C., and Tautz, J. *Zoology*, 105, 215–223, 2002. With permission.)

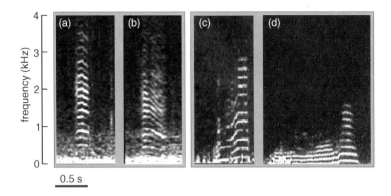

FIGURE 32.5 Worker piping in honeybees. (a, b) Sonograms of two piping signals emitted by tremble dancing nectar foragers showing the frequency modulation during the piping. (After Thom, C., Gilley, D. C., and Tautz, J. *Behav. Ecol. Sociobiol.*, 53, 199–205, 2003. With permission.) (c, d) Sonograms of two piping signals recorded from workers in a swarm shortly before lift-off. Note the rise in fundamental frequency and the appearance of numerous high frequency harmonics towards the end of each piping pulse. (After Seeley, T. D. and Tautz, J. *J. Comp. Physiol. A*, 187, 667–676, 2001. With permission.)

Kilpinen, 1998). Electrophysiologically, the highest sensitivity of the sensory cells is to vertical vibrations at frequencies between 150 and 900 Hz, with a response threshold of 0.06 to 0.15 mm/s peak–peak (Kilpinen and Storm, 1997). Indeed, both the fundamental frequency (honeybees: 200 to 500 Hz, Michelsen *et al.*, 1986a, 1986b; Kirchner, 1993b; Nieh and Tautz, 2000; stingless bees: 350 to 600 Hz, Hrncir *et al.*, 2000) and the amplitude of the substrate vibrations produced by bees (honey-bees: from 0.08 mm/s peak–peak, Nieh and Tautz, 2000; up to 6 mm/s peak–peak, Michelsen *et al.*, 1986b) overlap with this optimal range of the receptor. The average response threshold of the subgenual organ, however, increases by approximately 10 dB when the stimulus changes from vertical to horizontal (Rohrseitz and Kilpinen, 1997; Rohrseitz, 1998) (Figure 32.3a). Freely walking bees keep their legs in a variety of postures and the stimulus for the sensory cells (vibration in the axial direction of the tibia) certainly comprises both a vertical and a horizontal component (Figure 32.3a). Hence, the orientation of the leg of freely walking bees relative to the substrate (Figure 32.3a) is of great importance for a bee's sensitivity to substrate vibrations (Sandeman *et al.*, 1996; Rohrseitz and Kilpinen, 1997).

Airborne Sound

Transmission. Autrum (1936) pointed to the possibility that hair-like structures sensitive to weak air currents could serve as receptors of sound particle velocity. The bees' wings, which are not essential for the production of the "sounds" (Soltavalta, 1947; Lindauer and Kerr, 1958; Esch and Wilson, 1967; Schneider, 1975), are, however, of considerable importance for transforming thoracic vibrations into airborne sounds (Esch and Wilson, 1967; Michelsen *et al.*, 1987; Michelsen, 1993, 2003). It was demonstrated that the bees' wings act as an asymmetrical dipole emitter (Michelsen *et al.*, 1987; Michelsen, 1993): sound pressures above and below the plane of the wings are 180° out of phase, which generates pressure gradients of 1 Pa/mm (in the dorso-ventral direction) close to the edges of the wings. This creates a zone of intense air particle movements close to the abdomen of a bee (amplitude of particle movement: 1 m/s peak–peak, decreasing rapidly with distance from signalling bee; Michelsen *et al.*, 1987). Measurements of the air flow behind honeybees created during the waggle dance further indicate an intense motion of the air with peak velocities of about 15 cm/s at a distance of 1 mm from the dancer (Michelsen, 2003). As deduced from studies on a mechanical model of a dancing bee (for details, see Michelsen, 2003), an "air jet" is generated by the vibrating wings during "sound" production. Air moving out from the space between the wings and the abdomen during wing vibrations caused a jet air flow moving away behind the abdomen of the model bee. The direction of the air jet was constant and independent of the direction of the wing movement (Michelsen, 2003). In contrast to the rapid decrease in amplitude of the near field sound (Michelsen *et al.*, 1987), the amplitude of the air jet decreased more slowly and linearly with distance to the signal source (Michelsen, 2003).

Perception. Near the sound source most sound energy is in particle movement. Hence, flagellar structures such as hairs or antennae may function as particle velocity detectors (Bennet-Clark, 1971; Tautz, 1979, Barth, 1986; Eberl, 1999). It has long been demonstrated that bees are able to detect air currents by means of the Johnston's organ, a chordotonal organ in the pedicel of the antennae (McIndoo, 1922; Snodgrass, 1956). The sensory cells of this organ are stimulated by the deflection of the flagellum (distal part of the antenna). As in flies (Bässler, 1957), however, the Johnston's organ of bees was attributed primarily to the detection of flight speed (Heran, 1957, 1959). In a detailed analysis of the mechanical and physiological properties of the antennae of honeybees, Heran (1959) demonstrated that the Johnston's organ is most sensitive at oscillation frequencies of the antennae between 200 and 300 Hz. The author hypothesised that the bees might even be able to detect the recruitment sounds produced by dancing honeybees, which also range from 200 to 300 Hz (Esch, 1961b; Michelsen *et al.*, 1987).

The conversion of airborne acoustic energy into antennal vibrations, and the subsequent transmission of vibrations to Johnston's organs, has been described in flies (Eberl, 1999; Göpfert and Robert, 2001, 2002). Similar to the sound receiver system in flies (Göpfert and Robert, 2002), air particle oscillations deflect the flagellum of honeybees like a stiff rod (Kirchner, 1994) (Figure 32.3b). It has been suggested that airborne sound (air particle velocity) together with thoracic vibrations induce vibration of the receiver's antennae, and that the bees, similar to flies, do use the Johnston's organ to detect air particle oscillations evoked by the bees' "sounds". This mechanism of "sound" perception was proposed for honeybees (Kirchner *et al.*, 1991; Dreller and Kirchner, 1993, 1995), for stingless bees (Nieh *et al.*, 2003) and for bumble bees (Hensen and Kirchner, 2003).

BEHAVIOURAL RELEVANCE

As stated before, the term "acoustic" is misleading because "airborne sound" does not sufficiently describe the physical nature of the signals considered. On the other hand, the term "mechanical signals" would include too broad a variety of signals like tactile cues or even the honeybees' dance movements, which goes beyond the scope of this chapter. The signals described below (Table 32.1) have historically been denominated as "sounds" because they contain a component which the human ear can hear and which can be measured with a pressure microphone. We have already illustrated how thoracic oscillations are transformed into "signals" of a physical nature perceivable for bees: substrate vibrations and airborne sound (particle velocity or air flow). However, it has not been clarified yet whether only one (and which) of these two signal modalities is predominantly used by bees in the respective behavioural contexts.

ROYAL SIGNALS

One of the oldest accounts of acoustic signals in honeybees is on two easily distinguishable piping sounds produced inside the bees' nest (Butler, 1609). These signals are emitted by young queens during the process of swarming (Armbruster, 1922; Hansson, 1945). They are appreciably louder (at least to the human ear) than any other signal produced by bees inside the nest (von Frisch, 1965). Both signals are composed of a sequence of pulses and can be distinguished by different fundamental frequencies (Table 32.1) and different temporal structures (Hansson, 1945; Michelsen *et al.*, 1986b).

The piping emitted by emerged virgin queens is called "tooting" or "*Tüten*" (Armbruster, 1922; Hansson, 1945; Wenner, 1962a; von Frisch, 1965). In *Apis mellifera* tooting starts with one or two pulses of about 1 sec duration with an initial rise in both amplitude and frequency (Michelsen *et al.*, 1986b; Kirchner, 1993a, 1997). These first long pulses are followed by a variable number of short pulses of about 0.25 sec duration (Hansson, 1945; Michelsen *et al.*, 1986b; Kirchner, 1993a, 1997). The fundamental frequency rises from around 400 Hz on the day of emergence to more than 500 Hz 2 to 4 days after emergence, whereas the number of pulses decreases from about 17 to about 7 pulses per performance during the same period of time (Armbruster, 1922; Hansson, 1945; Michelsen *et al.*, 1986b).

When, after swarming, a surplus of queens is raised by the workers, the queen which first emerges from her cell announces her presence by tooting and also by release of pheromones (Winston, 1987). Mature queens still confined within their brood cells answer the tooting with a distinct piping sound, the so-called "quacking" or "*Quaken*". When several confined queens are present in the nest, a chorus of synchronised quacking follows each tooting (Armbruster, 1922; Hansson, 1945; Wenner, 1962a; von Frisch, 1965; Michelsen *et al.*, 1986b). The temporal structure of the quacking is very similar in all observed *Apis* species and has been described as a sequence of short pulses, typically less than 0.2 sec long (Kirchner, 1993a, 1997; Otis *et al.*, 1995). The fundamental frequency of quacking is approximately 350 Hz (Armbruster, 1922; Hansson, 1945; Michelsen *et al.*, 1986b) and thus lower than that of the tooting. Hansson (1945) offers two possible explanations for the lower frequency:

(i) The cell walls could have a damping effect on the thoracic vibrations.
(ii) The softness of the thoracic cuticle could influence its resonant properties and hardening of the bees' cuticle during the first days after hatching might also explain the increase in fundamental frequency of the queens' tooting with age (Hansson, 1945).

The behavioural context of tooting and quacking is the preparation for swarming (Simpson and Cherry, 1969; Winston, 1987). However, the exact messages carried by these signals and therefore their biological significance are still not fully understood (Michelsen, 1986b; Kirchner, 1997). The presence of a virgin queen in a colony usually suppresses the emergence of other queens (Winston, 1987). Substrate vibrations elicit a freezing reaction in worker bees which stop moving as long as the amplitude of the vibration is above threshold (Michelsen *et al.*, 1986a). This freezing reaction may also be observed within a few cm from a tooting queen (Michelsen *et al.*, 1986b). In the presence of a piping queen, workers stop chewing away the wax on capped queen cells preventing further queens from hatching (Winston, 1987). The piping dialogue may also inform both workers and emerged queens that other queens are present. A tooting queen may use the quacking response to evaluate both the number and viability of her competitors, and thus the risk of fighting and taking over the colony instead of leaving with a new swarm (Visscher, 1993).

COLONY DEFENCE

The successful defence of the nest, the brood and the stored resources is vital for colonies of social bees (Breed *et al.*, 2004) and requires the simultaneous action of many individuals (Michener, 1974; Sen Sarma *et al.*, 2002). Various mechanisms, some of which are highly sophisticated, have evolved to protect the nest against intruders (Michener, 1974).

In the behavioural context of alarm or colony defence, bumble bees emit a broadband hissing sound (Haas, 1961; Schneider, 1972, 1975; Heidelbach *et al.*, 1998; Kirchner and Röschard, 1999; Table 32.1) shown to function as an aposematic signal directed against potential predators and competitors for nesting sites such as mice (Kirchner and Röschard, 1999). The hissing is released by artificial vibrations of the nest and also by an increased CO_2 content of the air (Haas, 1961; Schneider, 1972; Kirchner and Röschard, 1999), indicating that it is a direct reaction to an animal intruder. In addition to hissing, bumble bees produce "sounds" like loud buzzing noises when defending their nest. These "sounds" were speculated to serve in communication between colony members (Wagner, 1907; Haas, 1961). However, conclusive experimental evidence for the biological significance of most "sounds" described in bumble bees is scarce (Dornhaus and Chittka, 2001).

In response to disturbances like a mechanical jolting or the appearance of a potential predator, colonies of various honeybee species (Table 32.1) also emit hissing sounds (Figure 32.4). They generate them by wing movements (Lindauer, 1956; Sakagami, 1960; Koeniger and Fuchs, 1972; Fuchs and Koeniger, 1974; Seeley *et al.*, 1982; Spangler, 1986; Sen Sarma *et al.*, 2002). The hissing behavior of honeybees has also frequently been called "shimmering" (Butler, 1954; Seeley *et al.*, 1982; Kastberger and Sharma, 2000), which describes the optical impression due to the coordinated, simultaneous wing movements of many individuals. The hissing sounds are broadband, mainly high frequency in nature (Koeniger and Fuchs, 1972; Fuchs and Koeniger, 1974; Spangler, 1986; Sen Sarma *et al.*, 2002) (Table 32.1; Figure 32.4). They were suggested to have evolved primarily for vertebrate ears and to represent warning signals emitted to deter or confuse potential predators (Schneider and Kloft, 1971; Fuchs and Königer, 1974; Sen Sarma *et al.*, 2002). In the giant honeybee (*A. dorsata*), the wing movements during hissing are also thought to hinder intruders, in particular wasps, from landing at the colony (Kastberger and Sharma, 2000). In *A. cerana* and *A. florea*, the temporary general suppression of colony activity often following hissing was proposed to make the colony less "suspicious" to predators (Fuchs and Koeniger, 1974; Sen Serma *et al.*, 2002). In order to activate and coordinate the simultaneous hissing of many individuals a trigger is required. Recently, it has been demonstrated that foragers of *A. florea*, upon their return to the colony, emit a warning piping (duration: 0.2 to 2.0 sec; fundamental frequency: 300 to 480 Hz) after having perceived a potential danger in the vicinity of a colony (Figure 32.4). Immediately after the piping, individuals close to the piping bee start hissing (Figure 32.4). The hissing action spreads rapidly to neighbours and finally involves most of the colony (Sen Sarma *et al.*, 2002). Both the close temporal relationship between piping by a forager and subsequent hissing by the entire colony and the absence of hissing when foragers were deterred from returning to the colony indicate that in *A. florea* the piping of a forager triggers the colony's hissing behaviour (Sen Sarma *et al.*, 2002).

WORKER PIPING IN HONEYBEES

"Worker piping" ("Piepen") was first described by Armbruster (1922), who noticed its similarity to the piping of honeybee queens (see above). However, the term "worker piping" does not describe a distinct signal, but rather represents a variety of obviously different signals produced by different types of workers in different behavioural contexts (Örösi-Pál, 1932; Esch, 1964; Ohtani and Kamada, 1980; Kirchner, 1993b; Nieh, 1993; Pratt *et al.*, 1996; Seeley and Tautz, 2001; Thom *et al.*, 2003; Table 32.1). We have already described the case of *A. florea* (see above). Here, we shall outline other situations in which behavioural responses to "worker piping" were observed.

Piping by dance followers. The first description of a behavioural reaction to worker piping was given by Esch (1964). He observed that hive bees attending the waggle dance of a honeybee forager stopped the dancing bee and surrounding dance followers by emitting a short piping pulse (duration: 0.1 to 0.2 sec; fundamental frequency: 300 to 400 Hz). The piper subsequently received a food sample from the dancing forager. Because of this behavioural response to the piping signal (delivering of a food sample), Michelsen *et al.* (1986a) referred to this form of worker piping by dance followers as a "begging signal". More recent studies (Nieh, 1993; Thom *et al.*, 2003) did not confirm the observation by Esch (1964) and argued that those piping dance followers might be nectar foragers which may have performed tremble dances (a behavioural context associated with worker piping; see below) at other times during their stay in the hive (Nieh, 1993; Thom *et al.*, 2003).

Piping by tremble dancing nectar foragers. According to a detailed study of the short piping signals of honeybee workers which explored the context of its occurrence and the identity of piping workers, almost 100% of the piping pulses (duration: 0.05 to 0.7 sec; fundamental frequency: 300 to 550 Hz; Figure 32.5a and b) inside a nest were shown to be emitted by tremble dancing nectar foragers (Thom *et al.*, 2003). Seeley (1992) demonstrated that nectar foragers, which had searched a long time for a food-storing bee inside the nest, performed tremble dances rather than waggle dances, which had a positive effect on the colony's nectar processing capacity by recruiting hive bees as additional nectar receivers. What could be the significance of the piping signals in the context of tremble dancing? Waggle dancing nectar foragers that received piping signals from tremble dancers stop dancing and leave the dance floor (Nieh, 1993). The piping signal therefore seems to inhibit waggle dancing, and thus the recruitment to nectar sources (Nieh, 1993; Kirchner, 1993b; Thom *et al.*, 2003). Yet, if piping serves merely to stop waggle dancing, then piping bees should produce their signals primarily on the dance floor, where the probability of encountering a waggle dancer is highest. However, about 50% the piping signals by tremble dancers are produced off the dance floor and predominantly directed at nonforaging bees (Thom *et al.*, 2003). The hypothesis is that piping also serves as a modulatory signal, which alters the response threshold of the signal recipient towards other stimuli (Hölldobler and Wilson, 1990). Hence, piping by tremble dancers might lower the response threshold of nestmates to the tremble dance itself and subsequently facilitate the recruitment of additional nectar receivers by the tremble dance (Thom *et al.*, 2003).

Piping by nest-site scouts. Prior to the coordinated takeoff of a swarm cluster for the new nest site, bees inside this cluster emit a high-pitched piping (Lindauer, 1955; Esch, 1964; Seeley *et al.*, 1979; Camazine *et al.*, 1999; Seeley and Tautz, 2001). The piping pulses were reported to have a duration between 0.1 sec and almost 2.0 sec and to be strongly frequency modulated, changing from a relatively pure, low-frequency tone (100 to 200 Hz) to a mixed, higher frequency sound (200 to 2000 Hz, including harmonics; Figure 32.5c and d). This piping signal that differs acoustically from the other forms of piping described above is predominantly emitted by nest-site scouts which press their thorax onto other bees during signalling (Seeley and Tautz, 2001). Camazine *et al.* (1999) suggested that these signals trigger the coordinated lift-off of the swarm cluster. Seeley and Tautz (2001), however, suggest a different significance for the piping. When piping bees were experimentally excluded from a swarm, the swarm bees did not warm up, which is essential for flight preparation prior to swarming (Heinrich, 1981). And indeed, the time course of the piping matches that of swarm warming, both starting at a low level about an hour before lift-off and building to a climax at lift-off (Seeley and Tautz, 2001). Therefore, the function of the piping by nest-site scouts may be to stimulate bees to prepare for lift-off rather than to trigger the coordinated lift-off.

RECRUITMENT TO FOOD SOURCES

A fascinating aspect of the biology of social insects is their ability to recruit nestmates to food sources (Wilson, 1971). Recruitment and communication mechanisms involved in this process increase the energetic expenses of the individual (Lewis, 1984), but also enhance the colony's efficiency in exploiting a food source and regulate its energy intake (Seeley, 1995). To understand a recruitment process, it is essential to know which signals are used to alert nestmates and to stimulate them to leave the nest, and which signals (if any) are used to guide recruits to a food source. Because the concept of symbolic communication during recruitment communication has received special interest over the past decades, the "sounds" produced during the recruitment process are among the best studied signals in bees (especially in honeybees and stingless bees; Table 32.1).

In bumble bees, an alerting mechanism which communicates the presence and the odour of a food source, but not its location, has been demonstrated in *Bombus terrestris* (Dornhaus and Chittka, 1999, 2001) and *B. transversalis* (Dornhaus and Cameron, 2003). Although food scent (Dornhaus and Chittka, 1999), pheromones (Dornhaus *et al.*, 2003) and agitated running by foragers (Dornhaus and Chittka, 2001) seem to be the signals predominantly responsible for alerting recruits, "sounds" are produced as well by foragers on entering and leaving the nest (Schneider, 1972; Heidelbach *et al.*, 1998; Oeynhausen and Kirchner, 2001). In *Bombus terrestris*, the number of "leaving sounds" correlates well with the sucrose concentration of the collected sugar solution (Oeynhausen and Kirchner, 2001). These "sounds" therefore possibly represent a recruitment signal, informing unemployed foragers about the existence of a valuable food source (Oeynhausen and Kirchner, 2001). However, only further studies will elucidate the true meaning of these "sounds" for the hive bees.

In many genera of stingless bees, "sounds" produced by foragers have been suggested to provide a mechanism of arousal used to activate nestmates to collect food. When returning from a food source foragers of *Nannotrigona testaceicornis* produce pulsed "sounds" while running agitatedly through the nest. Nestmates respond to this behaviour with buzzing and they leave the nest in a random (not directed) search for food (Lindauer and Kerr, 1958, 1960; Kerr and Esch, 1965; Kerr, 1969; Hubbel and Johnson, 1978). A similar arousal by "sounds" and an agitated zigzag running of foragers have been suggested for various other genera of stingless bees: *Tetragonisca*, *Cephalotrigona*, *Scaptotrigona*, *Plebeia* (Lindauer and Kerr, 1960), *Frieseomelitta*, *Leurotrigona*, *Meliponula* and *Hypotrigona* (Kerr and Esch, 1965; Kerr, 1969).

Bees of the genus *Melipona* are the most extensively studied stingless bees concerning recruitment behaviour and potential recruitment signals. As in the genera described above, foragers of *Melipona* emit "sounds" composed of a sequence of pulses when returning from a successful foraging trip (Lindauer and Kerr, 1958; Esch *et al.*, 1965; Esch, 1967a; Nieh, 1996; Nieh and Roubik, 1998; Hrncir *et al.*, 2000, 2004a, 2004b; Aguilar and Briceño, 2002; Nieh *et al.*, 2003) (Figure 32.2 and Figure 32.6). Because of the morphological resemblance of *Melipona* to honey-bees and the suggested evolution of the honeybee's language from a "more primitive" form of communication thought to be seen in the "sounds" of *Melipona* (Esch *et al.*, 1965; Kerr and Esch, 1965; Nieh *et al.*, 2003), investigations focused on the search for spatial information about food sources within these "sounds". Esch *et al.* (1965) and Esch (1967a) were the first to quantify the "sounds" produced by foragers of *M. quadrifasciata* and *M. seminigra*. According to these authors the duration of the "sound" pulses correlates positively with the distance of a food source. A similar correlation of pulse duration with food source distance was described in *M. panamica* (Nieh, 1996, cited in Kirchner, 1997; Nieh and Roubik, 1998). In addition, it has been speculated that these bees even code the elevation of a food source in their "sound" signals. The duration of the pause between two subsequent pulses (Nieh, 1996) and the duration of the longest pulse during a performance (Nieh and Roubik, 1998) correlated with the height of a food source above ground level. In spite of such correlations, however, there is no satisfying evidence that the "sounds" in *Melipona* are related causally to the location of a food source, nor that they are used by hive bees to locate the food source. So far, experiments have only demonstrated that the foragers' signals are followed by an increasing number of bees leaving the nest (Lindauer and Kerr, 1958; Esch, 1967a; Pereboom and Sommeijer, 1993) or by an increasing number of bees arriving at a food source (Lindauer and Kerr, 1958, 1960; Nieh and Roubik, 1995; Jarau *et al.*, 2000, 2003; Nieh *et al.*, 2003).

Recent studies strongly favour an alternative interpretation of the signals produced by foragers in stingless bees. These suggest that the temporal pattern of "sounds" represent the energy budget of the foraging trip rather than the location of a food source. In most *Melipona* species studied so far, the temporal pattern of the "sounds" emitted by foragers inside the nest (duration of single pulses, duration of pause between two subsequent pulses, duty cycle; Figure 32.2b) is strongly influenced by the foragers' energy intake (sugar concentration) at the food source (Aguilar and Briceño, 2002; Hrncir *et al.*, 2002, 2004b; Nieh *et al.*, 2003) (Figure 32.6b and c) and to some extent by the energy spent during a foraging trip (Hrncir *et al.*, 2002, 2004a). The view that the "sounds" of *Melipona* contain spatial information about a food source is additionally weakened by experiments with *M. seminigra* showing that the visual flow (lateral image motion that bees experience during flight)

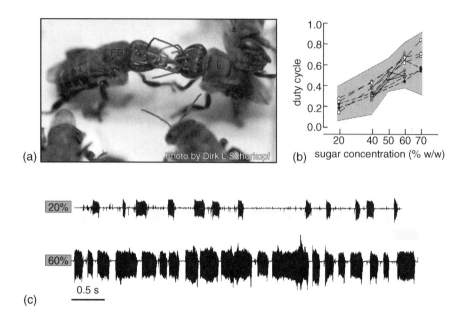

FIGURE 32.6 (a–c) Recruitment signals in stingless bees of the genus *Melipona*. (a) During trophallactic contacts (forager, F; food receiver, FR) foragers emit acoustic recruitment signals (photo showing a trophallaxis in *M. scutellaris*). (b, c) The temporal structure of these pulsed thoracic vibrations depends highly on the sugar concentration of the collected food. (b) Dependence of the duty cycle on the sugar concentration in different individuals (different symbols) in *M. seminigra*. (After Hrncir, M., Jarau, S., Zucchi, R., and Barth, F. G. *J. Comp. Physiol., 2004 A*, 190, 549–560. With permission.) (c) Differences in the temporal structure as a function of sugar concentration in *M. rufiventris*. Typical example of a recording of signals produced by an individual when collecting at a 20% and a 60% (weight on weight) sugar solution (Hrncir, unpublished data).

used by foragers to estimate the distance of a food source (Hrncir *et al.*, 2003) does not affect the temporal pattern of the "sounds" (Hrncir *et al.*, 2004a).

Stingless bees need information on the profitability of a food source for their decision on whether to forage or not (Biesmeijer *et al.*, 1998). Signals reflecting the energy budget of a foraging trip and thus the profitability of a food source therefore convey an important message. And indeed, the recruitment success of different *Melipona* species is influenced by the sugar concentration at the food source (Biesmeijer and Ermers, 1999; Jarau *et al.*, 2000; Nieh *et al.*, 2003). Yet it remains to be investigated whether the observed differences in the temporal pattern of the acoustic signals of bees collecting at food sources of different profitability really account for these differences in recruitment success.

In honeybees, the waggle dance offers an intriguing example for the study of a symbolic language in social insects. The orientation of the bees' wagging movements and their duration are highly correlated with the direction and distance of a food source (*e.g.* Lindauer, 1956; von Frisch, 1965; Dyer and Seeley, 1991). In those species which communicate in the dark (nesting in dark cavities: *A. mellifera*, *A. cerana*; nocturnal activity: *A. dorsata*), pulsed "sounds" accompany the waggle phase of the dance (Esch, 1961a, 1961b; Wenner, 1962b; Kirchner and Dreller, 1993). Two pulses are emitted during one waggle movement, predominantly at each turning point (Rohrseitz, 1998; Storm, 1998) (Figure 32.7). Hence, the duration of the waggle as well as the total number of "sound" pulses are correlated with the distance of a food source (Esch, 1961b; Wenner, 1962b; Kirchner and Dreller, 1993). It was therefore assumed that the signals emitted during waggling transmit information about the distance of a food source to the bees following the dance within the darkness of the hive (Esch, 1961b; Wenner, 1962b; Kirchner and Dreller, 1993; Kirchner

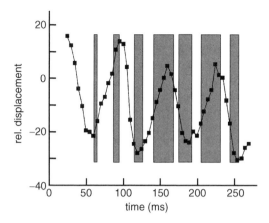

FIGURE 32.7 Acoustic signals during the waggle dance of honeybees. Sinusoid curve (filled squares) shows the displacement of waggle movements by a honeybee forager; shaded bars indicate the time of signal emission. It is demonstrated that dancing honeybees emit acoustic signals close to each turning point of the waggle movement, which agrees with the idea that the bees make use of the waggle movements for a better mechanical coupling to the substratum for the purpose of signal transmission (Rohrseitz and Storm, unpublished).

et al., 1996; Dyer, 2002). However, although the total time of the signal is clearly correlated with the distance of a food source, the temporal pattern of the "sounds" (duration of single pulses, duration of pauses, pulse sequence) is not (Esch, 1961b; Wenner, 1962b). It rather depends on the sugar concentration of the collected food (Esch, 1963; Waddington and Kirchner, 1992). Thus, the "sounds" produced during the honeybees' dance may have three different meanings to hive bees. First, they might serve to increase the attention of food receivers (Farina, 2000) and attract dance followers to the dancing forager and thus enhance the effectiveness of the dance itself (Tautz, 1996; Tautz and Rohrseitz, 1998; Dyer, 2002). Second, the "sounds" (total duration) which are strictly limited to the duration of the waggle phase may provide the dance followers with information about the distance to a food source. And third, the temporal pattern of the signal, which is strongly influenced by the energy intake of the foragers (similar to stingless bees), informs about the profitability of a food source. The "sounds" produced during the waggle dance clearly increase the recruitment success of a forager (Esch, 1963; Tautz, 1996). When foraging honeybees dance silently (without producing "sounds"), which occasionally happens, they fail to recruit nestmates to the food source where they are collecting (Esch, 1963). Therefore, these "sounds" emitted during dancing are significantly involved in the successful exploitation of a food source.

THE SENSORY WORLD OF BEES

Social insects show a high degree of coordination between individuals and a correspondingly complex signal system (Otte, 1974). Continuing communication, the spreading and exchanging of information among individuals is imperative for a colony to function as a single, purposeful unit (Hölldobler, 1984; Seeley, 1995, 1998). It has long fascinated mankind how it is at all possible that hundreds or even thousands of individuals can work together with such coherence. In bee colonies, chemical and tactile signals, substrate vibrations and airborne sounds interact in complex ways (Winston, 1987; Seeley, 1998). Chemical signals, like the queen pheromone, are suited to transmit messages to all colony members that persist over a long time period (*e.g.* the presence of the queen). Mechanical signals such as the substrate vibrations and airborne sounds illustrated in this chapter, however, provide important advantages over chemical signals in those behavioural contexts where a temporal coding of changing information is essential, as is the case during the exploitation of food sources. Also, the information contained in substrate vibrations and airborne sound signals remains restricted to those individuals which stay within a certain range around the signaller. In this way, signals are only perceived by those who are supposed to "listen", like potential recruits.

Knowledge has certainly been advanced considerably since the first description of "sounds" produced by bees (Butler, 1609). Yet, we are still only beginning to understand the significance of these substrate vibrations and airborne sound signals. In no case is the entire chain of information flow (sender–transmission channel–receiver) fully understood. A large area of the bees' acoustic world still waits to be discovered.

ACKNOWLEDGEMENTS

This work was financially supported by the Austrian Employment Office (MH), by grant P17530 of the Austrian Science Foundation (FGB) and the Deutsche Forschungsgemeinschaft, grants SFB 554, Ta 82/7-1 and 82/7-2 (JT).

References

Abbott, J. C. and Stewart, K. W. (1993). Male search behavior of the stonefly *Pteronarcella badia* (Hagen) (Plecoptera: Pteronarcyidae) in relation to drumming, *Journal of Insect Behaviour* **6:** 467–481.

Acker, T. S. (1966). Courtship and mating behavior in *Agulla* species (Neuroptera: Raphidiidae), *Annals of the Entomological Society of America* **59:** 1–6.

Adams, P. A. (1962). A stridulatory structure in Chrysopidae (Neuroptera), *Pan-Pacific Entomologist* **38:** 178–180.

Aguilar, I. and Briceño, D. (2002). Sounds in *Melipona costaricensis* (Apidae: Meliponini): effect of sugar concentration and nectar source distance, *Apidologie* **33:** 375–388.

Ahi, J., Kalmring, K., Ebendt, R., and Hellweg, J. (1993). Physiology and central projection of auditory receptor cells in the prothoracic ganglion of three related species of bushcrickets (Tettigoniidae, Insecta), *Journal of Experimental Zoology* **265:** 684–692.

Ahman, I. and Ekbom, B. S. (1981). Sexual behaviour of the greenhouse whitefly (*Trialeurodes vaporariorum*):Orientation and courtship, *Entomologia Experimentalis et Applicata* **29:** 330–338.

Aicher, B., Markl, H., Masters, W. M., and Kirschenlohr, H. L. (1983). Vibrational transmission through the walking legs of the fiddler crab, *Uca pugilator* (Brachyura, Ocypodidae) as measured by laser-Doppler vibrometry, *Journal of Comparative Physiology* **150:** 483–491.

Aicher, B. and Tautz, J. (1990). Vibrational communication in the fiddler crab *Uca pugilator*, *Journal of Comparative Physiology* **166:** 345–353.

Aidley, D. J. and White, D. S. C. (1969). Mechanical properties of glycerinated fibers from the tymbal muscles of a Brazilian cicada, *Journal of Physiology* **205:** 179–192.

Aiken, R. G. (1982). Sound production and mating in a waterboatman, *Palmacorixa nana* (Heteroptera: Corixidae), *Animal Behaviour* **30:** 54–61.

Aiken, R. B. (1985). Sound production by aquatic insects, *Biological Reviews of the Cambridge Philosophical Society* **60:** 163–211.

AKG Acoustics, The ABC's of AKG: Microphone Basics and Fundamentals of Usage.

Akingbohungbe, A. E. (1979). A new genus and four new species of Hyaliodinae (Heteroptera: Miridae) from Africa with comments on the status of the subfamily, *Revue de Zoologie Africaine* **93**(2)**:** 500–522.

Aldrich, J. R., Oliver, J. E., Lusby, W. R., Kochansky, J. P., and Lockwood, J. A. (1987). Pheromone strains of the cosmopolitan pest *Nezara viridula* (Heteroptera: Pentatomidae), *Journal of Experimental Zoology* **224:** 171–175.

Alexander, R. D. (1956). A comparative study of sound production in insects, with special reference to the singing Orthoptera and Cicadidae of the eastern United States, Unpubl. Ph.D. Thesis, Ohio State Univ. Columbus, p. 529.

Alexander, R. D. (1957a). Sound production and associated behavior in insects, *The Ohio Journal of Science* **57**(2)**:** 101–113.

Alexander, R. D. (1957b). The taxonomy of the field crickets of the eastern United States (Orthoptera: Gryllidae: Acheta), *Annals of the Entomological Society of America* **50:** 584–602.

Alexander, R. D. (1960). Sound communication in Orthoptera and Cicadidae, In *Animal Sounds and Communication*, Lanyon, W. and Tavolga, R., Eds., AIBS Publications, New York, pp. 38–92.

Alexander, R. D. (1967). Acoustical communication in arthropods, *Annual Review of Entomology* **12:** 495–526.

Alexander, R. D. and Bigelow, R. S. (1960). Allochronic speciation in field crickets, and a new species, *Acheta veletis*, *Evolution* **14:** 334–346.

Alexander, R. D. and Moore, T. E. (1958). Studies on the acoustical behavior of seventeen-year cicadas (Homoptera: Cicadidae: Magicicada), *The Ohio Journal of Science* **58:** 107–127.

Alexander, R. D. and Moore, T. E. (1962). The evolutionary relationships of 17-year and 13-year cicadas, and three new species (Homoptera, Cicadidae, Magicicada), *Miscellaneous Publications Museum of Zoology* **121:** 1–59.

Alexander, R. D., Moore, T. E., and Woodruff, R. E. (1963). The evolutionary differentiation of stridulatory signals in beetles, *Animal Behaviour* **11:** 111–115.

Alexander, K. D. and Stewart, K. W. (1996a). Description and theoretical considerations of mate finding and other adult behaviors in a Colorado population of *Claassenia sabulosa* (Banks) (Plecoptera: Perlidae), *Annals of the Entomological Society of America* **89:** 290–296.

Alexander, K. D. and Stewart, K. W. (1996b). The mate searching behavior of *Perlinella drymo* (Newman) (Plecoptera: Perlidae) in relation to drumming on a branched system, *Bulletin of the Swiss Entomological Society* **69:** 121–126.

Alexander, K. D. and Stewart, K. W. (1997). Further considerations of mate searching behavior and communication in adult stoneflies (Plecoptera); first report of tremulation in *Suwallia* (Chloroperlidae), In *Ephemeroptera and Plecoptera Biology–Ecology–Systematics*, Landolt, P. and Sartori, M., Eds., Mauron + Tinguely and Lachat S. A., Fribourg, pp. 107–112.

Allen, A. A. (1993). The site of stridulation in *Geotrupes* (Geotrupidae): an unsolved problem?, *Coleopterist* **2(2):** 56.

Allen, G. R. (1995). The biology of the phonotactic parasitoid, *Homotrixa* sp., (Diptera: Tachinidae), and its impact on the survival of male *Sciarasaga quadrata* (Orthoptera: Tettigoniidae) in the field, *Ecological Entomology* **20:** 103–110.

Allen, G. R. (1998). Deil calling and field survival of the bushcricket *Sciarasaga quadrata* (Orthoptera: Tettigoniidae): a role for sound locating parasitic flies?, *Ethology* **104:** 645–660.

Allen, G. R. and Hunt, J. (2000). Larval competition, adult fitness, and reproductive strategies in the acoustically orienting ormine *Homotrixa alleni* (Diptera: Tachinidae), *Journal of Insect Behaviour* **14:** 283–297.

Alonso-Pimentel, H., Spangler, H.G., and Heed, W. (1995). Courtship sounds and behaviour of the two sanguaro-breeding Drosophila and their relatives, *Animal Behaviour* **50:** 1031–1039.

Alpern, S. and Gal, S. (2003). *The Theory of Search Games Rendezvous*, Kluwer, Boston, p. 315.

Amon, T. (1990). Electrical brain stimulation elicits singing in the bug *Nezara viridula*, *Naturwissenschaften* **77:** 291–292.

Amon, T. and Čokl, A. (1990). Transmission of the vibratory song of the bug *Nezara viridula* (Pentatomidae, Heteroptera) on the *Hedera helix* plant. *Scopolia* **(Suppl. 1):** 133–141.

Arak, A. and Enquist, M. (1995). Conflict, receiver bias and the evolution of signal form, *Philosophical Transactions of the Royal Society of London B Biological Sciences* **349:** 337–344.

Armbruster, L. (1922). Über Bienentöne, Bienensprache und Bienengehör, *Archiv für Bienenkunde* **4:** 221–259.

Arrow, G. J. (1899). Notes on the classification of the coleopterous family Rutelidae, *Annals and Magazine of Natural History* **4:** 363–370.

Arrow, G. J. (1904). Sound production in Lamellicorn beetles, *Transactions of the Entomological Society of London* **IV:** 709–749.

Arrow, G. J. (1924). Vocal organs in the coleopterous families Dytiscidae, Erotylidae, and Endomychidae, *Transactions of the Entomological Society of London* **72:** 134–143.

Arrow, G. J. (1942). The origin of stridulation in beetles, *Proceedings of the Royal Entomological Society of London Series A* **17:** 83–86.

Arrow, G. J. (1951). *Horned Beetles — a Study of the Fantastic in Nature*, W. Junk, The Hague, p. 154.

Asada, N., Fujiwara, K., Ikeda, H., and Hihara, F. (1992). Mating behavior in three species of the *Drosophila hypocausta* subgroup, *Zoological Science* **9:** 397–404.

Ashlock, P. D. and Lattin, J. D. (1963). Stridulatory mechanisms in the Lygaeidae, with a new american genus of Orsillinae (Hemiptera: Heteroptera), *Annals of the Entomological Society of America* **56:** 693–703.

Aspi, J. (2000). Inbreeding and outbreeding depression in male courtship song characters in *Drosophila montana*, *Heredity* **84:** 273–282.

Aspöck, U. (2002). Phylogeny of the Neuropterida (Insecta: Holometabola), *Zoologica Scripta* **31:** 51–55.

Aspöck, U., Plant, J. D., and Nemeschkal, H. L. (2001). Cladistic analysis of Neuroptera and their systematic position within Neuropterida (Insecta: Holometabola: Neuropterida: Neuroptera), *Systematic Entomology* **26:** 73–86.

Assem, J. van. and Povel, G. D. E. (1973). Courtship behaviour of some *Muscidifurax* species (Hymenoptera: Pteromalidae): a possible example of a recently evolved ethological isolating mechanism, *Netherlands Journal of Zoology* **30**: 307–325.

Assem, J. van. and Werren, J. H. (1994). A comparison of the courtship and mating behaviour of three species of Nasonia (Hymenoptera: Pteromalidae), *Journal of Insect Behaviour* **7**: 53–66.

Au, E. C., (1969). The taxonomic value of the metathoracic wing in the Scutelleridae (Hemiptera: Heteroptera). MA Thesis, Ore. State Univ. p. 129.

Autrum, H. (1936). Über Lautäußerungen und Schallwahrnehmung bei Arthropoden. I. Untersuchungen an Ameisen. Eine allgemeine Theorie der Schallwahrnehmung bei Arthropoden, *Zeitschrift fur Vergleischende Physiologie* **23**: 332–373.

Autrum, H. (1941). Über Gehör- und Erschütterungssinn bei Locustiden, *Zeitschrift fur Vergleischende Physiologie* **28**: 580–637.

Autrum, H. and Schneider, W. (1948). Vergleichende Untersuchungen über den Erschütterungssinn der Insekten, *Zeitschrift fur Vergleischende Physiologie* **31**: 77–88.

Avise, J. C. and Walker, D. (1998). Pleistocene phylogeographic effects on avian populations and the speciation process, *Proceedings of the Royal Society of London B* **265**: 457–463.

Axen, A. H. and Pierce, N. E. (1998). Aggregation as a cost-reducing strategy for lycaenid larvae, *Behavioral Ecology* **9**: 109–115.

Bacher, S., Casas, J., and Dorn, S. (1996). Parasitoid vibrations as potential releasing stimulus of evasive behaviour in a leafminer, *Physiological Entomology* **21**: 33–43.

Bacher, S., Casas, J., and Dorn, S. (1997). Substrate vibrations elicit defensive behaviour in leafminer pupae, *Journal of Insect Physiology* **43**: 945–952.

Bailey, L. (1954). The respiratory currents in the tracheal system of the adult honey-bee, *Journal of Experimental Biology* **31**: 593–598.

Bailey, W. J. (1976). Species isolation and song types of the genus *Ruspolia* Schulthess in Uganda, *Journal of Natural History* **10**: 511–528.

Bailey, W. J. (1978). Resonant wing systems in the Australian whistling moth Hecatesia (Agarasidae, Lepidoptera), *Nature* **272**: 444–446.

Bailey, W. J. (1979). A review of Australian Copiphorini (Orthoptera: Tettigoniidae: Conocephalinae), *Australian Journal of Zoology* **27**: 1015–1049.

Bailey, W. J. (1985). Acoustic cues for female choice in bushcrickets (Tettigoniids), In *Acoustical and Vibrational Communication in Insects*, Kalmring, K. and Elsner, N., Eds., Paul Parey, Berlin, Hamburg, pp. 101–110.

Bailey, W. J. (1991). *The Acoustic Behaviour of Insects: An Evolutionary Perspective*, Chapman & Hall, London, p. 225.

Bailey, W. J. (2003). Insect duets: underlying mechanisms and their evolution, *Physiological Entomology* **28**: 157–174.

Bailey, W. J., Ager, E. I., O'Brien, E. K., and Watson, D. L. (2003). Visual and acoustic searching strategies of female bushcrickets (Orthoptera: Tettigoniidae): will females remain faithful to a male who stops calling?, *Physiological Entomology* **28**: 209–214.

Bailey, W. J. and Field, G. (2000). Acoustic satellite behaviour in the Australian bushcricket *Elephantodeta nobilis* (Phaneropterinae; Tettigoniidae; Orthoptera), *Animal Behaviour* **39**: 361–369.

Bailey, W. J. and Haythornthwaite, S. (1998). Risks of calling by the field cricket *Teleogryllus oceanicus*; potential predation by Australian long-eared bats, *Journal of Zoology London* **244**: 505–513.

Bailey, W. J. and Ridsdill-Smith, J. (1991). *Reproductive Behaviour of Insects*, Chapman and Hall, New York.

Bailey, W. J. and Robinson, D. (1971). Song as a possible isolating mechanism in the genus *Homorocoryphus* (Tettigonioidea, Orthoptera), *Animal Behaviour* **19**: 390–397.

Bailey, W. J., Withers, P. C., Endersby, M., and Gaull, K. (1993). The energetic costs of calling in the bushcricket *Requena verticalis* (Orthoptera: Tettigoniidae: Listroscelidinae), *Journal of Experimental Biology* **178**: 21–37.

Ball, E. and Field, L. H. (1981). Structure of the auditory system of the weta *Hemideina crassidens* (Blanchard, 1851) (Orthoptera, Ensifera, Gryllacridoidea, Stenopelmatidae), *Cell and Tissue Research* **217**: 321–343.

Balme, D. M., Ed. (1992). *Aristotle's De partibus animalium I and De generatione animalium I (with passages from II. 1-3). Translated with notes by D.M. Balme*, Reprint with new material. Clarendon Press, Oxford, p. 183.

Balthasar, V. (1964). *Monographie der Scarabaeidae und Aphodiidae der palaearktischen und orientalischen Region (Coleoptera: Lamellicornia)*, Vol. 3: Verlag der Tschechoslowakischen Akademie der Wissenschaften, Praha, pp. 1–652.

Bangert, M., Kalmring, K., Sickmann, T., Stephen, R., Jatho, M., and Lakes, R. (1998). Stimulus transmission in the auditory receptor organs of the foreleg of bushcrickets (Tettigoniidae). I. The role of the tympana, *Hearing Research* **115:** 27–38.

Barendse, W. (1990). Speciation in the genus Mygalopsis in southwestern Australia: A new look at distinguishing modes of speciation, In *The Tettigoniidae: Biology, Systematics and Evolution*, Bailey, W. J. and Rentz, D. C. F., Eds., Crawford House Press, Bathurst and Springer, Berlin, pp. 265–279.

Barr, T. C. Jr. (1968). Cave ecology and the evolution of troglobites, *Evolutionary Biology* **2:** 35–102.

Barr, B. A. (1969). Sound production in Scolytidae (Col.) with emphasis on genus *Ips*, *The Canadian Entomologist* **101**(6): 636–672.

Barrientos, L. and Den Hollander, J. (1994). Acoustic signals and taxonomy of Mexican Pterophylla (Orthoptera: Tettigoniidae: Pseudophyllinae), *Journal of Orthoptera Research* **2:** 35–40.

Barrientos-Lozano, L. (1998). Mate choice and hybridization experiments between allopatric populations of Pterophylla beltrani Bolivar and Bolivar and P. robertsi Hebard (Orthoptera: Tettigoniidae: Pseudophyllinae), *Journal of Orthoptera Research* **7:** 41–59.

Barth, F. G. (1986). Zur Organisation sensorischer Systeme: die cuticularen Mechanoreceptoren der Arthropoden, *Verhandlungen Deutsche Zoologische Gesellschaft* **79:** 69–90.

Barth, F. G. (1993). Sensory guidance in spider pre-copulatory behavior, *Comparative Biochemistry and Physiology,* **104A:** 717–733.

Barth, F. (1997). Vibratory communication in spiders: adaptation and compromise at many levels, In *Orientation and Communication in Arthropods*, Lehrer, M., Ed., Birkhauser Verlag, Basel, pp. 247–272.

Barth, F. G. (1998). The vibrational sense of spiders, In *Comparative Hearing: Insects*, Hoy, R. R., Popper, A. N., and Fay, R. R., Eds., Springer, New York, Berlin, Heidelberg, pp. 228–278.

Barth, F. (2001). *Sinne und Verhalten aus dem Leben einer Spinne*, Springer Verlag, Berlin, p. 423.

Barth, F. G. (2002a). *A Spider's World: Senses and Behavior*, Springer, Berlin, Heidelberg, chap. 18.

Barth, F. G. (2002b). *A Spider's World: Senses and Behavior*, Springer, Berlin, Heidelberg, chap. 20.

Barth, F. G., Bleckmann, H., Bohnenberger, J., and Seyfarth, E.-A. (1988). Spiders of the genus *Cupiennius* Simon 1891 (Araneae, Ctenidae) II. On the vibratory environment of a wandering spider, *Oecologia* **77:** 194–201.

Bartlett, A. C. and Gawel, N. J. (1993). Determining whitefly species, *Science* **261:** 1333–1334.

Bässler, U. (1957). Zur Funktion des Johnstonschen Organs bei der Orientierung der Stechmücken, *Naturwissenschaften.* **44:** 336.

Baudrimont, A. (1923a). Sur la «musique» du Hanneton du Pin (Col., Lamell.), *Procès-Verbaux Société. Linnéenne Bordeaux* **75:** 174–180.

Baudrimont, A. (1923b). Sur le Prione tanneur; sa facon de protester (Col., Longicorne), *Procès-Verbaux Société Linnéenne Bordeaux* **75:** 181–185.

Baudrimont, A. (1925). La musique du Hanneton du Pin. Son mecanisme, *Procès-Verbaux Société Linnéenne Bordeaux* **77:** 89–96.

Baudrimont, A. (1926). Note complémentaire sur la stridulation du Prione tanneur, *Procès-Verbaux Société Linnéenne Bordeaux* **77:** 111–113.

Bauer, T. (1976). Experimente zur Frage der biologischen Bedeutung des Stridulationsverhaltens von Käfern [Experiments concerning biological value of stridulation in Coleoptera], *Z. Tierpsychol.—J. Comp. Ethol.* **42:** 57–65.

Bauer, M. and von Helverson, O. (1987). Separate localization of sound recognizing and sound producing neural mechanisms in a grasshopper, *Journal of Comparative Physiology* **161A:** 95–101.

Baurecht, D. and Barth, F. G. (1992). Vibratory communication in Spiders. I. Representation of male courtship signals by female vibration receptor, *Journal of Comparative Physiology A* **171:** 231–243.

Beamer, R. H. (1928). Studies on the biology of Kansas Cicadidae, *The University of Kansas Science Bulletin* **18:** 155–263.

Beecher, M. D. (1988). Spectrographic analysis of animal vocalizations: implications of the «uncertainty principle», *Bioacoustics* **1:** 187–208.

Beecher, M. D. (1989). Signalling systems for individual recognition: An information theory approach, *Animal Behaviour* **38:** 248–261.

Beeman, K. (1998). Digital signal analysis, editing, and synthesis, In *Animal Acoustic Communication*, Hopp, S. L., Owren, M. J., and Evans, C. S., Eds., Springer, Berlin, pp. 59–104.

Bellows, T. S. Jr., Perring, T. M., Gill, R. J., and Headrick, D. H. (1994). Description of a species of *Bemisia* (Homoptera: Aleyrodidae) infesting North American agriculture, *Annals of the Entomological Society of America* **87:** 195–206.

Belwood, J. J. (1990). Anti-predator defences and ecology of neotropical forest katydids, especially the Pseudophyllinae, In *The Tettigoniidae: Biology, Systematics and Evolution*, Bailey, W. J. and Rentz, D. C. F., Eds., Crawford House Press, Bathurst, pp. 27–40.

Belwood, J. and Morris, G. K. (1987). Bat predation and its influence on calling behavior in Neotropical katydids, *Science* **238:** 64–67.

Benazzi-Lentati, G. (1966). Amphimixis and pseudogamy in fresh-water triclads: Experimental reconstitution of polyploid paseudogamic biotypes, *Chromosoma (Berlin)* **20:** 1–14.

Benedix, J. H. and Howard, D. J. (1991). Calling song displacement in a zone of overlap and hybridization, *Evolution* **45:** 1751–1759.

Bennet-Clark, H. C. (1970a). A new French mole cricket, differing in the song and morphology from *Gryllotalpa gryllotalpa* L. (Orthoptera: Gryllotalpidae), *Proceedings of the Royal Entomological Society of London Series B* **39:** 125–132.

Bennet-Clark, H. C. (1970b). The mechanism and efficiency of sound production in mole crickets, *Journal of Experimental Biology* **52:** 619–652.

Bennet-Clark, H. C. (1971). Acoustics of insect song, *Nature* **234:** 255–259.

Bennet-Clark, H. C. (1984a). A particle velocity microphone for the song of small insects and otheracoustic measurements, *Journal of Experimental Biology* **108:** 459–463.

Bennet-Clark, H. C. (1984b). Insect Hearing: Acoustics and transduction, In *Insect Communication*, Lewis, T., Ed., Academic Press, London, pp. 49–82.

Bennet-Clark, H. C. (1989). Songs and the physics of sound production, In *Cricket Behavior and Neurobiology*, Huber, F., Moore, T. E., and Loher, W., Eds., Cornell University Press, Ithaca, chap. 8.

Bennet-Clark, H. C. (1994). The world's noisiest insect — cicada song as a model of the biophysics of animal sound production, *Verhandlungen der Deutschen Zoologischen Gesellschaft* **87**(2)**:** 165–176.

Bennet-Clark, H. C. (1998a). Size and scale effects as constraints in insect sound communication, *Phil. Trans. R. Soc. Lond. B* **353:** 407–419.

Bennet-Clark, H. C. (1998b). Sound radiation by the bladder cicada *Cystosoma saudersii*, *Journal of Experimental Biology* **201:** 701–715.

Bennet-Clark, H. C. (1998c). How cicadas make their noise, *Scientific American* **278:** 36–39.

Bennet-Clark, H. C. (1999a). Resonators in insect sound production: how insects produce loud pure-tone songs, *Journal of Experimental Biology* **202:** 3347–3357.

Bennet-Clark, H. C. (1999b). Transduction of mechanical energy into sound energy in the cicada *Cyclochila australasiae*, *Journal of Experimental Biology* **202:** 1803–1817.

Bennet-Clark, H. C. and Daws, A. G. (1999). Transduction of mechanical energy into sound energy in the cicada *Cyclochila australasiae*, *Journal of Experimental Biology* **202:** 1803–1817.

Bennet-Clark, H. C. and Ewing, A. W. (1967). Stimuli provided by courtship of male *Drosophila melanogaster*, *Nature* **215:** 669–671.

Bennet-Clark, H. C. and Ewing, A. W. (1968). The wing mechanism involved in the courtship of *Drosophila*, *Journal of Experimental Biology* **49:** 117–128.

Bennet-Clark, H. C. and Ewing, A. W. (1969). Pulse interval as a critical parameter in the courtship song of *Drosophila melanogaster*, *Animal Behaviour* **17:** 755–759.

Bennet-Clark, H. C. and Leroy, Y. (1978). Regularity versus irregularity in specific songs of closely-related drosophilid flies, *Nature* **5644:** 442–444.

Bennet-Clark, H. C. and Young, D. (1992). A model of the mechanism of sound production in cicadas, *Journal of Experimental Biology* **173:** 123–153.

Bergroth, E. (1905). On stridulating Hemiptera of the subfamily Halyinae, with descriptions of new genera and new species, *Proceedings of the Zoological Society of London* **1905:** 145.

Bertram, S. M. (2000). The influence of age and size on temporal mate signalling behaviour, *Animal Behavior* **60**: 333–339.

Bessey, C. A. and Bessey, E. A. (1898). Further notes on thermometer crickets, *American Naturalist* **32**: 263–264.

Beutel, R. G. (2003). Coleoptera, In *Lehrbuch der Speziellen Zoologie, Part 5, Insecta*, Vol. I: Dathe, H. H., Ed., Spektrum Akad. Verl., Heidelberg, pp. 426–526.

Bickmeyer, U., Kalmring, K., Halex, H., and Mücke, A. (1992). The bimodal auditory-vibratory system of the thoracic ventral nerve cord in *Locusta migratoria* (Acrididae, Locustinae, Oedipodini), *Journal of Experimental Zoology* **264**: 381–394.

Biesmeijer, J. C. and Ermers, M. C. W. (1999). Social foraging in stingless bees: how colonies of *Melipona fasciata* choose among nectar sources, *Behavioral Ecology and Sociobiology* **46**: 129–140.

Biesmeijer, J. C., van Nieuwstadt, M. G. L., Lukács, S., and Sommeijer, M. J. (1998). The role of internal and external information in foraging decisions of *Melipona* workers (Hymenoptera: Meliponinae), *Behavioral Ecology and Sociobiology* **42**: 107–116.

Biström, O. (1997). Morphology and function of a possible stridulation apparatus in genus *Hydrovatus* (Coleoptera, Dytiscidae), *Entomologica Basiliensa* **19**: 43–50.

Blassioli-Moraes, M. C., Laumann, R. A., Čokl, A., and Borges, M., (2005). Vibratory signals of four Neotropical stink bug species. *Physiological Entomology* **30**: 175–188.

Bleckmann, H. (1994). *Reception of Hydrodynamic Stimuli in Aquatic and Semiaquatic Animals*, Progress in Zoology 41, Gustav Fischer, Stuttgart, Jena, New York.

Bleckmann, H. and Barth, F. G. (1984). Sensory ecology of a semi-aquatic spider (*Dolomedes triton*). II. The release of predatory behavior by water surface waves, *Behavioral Ecology and Sociobiology* **14**: 303–312.

Bleckmann, H., Borchardt, M., Horn, P., and Görner, P. (1994). Stimulus discrimination and wave source localization in fishing spiders (*Dolomedes triton* and *D. okefinokensis*), *Journal of Comparative Physiology* **174**: 305–316.

Blest, A. D. (1964). Protective display and sound productionin some New World arctiid and ctenuchid moths, *Zoologica* **49**: 151–164.

Blondheim, S. A. and Shulov, A. S. (1972). Acoustic communication and differences in the biology of two sibling species of grasshoppers, *Annuals of the Entomological Society of America* **65**: 17–24.

Booij, C. J. H. (1981). Biosystematics of the *Muellerianella* complex (Homoptera, Delphacidae), taxonomy, morphology and distribution, *Netherlands Journal of Zoology* **31**: 572–595.

Booij, C. J. H. (1982a). Biosystematics of the *Muellerianella* complex (Homoptera, Delphacidae), hostplants, habitats and phenology, *Ecological Entomology* **7**: 9–18.

Booij, C. J. H. (1982b). Biosystematics of the *Muellerianella* complex (Homoptera, Delphacidae), interspecific and geographic variation in acoustic behaviour, *Zeitschrift für Tierpsychologie* **58**: 31–52.

Booij, C. J. H. and Guldemond, A., (1984). Distributional and ecological differentiation between asexual gynogenetic planthoppers and related sexual species of the genus Muellerianella (Homoptera, Delphacidae), *Evolution* **38**: 163–175.

Borges, M., Jepson, P. C., and Howse, P. E. (1987). Long-range mate location and close-range courtship behavior of the green stink bug, *Nezara viridula* and its mediation by sex pheromones, *Entomologia Experimentalis et Applicata* **44**: 205–212.

Boulard, M. (1965). Comment vivent nos Cigales. *Sciences et Nature* **70**: 9–19.

Boulard, M. (1973a). Cigales de France, aspects de leur biologie. — FILM 16 mm, couleurs, son optique, 29 min., S.F.R.S. Prod., Paris.

Boulard, M. (1973b). Un type nouveau d'appareil stridulant chez les Cigales. *Comptes Rendus de l'Académie des Sciences*, Paris, sér. D, 277, 1487–1489, 1 Pl. H.T.

Boulard, M. (1974). *Spoerryana llewelyni* n.g., n. sp., une remarquable Cigale d'Afrique orientale (Hom. Cicadoidea), *Annales de la Socièté entomologique de France (N.S.)* **10**(3): 729–744, 35 fig.

Boulard, M., (1975). Les Cigales des savanes centrafricaines. Systématique, notes biologiques et biogéographiques, *Bulletin du Muséum national d'Histoire naturelle, Paris, 3e sér., 315, Zoologie* 222, pp. 869–928.

Boulard, M. (1976a). Sur une deuxième Cigale africaine dépourvue d'appareil sonore (Homoptera). *Bulletin de l' Institut Fondamental d'Afrique Noire*, 37, sér. A (3), pp. 629–636.

Boulard, M. (1976b). Un type nouveau d'appareil stridulant accessoire pour les Cicadoidea. Révision de la classification supérieure de la superfamille (Hom.), *Journal of Natural History* **10**(4)**:** 399–407.

Boulard, M. (1977). La cymbalisation coassante de certaines Cigales, *Bulletin de la Société Zoologique de France* **2**(Suppl.), 217–220.

Boulard, M. (1978). Description d'une troisième espèce du genre Plautilla Stål (Hom. Cicadoidea, Plautillidae). *Bulletin de la Société entomologique de France*, 82(9–10), 228–232, 1977, (1978).

Boulard, M. (1982a). Les Cigales du Portugal, contribution à leur étude, *Annales de la Société entomologique de France, (N.S.)* **18**(2): 181–198.

Boulard, M. (1982b). Une nouvelle Cigale néotropicale, halophile et crépusculaire (Hom. Cicadoidea). *Revue française d'Entomologie, (N.S.)* **4**(3)**:** 108–112.

Boulard, M. (1985). Appearance and Mimicry in Cicadasl, *Annales de la Societe entomologique de France (NS)* **90**(1–2)**:** 1016–1051.

Boulard, M. (1986). Cigales de la forêt de Taï (Côte d'Ivoire) et complément à la faune cicadéenne afrotropicale (Homoptera, Cicadoidea). *Revue française d'Entomologie*, (N.S.), 7(5), pp. 223–239.

Boulard, M. (1988). Les *Lyristes* d'Asie Mineure (Hom, Cicadidae).I.-.Sur deux formes éthospécifiques syntopiques et description de deux espèces nouvelles, *L'Entomologiste* **44**(3)**:** 153–167.

Boulard, M. (1990a). Contribution à l'Entomologie générale et appliquée. 2: Cicadaires (Homoptères Auchénorhynques), 1ère partie: Cicadoidea, *EPHE, Travaux Laboratoire Biologie et Evolution des Insectes* **3**: 55–245.

Boulard, M. (1990b). A new genus and species of cicada (Insecta, Homoptera, Tibicinidae) from Western Australia, In *Research in Shark bay. — Report of the France-Australe Bicentenary Expedition Committee*, Berry, P. F. *et al.*, Eds., pp. 237–243, Traduction partielle de Boulard 1989.

Boulard, M. (1991). Sur une nouvelle Cigale néocalédonienne et son étonnante cymbalisation (Homoptera Cicadoidea Tibicinidae), *EPHE, Travaux Laboratoire Bilogie et Evolution des Insectes* **4**: 73–82.

Boulard, M. (1992). Identité et Bio-Écologie de Cicadatra atra (Olivier, 1790), la Cigale noire in Gallia primordia patria (Homoptera, Cicadoidea, Cicadidae), *EPHE, Travaux Laboratoire Bilogie et Evolution des Insectes* **5**: 55–86.

Boulard, M. (1993a). *Pagiphora yanni*, Cigale anatolienne inédite. Description et premières informations biologiques (Cartes d'identité et d'éthologie sonores) (Homoptera, Cicadoidea, Tibicinidae). *Nouvelle Revue d' Entomologie (N.S.)*, **9(4):** 365–374.

Boulard, M. (1993b). *Bafutalna mirei*, nouveau genre, nouvelle espèce de Cigale acymbalique (Homoptera, Cicadoidea, Tibicinidae), *EPHE, Biologie et Evolution des Insectes* **6**: 87–92, 1 Pl.

Boulard, M. (1993c). Sur quatre nouvelles Cigales néocalédoniennes et leurs cymbalisations particulières (Homoptera, Cicadoidea, Tibicinidae), *EPHE, Travaux Laboratoire Biologie et Evolution des Insectes* **6**: 111–125.

Boulard, M. (1994). Postures de cymbalisation et cartes d'identité et d'éthologie acoustiques des Cigales. 1.— Généralités et espèces méditerranéennes, *EPHE, Biologie et Evolution des Insectes* **7–8:** 1–72.

Boulard, M. (1996a). Sur une Cigale wallisienne, crépusculaire et ombrophile (Homoptera, Cicadoidea, Tibicinidae), *Bulletin de la Société entomologique de France* **101**(2): 151–158.

Boulard, M. (1996b). Postures de cymbalisation et cartes d'identité et d'éthologie acoustiques des Cigales. 2.— Espèces forestières afro- et néotropicales, *EPHE, Biologie et Evolution des Insectes* **9**: 113–158.

Boulard, M. (1997). Nouvelles Cigales remarquables originaires de la Nouvelle-Calédonie (Homoptera, Cicadoidea, Tibicinidae), *Mémoires du Museum national d'Histoire naturelle* **171:** 179–196, Zoologica Neocaledonica, 4.

Boulard, M. (1998/1999). Postures de cymbalisation, cymbalisations et cartes d'identité acoustique des Cigales. 3.— Espèces tropicales des savanes et milieux ouverts, *EPHE, Biologie et Evolution des Insectes* **11/12:** 77–116.

Boulard, M. (2000a). Espèce, milieu et comportement, *EPHE, Biologie et Evolution des Insectes* **13:** 1–40.

Boulard, M. (2000b). Cymbalisation d'appel et cymbalisation de cour chez quatre Cigales thaïlandaises, *EPHE, Biologie et Evolution des Insectes* **13:** 49–59.

Boulard, M. (2000c). Appareils, productions et communications sonores chez les Insectes en général et chez les Cigales en particulier, *EPHE, Biologie et Evolution des Insectes* **13:** 75–110, 26 Pl.

Boulard, M. (2000d). *Tosena albata* (Distant, 1878) bonne espèce confirmée par l' acoustique (Auchenorhyncha, Cicadidae, Tosenini), *EPHE, Biologie et Evolution des Insectes* **13:** 119–126, 3 Pl.

Boulard, M. (2001a). Cartes d'Identité Acoustique et éthologie sonore de cinq espèces de Cigales thaïlandaises, dont deux sont nouvelles (Rhynchota, Auchenorhyncha, Cicadidae), *EPHE, Biologie et Evolution des Insectes* **14**: 49–71.

Boulard, M. (2001b). Éthologies sonore et larvaire de *Pomponia* pendleburyi n. sp., (précédé d'un historique taxonomique concernant les grandes *Pomponia*) (Cicadidae, Cicadinae, Pomponiini), *EPHE, Biologie et Evolution des Insectes* **14**: 80–107, 12 Pl.

Boulard, M. (2001c). Note rectificative: *Pomponia dolosa* nom. nov. pro P. fusca Boulard, 2000 [non Olivier, 1790]. Redescription de P. fusca (Olvr) (Rhynchota, Auchenorhyncha, Cicadidae), *EPHE, Biologie et Evolution des Insectes* **14**: 157–161.

Boulard, M. (2002). Éthologie sonore et Cartes d'Identité Acoustique de dix espèces de Cigales thaïlandaises, dont six restées jusqu'ici inédites, ou mal connues (Auchenorhyncha, Cicadoidea, Cicadidae), *Revue française d'Entomologie (N.S.)* **24**(1): 35–66.

Boulard, M. (2003a). Comportement sonore et Identité acoustique de six espèces de Cigales thaïlandaises, dont l'une est nouvelle (Rhynchota, Cicadomorpha, Cicadidae), *Bulletin de la Société entomologique de France* **108**(2): 185–200l.

Boulard, M. (2003b). Éthologie sonore et statut acoustique de quelques Cigales thaïlandaises, incluant la description de deux espèces nouvelles (Auchenorhyncha, Cicadoidea, Cicadidae), *Annales de la Société entomologique de France (N.S.)* **39**(2): 97–119.

Boulard, M. (2003c). Contribution à la connaissance des Cigales thaïlandaises, incluant la description de quatre espèces nouvelles (Rhynchota, Cicadoidea, Cicadidae), *Revue française d'Entomologie (N.S.)* **25**(4): 171–201.

Boulard, M. (2004a). Tibicen Latreille, 1825, «Fatal error», *Nouvelle Revue d'Entomologie* **20**(4): 371–372.

Boulard, M. (2004b). Données statutaires et éthologiques sur des Cigales thaïlandaises, incluant la description de huit espèces nouvelles, ou mal connues (Rhynchota, Cicadoidea, Cicadidae), *EPHE (École Pratique des Hautes Études), Biolologie et Evolution des Insectes* **15**: 5–57.

Boulard, M. (2004c). *Mimétisme, usages biologiques du paraître*, éditions Boubée, Paris.

Boulard, M. (2004d). Description de six nouvelles espèces de Cigales d'Asie continento-tropicale (Rhynchota, Cicadoidea, Cicadidae), *EPHE, Biologie et Evolution des Insectes* **15/16**: 57–74.

Boulard, M. (2004e). Statut taxonomique et acoustique de quatre Cigales thaïlandaises, dont deux restées inédites jusqu'ici (Rhynchota, Cicadoidea, Cicadidae), *Nouvelle Revue d'Entomologie* **20**(3): 2003/2004, 259–279.

Boulard, M. and Lodos, N. (1987). A study on the sounds as a taxonomic characteristics in some species of Cicadidae (Homoptera: Auchenorhyncha), *Proc. First Turk. Natl. Congr. Entomol., Izmir, 13–16 October* 643–648, 1987, 2 fig.

Boulard, M. and Mondon, B. (1995). *Vies et Mémoires de Cigales*. Éditions de l'Équinoxe, 157p. + un CD: «Chants de Cigales méditerranéennes» (17min.).

Boyan, G. S. (1984). What is an "auditory" neurone?, *Naturwissenschaften* **71**: 482.

Boyle, W. W. (1956). A revision of the Erotylidae of America north of Mexico (Col.), *Bulletin of the American Museum of Natural History* **110**: 67–172.

Brach, V. (1975). A case of active brood defense in the thornbug, Umbonia crassicornis (Homoptera: Membracidae), *Bulletin of the Southern California Academy of Science* **74**: 163–164.

Bradbury, J. W., and Vehrencamp, S. L. (1998). *Principles of Animal Communication*, Sinauer Associates, Massachusetts, ISBN 0-87893-100-107.

Bredohl, R. (1984). Zur Bioakustik mitteleuropäischer Totengräber (Coleoptera: Silphidae: *Necrophorus*), *Entomologia Generalis* **10**: 11–25.

Breed, M. D., Guzmán-Novoa, E., and Hunt, G. J. (2004). Defensive behavior of honey bees: organization, genetics, and comparisons with other bees, *Annual Review of Entomology* **49**: 271–298.

Breidbach, O. (1986). Studies on the stridulation of *Hylotrupes bajulus* (L.) (Cerambycidae, Col.): communication through support vibration — morphology and mechanics of the signal, *Behavioural Processes* **12**: 169–186.

Breidbach, O. (1988). Zur Stridulation der Bockkäfer (Cerambycidae, Col.), *Deutsche Entomologische Zeitschrift* **35**: 417–425.

Briggs, C. A. (1897). A curious habit in certain male Perlidae, *Entomologist's Monthly Magazine* **8**: 207–208.

Brinck, P. (1949). Studies on Swedish stoneflies (Plecoptera), *Opuscula Entomologica* (**Suppl. XI**): 1–250, Lund.

Brogdon, W. G. (1994). Measurement of flight tone differences between female *Aedes aegypti* and *A. albopictus* (Diptera: Culicidae), *Journal of Medical Entomology* **31**: 700–703.

Brogdon, W. G. (1998). Measurement of flight tone differentiates among members of the *Anopheles gambiae* species complex (Diptera, Culicidae), *J. Med. Ent.* **35**: 681–684.

Brooks, M. W. (1882). Influence of temperature on the chirp rate of the cricket, *Popular Science Monthly* **20**: 268.

Brooks, S. J. (1987). Stridulatory structures in three green lacewings (Neuroptera: Chrysopidae), *International Journal of Insect Morphology and Embryology* **16**: 237–244.

Brooks, S. J. (1994). A taxonomic review of the common green lacewing genus *Chrysoperla* (Neuroptera: Chrysopidae), *Bulletin of the British Museum of Natural History (Entomology)* **63**: 137–210.

Broughton, W. B. (1963). Method in bio-acoustic terminology, In *Acoustic Behaviour of Animals*, Busnel, R. G., Ed., Elsevier, London, pp. 3–24.

Brown, H. P. (1952). The life history of *Climacia areolaris* (Hagen), a neuropterous "parasite" of fresh-water sponges, *American Midland Naturalist* **47**: 130–160.

Brown, J. K., Coats, S. A., Bedford, I. D., Markham, P. G., Bird, J., and Frohlich, D. R. (1995a). Characterization and distribution of esterase electromorphs in the whitefly, *Bemisia tabaci* (Genn.) (Homoptera:Aleyrodidae), *Biochemical Genetics* **33**: 205–214.

Brown, J. K., Frohlich, D. R., and Rosell, R. C. (1995b). The sweetpotato or silverleaf whiteflies: biotypes of *Bemisia tabaci* or a species complex?, *Annual Review of Entomology* **40**: 511–534.

Brown, J. K., Perring, T. M., Cooper, A. D., Bedford, I. D., and Markham, P. G. (2000). Genetic analysis of *Bemisia* (Homoptera: Aleyrodidae) populations by isoelectric focusing electrophoresis, *Biochemical Genetics* **38**: 13–25.

Browne, J. and Scholtz, C. H. (1999). A phylogeny of the families of Scarabaeoidea (Col.), *Systematic Entomology* **24**: 51–84.

Brownell, P. H. (1977). Compressional and surface waves in sand: used by desert scorpions to locate prey, *Science* **197**: 479–482.

Brownell, P. and Farley, R. D. (1979a). Prey-localizing behaviour of the nocturnal desert scorpion *Paruroctonus mesaensis*: orientation to substrate vibrations, *Animal Behaviour* **27**: 185–193.

Brownell, P. and Farley, R. D. (1979b). Detection of vibrations by tarsal sense organs of the nocturnal scorpion *Paruroctonus mesaensis*, *Journal of Comparative Physiology A* **131**: 23–30.

Brownell, P. and Farley, R. D. (1979c). Orientation to vibrations in sand by the nocturnal scorpion *Paruroctonus mesaensis*: mechanism of target location, *Journal of Comparative Physiology A* **131**: 31–38.

Brownell, P. H. and van Hemmen, J. L. (2001). Vibration sensitivity and a computational theory for prey-localizing behavior in sand scorpion, *American Zoologist* **41**: 1229–1240.

Broza, M., Blondheim, S., and Nevo, E. (1998). New species of mole crickets of the *Gryllotalpa gryllotalpa* group (Orthoptera: Gryllotalpidae) from Israel, based on morphology, song recordings, chromosomes and cuticular hydrocarbons, with comments on the distribution of the group in Europe and the Mediterranean region, *Systematic Entomology* **23**: 125–135.

Buhk, F. (1910). Stridulationsapparat bei *Spercheus emarginatus* Schall, *Zeitschrift fur Wissenschaftliche Insektenbiologie* **6**: 342–346.

Bukhvalova, M. A. and Vedenina, V. Yu. (1998). Contributions to the study of acoustic signals of grasshoppers (Orthoptera: Acrididae: Gomphocerinae) from Russia and adjacent countries. I. New recordings of the calling songs of grasshoppers from Russia and adjacent countries, *Russian Entomological Journal* **7**(3–4)**:** 109–125.

Burbidge, A. and Iannantuoni, M. *et al.* (1997). *An Introduction to Wildlife Sound Recording*, Wildlife Sound Recording Society, p. 26.

Burmeister, H. (1832–1847). *Handbuch der Entomologie*, 5 Vols.: G. Reimer, Berlin.

Burrows, M. (1996). *The Neurobiology of an Insect Brain*, Oxford University Press Inc., Oxford.

Bush, G. L. (1994). Sympatric speciation, *Trends in Ecology and Evolution* **9**: 285–288.

Busnel, R. G., Ed. (1963). *Acoustic Behaviour of Animals*, Elsevier, Amsterdam, p. 933.

Busnel, R. G. and Dumortier, B. (1959). Vérification par les methods d'analyse acoustique des hypothèses sur l'origine du cri du Sphinx, *Acherontia atropos* (Linné). *Bulletin de la Société entomologique de France* **64**: 44–58.

Busnel, R. G., Dumortier, B., and Busnel, M. C. (1956). Recherches sur le comportement acoustique des Ephippigeres, *Le Bulletin Biologique de la France et de la Belgique* **3**: 219–228.

Butler, C. (1609). *The Feminine Monarchy*, Joseph Barnes, Oxford.

Butler, C. G. (1954). *The World of the Honeybee*, Collins, London, p. 226.

Butlin, R. K. (1989). Reinforcement of premating isolation, In *Speciation and Its Consequences*, Otte, D. and Endler, J. A., Eds., Sinauer Associates, Sunderland, MA, pp. 158–179.

Butlin, R. K. (1995). Genetic variation in mating signals and responses, In *Speciation and the Recognition Concept*, Lambert, D. M. and Spencer, H. G., Eds., Johns Hopkins University Press, Baltimore, Maryland, pp. 327–366.

Butlin, R. K., Hewitt, G. M., and Webb, S. F. (1985). Sexual selection for intermediate optimum in *Chorthippus brunneus* (Orthoptera: Acrididae), *Animal Behavior* **33**: 1281–1292.

Cade, W. H. (1975). Acoustically orienting parasitoids: fly phonotaxis to cricket song, *Science* **190**: 1312–1313.

Cade, W. H. (1979). The evolution of alternative male reproductive strategies in field crickets, In *Sexual Selection and Reproductive Competition in Insects*, Blum, M. S. and Blum, N. A., Eds., Academic Press, New York, pp. 343–349.

Cade, W. H. and Otte, D. (2000). *Gryllus texensis* n. sp.: a widely studied field cricket (Orthoptera; Gryllidae) from the southern United States, *Transactions of the American Entomological Society* **126**: 117–123.

Camazine, S., Visscher, P. K., Finley, J., and Vetter, R. S. (1999). House hunting by honey bee swarms: collective decision and individual behaviors, *Insectes Sociaux* **46**: 348–360.

Campbell, K. G. (1964). Sound production by Psyllidae (Hemiptera), *Journal of the Australian Entomological Society (N. S. W.)* **1**: 3–4.

Campesan, S., Dubrova, Y., Hall, J. C., and Kyriacou, C. P. (2001). The *nonA* gene in *Drosophila* conveys species-specific behavioral characteristics, *Genetics* **158**: 1535–1543.

Canary 1.2, (1995). User's Manual. Appendix A Digital Representation of Sound, Cornell Laboratory of Ornithology.

Carisio, L., Palestrini, C., and Rolando, A. (2004). Stridulation variability and morphology: an examination in dung beetles of the genus *Trypocopris* (Coleoptera, Geotrupidae), *Population Ecology* **46**: 27–37.

Carlet, G. (1877a). Mémoire sur l'appareil musical de la cigale, *Annales des Science Naturelles, Paris* **5**(6): 1–35, Pl. 2.

Carlet, G. (1877b). Le chant de la cigale, *Revue Scientifique, Paris* **13**(2): 516–519.

Carne, P. B. (1962). The characteristics and behaviour of the sawfly *Perga affinis affinis* (Hymenoptera), *Australian Journal of Zoology* **10**: 1–34.

Carson, H. L. (1985). Genetic variation in a courtship-related male character in Drosophila *silvestris* from a single Hawaiian locality, *Evolution* **39**: 678–786.

Carver, M. (1987). Distinctive motory behaviour in some adult psyllids (Homoptera: Psylloidea), *Journal of the Australian Entomological Society* **26**: 369–372.

Casas, J. (1989). Foraging behaviour of a leafminer parasitoid in the field, *Ecological Entomology* **14**: 257–265.

Casas, J., Bacher, S., Tautz, J., Meyhöfer, R., and Pierre, D. (1998). Leaf vibrations and air movements in a leafminer-parasitoid system, *Biocontrol* **11**: 147–153.

Casas, J. and Baumgärtner, J. (1990). Liste des parasitoïdes de *Phyllonorycter blancardella* (F.) (Lepidoptera, Gracillariidae) en Suisse, *Bulletin of the Swiss Entomological Society* **63**: 299–302.

Casas, J., Gurney, W. C. S., Nisbet, R., and Roux, O. (1993). A probabilistic model for the functional response of a parasitoid at the behavioural time scale, *Journal of Animal Ecology* **63**: 194–204.

Casserius, J. (1600). De vocis auditusque organis historia anatomica singulari fide methode ac industria concinnata tractatibus duobus explicata ac variis iconibus aere excusis. Ferrara, 1 vol., Tab. I-XXII + I-XII.

Chapman, R. F. (1998). *The Insects: Structure and Function*, 4th ed., University Press, Cambridge.

Cheng, J. (1991). Sound production in longhorned beetles: stridulation and associated behaviour of the adult (Coleoptera: Cerambycidae), *Scientia Silvae Sinicae* **27**: 234–237.

Cheng, J. (1993). A study on the acoustical properties of thoracic stridulation and elytral vibration sounding in beetle *Anoplophora horsfieldi* (Hope) (Coleoptera: Cerambycidae), *Acta Entomologica Sinica* **36**: 150–157.

Chesmore, E. D. (2001). Application of time domain signal coding and artificial neural networks to passive acoustical identification of animals, *Applied Acoustics* **62:** 1359–1374.

China, W. E. and Fennah, R. G. (1952). A remarkable new genus and species of Fulgoroidea (Homoptera) representing a new family, *Annals and Magazine of Natural History* **12**(5)**:** 189–199.

Chou, I., Lei, Z., Li, L., Lu, X., and Yao, W. (1997). *The Cicadidae of China (Homoptera: Cicadoidea)*, Tianze Eldoneio, Hong Kong, 380pp.

Chowdhury, V. and Parr, M. J. (1981). The 'switch mechanism' and sound production in tsetse flies (Diptera: Glossinidae), *Journal of Natural History* **15:** 87–95.

Christensen, B. and O'Connor, F. B. (1958). Pseudofertilization in the genus *Lumbricillus*, *Nature* **181:** 1085–1086.

Chvala, M., Doskocil, J., Mook, J. H., and Pokorny, V. (1974). The genus *Lipara* Meigen (Diptera, Chloropidae), systematics, morphology, behaviour, and ecology, *Tijdschrift voor Entomologie* **117:** 1–25.

Ciceran, M., Murray, A. M., and Rowell, G. (1994). Natural variation in the temporal patterning of calling song structure in the field cricket *Gryllus pennsylvanicus*: effects of temperature, age, mass, time of day, and nearest neighbor, *Canadian Journal of Zoology* **72:** 38–42.

Çiplak, B. (2000). Systematics and phylogeny of Parapholidoptera (Orthoptera: Tettigoniidae: Tettigoniinae), *Systematic Entomology* **25:** 411–436.

Çiplak, B. (2004). Systematics, phylogeny and biogeography of *Anterastes* (Orthoptera, Tettigoniidae, Tettigoniinae): evolution within a refugium, *Zoologica Scripta* **3:** 19–44.

Çiplak, B. and Heller, K.-G. (2001). Notes on the song of *Bolua turkiyae* and on the phylogeny of the genus *Bolua* (Orthoptera, Tettigoniidae, Tettigoniinae), *Israel Journal of Zoology* **47:** 233–242.

Claridge, L. C. (1968). Sound production in species of *Rhynchaenus* (= *Orchestes*) (Col., Curculionidae), *Transactions of the Royal Entomological Society of London* **120:** 287.

Claridge, M. F. (1974). Stridulation and defensive behavior in ground beetle, *Cychrus caraboides* (L.), *Journal of Entomology Series A - Physiology & Behaviour* **49:** 7–15.

Claridge, M. F. (1985a). Acoustic behaviour of Leafhoppers and Planthoppers: species problems and speciation, In *The Leafhoppers and Planthoppers*, Nault, L. R. and Rodriguez, J. G., Eds., John Wiley, Berlin, pp. 103–125.

Claridge, M. F. (1985b). Acoustic signals in the homoptera: behaviour taxonomy, and evolution, *Annual Review of Entomology* **30:** 297–317.

Claridge, M. F. (1988). Species concepts and speciation in parasites, In *Prospects in Systematics*, Hawksworth, D. L., Ed., Clarendon Press, Oxford.

Claridge, M. F. (1990). Acoustic recognition signals: barriers to hybridization in Homoptera Auchenorrhyncha, *Canadian Journal of Zoology* **68:** 1741–1746.

Claridge, M. F. (1991). Genetic and biological diversity of insect pests and their natural enemies, In *The Biodiversity of Microorganisms and Invertebrates: Its Role in Sustainable Agriculture*, Hawksworth, D. L., Ed., CAB International, Wallingford, UK.

Claridge, M. F. (1993). Speciation in insect herbivores — the role of acoustic signals in leafhoppers and planthoppers, In *Evolutionary Patterns and Processes*, Lees, D. R. and Edwards, D., Eds., Linnean Society of London, Academic Press, London, pp. 285–297.

Claridge, M. F., Dawah, H. A., and Wilson, M. R. (1997a). Practical approaches to species concepts for living organisms, In *Species: the units of biodiversity*, Claridge, M. F., Dawah, H. A., and Wilson, M. R., Eds., Chapman and Hall, London, pp. 1–15.

Claridge, M. F., Dawah, H. A., and Wilson, M. R. (1997b). Species in insect herbivores and parasitoids — sibling species, host races and biotypes, In *Species: The Units of Biodiversity*, Claridge, M. F., Dawah, H. A., and Wilson, M. R., Eds., Chapman and Hall, London.

Claridge, M. F., Den Hollander, J., and Morgan, J. C. (1985a). Variation in courtship signals and hybridization between geographically definable populations of the rice Brown planthopper, *Nilaparvata lugens* (Stål), *Biological Journal of the Linnean Society* **24:** 35–49.

Claridge, M. F., Den Hollander, J., and Morgan, J. C. (1985b). The status of weed-associated populations of the brown planthopper, *Nilaparvata lugens* (Stål) — host race or biological species?, *Zoological Journal of the Linnean Society* **84:** 77–90.

Claridge, M. F., Den Hollander, J., and Morgan, J. C. (1988). Variation in hostplant relations and courtship signals of weed-associated populations of the rice Brown planthopper, *Nilaparvata lugens* (Stål), from

Australia and Asia: a test of the recognition concept, *Biological Journal of the Linnean Society* **35:** 79–93.

Claridge, M. F. and de Vrijer, P. W. F. (1994). Reproductive behavior: the role of acoustic signals in species recognition and speciation, In *Planthoppers: Their Ecology and Management*, Denno, R. F. and Perfect, T. J., Eds., Chapman & Hall, New York, pp. 216–233.

Claridge, M. F. and Morgan, J. C. (1987). The Brown Planthopper, *Nilaparvata lugens* (Stal), and some related species: a biotaxonomic approach, In *Proceedings of the 2nd International Workshop on Leafhoppers and Planthoppers of Economic Importance*, Wilson, M. R. and Nault, L. R., Eds., CAB International Institute of Entomology, London.

Claridge, M. F. and Morgan, J. C. (1993). Geographical variation in acoustic signals of the planthopper, *Nilaparvata bakeri* (Muir), in Asia: species recognition and sexual selection, *Biological Journal of the Linnean Society* **48:** 267–281.

Claridge, M. F. and Nixon, G. A. (1986). *Oncopsis flavicollis* (L.) associated with tree birches (*Betula*): a complex of biological species or a host plant utilization polymorphism?, *Biological Journal of the Linnean Society* **48:** 267–281.

Claridge, M. F., Wilson, M. R., and Singhrao, J. S. (1979). The songs and calling site of two European cicadas, *Ecological Entomology* **4:** 225–229.

Clements, M. (1998). Digital signal acquisition and representation, In *Animal Acoustic Communication*, Hopp, S. L., Owren, M. J., and Evans, C. S., Eds., Springer, Berlin, pp. 27–58.

Cocroft, R. B. (1996). Insect vibrational defence signals, *Nature* **382:** 679–680.

Cocroft, R. B. (1999). Parent-offspring communication in response to predators in a subsocial treehopper (Hemiptera: Membracidae: *Umbonia crassicornis*), *Ethology* **105:** 553–568.

Cocroft, R. B. (2001). Vibrational communication and the ecology of group-living herbivorous insects, *American Zoologist* **41:** 1215–1221.

Cocroft, R. B. (2002). Maternal defense as a limited resource: unequal predation risk in broods of an insect with maternal care, *Behavioral Ecology* **13:** 125–133.

Cocroft, R. B. (2003). The social environment of an aggregating, ant-attended treehopper, *Journal of Insect Behaviour* **16:** 79–95.

Cocroft, R. B. (2005). Vibrational communication facilitates cooperative foraging in a phloem-feeding insect, *Proceedings of the Royal Society of London Series B* in press.

Cocroft, R. B. and Pogue, M. (1996). Social behavior and communication in the neotropical cicada *Fidicina mannifera* (Fabricius) (Homoptera: Cicadidae), *Journal of the Kansas Entomological Society* **69**(4): (Suppl.), 85–97.

Cocroft, R. B. and Rodríguez, R. L. (2005). The behavioral ecology of vibrational communication in insects, *BioScience* **55:** 323–334.

Cocroft, R. B., Tieu, T. D., Hoy, R. R., and Miles, R. N. (2000). Directionality in the mechanical response to substrate vibration in a treehopper (Hemiptera, Membracidae: *Umbonia crassicornis*), *Journal of Comparative Physiology A* **186:** 695–705.

Cohen, L. (1995). *Time-Frequency Analysis*, Prentice-Hall, Englewood Cliffs, NJ.

Čokl, A. (1983). Functional properties of vibroreceptors in the legs of *Nezara viridula* (L.) (Heteroptera: Pentatomidae), *Journal of Comparative Physiology A* **150:** 261–269.

Čokl, A. (1988). Vibratory signal transmission in plants as measured by laser vibrometer, *Periodicum Biologorum* **90:** 193–196.

Čokl, A. and Amon, T. (1980). Vibratory interneurons in central nervous system of *Nezara viridula* L. (Pentatomidae, Heteroptera), *Journal of Comparative Physiology A* **139:** 87–95.

Čokl, A. and Bogataj, E. (1982). Factors affecting vibrational communication in *Nezara viridula* L. (Heteroptera, Pentatomidae), *Biološki Vestnik (Ljubljana)* **30:** 1–20.

Čokl, A., Gogala, M., and Blaževič, A. (1978). Principles of sound recognition in three pentatomide bug species (Heteroptera), *Biološki vestnik (Ljubljana)* **26:** 81–94.

Čokl, A., Gogala, M., and Jež, M. (1972). The analysis of the acoustic signals of the bug Nezara viridula. *Biološki vestnik (Ljubljana)* **20:** 47–53, in Slovenian? Analiza zvočnih signalov stenice *Nezara viridula* (L.).

Čokl, A., Kalmring, K., and Rössler, W. (1995). Physiology of atympanate tibial organs in forelegs and midlegs of the cave-living Ensifera, *Troglophilus neglectus* (Raphidophoridae, Gryllacridoidea), *Journal of Experimental Zoology* **273:** 376–388.

Čokl, A., Kalmring, K., and Wittig, H. (1977). The responses of auditory ventral-cord neurons of *Locusta migratoria* to vibration stimuli. *Journal of Comparative Physiology* **120:** 161–172.

Čokl, A., McBrien, H. L., and Millar, J. G. (2001). Comparison of substrate-borne vibrational signals of two stink bug species *Acrosternum hilare* and *Nezara viridula* (Heteroptera: Pentatomidae), *Annals of the Entomological Society of America* **94:** 471–479.

Čokl, A., Otto, C., and Kalmring, K. (1985). The processing of directional vibratory signals in the ventral nerve cord of Locusta migratoria, *Journal of Comparative Physiology A* **156:** 45–52.

Čokl, A., Prešern, J., Virant-Doberlet, M., Bagwell, G. J., and Millar, J. G. (2004a). Vibratory signals of the harlequin bug and their transmission through plants, *Physiological Entomology* **29:** 372–380.

Čokl, A. and Virant-Doberlet, M. (1997). Tuning of tibial organ receptor cells in *Periplaneta americana* L, *Journal of Experimental Zoology* **278:** 395–404.

Čokl, A. and Virant-Doberlet, M. (2003). Communication with substrate-borne signals in small plant-dwelling insects, *Annual Review of Entomology* **48:** 29–50.

Čokl, A., Virant-Doberlet, M., and McDowell, A. (1999). Vibrational directionality in the southern green stink bug *Nezara viridula* (L.) is mediated by female song, *Animal Behaviour* **58:** 1277–1283.

Čokl, A., Virant-Doberlet, M., and Stritih, N. (2000). The structure and function of songs emitted by southern green stink bugs from Brazil, Florida, Italy and Slovenia, *Physiological Entomology* **25:** 196–205.

Čokl, A., Zorović, M., Žunič, A., and Virant-Doberlet, M. (2005). Tuning of host plants with vibratory songs of *Nezara viridula* L (Heteroptera: Pentatomide), *Journal of Experimental Biology* **208:** 1481–1488.

Colegrave, N., Hollocher, H., Hinton, K., and Ritchie, M. G. (2000). The courtship song of African *Drosophila melanogaster*, *Journal of Evolutionary Biology* **13:** 143–150.

Collyer, E. (1951). The separation of *Conwentzia pineticola* End. from *Conwentzia psociformis* (Curt.) and notes on their biology, *Bulletin of Entomological Research* **42:** 555–564.

Conner, W.E. (1999). 'Un chat d'appel amoureux' acoustic communication in moths. *Journal of Experimental Biology* **202:** 1711–1723.

Contreras-Ramos, A. (1999). Mating behavior of *Platyneuromus* (Megaloptera: Corydalidae), with life history notes on dobsonflies from Mexico and Costa Rica, *Entomological News* **110:** 125–135.

Cook, M. A. and Scoble, M. J. (1992). Tympanal organs of geometrid moths: a review of their morphology, function, and systematic importance, *Systematic Entomology* **17:** 219–232.

Cooley, J. R. and Marshall, D. C. (2001). Sexual signaling in periodical cicadas, *Magicicada* spp., (Hemiptera: Cicadidae), *Behaviour* **138:** 827–855.

Cooley, J. R., Simon, C., and Marshall, D. C. (2003). Temporal separation and speciation in periodical cicadas, *BioScience* **53:** 151–157.

Costa, C. T. A. and Sene, F. M. (2002). Characterizations of courtship sounds of species of the subgroup *fasciola* (Diptera, Drosophilidae, *Drosophila repleta* group): interspecific and interpopulational analysis. *Brazilian Journal of Biology* **62:** 573–583.

Costa, H. S. and Brown, J. K. (1991). Variation in biological characteristics and esterase patterns among populations of *Bemisia tabaci* (Genn.) and the association of one population with silverleaf symptom development, *Entomologia Experimentalis et Applicata* **61:** 211–219.

Cowling, D. E. (1980). The genetics of *Drosophila melanogaster* courtship song — diallel analysis, *Heredity* **45:** 401–403.

Cowling, D. E. and Burnet, B. (1981). Courtship songs and genetic control of their acoustic characteristics in sibling species of the *Drosophila melanogaster* subgroup, *Animal Behaviour* **29:** 924–935.

Coyne, J. A. and Orr, H. A. (2004). *Speciation*, Sinauer Assoc., Inc., Sunderland, MA.

Crawford, C. S. and Dadone, M. M. (1979). Onset of evening chorus in *Tibicen marginalis* (Homoptera: Cicadidae), *Environmental Entomology* **8:** 1157–1160.

Cremer, L. and Heckl, M. (1973). *Structure-Borne Sound*, Springer, Berlin.

Crossley, S. A. (1986). Courtship sounds and behaviour in the four species of the *Drosophila bipectinata* complex, *Animal Behaviour* **34:** 1146–1159.

Crowcroft, P. (1957). *The Life of the Shrew*, Max Reinhardt, London, p. 166.

Crowson, R. A. (1981). *The Biology of the Coleoptera*, Academic Press, London, p. 802.

Dadour, I. and Bailey, W. J. (1985). Male agonistic behaviour of the bush cricket *Mygalopsis marki* Bailey in response to conspecific song (Orthoptera, Tettigoniidae), *Zeitschrifte fur Tierpsychologie* **70:** 320–330.

Dadour, I. R. and Johnson, M. S. (1983). Genetic differentiation, hybridization and reproductive isolation in *Mygalopsis marki* Bailey (Orthoptera: Tettigoniidae), *Australian Journal of Zoology* **31**: 353–360.

Dagley, J. R., Butlin, R. K., and Hewitt, G. M. (1994)., Divergence in morphology and mating signals, and assortative mating among populations of *Chorthippus parallelus* (Orthoptera: Acrididae), *Evolution* **48**(4): 1202–1210.

Dambach, M. (1972). Der Vibrationssinn der Grillen. II. Antworten von Neuronen im Bauchmark, *Journal of Comparative Physiology* **79**: 305–324.

Dambach, M. (1989). Vibrational responses, In *Cricket Behavior and Neurobiology*, Huber, F., Moore, J. E., and Loher, W., Eds., Cornell University Press, Ithaca, chap. 6.

Danchin, E., Giraldeau, L. A., Valone, T. J., and Wagner, R. H. (2004). Public information: from noisy neighbors to cultural evolution, *Science* **305**: 487–491.

Darwin, C. R. (1877). *The Descent of Man in Relation to Sex*, 2nd ed., John Murray, London.

Davis, W. T. (1894a). Staten island harvest flies, *American Naturalist* **28**: 363–364.

Davis, W. T. (1894b). Staten island harvest flies, *Proceedings of the National Science Association of Staten Island* **4**: 9–10.

Davis, W. T. (1943). Two ways of song communication among our North America cicadas, *Journal of the New York Entomological Society* **51**: 185–190.

Daws, A. G., Henning, R. M., and Young, D. (1997). Phonotaxis in the cicadas *Cystosoma saundersii* and *Cyclochila australasiae*, *Bioacoustics* **7**: 173–188.

Debaisieux, P. (1938). Organes scolopidiaux des pattes d'insectes II, *Cellule* **47**: 77–202.

Dechambre, R. P. (1984). Contribution a l'etude phylogenetique des *Oryctes* de la region malgache; utilisation des stries stridulatoires propygidiales (Col., Dynastidae), *Bulletin de la Societe Entomologique de France* **88**: 436–448.

De Luca, P. A. and Morris, G. K. (1998). Courtship communication in meadow katydid: female preference for large male vibrations, *Behaviour* **135**: 777–793.

den Bieman, C. F. M. (1986). Acoustic differentiation and variation in planthoppers of the genus *Ribautodelphax* (Homoptera, Delphacidae), *Neth. J. Zool.* **36**: 461–480.

den Bieman, C. F. M. (1987). Biological and taxonomic differentiation in the *Ribautodelphax collinus* complex (Homoptera, Delphacidae), PhD Thesis, Ponsen & Looijen, Wageningen, p. 163.

den Bieman, C. F. M. and de Vrijer, P. W. F. (1987). True parthenogenesis for the first time demonstrated in planthoppers (Homoptera, Delphacidae), *Annls. Soc. Entomol.Fr., (n.s.)* **23**: 3–9.

den Bieman, C. F. M. and Eggers-Schumacher, H. A. (1987). Allozyme polymorphism in planthoppers of the genus *Ribautodelphax* (Homoptera, Delphacidae) and yhe origin of the pseudogamous triploid forms, *Neth. J. Zool.* **37**: 239–254.

Den Hollander, J. (1995). Acoustic signals as specific-mate recognition signals in leafhoppers (Cicadellidae) and planthoppers (Delphacidae) (Homoptera: Auchenorrhyncha), In *Speciation and the Recognition Concept. Theory and Application*, Lambert, D. M. and Spencer, H. G., Eds., The John Hopkins University Press, Baltimore and London, pp. 440–463.

Den Hollander, J. and Barrientos, L. (1994). Acoustic and morphometric differences between allopatric populations of *Pterophylla beltrani* (Orthoptera: Tettigoniidae: Pseudophyllinae), *Journal of Orthoptera Research* **2**: 29–34.

Desutter-Grandcolas, L. (2003). Phylogeny and the evolution of acoustic communication in extant Ensifera (Insecta, Orthoptera), *Zoologica Scripta* **32**: 525–561.

Desutter-Grandcolas, L. and Robillard, T. (2003). Phylogeny and the evolution of calling songs in *Gryllus* (Insecta, Orthoptera, Gryllidae), *Zoologica Scripta* **32**: 172–183.

Devetak, D. (1985). Detection of substrate vibrations in the antlion larva *Myrmeleon formicarius* (Neuropetra: Myrmeleontidae), *Bioloski Vestnik* **33**: 11–22.

Devetak, D. (1998). Detection of substrate vibration in Neuropteroidea: a review, *Acta Zoologica Fennica* **209**: 87–94.

Devetak, D. and Amon, T. (1997). Substrate vibration sensitivity of the leg scolopidial organs in the green lacewing, *Chrysoperla carnea*, *Journal of Insect Physiology* **43**: 433–437.

Devetak, D., Gogala, M., and Čokl, A. (1978). A contribution to the physiology of the vibration receptors in bugs of the family Cydnidae (Heteroptera), *Bioloski Vestnik (Ljubljana)* **36**: 131–139.

Devetak, D. and Pabst, M. A. (1994). Structure of the subgenual organ in the green lacewing, *Chrysoperla carnea*, *Tissue Cell* **26**: 249–257.

DeVries, P. J. (1991). Call production by myrmecophilous riodinid and lycaenid butterfly caterpillars (Lepidoptera); morphological, acoustical, functional and evolutionary patterns, *American Museum Novitates* **3025**: 1–23.

DeVries, P. J., Cocroft, R. B., and Thomas, J. (1993). Comparison of acoustical signals in *Maculinea* butterfly caterpillars and their obligate host *Myrmica* ants, *Zoological Journal of the Linnean Society* **49**: 229–238.

De Vrijer, P. W. F. (1984). Variability in calling signals of the planthopper *Javesella pellucida* (F.) (Homoptera: Delphacidae) in relation to temperature, and consequences for species recognition during distant communication, *Netherlands Journal of Zoology* **34**: 388–406.

De Vrijer, P. W. F. (1986). Species distinctiveness and variability of acoustic calling signals in the planthopper genus *Javesella* (Homoptera: Delphacidae), *Netherlands Journal of Zoology* **36**(1): 162–175.

De Winter, A. J. (1992). The genetics basis and evolution of acoustic mate recognition signals in a *Ribautodelphax* planthopper (Homoptera, Delphacidae). 1. The female call, *Journal of Evolutionary Biology* **5**: 249–265.

De Winter, A. J. and Rollenhagen, T. (1990). The importance of male and female acoustic behaviour for reproductive isolation in *Ribautodelphax* planthoppers (Homoptera: Delphacidae), *Biological Journal of the Linnean Society* **40**: 191–206.

Dias, J. C. P. and Schofield, C. J. (1999). The Evolution of Chagas Disease (American Trypanosomiasis) Control after 90 years since Carlos Chagas discovery, *Memorias do Instituto Oswaldo Cruz* **94**(Suppl. 1): 103–121.

Dietrich, C. H. and Deitz, L. L. (1991). Revision of the Neotropical treehopper tribe Aconophorini (Homoptera: Membracidae), *North Carolina Agricultural Research Service Technical Bulletin* **293**: 1–134.

Dietrich, C. H. and Deitz, L. L. (1993). Superfamily Membracoidea (Homoptera: Auchenorrhyncha). II. Cladistic analysis and conclusions, *Systematic Entomology* **18**: 297–311.

Dietrich, C. H. and McKamey, S. H. (1990). Three new idiocerine leafhoppers (Homoptera: Cicadellidae) from Guyana with notes on ant-mutualism and subsociality, *Proceedings of the Entomological Society of Washington* **92**: 214–223.

Dietrich, C. H., Rakitov, R. A., Holmes, J. L., and Black, W. C., IV (2001). Phylogeny of the mayor lineages of Membracoidea (Hemiptera: Cicadomorpha) based on 28S rDNA sequences, *Molecular Phylogenetics and Evolution* **18**(2): 293–305.

Di Luciano, V. S. (1981). Morphology of the stridulatory groove of *Triatoma infestans* (Hemiptera: Reduviidae), *Journal of Medical Entomology* **18**: 24–32.

Dingler, M. (1932a). Das Stridulationsorgan von *Crioceris, Biologisches Zentralblatt* **52**: 705–709.

Dingler, M. (1932b). Über unsere beiden Spargelkäfer (*Crioceris duodecimpunctata* L. und *C. asparagi* L.), *Zeitschrift fur Angewandte Entomologie* **21**: 415–442.

Distant, W. L. (1897). Stridulation and habits of Cicadidae, *Zoologist* **4**(1): 520–521.

Djemai, I., Meyhöfer, R., and Casas, J. (2000). Geometrical games in a host-parasitoid system, *American Naturalist* **156**: 257–265.

Djemai, I., Casas, J., and Magal, C. (2001). Matching host reactions to parasitoid wasp vibrations, *Proceedings of the Royal Society of London Series B* **268**: 2403–2408.

Djemai, I., Casas, J., and Magal, C. (2004). Parasitoid foraging decisions mediated by artificial vibrations, *Animal Behaviour* **67**: 567–571.

Dlabola, J. (1984). Neue Zikadenarten aus Mediterraneum und dem Iran mit weiteren beitraegen zur Iranischen Fauna (Homoptera — Auchenorrhyncha), *Acta Musei Nationalis Pragae* **40B**: 21–63.

Doherty, J. A. (1985a). Phonotaxis in the cricket, *Gryllus bimaculatus* DeGeer: comparison of choice and no-choice paradigms, *Journal of Comparative Physiology A* **157**: 279–289.

Doherty, J. A. (1985b). Temperature coupling and 'trade-off' phenomena in the acoustic communication system of the cricket, *Gryllus bimaculatus* De Geer (Gryllidae), *Journal of Experimental Biology* **114**: 17–35.

Doherty, J. A. and Callos, J. D. (1991). Acoustic communication in the trilling field cricket, *Gryllus rubens* (Orthoptera, Gryllidae), *Journal of Insect Behaviour* **4**: 67–82.

Doherty, J. A. and Howard, D. J. (1996). Lack of preference for conspecific calling songs in female crickets, *Animal Behaviour* **51**: 981–990.

Doherty, J. and Huber, R. R. (1983). Temperature effects on acoustic communication in the cricket *Gryllus bimaculatus* De Geer, *Verhandlungen Deutsche Zoologische Gesellschaft* **1983:** 188.

Dolbear, A. E. (1897). The cricket as a thermometer, *American Naturalist* **31:** 970–971.

Doolan, J. M. and Young, D. (1989). Relative importance of song parameters during flight phonotaxis and courtship in the bladder cicada *Cystosoma saundersii*, *Journal of Experimental Biology* **141:** 113–131.

Dornhaus, A., Brockmann, A., and Chittka, L. (2003). Bumble bees alert to food with pheromone from tergal gland, *Journal of Comparative Physiology A* **189:** 47–51.

Dornhaus, A. and Cameron, S. (2003). A scientific note on food alert in *Bombus transversalis*, *Apidologie* **34:** 87–88.

Dornhaus, A. and Chittka, L. (1999). Evolutionary origins of bee dances, *Nature* **401:** 38.

Dornhaus, A. and Chittka, L. (2001). Food alert in bumblebees (*Bombus terrestris*): possible mechanisms and evolutionary implications, *Behavioral Ecology and Sociobiology* **50:** 570–576.

Drašlar, K. and Gogala, M. (1976). Struktura stridulacijskih organov pri žuželkah iz družine Cydnidae (Heteroptera), *Bioloski Vestnik (Ljubljana)* **24:** 175–200.

Dreller, C. and Kirchner, W. H. (1993). Hearing in honeybees: localization of the auditory sense organ, *Journal of Comparative Physiology A* **173:** 275–279.

Dreller, C. and Kirchner, W. H. (1995). The sense of hearing in honey bees, *Bee World* **76:** 6–17.

Drosopoulos, S. (1976). Triploid pseudogamous biotype of the leafhopper *Muellerianella fairmairei*, *Nature* **236:** 499–500.

Drosopoulos, S. (1977). Biosystematic studies on the *Muellerianella* complex (Delphacidae, Homoptera, Auchenorrhyncha), *Meded. Landbouwhogeschool Wageningen* **77:** 1–133.

Drosopoulos, S. (1978). Laboratory synthesis of a pseudogamous triploid "species" of the genus *Muellerianella* (Homoptera, Delphacidae), *Evolution* **32:** 916–920.

Drosopoulos, S. (1983). Some notes on the genera *Muellerianella* and *Florodelphax* from Greece (Homoptera: Delphacidae) with a description of *Florodelphax mourikisi* n. sp. From Ikaria Island, *Entomologische Berichten* **43:** 72–75.

Drosopoulos, S. (1985). Acoustic communication and mating behaviour in *Muellerianella* complex (Homoptera, Delphacidae), *Behaviour* **94:** 183–201.

Drosopoulos, S. (1990). The family Issidae (Homoptera, Auchenorrhyncha) in Greece: endemism and speciation, *Scopolia Suppl.* **1:** 89–92.

Drosopoulos, S. (2003). New data on the nature and origin of colour polymorphism in the spittlebug genus *Philaenus* (Hemiptera: Aphrophoridae), *Annales de la Societe entomologique de France (NS)* **39:** 31–42.

Drosopoulos, S., Asche, M., and Hoch, H. (1985). Contribution to the planthopper fauna of Greece (Homoptera, Auchenorrhyncha, Fulgoromorpha, Delphacidae), *Annals de l' Institut Phytopathologique Benaki* **14:** 35–88.

Drosopoulos, S. and Loukas, M. (1988). Genetic differentiation between coexisting color types of the *Alebra albostriella* group (Homoptera: Cicadellidae), *Journal of Heredity* **79:** 434–438.

Du Bose, W. P. (1960). The genus *Delphacodes* Fieber in North Carolina (Homoptera: Delphacidae), *Journal of the Elisha Mitchell Scientific Society* **76:** 36–63.

Dudich, E. (1920). Über den Stridulationsapparat einiger Käfer, *Entomologische Blätter* **16:** 146–161.

Duelli, P. and Johnson, J. B. (1982). Behavioral origin of tremulation, and possible stridulation, in green lacewings (Neuroptera: Chrysopidae), *Psyche* **88:** 375–381.

Duffels, J. P. (1988). The Cicadas of the Fiji, Samoa and Tonga Islands, their Taxonomy and Biogeography (Homoptera, Cicadoidea), *Entomonograph*, Vol. 10: Scandinavian Sience Press, Leiden, p. 108.

Duffels, J. P. (1993). The systematic position of *Moana expansa* (Homoptera, Cicadidae), with reference to sound organs and the higher classification of the superfamily Cicadoidea, *Journal of Natural History* **27:** 1223–1237.

Dumortier, B. (1963a). Morphology of sound emission apparatus in Arthropoda, In *Acoustic Behavior of Animals*, Busnel, R. G., Ed., Elsevier, Amsterdam, pp. 277–345.

Dumortier, B. (1963b). The physical characteristics of sound production in Arthropoda, In *Acoustic Behaviour of Animals*, Busnel, R. G., Ed., Elsevier, Amsterdam, pp. 346–373.

Dumortier, B. (1963c). Ethological and physiological study of sound emissions in Arthropoda, In *Acoustic Behaviour of Animals*, Busnel, R. G., Ed., Elsevier, Amsterdam, pp. 583–654.

Dunning, D. C., Byers, J. A., and Zanger, C. D. (1979). Courtship in two species of periodical cicadas, *Magicicada septendecim and Magicicada cassini. Animal Behaviour* **27:** 1073–1090.

Dupuis, C. (1953). Notes, remarques et observations diverses sur les Hémiptères Note VI - Appareil stridulatoire et stridulation des Cydnidae et Tessaratomidae (Heteroptera, Pentatomidae), *Cahiers des Naturalistes, Bulletin des N.P. n.s.* **8**(3–4)**:** 25–27.

Dyer, F. C. (2002). The biology of the dance language, *Annual Review of Entomology* **47:** 917–949.

Dyer, F. C. and Seeley, T. D. (1991). Dance dialects and foraging range in three Asian honey bee species, *Behavioral Ecology and Sociobiology* **28:** 227–233.

Ebendt, R., Friedel, J., and Kalmring, K. (1994). Central projection of auditory receptors in the prothoracic ganglion of the bushcricket Psorodonotus illyricus (Tettigoniidae): computer-aided analysis of the end branch pattern, *Journal of Neurobiology* **25:** 35–49.

Eberhard, W. G. (1985). *Sexual Selection and Animal Genitalia*, Harvard University Press, Cambridge.

Eberhard, W. G. (1986). Possible mutualism between females of the subsocial membracid *Polyglypta dispar* (Homoptera), *Behavioral Ecology and Sociobiology* **19:** 447–453.

Eberhard, W. G. (1991). Copulatory courtship and cryptic female choice in insects, *Biological Reviews of the Cambridge Philosophical Society* **66:** 1–31.

Eberhard, W. G. (1994). Evidence for widespread courtship during copulation in 131 species of insects and spiders, and implications for cryptic female choice, *Evolution* **48:** 711–733.

Eberhard, W. G. (1996). *Female Control: Sexual Selection by Cryptic Female Choice*, Princeton University Press, Princeton.

Eberhard, W. G. (2003). Sexual behavior and morphology of *Themira minor* (Diptera:Sepsidae) males and the evolution of male sternal lobes and genitalic surstyli, *Canadian Entomologist* **135:** 569–581.

Eberl, D. F. (1999). Feeling the vibes: chordotonal mechanisms in insect hearing, *Current Opinion in Neurobiology* **9:** 389–393.

Economides, P. S. (1991). *Check List of Fresh Water Fishes of Greece*, Hellenic Soc. for the Protection of Nature, Athens.

Eglin, W. (1939). Zur Biology und Morphologie der Raphidien und Myrmeleoniden (Neuropteroidea) von Basel und Umgegung, *Verhandlungen der Naturforschenden Gesellschaft, Basel* **50:** 163–220.

Eibl, E. and Huber, F. (1979). Central projections of tibial sensory fibers within the three thoracic ganglia of the crickets (*Gryllus campestris* L., *Gryllus bimaculatus* DeGeer), *Zoomorphologie* **92:** 1–17.

Eichele, G. and Villiger, W. (1974). Untersuchungen an den Stridulationsorganen de Florfliege, *Chrysopa carnea* (St.) (Neuroptera: Chrysopidae), *International Journal of Insect Morphology and Embryology* **3:** 41–46.

Eiriksson, T. (1993). Female preference for specific pulse duration of male songs in the grasshopper, *Omocestus viridulus, Animal Behaviour* **45**(3)**:** 471–477.

Elliot, C. J. and Koch, U. T. (1985). The clockwork cricket, *Naturwissenschaften* **72:** 150–153.

Elsner, N. and Popov, A. V. (1978). Neuroethology of acoustic communication, *Advances in Insect Physiology* **13:** 229–353.

Embleton, T. F. W., Piercy, J. E., and Olson, N. (1976). Outdoor sound propagation over ground of finite impedance, *Journal of the Acoustical Society of America* **35:** 1119–1125.

Emelyanov, A. F. and Kirillova, V. I. (1989). [Trends and modes of karyotype evolution in the Cicadinea (Homoptera)], *Entomologicheskoe Obozrenie*, 68(3), 587–603, in Russian, with English summary.

Endler, J. A. (1992). Signals, signal conditions, and the direction of evolution, *American Naturalist* **139:** S125–S153.

Endler, J. A. and Basolo, A. L. (1998). Sensory ecology, receiver biases and sexual selection, *Trends in Ecology and Evolution* **13:** 415–420.

Erbe, P. and Hoch, H. (2004). Two new species of the Australian planthopper genus *Solonaima* Kirkaldy (Hemiptera: Fulgoromorpha: Cixiidae), *Zootaxa* **536:** 1–7.

Erichson, W. F. (1837). *Die Käfer der Mark Brandenburg*, 1. Band, 1, Abt. Morin, Berlin, p. 740.

Erichson, W. F. (1848). *Naturgeschichte der Insecten Deutschlands I. Coloptera, 3*. Berlin, 1–968.

Esch, H. (1961a). Ein neuer Bewegungstyp im Schwänzeltanz der Bienen, *Naturwissenschaften* **48:** 140–141.

Esch, H. (1961b). Über die Schallerzeugung beim Werbetanz der Honigbiene, *Zeitschrift fur Vergleichende Physiologie* **45:** 1–11.

Esch, H. (1963). Über die Auswirkung der Futterplatzqualität auf die Schallerzeugung im Werbetanz der Honigbiene (*Apis mellifica*), *Zoologischer Anzeiger* (Suppl. 26), 302–309.

Esch, H. (1964). Beiträge zum Problem der Entfernungsweisung im Schwänzeltanz der Honigbiene, *Zeitschrift fur Vergleichende Physiologie* **48:** 534–546.

Esch, H. (1967a). Die Bedeutung der Lauterzeugung für die Verständigung der stachellosen Bienen, *Zeitschrift fur Vergleichende Physiologie* **56:** 199–220.

Esch, H. (1967b). The sounds produced by swarming honey bees, *Zeitschrift fur Vergleichende Physiologie* **56:** 408–411.

Esch, H., Esch, I., and Kerr, W. E. (1965). Sound: an element common to communication of stingless bees and to dances of honey bees, *Science* **149:** 320–321.

Esch, H. and Goller, F. (1991). Neural control of fibrillar muscles in bees during shivering and flight, *Journal of Experimental Biology* **159:** 419–431.

Esch, H., Huber, F., and Wohlers, D. W. (1980). Primary auditory neurons in crickets: Physiology and central projections, *Journal of Comparative Physiology* **137:** 27–38.

Esch, H. and Wilson, D. (1967). The sounds produced by flies and bees, *Zeitschrift fur Vergleichende Physiologie* **54:** 256–267.

Evans, J. W. (1946). A natural classification of leafhoppers, Pt. 2, *Proceedings of the Royal Entomological Society of London Series A* **97:** 39.

Evans, J. W. (1957). Some aspects of the morphology and inter-relationships of extinct and recent Homoptera, *Proceedings of the Royal Entomological Society of London Series A* **109:** 275–294.

Ewart, A. (1989). Cicada songs — song production, structures, variation and uniqueness within species. New Bulletin, *The Entomological Society of Queensland* **17**(7): 75–82.

Ewing, A. W. (1969). The genetic basis of sound production in *Drosophila pseudoobscura* and *D. persimilis*, *Animal Behaviour* **17:** 555–560.

Ewing, A. W. (1970). The evolution of courtship songs in *Drosophila*, *Rev. Comp. Anim.* **4:** 3–8.

Ewing, A. W. (1977). The neuromuscular basis of courtship song in *Drosophila*: the role of the indrect muscles, *Journal of Comparative Physiology* **119:** 249–265.

Ewing, A. W. (1978). The antenna of *Drosophila* as a 'love song' receptor, *Physiological Entomology* **3:** 33–36.

Ewing, A. W. (1979a). Complex courtship songs in the *Drosophila funebris* species group: escape from an evolutionary bottleneck, *Animal Behaviour* **27:** 343–349.

Ewing, A. W. (1979b). The neuromuscular basis of courtship song in *Drosophila*: the role of the direct and axillary wing muscles, *Journal of Comparative Physiology* **130:** 87–93.

Ewing, A. W. (1989). *Arthropod Bioacoustics: Neurobiology and Behaviour*, Edinburgh University Press, Edinburgh, p. 260.

Ewing, A. W. and Bennet-Clark, H. C. (1968). The courtship songs of *Drosophila*, *Behaviour* **31:** 288–301.

Ewing, A. W. and Miyan, J. A. (1986). Sexual selection, sexual isolation and the evolution of song in the *Drosophila repleta* group of species, *Animal Behaviour* **34:** 421–429.

Fabre, J. H. (1917). *Souvenirs entomologiques. Ètudes sur l'instinct et les mœurs des insectes*, (Sixième Série) Dixième édition. Librairie Delgrave, Paris, p. 418.

Farina, W. M. (2000). The interplay between dancing and trophallactic behavior in the honey bee *Apis mellifera*, *Journal of Comparative Physiology A* **186:** 239–245.

Fennah, R. G. (1973a). The cavernicolous fauna of Hawaiian lava tubes. 4. Two new blind *Oliarus* (Fulgoroidea Cixiidae), *Pacific Insects* **15**(1): 181–184.

Fennah, R. G. (1973b). Three new cavernicolous species of Fulgoroidea (Homoptera) from Mexico and Western Australia, *Proceedings of the Biological Society, Washington* **86**(38): 439–446.

Fennah, R. G. (1975). New cavernicolous cixiid from New Zealand (Homoptera Fulgoroidea), *New Zealand Journal of Zoology* **2**(3): 377–380.

Fennah, R. G. (1980a). A cavernicolous new species of Notuchus from New Caledonia (Homoptera: Fulgoroidea: Delphacidae), *Revue Suisse Zoologie* **87**(3): 757–759.

Fennah, R. G. (1980b). New and little-known neotropical Kinnaridae (Homoptera Fulgoroidea), *Proceedings of the Biological Society, Washington* **93**(3): 674–696.

Ferreira, M. and Ferguson, J. W. H. (2002). Geographic variation in the calling song of the field cricket *Gryllus bimaculatus* (Orthoptera: Gryllidae) and its relevance to mate recognition and mate choice, *Journal of Zoology* **257:** 163–170.

Finn, W. E., Payne, T. L., and Mastro, V. C. (1972). Stridulatory apparatus and analyses of acoustics of 4 species of subfamily Lamiinae (Col., Cerambycidae), *Annals of the Entomological Society of America* **65:** 644ff.

Fish, J. and Alcock, J. (1973). The behavior of *Chlorochroa ligata* (Say) and *Cosmopepla bimaculata* (Thomas) (Hemiptera: Pentatomidae), *Entomological News* **84:** 250–268.

Fitzpatrick, M. J. and Gray, D. A. (2001). Divergence between the courtship songs of the field crickets *Gryllus texensis* and *Gryllus rubens* (Orthoptera, Gryllidae), *Ethology* **107:** 1075–1085.

Flemming, C. A. (1971). A new species of cicada from rock fans in southern Wellington, with a review of three species with similar songs and habitat, *New Zealand Journal of Science* **14:** 443–449.

Fletcher, B. S. (1968). The storage and release of a sex pheromone by the Queensland fruit fly, *Dacus ctryoni* (Diptera: Tephritidae), *Nature* **219:** 631–632.

Fletcher, B. S. (1969). The structure and function of the sex pheromone glands of the male Queensland fruit fly, *Dacus tryoni*, *Journal of Insect Physiology* **15:** 1309–1322.

Fletcher, N. H. (2004). A simple frequency-scaling rule for animal communication, *Journal of the Acoustical Society of America* **115:** 2334–2338.

Fonseca, P. J. (1993). Directional hearing of a cicada: biophysical aspect, *Journal of Comparative Physiology A* **172:** 767–774.

Fonseca, P. J. and Bennet-Clark, H. C. (1998). Assymmetry of tymbal action and structure in a cicada: a possible role in the production of complex songs. *Journal of Experimental Biology* **201:** 717–730.

Fonseca, P. J. and Revez, M. A. (2002). Temperature dependence of cicada songs (Homoptera, Cicadoidea), *Journal of Comparative Physiology* **187A:** 971–976.

Fontana, P. (2001). Identità e bioacustica di *Roeseliana brunneri* Ramme, 1951, un endemita da tutelare (Insecta Orthoptera) [Italian], *Boll. Mus. Civ. Stor. Nat. Venezia* **52:** 59–75.

Fontana, P., Ode, B., and Malagnini, V. (1999). On the identity of *Decticus loudoni* Ramme, 1933 (Insecta Orthoptera Tettigoniidae), *Boll. Ist. Entomol. Guido Grandi Univ. Studi Bologna* **53:** 71–85.

Forrest, T. G. (1982). Acoustic communication and bafling behaviors of cricket, *Florida Entomologist* **65:** 33–44.

Forrest, T. G. (1983). Calling songs and mate choice in mole crickets, in orthopteran mating systems: sexual selection, In *A Diverse Group of Insects*, Gwynne, D. T. and Morris, G. K., Eds., Westview Press, Boulder, Colorado, chap. 9.

Forrest, T. G., Read, M. P., Farris, H. E., and Hoy, R. R. (1997). A tympanal hearing organ in scarab beetles, *Journal of Experimental Biology* **200:** 601–606.

Forsythe, T. G. (1978). Preliminary investigation into the stridulation mechanisms of the genus *Elaphrus* (Coleoptera: Carabidae) *Elaphrus cupreus* Duftschmid, *Coleopterists Bulletin* **32:** 41–46.

Forsythe, T. G. (1979). Preliminary investigation into the stridulatory mechanism of *Platyderus ruficollis* (Marsham) (Coleoptera: Carabidae), *Coleopterists Bulletin* **33:** 351–356.

Forsythe, T. G. (1980). Sound production in *Amara familiaris* Duft with a review of sound production in British Carabidae, *Entomologist's Monthly Magazine* **115:** 177–179.

Frantsevich, L. and Gorb, S. (1998). The probable purpose of the mid-coxal prong in Brachycera (Diptera), *Naturwissenschaften* **85:** 31–33.

Freitag, R. and Lee, S. K. (1972). Sound producing structures in adult *Cicindela tranquebarica* (Col., Cicindelidae) including a list of tiger beetles and ground beetles with flight wing files, *Canadian Entomologist* **104:** 851.

Frings, H. and Frings, M. (1957). The effects of temperature on chirp-rate of male conehead grasshoppers, *Neoconocephalus ensiger*, *Journal of Experimental Zoology* **134:** 411–425.

Frisch, J. L. (1724). *Beschreibung von allerley Insecten in Deutschland*, Fünfter Theil. C.G. Nicolai, Berlin, p. 55.

Frohlich, D. R., Torres-Jerez, I., Bedford, I. D., Markham, P. G., and Brown, J. K. (1999). A phylogeographic analysis of the *Bemisia tabaci* species complex based on mitochondrial DNA markers, *Molecular Ecology* **8:** 1683–1691.

Fuchs, S. and Koeniger, N. (1974). Schallerzeugung im Dienst der Verteidigung des Bienenvolkes (*Apis cerana* Fabr.), *Apidologie* **5:** 271–287.

Fullard, J. H. (1977). Phenology of sound producing arctiid moths and the activity of insectivorous bats, *Nature* **267:** 42–43.

Fullard, J. H. (1992). The neuroethology of sound production in tiger moths (Lepidoptera Arctiiddae). I. Rhytmicity and central contrôle, *Journal of Comparative Physiology A* **170:** 575–588.

Fullard, J. H. (1998). The sensory coevolution of moths and bats, In *Comparative Hearing: Insects*, Hoy, R. R., Popper, A. N., and Fay, R. R., Eds., Springer Verlag, New York, pp. 279–326.

Funkhouser, W. D. (1917). Biology of the Membracidae of the Cayuga Lake basin, *Memoirs of the Cornell University Agricultural Experiment Station* **2:** 177–445.

Gahan, C. J. (1900). Stridulation organs in Coleoptera, *Trans. Entomol. Soc. London* **48:** 433–452.

Galliart, P. L. and Shaw, K. C. (1992). The relation of male and female acoustic parameters to female phonotaxis in the katydid *Amblycorypha parvipennis*, *Journal of Orthopteran Research* **1:** 110–115.

Galliart, P. L. and Shaw, K. C. (1996). The effect of variation on parameters of the male calling song of the katydid, *Amblycorypha parvipennis* (Orthoptera: Tettigoniidae), on female phonotaxis and phonoresponse, *Journal of Insect Behaviour* **9:** 841–855.

Gaumont, J. (1976). L'appareil digestif des larves de Planipenes, *Annales des Sciences Naturelles — Zoologie et Biologie Animale* **18:** 145–250.

Gawel, N. J. and Bartlett, A. C. (1993). Characterization of differences between whiteflies using APD-PCR, *Insect Molecular Biology* **2:** 33–38.

Gerhardt, H. C. (1978). Temperature coupling in the vocal communication system of the gray treefrog, *Hyla versicolor, Science* **199:** 992–994.

Gerhardt, H. C. (1998). Acoustic signals of animals: recording, field measurements, analysis and description, In *Animal Acoustic Communication*, Hopp, S. L., Owren, M. J., and Evans, C. S., Eds., Springer, Berlin, pp. 1–25.

Gerhardt, H. C. and Huber, F. (2002). *Acoustic Communication in Insects and Anurans. Common Problems and Diverse Solutions*, University of Chicago Press, Chicago, IL.

Gilham, M. C. and de Vrijer, P. W. F. (1984). Patterns of variation in the acoustic calling signals of *Chloriona* planthoppers (Homoptera: Delphacidae) coexisting on common reed *Phragmites australis*, *Biological Journal of the Linnean Society* **54:** 245–269.

Gleason, J. M., Nuzhdin, S. V., and Ritchie, M. G. (2002). Quantitative trait loci affecting a courtship signal in *Drosophila melanogaster*, *Heredity* **89:** 1–6.

Gleason, J. M. and Ritchie, M. G. (2004). Do quantitative trait loci (QTL) for a courtship song difference between *Drosophila simulans* and *D. sechellia* coincide with candidate genes and intraspecific QTL?, *Genetics* **166:** 1303–1311.

Gogala, M. (1969). Die akustische Kommunikation bei der Wanze Tritomegas bicolor (L.) (Heteroptera, Cydnidae), *Zeitschrift fur Vergleichende Physiologie* **63:** 379–391.

Gogala, M. (1970). Artspezifität der Lautäusserungen bei Erdwanzen (Heteroptera, Cydnidae), *Zeitschrift fur Vergleichende Physiologie* **70:** 20–28.

Gogala, M. (1978a). Akustični signali štirih vrst iz družine Cydnidae (Heteroptera), *Bioloski Vestnik, (Ljubljana)* **26:** 153–168.

Gogala, M. (1978b). Ecosensory functions in insects (with remarks on Arachnida), In *Sensory Ecology*, Ali, M. A., Ed., Plenum Press, New York, London, pp. 123–153.

Gogala, M. (1984). Vibration producing structures and songs of terrestrial Heteroptera as systematic character, *Bioloski Vestnik, (Ljubljana)* **32:** 19–36.

Gogala, M. (1985a). Vibrational communication in insects (biophysical and behavioural aspects), In *Acoustic and Vibrational Communication in Insects*, Kalmring, K. and Elsner, N., Eds., Paul Parey, Berlin, Hamburg, pp. 117–126, (Proc. XVII. Intern. Congress Entomol. Hamburg 1984).

Gogala, M. (1985b). Vibrational songs of land bugs and their production, In *Acoustic and Vibrational Communication in Insects*, Kalmring, K. and Elsner, N., Eds., Paul Parey, Berlin, Hamburg, pp. 143–150, (Proc. XVII. Intern. Congress Entomol. Hamburg 1984).

Gogala, M. (1990). Distribution of low frequency vibrational songs in local Heteroptera, *Scopolia* **1**(Suppl.)**:** 125–132.

Gogala, M. (1995). Songs of four cicada species from Thailand, *Bioacoustics* **6:** 101–116.

Gogala, M. and Čokl, A. (1983). The acoustic behaviour of the bug Phymata crassipes (F.) (Heteroptera), *Revue Canadienne de Biologie Experimentale* **42:** 249–256.

Gogala, M., Čokl, A., Drašlar, K., and Blaževič, A. (1974). Substrate-borne sound communication in Cydnidae (Heteroptera), *Journal of Comparative Physiology* **94:** 25–31.

Gogala, M. and Hočevar, I. (1990). Vibrational songs in three sympatric species of Tritomegas, *Scopolia* **1**(Suppl.): 117–123.

Gogala, M., Popov, A. V., and Ribaric, D. (1996). Bioacoustics of singing cicadas of the western Palaearctic: *Cicadetta tibialis* (Panzer)(Cicadoidea: Tibicinidae), *Acta Entomologica Slovenica* **4**: 45–62.

Gogala, M. and Razpotnik, R. (1974). Metoda oscilografske sonagrafije za bioakustične raziskave, *Bioloski Vestnik (Ljubljana)* **22**: 209–216.

Gogala, M. and Riede, K. (1995). Time sharing of song activity by cicadas in Tremengor forest reserve, Hulu Perak, and Sabah, Malaysia, *Malaysian Nature Journal* **48**: 297–305.

Gogala, M. and Trilar, T. (1998). The song structure of *Cicadetta montana macedonica* Schedl with remarks on songs of related singing cicadas (Hemiptera: Auchenorhyncha: Cicadomorpha: Tibicinidae), *Reichenbachia, Dresden* **33**(11): 91–97.

Gogala, M. and Trilar, T. (2003). Video analysis of wing clicking in cicadas of the genera Cicadatra and Pagiphora (Homoptera: Auchenorrhyncha: Cicadoidea), *Acta Entomologica Slovenica* **11**(1): 5–15.

Gogala, M. and Trilar, T. (2004). Bioacoustic investigations and taxonomic considerations on the *Cicadetta montana* species complex (Homoptera: Cicadoidea: Tibicinidae), *Anais da Academia Brasileira de Ciências* **76**(2): 316–324.

Gogala, M., Trilar, T., Kozina, U., and Duffels, H. (2004). Frequency modulated song of the cicada *Maua albigutta* (Walker 1856) (Auchenorrhyncha: Cicadoidea) from South East Asia, *Scopolia* **54**: 1–15.

Gogala, M., Virant, M., and Blejec, A. (1984). Mocking bug Phymata crassipes (Heteroptera), *Acoustics Letters* **8**: 44–51.

Göpfert, M. C. and Robert, D. (2001). Turning the key of *Drosophila* audition, *Nature* **411**: 908.

Göpfert, M. C. and Robert, D. (2002). The mechanical basis of *Drosophila* audition, *Journal of Experimental Biology* **205**: 1199–1208.

Goulson, D., Birch, M. C., and Wyatt, T. D. (1994). Mate location in the deathwatch beetle *Xestobium rufovillosum* De Geer (Anobiidae): orientation to substrate vibrations, *Animal Behaviour* **47**: 899–907.

Goureau, M. (1837). Essay sur la stridulation des insectes, *Annales de la Societe entomologique de France (NS)* **6**: 31–75.

Gray, D. A. (1997). Female house crickets, *Acheta domesticus*, prefer the chirps of large males, *Animal Behaviour* **54**: 1553–1562.

Gray, D. A. and Cade, W. H. (1999). Sex, death and genetic variation: natural and sexual selection on cricket song, *Proceedings of the National Academy of Science, USA B* 14449–14454.

Gray, D. A. and Cade, W. H. (2000). Sexual selection and speciation in field crickets, *Proceedings of the National Academy of Sciences of the United States of America* **97**: 14449–14454.

Gray, D. A. and Cade, W. H. (2003). Sex, death and genetic variation: natural and sexual selection on cricket song, *Proceedings of the Royal Society of London Series B* **266**: 707–709.

Gray, D. A. and Eckhardt, G. (2001). Is cricket courtship song condition dependent?, *Animal Behaviour* **62**: 871–877.

Greenfield, M. D. (1983). Unsynchronized chorusing in the coneheaded katydid *Neoconocephalus affinis* (Beauvois), *Animal Behaviour* **31**: 102–112.

Greenfield, M. D. (1988). Interspecific acoustic interactions among katydids (*Neoconocephalus*): inhibition-induced shifts in diel periodicity, *Animal Behaviour* **36**: 684–690.

Greenfield, M. D. (1994). Synchronous and alternating choruses in insects and Anurans: common mechanisms and diverse functions, *American Zoologist* **34**: 605–615.

Greenfield, M. D. (1997). Acoustic communication in Orthoptera, In *Bionomics of Grasshoppers, Katydids and Their Kin*, Gangwere, S. K., Muralirangan, M. C., and Muralirangan, M., Eds., CAB International, Wallingford, Oxon, UK, pp. 197–230.

Greenfield, M. D. (2002). *Signalers and Receivers: Mechanisms and Evolution of Arthropod Communication*, Oxford University Press, Oxford.

Greenfield, M. D., Alkaslassy, E., Wang, G., and Shelly, T. E. (1989). Long-term memory in territorial grasshoppers, *Experientia* **45**: 775–777.

Greenfield, M. D. and Roizen, I. (1993). Katydid synchronous chorusing is an evolutionary stable outcome of female choice, *Nature* **364**: 618–620.

Gregory, P. G. and Howard, D. J. (1994). A postinsemination barrier to fertilization isolates two closely related ground crickets, *Evolution* **48**: 705–710.

Grolous, J. (1880). *De l'origine et des Métamorphoses des Insectes*, Lubbock, J., Ed., Reinwald, Paris, p. 130, [30, footnote].

Gross, G. (1976). *Plant-feeding and Other Bugs (Hemiptera) of South Australia, Heteroptera — Part I*, A.B. James, South Australia.

Groot, A. T. (2000). *Sexual Behaviour of the Green Capside Bug*, Ponsen & Looyen BV, Wageningen.

Guerra, P. A. and Morris, G. K. (2002). Calling communication in meadow katydids (Orthoptera, Tettigoniidae): female preferences for species-specific wingstroke rates, *Behaviour* **139**: 23–43.

Gullan, P. J. and Cranston, P. S. (2005). *The Insects. An Outline of Entomology*, 3rd ed., Blackwell, Oxford.

Gwynne, D. T. (1984). Courtship feeding increases female reproductive success in bushcrickets, *Nature* **307**: 361–363.

Gwynne, D. T. (1987). Sex-biased predation and the risky mate-locating behaviour of male tick-tock cicadas (Homoptera: Cicadidae), *Animal Behaviour* **35**: 571–576.

Gwynne, D. T. (2001). *Katydids and Bush-Crickets: Reproductive Behavior and Evolution of the Tettigoniidae*, Cornell University Press, New York.

Gwynne, D. T. and Bailey, W. J. (1988). Mating system, mate choice and ultrasonic calling in a zaprochiline katydid (Orthoptera: Tettigoniidae), *Behaviour* **105**: 202–223.

Gwynne, D. T. and Bailey, W. J. (1999). Female–female competition in katydids: sexual selection for increased female to a male signal?, *Evolution* **53**: 546–555.

Haas, A. (1961). Das Rätsel des Hummeltrompeters: Lichtalarm — 1. Bericht über Verhaltensstudien an einem kleinen Nest von *Bombus hypnorum* mit Arbeiter-Königin, *Z. Tierpsychol.* **18**: 129–138.

Hamilton, K. G. A. (1975). Review of the tribal classification of the leafhopper subfamily Aphrodinae (Deltocephalinae of authors) of the Holarctic region (Rhynchota: Homoptera: Cicadellidae), *Canadian Entomologist* **107**(5–6): 477–498.

Hamilton, K. G. A. (1981). Morphology and evolution of the Rhychotan head (Insecta: Hemiptera, Homoptera), *Canadian Entomologist* **113**: 953–974.

Hammond, T. J. and Bailey, W. J. (2003). Eavesdropping and defensive auditory masking in an Australian bushcricket *Caedicia* (Phaneropterinae: Tettigoniidae: Orthoptera), *Behaviour* **140**: 79–95.

Handlirsch, A. (1900a). Zur Kenntnis der Stridulationsorgane bei den Rhynchota, *Annalen des Kaiserlich-Koniglichen Naturhistorischen Hofmuseums Wien* **15**: 127–141.

Handlirsch, A. (1900b). Neue Beiträge zur Kenntnis der Stridulationsorgane bei den Rhynchoten, *Verhandlungen der Kaiserlich-Koniglichen Zoologisch-Botanischen Gesellschaft in Wien* **50**: 555–560.

Hanrahan, S. A. and Kirchner, W. H. (1994). Acoustic orientation and communication in desert tenebrionid beetles in sand dunes, *Ethology* **97**: 26–32.

Hanski, I. (1991). In *Dung Beetle Ecology*, Cambefort, Y. and Hanski, I., Eds., Princeton University Press, Princeton, New Jersey, pp. 75–96.

Hansson, Å. (1945). Lauterzeugung und Lautauffassungsvermögen der Bienen, *Opuscula Entomologica* **VI**: 1–124.

Haring, E. and Aspöck, U. (2004). Phylogeny of the Neuropterida: a first molecular approach, *Systematic Entomology* **29**: 415–430.

Harman, D. M. and Harman, A. L. (1984). Comparison of stridulatory structures in North American Pissodes spp. (Coleoptera: Curculionidae), *Proceedings of the Entomological Society of Washington* **86**: 228–238.

Harris, W. H. (1903). Remarks on the emission of musical notes and on the hovering habit of *Eristalis tenax*, *Journal of the Quekett Microscopical Club* **8**: 513–520, 2nd Ser.

Harris, V. E. and Todd, J. W. (1980). Temporal and numerical patterns of reproduction behavior in the southern green stink bug, *Nezara viridula* (Hemiptera: Pentatomidae), *Entomologia Experimentalis et Applicata* **27**: 105–116.

Harz, K. (1969). *Die Orthopteren Europas 1*, Series entomologica 5, Dr. W. Junk N.V., The Hague.

Harz, K. (1975). *The Orthoptera of Europe II*, Series entomologica 11, Dr. W. Junk N.V., The Hague.

Hase, A. (1933). Über die Lauterzeugung sowie deren mutmassliche Bedeutung bei die Wanze *Panstrongylus*, *Biologisches Zentralblatt* **53**: 607–614.

Haskell, P. T. (1957). Stridulation and its analysis in certain Geocorisae (Hemiptera, Heteroptera), *Proceedings of the Zoological Society of London* **129**: 351–358.

Haskell, P. T. (1961). *Insect Sounds*, H. F. and G. Witherby Ltd, London, p. 189.

Hatch, M. (1991). *The Beetles of the Pacific North West*, Part V, Vol. 16: University of Washington Publications in Biology, pp. 433–533.

Hayashi, F. (1993). Male mating costs in two insect species (*Protohermes*, Megaloptera) that produce large spermatophores, *Animal Behaviour* **45**: 343–349.

Hayashi, F. (1996). Insemination through an externally attached spermatophore: Bundled sperm and post-copulatory mate guarding by male fishflies (Megaloptera: Corydalidae), *Journal of Insect Physiology* **42**: 859–866.

Heady, S. E. and Nault, L. R. (1991). Acoustic signals of *Graminella nigrifons* (Homoptera: Cicadellidae). *Great Lakes Entomology* **24**: 9–16.

Heady, S. E., Nault, L. R., Shambaugh, G. F. and Fairchild, L. (1986). Acoustic and mating behavior of *Dalbulus* leafhoppers (Homoptera: Cicadellidae). *Annals of the Entomological Society of America* **79**: 727–736.

Heath, J. E. (1967). Temperature responses of the periodical 17-year cicada, *Magicicada cassini* (Homoptera, Cicadidae), *American Midland Naturalist* **77**: 64–67.

Heath, M. S. (1972). Temperature requirements of the cicada *Okanagana striatipes beameri*: a study from Flagstaff, Arizona, *Plateau* **45**: 31–40.

Heath, J. E., Hanagan, J. L., Wilkin, P. J., and Heath, M. S. (1971). Adaptation of the thermal responses of insects, *American Zoologist* **11**: 147–158.

Heath, J. E. and Josephson, R. K. (1970). Body temperature and singing in the katydid, *Neoconocephallus robustus* (Orthoptera, Tettigoniidae), *Biological Bulletin* **138**(3): 272–285.

Heath, J. E. and Wilkin, P. J. (1970). Temperature responses of the desert cicada, *Diceroprocta apache* (Homoptera, Cicadidae), *Physiological Zoology* **43**: 145–154.

Heath, J. E., Wilkin, P. J., and Heath, M. S. (1972). Temperature responses of the cactus dodger, *Cacama valvata* (Homoptera, Cicadidae), *Physiological Zoology* **45**: 238–246.

Hedrick, A. V. and Weber, T. (1998). Variance in female responses to the fine structure of male song in the field cricket, *Gryllus integer. Behavioural Ecology* **9**: 582–591.

Heidelbach, J., Böhm, H., and Kirchner, W. H. (1998). Sound and vibration signals in a bumblebee colony [*Bombus terrestris*], *Zoology* **101**(Suppl. 1): 82.

Heinrich, B. (1981). The mechanisms and energetics of honeybee swarm temperature regulation, *Journal of Experimental Biology* **80**: 217–229.

Heinrich, R., Jatho, M., and Kalmring, K. (1993). Acoustic transmission characteristics of the tympanal tracheae of bushcrickets (Tettigoniidae). II: Comparative studies of the tracheae of seven species, *Journal of the Acoustical Society of America* **93**: 3481–3489.

Heller, K.-G. (1984). Zur Bioakustik und Phylogenie der Gattung *Poecilimon* (Orthoptera, Tettigoniidae, Phaneropterinae), *Zoologische Jahrbucher, Abteilung Systematik* **111**: 69–117.

Heller, K.-G. (1986). Warm-up and stridulation in the bushcricket *Hexacentrus unicolor* Serville (Orthoptera, Conocephalidae, Listroscelinae), *Journal of Experimental Biology* **126**: 97–109.

Heller, K.-G. (1988). *Bioakustik der Europäischen Laubheuschrecken. Ökologie in Forschung und Anwendung*, Verlag Josef Margraf, Weikersheim.

Heller, K.-G. (1990). Evolution of song pattern in east Mediterranean Phaneropterinae: Constraints by the communication system, In *The Tettigoniidae: Biology, Systematics and Evolution*, Bailey, W. J. and Rentz, D. C. F., Eds., Crawford House Press, Bathurst and Springer, Berlin, pp. 130–151.

Heller, K.-G. (1992). Risk shift between males and females in the pair-forming behaviour of bushcrickets, *Naturwissenschaften* **79**: 89–91.

Heller, K.-G. (1995). Acoustic signalling in palaeotropical bushcrickets (Orthoptera: Tettigonioidea: Pseudophyllidae): does predation pressure by eavesdropping enemies differ in the Palaeo- and Neotropics?, *Journal of Zoology London* **237**: 469–485.

Heller, K.-G. (2003). Calling songs of North African bush-crickets, recorded by Albrecht Faber in 1965, *Articulata* **18**: 1–9.

Heller, K.-G. (2004). *Poecilimon martinae sp. n* and *P. inflatus* Brunner von Wattenwyl, 1891 (Orthoptera, Tettigonioidea, Phaneropteridae), two bush-cricket species endemic to southwest Anatolia: morphology, bioacoustics and systematics, *Articulata* **19**: 1–17.

Heller, K.-G. and Lehmann, A. (2004). Taxonomic revision of the European species of the *Poecilimon ampliatus*-group (Orthoptera: Tettigonioidea: Phaneropteridae), *Memorie della Societa Entomologia Italiana* **82**: 403–422.

Heller, K.-G. and Reinhold, K. (1992). A new bushcricket of the genus *Poecilimon* from the Greek islands (Orthoptera: Phaneropterinae), *Tijdschrift voor Entomologie* **135**: 163–168.

Heller, K.-G. and von Helversen, D. (1986). Acoustic communication in phaneropterid bushcrickets: species-specific delay of female stridulatory response and matching male sensory time window, *Behavioral Ecology and Sociobiology* **18**: 189–198.

Heller, K.-G. and von Helversen, D. (1993). Calling behaviour in bushcrickets of the genus Poecilimon with differing communication systems (Orthoptera: Tettigonioidea: Phaneropteridae), *Journal of Insect Behaviour* **6**: 361–377.

Heller, K.-G. and Willemse, F. (1989). Two new bush-crickets from Greece, *Leptophyes lisae* sp. nov. *Platycleis (Parnassiana) tenuis* sp. nov. (Orthoptera: Tettigoniidae), *Entomologische Berichten* **49**: 144–156.

Hennig, R. M., Weber, T., Moore, T. E., Huber, F., Kleindienst, H.-U., and Popov, A. V. (1994). Function of the tensor muscle in the cicada *Tibicen linnei*, *Journal of Experimental Biology* **187**: 33–44.

Henry, T. (1976). *Aleuropteryx juniperi*: a European scale predator established in North America (Neuroptera: Coniopterygidae), *Proceedings of the Entomological Society of Washington* **78**: 195–201.

Henry, C. S. (1979). Acoustical communication during courtship and mating in the green lacewing *Chrysopa carnea* (Neuroptera: Chrysopidae), *Annals of the Entomological Society of America* **72**: 68–79.

Henry, C. S. (1982a). Reproductive and calling behavior in two closely related, sympatric lacewing species, *Chrysopa oculata* and *C. chi* (Neuroptera: Chrysopidae), *Proceedings of the Entomological Society of Washington* **84**: 191–203.

Henry, C. S. (1982b). Neuroptera, In *McGraw Hill Synopsis and Classification of Living Organisms*, Parker, S., Ed., McGraw Hill, New York, pp. 470–482.

Henry, C. S. (1985). Sibling species, call differences, and speciation in green lacewings (Neuroptera: Chrysopidae: *Chrysoperla*), *Evolution* **39**: 965–984.

Henry, C. S. (1989). The unique purring song of *Chrysoperla comanche* (Neuroptera: Chrysopidae), a western sibling of *C. rufilabris*, *Proceedings of the Entomological Society of Washington* **91**: 133–142.

Henry, C. S. (1993). *Chrysoperla johnsoni* (Neuroptera: Chrysopidae): acoustic evidence for full species status, *Annals of the Entomological Society of America* **86**: 14–25.

Henry, C. S. (1994). Singing and cryptic speciation in insects, *Trends in Ecology and Evolution* **9**: 388–392.

Henry, C. S. (1997). Modern mating systems in archaic Holometabola: sexuality in neuropterid insects, In *The Evolution of Mating Systems in Insects and Arachnids*, Choe, J. C. and Crespi, B. J., Eds., Cambridge University Press, Cambridge, pp. 193–210.

Henry, C. S., Brooks, S. J., Johnson, J. B., and Duelli, P. (1996). *Chrysoperla lucasina* (Lacroix): a distinct species of green lacewing, confirmed by acoustical analysis (Neuroptera: Chrysopidae), *Systematic Entomology* **21**: 205–218.

Henry, C. S., Brooks, S. J., Johnson, J. B., and Duelli, P. (1999). Revised concept *Chrysoperla mediterranea* (Hölzel), a green lacewing associated with conifers: courtship song across 2800 kilometres of Europe (Neuroptera: Chrysopidae), *Systematic Entomology* **24**: 335–350.

Henry, C. S., Brooks, S. J., Thierry, D., Duelli, P., and Johnson, J. B. (2001). The common green lacewing (*Chrysoperla carnea s. lat.*) and the sibling species problem, In *Lacewings in the Crop Environment*, McEwen, P. K., New, T. R., and Whittington, A. E., Eds., Cambridge University Press, Cambridge, pp. 29–42.

Henry, C. S., Brooks, S. J., Duelli, P., and Johnson, J. B. (2002a). Discovering the true *Chrysoperla carnea* (Stephens) (Insecta: Neuroptera: Chrysopidae) using song analysis, morphology, and ecology, *Annals of the Entomological Society of America* **95**: 172–191.

Henry, C. S., Brooks, S. J., Duelli, P., and Johnson, J. B. (2003). A lacewing with the wanderlust: the European song species 'Maltese', *Chrysoperla agilis* sp.n., of the *carnea* group of *Chrysoperla* (Neuroptera: Chrysopidae), *Systematic Entomology* **28**: 131–148.

Henry, C. S. and Johnson, J. B. (1989). Sexual singing in a nonchrysoperlan green lacewing, *Chrysopiella minora* Banks, *Canadian Journal of Zoology* **67**: 1439–1446.

Henry, C. S., Martínez, M. L., and Holsinger, K. E. (2002). The inheritance of mating songs in two cryptic, sibling lacewing species (Neuroptera: Chrysopidae: Chrysoperla), *Genetica* **116**: 269–289.

Henry, C. S. and Wells, M. L. M. (1990). Geographical variation in the song of *Chrysoperla plorabunda* (Neuroptera: Chrysopidae) in North America. *Annals of the Entomological Society of America* **83:** 317–325.

Henry, C. S. and Wells, M. L. M. (2004). Adaptation or random change? The evolutionary response of songs to substrate properties in lacewings (Neuroptera: Chrysopidae: *Chrysoperla*), *Animal Behaviour* **68:** 879–895.

Henry, C. S., Wells, M. M., and Pupedis, R. J. (1993). Hidden taxonomic diversity within *Chrysoperla plorabunda* (Neuroptera: Chrysopidae): two new species based on courtship songs, *Annals of the Entomological Society of America* **86:** 1–13.

Henry, C. S., Wells, M. L. M., and Simon, C. M. (1999). Convergent evolution of courtship songs among cryptic species of the *carnea*-group of green lacewings (Neuroptera: Chrysopidae: *Chrysoperla*), *Evolution* **53:** 1165–1179.

Hensen, K. and Kirchner, W. H. (2003). *Wahrnehmung von luftgetragenem Nahfeldschall bei Hummeln (Bombus terrestris), Proceedings of 18th German IUSSI-Meeting,* Regensburg, Germany, p. 24.

Henwood, K. and Fabrick, A. (1979). A quantitative analysis of the dawn chorus: temporal selection for communicatory optimization, *American Naturalist* **114:** 260–274.

Heran, H. (1957). Die Bienenantenne als Meßorgan der Flugeigengeschwindigkeit, *Naturwissenschaften* **44:** 475.

Heran, H. (1959). Wahrnehmung zur Regelung der Flugeigengeschwindigkeit bei *Apis mellifica* L., *Zeitschrift fur Vergleichende Physiologie* **42:** 103–163.

Hergenröder, R. and Barth, F. G. (1983). Vibratory signals and spider behavior: how do the sensory inputs from the eight legs interact in orientation?, *Journal of Comparative Physiology A* **152:** 361–371.

Heslop-Harrison, G. (1960). Sound production in the Homoptera with special reference to sound producing mechanisms in the Psyllidae, *Annals and Magazine of Natural History* **3**(34)**:** 633–640, Ser. 13.

Hesse, A. J. (1936). The sound-producing or stridulating organs of a few Peninsula insects, *Cape Naturalist (Cape Town)* **1:** 70–76.

Hesse, A. J. (1948). Notes on the sub-family Acanthocerinae (Col., Scarabaeidae) and description of new species from Natal and Zululand, *Annals of the Natal Museum* **11:** 377–383.

Hewitt, G. M. (2000). The genetic legacy of the Quaternary ice ages, *Nature* **405:** 907–913.

Higgins, L. A. and Waugaman, R. D. (2004). Sexual selection and variation: a multivariate approach to species-specific calls and preferences, *Animal Behaviour* **68:** 1139–1153.

Hill, P. S. M. (1998). Environmental and social influences on calling effort in the prairie mole cricket (*Gryllotalpa major*), *Behavioral Ecology* **9:** 101–108.

Hill, K. G. and Oldfield, B. P. (1981). Auditory function in Tettigoniidae (Orthoptera: Ensifera), *Journal of Comparative Physiology* **142:** 169–180.

Hill, P. S. M. and Shadley, J. R. (2001). Talking back: sending soil vibration signals to lekking prairie mole cricket males, *American Zoologist* **41:** 1200–1214.

Hinton, H. E. (1946). The "Gin-traps" of some beetle pupae; a protective device which appears to be unknown, *Proceedings of the Royal Entomological Society of London Series A* **97:** 473–496.

Hinton, H. E. (1955). Protective devices of endopterygote pupae, *Transactions of the Society for British Entomology* **12:** 49–92.

Hirschberger, P. (1998). Spatial distribution, resource utilisation and intraspecific competition in the dung beetle *Aphodius ater*, *Oecologia* **116:** 136–142.

Hirschberger, P. (2001). Stridulation in *Aphodius* dung beetles: Behavioral context and intraspecific variability of song patterns in *Aphodius ater* (Scarabaeidae), *Journal of Insect Behaviour* **14:** 69–88.

Hirschberger, P. and Rohrseitz, K. (1995). Stridulation in the adult dung beetle *Aphodius ater* (Col., Aphodiidae), *Zoology* **99:** 97–102.

Hlawatsch, F. and Boudreaux-Bartels, G. F. (1992). Linear and quadratic time-frequency signal representations, *IEEE Signal Process. Mag.* **9**(2)**:** 21–67.

Hoch, H. (1988a). A new cavernicolous planthopper species (Homoptera: Fulgoroidea: Cixiidae) from Mexico, *Mitteilungen der Schweizerischen Entomologischen Gesellschaft* **61:** 295–302.

Hoch, H. (1988b). Five new epigean species of the Australian planthopper genus Solonaima Kirkaldy (Homoptera: Fulgoroidea: Cixiidae), *The Beagle, Rec. North. Territ. Mus. Art. Sci.* **5**(1)**:** 125–133.

Hoch, H. (1990). Cavernicolous Meenoplidae (Homoptera Fulgoroidea) from Australia, *Occasional Papers of the Bishop Museum* **30:** 188–203.

Hoch, H. (1991). Cave-dwelling Cixiidae (Homoptera Fulgoroidea) from the Azores, *Bocagiana* **149**: 1–9.

Hoch, H. (1993). A new troglobitic planthopper species (Hemiptera: Fulgoroidea: Meenoplidae) from Western Australia, *Records of the Western Australian Museum* **16**(3): 393–398.

Hoch, H. (1994). Homoptera (Auchenorrhyncha Fulgoroidea), In *Encyclopaedia Biospeologica*, Tome, I., Juberthie, C., and Decu, V., Eds., Moulis-Bucarest, pp. 313–325, See also p. 834.

Hoch, H. (1996). A new cavernicolous planthopper of the family Meenoplidae from New Caledonia (Hemiptera: Fulgoroidea), *Records of the Western Australian Museum* **17**: 451–454.

Hoch, H. (1997). The Hawaiian cave planthoppers (Homoptera: Fulgoroidea: Cixiidae) — a model for rapid subterranean speciation?, *International Journal of Speleology* **26**(1–2): 21–31.

Hoch, H. (2000). Acoustic communication in darkness. 211-219, In *Ecosystems of the World*, Subterranean Ecosystems, Vol. 30: Wilkens, H., Culver, D. C., and Humphreys, W. F., Eds., Elsevier, Amsterdam, p. 791.

Hoch, H. (2002). Hidden from the light of day: planthoppers in subterranean habitats (Hemiptera: Auchenorrhyncha: Fulgoromorpha), In *Zikaden: Leafhoppers, Planthoppers and Cicadas (Insecta: Hemiptera: Auchenorrhyncha)*, Holzinger, W. E., Ed., (Denisia 4), Katalog des Oberösterreichischen Landesmuseums, Lindz, Austria, 139–146.

Hoch, H. and Asche, M. (1988). A new troglobitic meenoplid from a lava tube in Western Samoa (Homoptera Fulgoroidea Meenoplidae), *Journal of Natural History* **22**: 1489–1494.

Hoch, H. and Asche, M. (1993). Evolution and speciation of cave-dwelling Fulgoroidea in the Canary Islands (Homoptera: Cixiidae and Meenoplidae), *Zoological Journal of the Linnean Society* **109**: 53–101.

Hoch, H., Bonfils, J., Reynaud, B., and Attié, M. (2003). First record of troglobitic Hemiptera from La Réunion (Fulgoromorpha: Cixiidae), *Annales de la Société Entomologique de France (n.s.)* **39**(3): 265–270.

Hoch, H. and Howarth, F. G. (1989a). Reductive evolutionary trends in two new cavernicolous species of a new Australian cixiid genus (Homoptera Fulgoroidea), *Systematic Entomology* **14**: 179–196.

Hoch, H. and Howarth, F. G. (1989b). Six new cavernicolous cixiid planthoppers in the genus *Solonaima* from Australia (Homoptera Fulgoroidea), *Systematic Entomology* **14**: 377–402.

Hoch, H. and Howarth, F. G. (1993). Evolutionary dynamics of behavioral divergence among populations of the Hawaiian cave-dwelling planthopper *Oliarus polyphemus* (Homoptera: Fulgoroidea: Cixiidae), *Pacific Science* **47**(4): 303–318.

Hoch, H. and Howarth, F. G. (1999). Multiple cave invasions by species of the planthopper genus *Oliarus* in Hawaii (Homoptera: Fulgoroidea: Cixiidae), *Zoological Journal of the Linnean Society* **127**: 453–475.

Hoch, H. and Izquierdo, I. (1996). A cavernicolous planthopper in the Galápagos Islands (Homoptera Auchenorrhyncha Cixiidae), *Journal of Natural History* **30**: 1495–1502.

Hoch, H., Oromi, P. and Arechavaleta, M. (1999). *Nisia subfogo* sp.n, a new cave-dwelling planthopper from the Cape Verde Islands (Hemiptera: Fulgoromorpha: Meenoplidae), *Revista de la Academia Canaria de Ciencias* **11**(3–4): 189–199.

Hödl, W. (1977). Call differences and calling site segregation in anuran species from Central Amazonian floating meadows, *Oecologia* **28**: 351–363.

Hoffmann, E. and Jatho, M. (1995). The acoustic trachea of tettigoniids as an exponential horn: theoretical calculations and bioacoustical measurements, *Journal of the Acoustical Society of America* **98**: 1845–1851.

Hograefe, T. (1984). Substrat-Stridulation bei den koloniebildenden Blattwespenlarven von *Hemichroa crocea* (Geoff) (Hymenoptera: Tenthredinidae), *Zoologischer Anzeiger* **213**: 234–241.

Hoikkala, A. (1985). Genetic variation in the male courtship sound of *Drosophila littoralis*, *Behavioral Genetics* **15**: 135–142.

Hoikkala, A. and Crossley, S. A. (2000). Copulatory courtship in *Drosophila*: behaviour and songs of *D. birchii* and *D. serrata*, *Journal of Insect Behaviour* **13**: 71–86.

Hoikkala, A., Crossley, S., and Castillo-Melendez, C. (2000a). Copulatory courtship in *Drosophila birchii* and *D. serrata*, species recognition and sexual selection, *Journal of Insect Behaviour* **13**: 361–373.

Hoikkala, A., Hoy, R. R., and Kaneshiro, K. Y. (1989). High-frequency clicks of Hawaiian picture-winged *Drosophila* species, *Animal Behaviour* **37**: 927–934.

Hoikkala, A., Kaneshiro, K. Y., and Hoy, R. R. (1994). Courtship songs of the picture-winged *Drosophila planitibia* subgroup species, *Animal Behaviour* **47**: 1363–1374.

Hoikkala, A. and Lumme, J. (1987). The genetic basis of evolution of male courtship sounds in the *Drosophila virilis* group, *Evolution* **4:** 827–845.

Hoikkala, A., Päällysaho, S., Aspi, J., and Lumme, J. (2000b). Localization of genes affecting species differences in male courtship song between *Drosophila virilis* and *D. littoralis*, *Genetical Research, Cambridge* **75:** 37–45.

Hölldobler, B. (1984). Evolution of insect communication, In *Insect Communication*, Lewis, T., Ed., Academic Press, London, pp. 349–377.

Hölldobler, B. and Wilson, E. O. (1990). *The Ants*, Springer, Berlin, p. 732.

Holzapfel, C. M. and Cantrall, I. J. (1972). Evolution in the Canary Islands. The genus Calliphona (Orthoptera: Tettigoniidae), *Occasional Papers of the Museum of Zoology, The University of Michigan* **663:** 1–22.

Hopp, S. L. and Morton, E. S. (1998). Sound playback studies, In *Animal Acoustic Communication*, Hopp, S. L., Owren, M. J., and Evans, C. S., Eds., Springer, Berlin, pp. 323–352.

Hopp, S. L., Owren, M. J., Evans, C. S., Eds. (1998). *Animal Acoustical Communication: Sound Analysis and Research Methods*, Springer-Verlag, Berlin/Heidelberg, ISBN 3-540-53353-2.

Horn, G. H. (1876). Revision of the United States species *Ochodaeus* and other genera, *Transactions of the American Entomological Society* **5:** 177ff.

Howard, D. J. (1993). Reinforcement: Origin, dynamics, and fate of an evolutionary hypothesis, In *Hybrid Zones and the Evolutionary Process*, Harrison, R. G., Ed., Oxford University Press, New York, pp. 46–69.

Howard, D. J. and Furth, D. G. (1986). Review of the *Allonemobius fasciatus* (Orthoptera: Gryllidae) complex with the description of two new species separated by electrophoresis, songs, and morphometrics, *Annals of the Entomological Society of America* **79:** 472–481.

Howarth, F. G. (1981). Community structure and niche differentiation in Hawaiian lava tubes, In *Island Ecosystems*, US/IBP Synthesis Series 15, Mueller-Dombois, D., Bridges, K. W., and Carson, H. L., Eds., pp. 318–336.

Howarth, F. G. (1983). Ecology of cave arthropods, *Annual Review of Entomology* **28:** 365–389.

Howarth, F. G. (1986). The tropical environment and the evolution of troglobites, Proceedings of the 9th International Congress of Speleology, Vol. II, Barcelona, Spain, pp. 153–155.

Howarth, F. G. and Hoch, H. (2004). Adaptive shifts, In *Encyclopedia of Caves*, Culver, D. C. and White, W. B., Eds., Elsevier Academic Press, New York, p. 696.

Howarth, F. G. and Stone, F. D. (1990). Elevated carbon dioxide levels in Bayliss Cave, Australia: implications for the evolution of obligate cave species, *Pacific Science* **44**(3): 207–218.

Howse, P. E. (1964). The significance of sound produced by the termite *Zootermopsis angusticollis* (Hagen), *Animal Behaviour* **12:** 284–300.

Hoy, R. R. (1974). Genetic control of acoustic behavior in crickets, *American Zoologist* **14:** 1067–1080.

Hoy, R. R., Hoikkala, A., and Kaneshiro, K. (1988). Hawaiian courtship songs: evolutionary innovation in communication signals of *Drosophila*, *Science* **240:** 217–219.

Hrncir, M., Jarau, S., Zucchi, R., and Barth, F. G. (2000). Recruitment behavior in stingless bees, *Melipona scutellaris* and *M. quadrifasciata*. II. Possible mechanisms of communication, *Apidologie* **31:** 93–113.

Hrncir, M., Jarau, S., Zucchi, R., and Barth, F. G. (2003). A stingless bee (*Melipona seminigra*) uses optic flow to estimate flight distances, *Journal of Comparative Physiology A* **189:** 761–768.

Hrncir, M., Jarau, S., Zucchi, R., and Barth, F. G. (2004a). Thorax vibrations in stingless bees (*Melipona seminigra*). I. No influence of visual flow, *Journal of Comparative Physiology A* **190:** 539–548.

Hrncir, M., Jarau, S., Zucchi, R., and Barth, F. G. (2004b). Thorax vibrations in stingless bees (*Melipona seminigra*). II. Dependence on sugar concentration, *Journal of Comparative Physiology A* **190:** 549–560.

Hrncir, M., Zucchi, R., and Barth, F. G. (2002). Mechanical recruitment signals in *Melipona* vary with gains and costs at the food source, *Anais do V Encontro Sobre Abelhas*, Ribeirão Preto, Brasil, pp. 172–175.

Huang, Y., Ortí, G., Sutherlin, M., Duhachek, A., and Zera, A. (2000). Phylogenetic relationships of North American field crickets inferred from mitochondrial DNA data, *Molecular Phylogenetics and Evolution* **17:** 48–57.

Hubbel, S. P. and Johnson, L. K. (1978). Comparative foraging behavior of six stingless bee species exploiting a standardized resource, *Ecology* **59:** 1123–1136.

Huber, F. (1964). The insect and the interneural environment-homeostasis, In *Physiology of Insecta*, Vol. 2: Rockstein, M., Ed., Academic Press, New York, pp. 334–400.

Huber, F., Kleindienst, H.-U., Moore, T. E., Schildberger, and Weber, T. (1990). *Acoustic* communication in periodical cicadas: neuronal responses to songs of sympatric species, In *Sensory Systems and Communication in Arthropods*, Birkhäuser Verlag, Basel, pp. 217–228.

Huerta, C., Halffter, G., and Fresneau, D. (1992). Inhibition of stridulation in Necrophorus (Coleoptera: Silphidae): consequences for reproduction, *Elytron* **6:** 151–157.

Hughes-Schrader, S. (1975). Segregational mechanisms of sex chromosomes in spongilla-flies (Neuroptera: Sisyridae), *Chromosoma* **52:** 1–10.

Hunt, R. E. (1993). Role of vibrational signals in mating behavior of *Spissistilus festinus* (Homoptera: Membracidae), *Annals of the Entomological Society of America* **86:** 356–361.

Hunt, R. E. (1994). Vibrational signals associated with mating behavior in the treehopper, *Enchenopa binotata* Say (Hemiptera: Homoptera: Membracidae), *Journal of the New York Entomological Society* **102:** 266–270.

Hunt, R. E. and Morton, T. L. (2001). Regulation of chorusing in the vibrational communication system of the leafhopper *Graminella nigrifrons*, *American Zoologist* **41:** 1222–1228.

Hunt, R. E. and Nault, R. R. (1991). Roles of interplant movement, acoustic communication and phototaxis in mate-location behavior of the leafhopper *Graminella nigrifrons*, *Behavioral Ecology and Sociobiology* **28:** 315–320.

Huttunen, S. and Aspi, J. (2003). Complex inheritance of male courtship song characters in *Drosophila virilis*, *Behavioral Genetics* **33:** 17–24.

Huttunen, S., Aspi, A., Hoikkala, A., and Schlötterer, C. (2004). QTL analysis of variation in male courtship song characters in *Drosophila virilis*, *Heredity* **92:** 263–269.

Huttunen, S., Campesan, S., and Hoikkala, A. (2002a). Nucleotide variation at the *no-on-transient A* gene in *Drosophila littoralis*, *Heredity* **88:** 39–45.

Huttunen, S., Vieira, J., and Hoikkala, A. (2002b). Nucleotide and repeat length variation at the *nonA* gene of the *Drosophila virilis* group species and its effects on male courtship song, *Genetica* **115:** 159–167.

Ichikawa, T. (1976). Mutual communication by substrate vibration in the mating behaviour of planthoppers (Homoptera: Delphacidae), *Applied Entomology and Zoology* **11:** 8–23.

Ichikawa, T. (1982). Density-related changes in male-male competitive behaviour in the rice brown planthopper, *Nilaparvata lugens* (Stål) (Homoptera: Delphacidae), *Aplied Entomolgy and Zoology* **17**(4)**:** 439–452.

Ichikawa, T. and Ishii, S. (1974). Mating signal of the brown planthopper, *Niloparvata lugens* STÅL (Homoptera: Delphacidae): vibration of the substrate, *Aplied Entomolgy and Zoology* **9:** 196–198.

Illies, J. (1966). Katalog. der rezenten Plecoptera, *Das Tierreich*, Vol. 82: Walter de Gruyter, Berlin.

Inagaki, H., Matsuura, I., and Sugimoto, T. (1986). Deux espèces jumelles acoustiques de Hexacentrus (Orthoptera, Tettigoniidae), *H. japonicus* Karny et *H. unicolor* Serville, et leur ségrégation d' habitat, *Comptes Rendus de Seances de la Societe de Biologie* **180:** 589–592.

Inagaki, H. and Sugimoto, T. (1994). Interspezifische Mikrodivergenz der männlichen Genitalapparate bei zwei akustisch unterscheidbaren Geschwisterarten der Gattung *Hexacentrus* (Orthoptera: Tettigoniidae), *Entomologia Generalis* **18:** 165–170.

Inagaki, H., Sugimoto, T., and Irisawa, S. (1990). Comparaison de la pars stridens de deux espèces jumelles acoustique de Hexacentrus (Orthoptera, Tettigoniidae), *H. japonicus* Karny et *H. unicolor* Serville, *Comptes Rendus Hebdomadaires des Seances de l'Academie des* **310:** 515–519.

Ingard, U. (1953). A review of the influence of metereological conditions on sound propagation, *Journal of the Acoustical Society of America* **25:** 405–411.

Ingrisch, S., Willemse, F., and Heller, K.-G. (1992). Eine neue Unterart des Warzenbeißers *Decticus verrucivorus* (Linnaeus 1758) aus Spanien (Ensifera: Tettigoniidae), *Entomologische Zeitschrift* **102:** 173–192.

Isaac, N. J. B., Mallet, J., and Mace, G. M. (2004). Taxonomic inflation: its influence on macroecology and conservation, *Trends in Ecology & Evolution* **19:** 464–469.

Jackson, R. R. and Walls, E. I. (1998). Predatory and scavenging behaviour of *Microvelia macgregori* (Hemipetra: Vellidae), a water-surface bug from New Zealand, *New Zealand Journal of Zoology* **25:** 23–28.

Jacobi, A. (1907). Ein Schrillapparat bei Singcicaden, *Zoologischer Anzeiger* **32:** 67–71.

Jacobs, M. E. (1953). Observations on the two forms of the periodical cicada *Magicicada septendecim* (L.), *Proceedings of the Indian Academy of Sciences* **63:** 177–179.

Jaffe, D. A. (1987). Spectrum analysis tutorial. Part 1, The Discrete Fourier Transform; Part 2, Properties and applications of the Discrete Fourier Transform, *Computer Music J.* **11**(3)**:** 9–35.

Jarau, S., Hrncir, M., Schmidt, V. M., Zucchi, R., and Barth, F. G. (2003). Effectiveness of recruitment behavior in stingless bees (Apidae, Meliponini), *Insectes Sociaux* **50:** 365–374.

Jarau, S., Hrncir, M., Zucchi, R., and Barth, F. G. (2000). Recruitment behavior in stingless bees, *Melipona scutellaris* and *Melipona quadrifasciata*. I. Foraging at food sources differing in direction and distance, *Apidologie* **31:** 81–91.

Jatho, M. (1995). *Untersuchungen zur Schallproduktion und zum phonotaktischen Verhalten von Laubheuschrecken (Orthoptera: Tettigoniidae)*, Cuvillier Verlag, Göttingen.

Jatho, M., Schul, J., Stiedl, O., and Kalmring, K. (1994). Specific differences in sound production and pattern recognition in tettigoniids, *Behavioural Processes* **31:** 293–300.

Jatho, M., Weidemann, S., and Kretzen, D. (1992). Species-specific sound production in 3 Ephippigerine bushcrickets, *Behavioural Processes* **26:** 31–42.

Jeram, S. (1993). Anatomical and physiological properties of antennal mechanoreceptors of the bug species *Nezara viridula* (L.) (Pentatomidae, Heteroptera), Ms.C. thesis, University of Ljubljana, Ljubljana.

Jeram, S. (1996). Structure and function of Johnston's organ in bug species *Nezara viridula*, Ph.D. thesis, University of Ljubljana, Ljubljana.

Jeram, S. and Čokl, A. (1996). Mechanoreceptors in insects: Johnston's organ in *Nezara viridula* (L.) (Pentatomidae, Heteroptera), *Pflugers Archiv* **431**(Suppl.)**:** R281.

Jeram, S. and Pabst, M. A. (1996). Johnston's organ and central organ in *Nezara viridula* (L.) (Heteroptera, Pentatomidae), *Tissue Cell* **28:** 227–235.

Jocqué, R. (2002). Genitalic polymorphism—a chalange for taxonomy, *The Journal of Arachnology* **30:** 298–306.

Johnson, V. and Morrison, W. P. (1979). Mating behavior of three species of Coniopterygidae (Neuroptera), *Psyche* **86:** 395–398.

Jones, A. E., Ten Cate, C., and Kijleveld, C. J. H. (2001). The interobserver reliability of scoring sonograms by eye: a study on methods, illustrated on zebra finch songs, *Animal Behaviour* **62:** 791–801.

Jordan, K. H. C. (1958). Lautäusserungen bei den Hemipteren-Familien der Cydnidae, Pentatomidae und Acanthosomidae. *Zoologischer Anzeiger* **161**(5/6)**:** 130–144.

Josephson, R. K. (1973). Contraction kinetics of the fast muscles used in singing by a katydid, *Journal of Experimental Biology* **59:** 781–801.

Josephson, R. K. (1975). Extensive and intensive factors determining the performance of striated muscle, *Journal of Experimental Zoology* **94:** 135–153.

Josephson, R. K. (1981). Temperature and the mechanical performance of insect muscle, In *Insect Thermoregulation*, Heinrich, B., Ed., Wiley, New York, chap. 2.

Josephson, R. K. (1984). Contraction dynamics of flight and stridulatory muscles of Tettigoniid insects, *Journal of Experimental Biology* **108:** 77–96.

Josephson, R. K. and Halverson, R. C. (1971). High frequency muscles used in sound production by a katydid. I. Organisation of the motor system, *Biological Bulletin* **141:** 411–433.

Josephson, R. K. and Young, D. (1979). Body temperature and singing in the bladder cicada, *Cystosoma saundersii*, *Journal of Experimental Biology* **80:** 69–81.

Josephson, R. K. and Young, D. (1981). Synchronous and asynchronous muscles in cicadas, *Journal of Experimental Biology* **91:** 219–237.

Josephson, R. K. and Young, D. (1985). A synchronous insect muscle with an operating frequency greater than 500 Hz, *Journal of Experimental Biology* **118:** 185–208.

Josephson, R. K. and Young, D. (1987). Fiber ultrastructure and contraction kinetics in insect fast muscles, *American Zoologist* **27:** 991–1000.

Kahn, M. C. and Offenhauser, W. Jr. (1949). The first field tests of recorded mosquito sounds used for mosquito destruction, *American Journal of Tropical Medicine* **29:** 811–825.

Kalmring, K. (1985). Vibrational communication in insects (reception and integration of vibratory information), In *Acoustic and Vibrational Communication in Insects*, Kalmring, K. and Elsner, N., Eds., Paul Parey, Berlin, pp. 127–134.

Kalmring, K., Hoffmann, E., Jatho, M., Sickmann, T., and Grossbach, M. (1996). Auditory-vibratory sensory system of the bushcricket *Polysarcus denticauda* (Phaneropterinae, Tettigoniidae). II. Physiology of receptor cells, *Journal of Experimental Zoology* **276**: 315–329.

Kalmring, K. and Kühne, R. (1980). The coding of airborne-sound and vibration signals in bimodal ventral-cord neurons of the grasshopper *Tettigonia cantans*, *Journal of Comparative Physiology, A* **139**: 267–275.

Kalmring, K., Lewis, B., and Eichendorf, A. (1978). The physiological characteristics of the primary sensory neurons of the complex tibial organ of *Decticus verrucivorus* L (Orthoptera, Ensifera), *Journal of Comparative Physiology* **127**: 109–121.

Kalmring, K., Reitböck, H. J., Rössler, W., Schröder, J., and Bailey, W. J. (1991). Synchronous activity in neuronal assemblies as a coding principle for pattern recognition in sensory systems of insects, In *Trends in Biological Cybernetics*, Research Trends, Vol. 1: Menon, J., Ed., Council of Scientific Research Integration, Trivandrum, pp. 45–64.

Kalmring, K., Rössler, W., Ebendt, R., Ahi, J., and Lakes, R. (1993). The auditory receptor organs in the forelegs of bushcrickets: physiology, receptor cell arrangement, and morphology of the tympanal and intermediate organs of three closely related species, *Zoologische Jahrbuecher Physiologie* **97**: 75–94.

Kalmring, K., Rössler, W., Hoffmann, E., Jatho, M., and Unrast, C. (1995a). Causes of the differences in detection of low frequencies in the auditory receptor organs of two species of bushcrickets, *Journal of Experimental Zoology* **272**: 103–115.

Kalmring, K., Rössler, W., Jatho, M., and Hoffmann, E. (1995b). Comparison of song frequency and receptor tuning in two closely related bushcricket species, *Acta Biologica Hungarica* **46**: 457–469.

Kalmring, K., Rössler, W., and Unrast, C. (1994). Complex tibial organs in the forelegs, midlegs, and hindlegs of the bushcricket *Gampsocleis gratiosa* (Tettigoniidae): comparison of the physiology of the organs, *Journal of Experimental Zoology* **270**: 155–161.

Kalmring, K., Sickmann, T., Jatho, M., Zhantiev, R., and Grossbach, M. (1997). The auditory-vibratory sensory system of *Polysarcus denticauda* (Phaneropterinae, Tettigoniidae). III. Physiology of the ventral cord neurons ascending to the head ganglia, *Journal of Experimental Zoology* **278**: 9–38.

Kanmiya, K. (1981a). Life of *Lipara* species, *Insectarium* **18**(6): 140–148, in Japanese.

Kanmiya, K. (1981b). Mating behavior in the genus *Lipara* Meigen:mutual communication by transmission of substrate vibration of the reed stem. Part I, *Insectarium* **18**(9): 224–229, in Japanese.

Kanmiya, K. (1981c). Mating behavior in the genus *Lipara* Meigen:mutual communication by transmission of substrate vibration of the reed stem. Part II, *Insectarium* **18**(10): 260–267, in Japanese.

Kanmiya, K. (1983). A systematic study of the Japanese Chloropidae (Diptera), *Memoirs of the Entomological Society of Washington* **11**: 1–370.

Kanmiya, K. (1985). Record of courtship sound in a phorid species, *Puliciphora tokyoensis* Kinoshita, with notes on its acoustic properties, *Acta Dipterologica* **13**: 1–7, in Japanese.

Kanmiya, K. (1986). Speciation of the genus *Lipara*: Life-form and mating behavior, In *Biogeography in Japan*, Kimoto, S., Ed., Tokai Univ. Press, Tokyo, pp. 66–76, in Japanese.

Kanmiya, K. (1988). Acoustic studies on the mechanism of sound production in the mating songs of the melon fly, *Dacus cucurbitae* Coquillett (Diptera, Tephritidae), *Journal of Ethology* **6**: 143–151.

Kanmiya, K. (1990). Acoustic properties and geographic variation in the vibratory courtship signals of the European Chloropid fly, *Lipara lucens* Meigen (Diptera, Chloropidae), *Journal of Ethology* **8**: 105–120.

Kanmiya, K. (1996a). Discovery of male acoustic sounds in the Greenhouse whitefly, *Trialeurodes vaporariourum* (Westwood)(Homoptera:Aleyrodidae), *Aplied Entomolgy and Zoology* **31**: 255–262.

Kanmiya, K. (1996b). Discovery of male acoustic sounds in whitefly genera, *Bemisia*, *Trialeurodes* and *Pealius* (Homoptera:Aleyrodidae), *Proc. XX Intern. Congr. Entomol.* **1996**: 475.

Kanmiya, K. (1996c). Mating behavior of *Tinearia alternata* (Say), and acoustic signals by wing fanning (Diptera, Psychodidae), *Acta Dipterologica* **19**: 40–49, in Japanese.

Kanmiya, K. (1998). Mating behavior and vibrational sounds on whiteflies, *Plant Protection* **52**: 17–22, in Japanese.

Kanmiya, K. (1999). Insect acoustic communication, In *Environmental Entomology, Behavior, Physiology and Chemical Ecology*, Hidaka, T. *et al.*, Eds., University of Tokyo Press, Tokyo, pp. 495–509, in Japanese.

Kanmiya, K. (2001). Phylogeny and origin of the Reed-flies, In *The Natural History of Flies*, Shinonaga, S. and Shima, H., Eds., Tokai University Press, Tokyo, pp. 215–243, in Japanese.

Kanmiya, K., Kaneshiro, K. Y., Kanegawa, K. M., and Whittier, T. S. (1991). Time-domain acoustic analyses on the male melon fly mating songs between populations of Hawaii and Japan, *Proc. Intern. Symp. Biol. Contr. Fruit Flies* **1991:** 297–301.

Kanmiya, K. and Sonobe, R. (2002). Records of two citrus whiteflies in Japan with special reference to their mating sounds (Homoptera:Aleyrodidae), *Aplied Entomolgy and Zoology* **37:** 487–495.

Kanmiya, K., Tanaka, A., Kamiwada, H., Nakagawa, K., and Nishioka, T. (1987). Time-domain analysis of the male courtship songs produced by wild, mass-reared, and by irradiated melon flies, *Dacus cucurbitae* Coquillett (Diptera:Tephritidae), *Aplied Entomolgy and Zoology* **22:** 181–194.

Kasper, J. and Hirschberger, P. (2005). Stridulation in *Aphodius* dung beetles: songs and morphology of stridulatory organs in North American Aphodius species (Scarabaeidae), *Journal of Natural History* **39:** 91–99.

Kastberger, G. and Sharma, D. K. (2000). The predator-prey interaction between blue-bearded bee eaters (*Nyctyornis athertoni* Jardine and Selby 1830) and giant honeybees (*Apis dorsata* Fabricius 1798), *Apidologie* **31:** 727–736.

Kaszab, Z. (1936/37). Morphologische und systematische Untersuchungen über das Stridulationsorgan der Blumenbockkäfer (Lepturina). *Festschrift Embrik Strand.* **4:** 149–163.

Kavanach, M. W. (1987). The efficiency of sound production in two cricket species, *Journal of Experimental Biology* **130:** 107–119.

Kawanishi, M. and Watanabe, T. K. (1980). Genetic variation of courtship song of *Drosophila melanogaster* and *D. simulans*, *Japanese Journal of Genetics* **55:** 235–240.

Kawanishi, M. and Watanabe, T. K. (1981). Genes affecting courtship song and mating preference in *Drosophila melanogaster*, *Drosophila simulans* and their hybrids, *Evolution* **35:** 1128–1133.

Keller, E. (1994). *Signalyse, analyse du signal pour la parole et le son, InfoSignal*, Lausanne, Suisse, p. 270.

Kerr, W. E. (1969). Some aspects of the evolution of social bees (Apidae), *Evolutionary Biology* **3:** 119–175.

Kerr, W. E. and Esch, H. (1965). Communicação entre as abelhas sociais brasileiras e sua contribuição para o entendimento da sua evolução, *Ciência e Cultura* **17:** 529–538.

Kettle, R. and Vieillard, J. M. E. (1991). Documentation standards for wildlife sound recordings, *Bioacoustics* **3:** 235–238.

Keuper, A. (1981). *Biophysikalische Untersuchungen zur Luft- und Substartschallausbreitung im Biotop von Heuschrecken, Diplomarbeit Naturwiss*, Fak. Philipps-Universität Marburg, Marburg.

Keuper, A., Kalmring, K., Schatral, A., Latimer, W., and Kaiser, W. (1986). Behavioural adaptions of ground living bushcrickets to the properties of sound propagation in low grassland, *Oecologia* **70:** 414–422.

Keuper, A. and Kühne, R. (1983). The acoustic behaviour of the bushcricket *Tettigonia cantans*. II. Transmission of airborne-sound and vibration signals in the biotope, *Behavioural Processes* **8:** 125–145.

Keuper, A., Otto, C. W., and Schatral, A. (1985). Airborne sound and vibration signals of bush crickets and locusts. Their importance for the behaviour in the biotope, In *Acoustical and Vibrational Communication in Insects*, Kalmring, K. and Elsner, N., Eds., Paul Parey, Berlin, pp. 135–142.

Keuper, A., Weidemann, S., Kalmring, K., and Kaminski, D. (1988). Sound production and sound emission in seven species of european Tettigoniidae. Part I: The different parameters of the song; their relations to the morphology of the bushcricket. Part II: Wing morphology and the frequency content of the song, *Bioacoustics* **1:** 31–48, See also pp. 171–186.

Khanna, H., Gaunt, S. L. L., and McCallum, D. A. (1997). Digital spectrographic cross-correlation: tests of sensitivity, *Bioacoustics* **7:** 209–234.

Kidd, D. M. and Ritchie, M. G. (2000). Inferring the patterns and causes of geographic variation in *Ephippiger ephippiger* (Orthoptera, Tettigoniidea) using Geographical Information Systems (GIS), *Biological Journal of the Linnean Society* **71:** 269–295.

Killington, F. J. (1932). On the pairing of *Sialis fuliginosa* Pict, *Entomologist* **65:** 66–67.

Killington, F. J. (1936). *A Monograph of the British Neuroptera*, Vol. I: Ray Society, London.

Kilpinen, O. and Storm, J. (1997). Biophysics of the subgenual organ of the honeybee, *Apis mellifera, Journal of Comparative Physiology A* **181:** 309–318.

King, I. M. (1999). Species-specific sounds in water bugs of the genus *Micronecta*. Part 1. Sound analysis, *Bioacoustics* **9:** 297–323.

King, M. J., Buchmann, S. L., and Spangler, H. (1996). Activity of asynchronous flight muscle from two bee families during sonication (buzzing), *Journal of Experimental Biology* **199**: 2317–2321.

Kingsolver, J. M., Romero, N. J., and Johnson, C. D. (1993). Files and scrapers: Circumstantial evidence for stridulation in three species of Amblycerus, one new (Coleoptera: Bruchidae), *Pan-Pacific Entomologist* **69**: 122–132.

Kirchner, W. H. (1993a). Acoustical communication in honeybees, *Apidologie* **24**: 297–307.

Kirchner, W. H. (1993b). Vibrational signals in the tremble dance of the honeybee, *Apis mellifera*, *Behavioral Ecology and Sociobiology* **33**: 169–172.

Kirchner, W. H. (1994). Hearing in honeybees: the mechanical response of the bee's antenna to near field sound, *Journal of Comparative Physiology A* **175**: 261–265.

Kirchner, W. H. (1997). Acoustical communication in social insects, In *Orientation and Communication in Arthropods*, Lehrer, M., Ed., Birkhäuser Verlag, Basel Switzerland, pp. 273–300.

Kirchner, W. H. and Dreller, C. (1993). Acoustic signals in the dance language of the giant honeybee, *Apis dorsata*, *Behavioral Ecology and Sociobiology* **33**: 67–72.

Kirchner, W. H., Dreller, C., Grasser, A., and Baidya, D. (1996). The silent dances of the Himalayan honeybee, *Apis laborosa*. *Apidologie* **27**: 331–339.

Kirchner, W. H., Dreller, C., and Towne, W. F. (1991). Hearing in honeybees: operant conditioning and spontaneous reactions to airborne sound, *Journal of Comparative Physiology A* **168**: 85–89.

Kirchner, W. H. and Röschard, J. (1999). Hissing in bumblebees: an interspecific defence signal, *Insectes Sociaux* **46**: 239–243.

Kiritani, K., Hokyo, K. N., and Yukawa, J. (1963). Coexistence of the two related stink bugs *Nezara viridula* and *Nezara antennata* under natural conditions, *Research on Population Ecology* **5**: 11–22.

Kirk, A. A., Lacey, L. A., Brown, J. K., Ciomperlik, M. A., Goolsby, J. A., Vacek, D. C., Wendel, L. E., and Napompeth, B. (2000). Variation in the *Bemisia tabaci* s. l. species complex (Hemiptera: Aleyrodidae) and its natural enemies leading to successful biological control of *Bemisia* biotype B in the USA, *Bulletin of Entomological Research* **90**: 317–327.

Klappert, K. and Reinhold, K. (2003). Acoustic preference functions and sexual selection on the male calling song in the grasshopper *Chorthippus biguttulus*, *Animal Behaviour* **65**: 225–233.

Kleine, R. (1918). Über den Stridulationsapparat der Brenthidae, *Arch. Naturgesch., Abt. A* **84**: 1–84.

Kleine, R. (1920). Über den Stridulationsapparat der Familie Nemonychidae Desbr, *Entomologische Blatter* **16**: 20–22.

Kleine, R. (1932). Der Stridulationsapparat der Ipidae. III, *Entomologische Rdsch* **49**: 7–11.

Klingstedt, H. (1937). Chromosome behavior and phylogeny in the Neuroptera, *Nature* **139**: 468.

Koch, U. T., Elliott, C. J. H., Schäffner, K.-H., and Kleindienst, H.-U. (1988). The mechanics of stridulation of the cricket *Gryllus campestris*, *Journal of Comparative Physiology* **162A**: 213–223.

Koenig, W., Dunn, H. K., and Lacy, L. Y. (1946). The sound spectrograph, *Journal of the Acoustical Society of America* **18**(1): 19–49.

Koeniger, N. and Fuchs, S. (1972). Kommunikative Schallerzeugung *von Apis cerana* Fabr. im Bienenvolk, *Naturwissenschaften* **59**: 169.

Kon, M., Akemi, O. E., Numata, H., and Hidaka, T. (1988). Comparison of the mating behaviour between two sympatric species. *Nezara antennata* and *Nezara viridula* (Heteroptera: Pentatomidae), with special reference to sound emission, *Journal of Ethology* **2**: 91–98.

Kon, M., Oe, A., and Numata, H. (1993). Intra- and interspecific copulation in the two congeneric green stink bugs. *Nezara antennata* and *Nezara viridula* (Heteroptera: Pentatomidae), with reference to postcopulatory changes in the spermatheca, *Journal of Ethology* **11**: 63–89.

Kon, M., Oe, A., and Numata, H. (1994). Ethological isolation between two congeneric green stink bugs *Nezara antennata* and *N. viridula* (Heteroptera: Pentatomidae), *Journal of Ethology* **12**: 67–71.

Kon, M., Oe, A., Numata, H., and Hidaka, T. (1988). Comparison of the mating behavior between the sympatric species *Nezara antennata* and *N. viridula* (Heteroptera: Pentatomidae) with special reference to sound emission, *Journal of Ethology* **6**: 91–98.

Kotiaho, J. S., Alatalo, R. V., Mappes, J., Parri, S., and Rivero, A. (1998). Energetic costs of size and signalling in a wolf spider, *Proceedings of the Royal Society of London B* **265**: 2203–2209.

Kovarik, P. W., Burke, H. R., and Agnew, C. W. (1991). Development and behavior of a snakefly, *Raphidia bicolor* Albarda (Neuroptera, Raphidiidae), *Southwestern Entomologist* **16**: 353–364.

Kraus, W. F. (1989). Surface wave communication during courtship in the giant water bug *Abedus indentatus* (Heteroptera: Belostomatidae), *Journal of the Kansas Entomological Society* **62**: 316–328.

Krell, F.-T. and Fery, H. (1992). Trogidae, Geotrupidae, Scarabaeidae, Lucanidae, In *Die Käfer Mitteleuropas*, Vol. 13: Lohse, G. A. and Lucht, W. H., Eds., Goecke and Evers, Krefeld, pp. 201–253.

Kriegbaum, H. (1989). Female choice in the grasshopper *Chorthippus biguttulus*, *Naturwissenschaften* **76**: 81–82.

Kriegbaum, H. and von Helversen, O. (1992). Influence of male songs on female mating behavior in the grasshopper *Chorthippus biguttulus* (Orthoptera: Acrididae), *Ethology* **91**: 248–254.

Kristensen, N. P. (1999). Phylogeny of endopterygote insects, the most successful lineage of living organisms, *European Journal of Entomology* **96**: 237–253.

Kroodsma, D. E., Budney, G. F., Grotke, R. W., Vieillard, J. M. E., Gaunt, S. L. L., Ranft, R., and Veprintseva, O. D. (1996). Natural sound archives: guidance for recordists and a request for cooperation, In *Ecology and Evolution of Acoustic Communication in Birds*, Kroodsma, D. E. and Miller, E. H., Eds., Cornell University Press, New York, pp. 474–486.

Kühne, R. (1982a). Neurophysiology of the vibration sense in locusts and bushcrickets: response characteristics of single receptor units, *Journal of Insect Physiology* **28**: 155–163.

Kühne, R. (1982b). Neurophysiology of the vibration sense in locusts and bushcrickets: the responses of ventral-cord neurons, *Journal of Insect Physiology* **28**: 615–623.

Kühne, R., Silver, S., and Lewis, B. (1984). Processing of vibratory and acoustic signals by ventral cord neurons in the cricket *Gryllus campestris*, *Journal of Insect Physiology* **30**: 575–585.

Kühne, R., Silver, S., and Lewis, B. (1985). Processing of vibratory signals in the central nervous system of the cricket, In *Acoustic and Vibrational Communication in Insects*, Kalmring, K. and Elsner, N., Eds., Paul Parey, Berlin, pp. 183–192.

Kühnelt, W. (1986). Contribution to the knowledge of historical biogeography of the Balkan Peninsula especially of Greece, *Biologia Gallo-hellenica* **12**: 71–84.

Kulkarni, S. J. and Hall, J. C. (1987). Behavioral and cytogenetic analysis of the *cacophony* courtship song mutant and interacting genetic variants in *Drosophila melanogaster*, *Genetics* **116**: 461–475.

Kulkarni, S. J., Steinlauf, A. F., and Hall, J. C. (1988). The *dissonance* mutant of *Drosophila melanogaster*: isolation, behavior and cytogenetics, *Genetics* **118**: 267–285.

Kunze, L. (1959). Die funktionsanatomischen Grundlagen der Kopulation der Zwergzikaden, untersucht an *Euscelis plebejus* (FALL.) und einigen Typhlocybinen (Homoptwera Auchenorrhyncha), *Deutsche Entomologische Zeitschrift (N.F.)* **6**(IV): 322–387.

Kuštor, V. (1989). Activity of vibratory organ muscles in the bug *Nezara viridula* (L.) (in slovene), Ms.C. thesis, University of Ljubljana, Ljubljana, p. 68.

Kyriacou, C. P. and Hall, J. C. (1980). Circadian rhythm mutations in *Drosophila melanogaster* affect short-term fluctuations in the male's courtship song, *Proceedings of the National Academy of Sciences of the United States of America* **77**: 6729–6733.

Kyriacou, C. P. and Hall, J. C. (1986). Interspecific genetic control of courtship song production and reception in *Drosophila*, *Science* **232**: 494–497.

Lakes, R., Kalmring, K., and Engelhard, K. H. (1990). Changes in the auditory system of locusts (*Locusta migratoria* and *Schistocerca gregaria*) after deafferentiation, *Journal of Comparative Physiology* **166**: 553–563.

Lalley, S. and Robbins, H. (1988). Stochastic search in a convex region, *Probability Theory and Related Fields* **77**: 99–116.

Landois, H. 1867 Die Ton- und Stimmapparate der Insekten in anatomischphysiologischer und akustischer Beziehung, *Zeitschrift fur Wissenschaftliche Zoologie* **17**: 105–186.

Landois, H. (1874). *Thierstimmen*, Herder'sche Verlagsbuchhandlung, Freiburg/Br., p. 229.

Lane, D. H. (1995). The recognition concept of species applied in an analysis of pupative hybridization in New Zealand cicadas of the genus *Kikihia*, In *Speciation and the Recognition Concept, Theory and Application*, Lambert, D. M. and Spencer, H. G., Eds., The John Hopkins University Press, Baltimore and London, pp. 367–421.

Lane, C. and Rothschild, M. (1965). A case of Mullerian mimicry of sound, *Proceedings of the Royal Entomological Society of London Series A* **40**: 156–158.

Lang, F. (2000). Acoustic communication distances of a gomphocerine grasshopper, *Bioacoustics* **10**: 233–258.

Larsen, O.-N. (1981). Mechanical time resolution in some insect ears. II. Impulse sound transmission in acoustic tracheal tubes, *Journal of Comparative Physiology A* **143**: 297–304.

Larsen, O.-N., Gleffe, G., and Tengö, J. (1986). Vibration and sound communication in solitary wasps, *Physiological Entomology* **11**: 287–296.

Las, A. (1979). Male courtship persistence in the greenhouse whitefly, *Trialeurodes vaporariorum* Westwood (Homoptera:Aleyrodidae), *Behavior* **72**: 107–125.

Latimer, W. and Schatral, A. (1983). The acoustic behaviour of the bushcricket *Tettigonia cantans*. I. Behavioural responses to sound and vibration, *Behavioural Processes* **8**: 113–124.

Lattin, J. D. (1958). A stridulatory mechanism in Arhaphe cicindeloides Walker (Hemiptera: Heteroptera: Pyrrhocoridae), *The Pan-Pacific Entomologist* **34**: 217–219.

Lawrence, P. O. (1981). Host vibration — a cue to host location by the parasite *Biosteres longicaudatus*, *Oecologia* **48**: 249–251.

Lawson, F. A. and Chu, J. (1971). A scanning electron microscopy study of stridulating organs in two Hemiptera, *Journal of the Kansas Entomological Society* **44**: 245–253.

LeConte, J. L. (1878). Stridulation of Coleoptera, *Psyche* **2**: 126.

Lehmann, A. (1998). Speciation, acoustic communication and sexual selection in Greek bush-crickets of the Poecilimon propinquus group (Tettigonioidea, Phaneropteridae), Dissertation, University Erlangen-Nürnberg, in German.

Lehr, W. (1914). Die Sinnesorgane der beiden Flügelpaare von *Dytiscus marginalis*, *Zeitschrift fur Wissenschaftliche Zoologie* **110**: 87–150.

Leighton, J. R. B. (1987). Cost of tokking: the energetics of substrate communication in the tok–tok beetle, *Psammodes striatus*, *Journal of Comparative Physiology B.* **157**: 11–20.

Lennan Mc, D. A. and Ryan, M. J. (1997). Responses to conspecific and heterospecific olfactory cues in the swordtail *Xiphophorus cortezi*, *Animal Behaviour* **54**: 1077–1088.

Leroy, Y. (1978). Signal sonore et systématique animale, *Journal of Psychology* **2**: 197–229.

Leroy, Y. (1979). *L'univers sonore animal, rôles et évolution de la communication acoustique*, Paris: Gauthiers-Villars, p. 350.

Leroy, Y. (1980). Les critères acoustiques de l'espèce, *Mém. Soc. Zool. Fr.* **40**: 151–198.

Lesser, F.C. (1740). *Insecto-Theologia oder Vernunfft- und schrifftmäßiger Versuch, wie ein Mensch durch aufmercksame Betrachtung derer sonst wenig geachteten Insecten zu lebendiger Erkänntniß und Bewunderung der Allmacht, Weißheit, der Güte und Gerechtigkeit des grossen GOttes gelangen könne*, 2nd ed., Blochberger, M., Franckfurt & Leipzig, p. 512.

Leston, D. (1954). Strigils and stridulation in Pentatomoidea (Hem.): some new data and a review, *Entomologist's Monthly Magazine* **90**: 49–56.

Leston, D. (1957). The stridulatory mechanisms in terrestrial species of Hemiptera Heteroptera, *Proceedings of the Zoological Society of London* **128**: 369–386.

Leston, D. and Pringle, J. W. S. (1963a). Acoustical behaviour of Hemiptera, In *Acoustic Behaviour of Animals*, Busnell, R. G., Ed., Elsevier, Amsterdam, pp. 391–411.

Leston, D.-and Pringle, J. W. S. (1963b). Acoustic behaviour of Hemiptera, In *Acoustic Behaviour of Animals*, Busnel, R. G., Ed., Elsevier, Amsterdam, chap. 14.

Lewis, D. B. (1974). The physiology of the tettigoniid ear. I–IV, *Journal of Experimental Biology* **60**: 821–869.

Lewis, B. (1983). *Bioacoustics: A Comparative Approach*, Academic Press, London, p. 493.

Lewis, T. (1984). The elements and frontiers of insect communication, In *Insect Communication*, Lewis, T., Ed., Academic Press, London, pp. 1–27.

Lewis, E. E. and Cane, J. H. (1990). Stridulation as a primary anti-predator defence of a beetle, *Animal Behaviour* **40**: 1003–1004.

Lewis, E. E. and Cane, J. H. (1992). Inefficacy of courtship stridulation as a premating ethological barrier for Ips bark beetles (Coleoptera: Scolytidae), *Annals of the Entomological Society of America* **85**: 517–524.

Li, T.-Y. and Maschwitz, U. (1985). Sexual behavior of whitefly *Trialeurodes vaporariorum*, *Acta Entomologica Sinica* **28**: 233–235.

Li, T.-Y., Vinson, S. B., and Gerling, D. (1989). Courtship and mating behavior of *Bemisia tabaci* (Homoptera; Aleyrodidae), *Environmental Entomology* **18**: 800–806.

Lin, C. P., Danforth, B. P., and Wood, T. K. (2004). Molecular phylogenetics and evolution of maternal care in membracine treehoppers, *Systematic Biology* **53**: 400–421.

Lin, Y., Kalmring, K., Jatho, M., Sickmann, T., and Rössler, W. (1993). Auditory receptor organs in the forelegs of *Gampsocleis gratiosa* (Tettigoniidae): morphology and function of the organs in comparison to the frequency parameters of the conspecific song, *Journal of Experimental Zoology* **267**: 377–388.

Lin, Y., Rössler, W., and Kalmring, K. (1995). Morphology of the tibial organs of Acrididae: comparison of the subgenual- and distal-organs of the for-, mid- and hindlegs of *Schistocerca gregaria* (Acrididae, Catantopinae) and Locusta migratoria/Acrididae, Oedipodinae), *Journal of Morphology* **226**: 351–360.

Lin, Y., Rössler, W., and Kalmring, K. (1994). Complex tibial organs in the fore-, mid- and hindlegs of the bushcricket *Gampsocleis gratiosa* (Tettigoniidae): comparison of the morphology of the organs, *Journal of Morphology* **221**: 191–198.

Lindauer, M. (1955). Schwarmbienen auf Wohnungssuche, *Zeitschrift fur Vergleichende Physiologie* **37**: 263–324.

Lindauer, M. (1956). Über die Verständigung bei indischen Bienen, *Zeitschrift fur Vergleichende Physiologie* **38**: 521–557.

Lindauer, M. and Kerr, W. E. (1958). Die gegenseitige Verständigung bei den stachellosen Bienen, *Zeitschrift fur Vergleichende Physiologie* **41**: 405–434.

Lindauer, M. and Kerr, W. E. (1960). Communication between workers of stingless bees, *Bee World* **41**: 29–41, See also pp. 65–71.

Liou, L. W. and Price, T. D. (1994). Speciation by reinforcement of premating isolation, *Evolution* **48**: 1451–1459.

Loher, W. and Wiedenmann, G. (1981). Temperature dependent changes in circadian patterns of cricket premating behaviour, *Physiological Entomology* **6**: 35–43.

Loukas, M. and Drosopoulos, S. (1992a). Population genetic studies of leafhopper (*Empoasca*) species, *Entomologia Experimentalis et Applicata* **63**: 71–79.

Loukas, M. and Drosopoulos, S. (1992b). Population genetics of the spittlebug genus *Philaenus* (Homoptera: Cercopidae) in Greece, *Biological Journal of the Linnean Society* **46**: 403–413.

Lyal, C. H. C. and King, T. (1996). Elytro-tergal stridulation in weevils (Insecta: Coleoptera: Curculionoidea), *Journal of Natural History* **30**: 703–773.

MacGregor, H. C. and Uzzell, T. M. (1964). Gynogenesis in salamanders related to *Ambystoma jeffersoniamum*, *Science* **143**: 1043–1045.

MacLeod, E. G. and Adams, P. A. (1967). A review of the taxonomy and morphology of the Berothidae, with the description of a new subfamily from Chile (Neuroptera), *Psyche* **74**: 237–265.

MacNamara, C. (1926). The "drumming" of stoneflies (Plecoptera), *Can. Ent.* **58**: 53–54.

Magal, C., Schöller, M., Tautz, J., and Casas, J. (2000). The role of leaf structure in vibratory signal propagation, *Journal of the Acoustical Society of America* **108**: 2412–2418.

Maillard, Y. P., Sellier, R. (1970). *La pars stridens des Hydrophilidae (Ins., Col.) etude au microscope electronique a balayage*, Vol. 270: *Comptes Rendus de l'Acadamie des Sciences, Paris*, pp. 2969–2972.

Maketon, M. and Stewart, K. W. (1984). Further Studies of drumming behavior of North American Perlidae (Plecoptera), *Annals of the Entomological Society of America* **77**: 770–778.

Maketon, M. and Stewart, K. W. (1988). Patterns and evolution of drumming behavior in the stonefly families Perlidae and Peltoperlidae, *Aquatic Insects* **10**: 77–98.

Maketon, M., Stewart, K. W., Kondratieff, B. C., and Kirchner, R. F. (1988). New descriptions of drumming and evolution of the behavior in North American Perlodidae (Plecoptera), *Journal of the Kansas Entomological Society* **61**: 161–168.

Manrique, G. and Lazzari, C. R. (1994). Sexual behaviour and stridulation during mating in *Triatoma infestans* (Hemiptera: Reduviidae), *Memorias do Instituto Oswaldo Cruz* **89**: 629–633.

Manrique, G. and Schilman, P. E. (2000). Two different vibratory signals in *Rhodnius prolixus* (Hemiptera: Reduviidae), *Acta Tropica* **77**: 271–278.

Marcu, O. (1930a). Beitrag zur Kenntnis der Stridulationsorgane von *Prionus coriarus* L, *Zoologischer Anzeiger* **92**: 65–66.

Marcu, O. (1930b). Beiträge zur Kenntnis der Stridulationsorgane der Curculioniden, *Zoologischer Anzeiger* **87:** 283–289.

Marcu, O. (1932a). Beitrag zur Kenntnis der Stridulationsorgane von *Hydrophilus* und *Hydrobius*, *Zoologischer Anzeiger* **100:** 80–81.

Marcu, O. (1932b). Zur Kenntnis der Stridulationsorgane der Gattung *Limnoxenus* (Hydrophilidae), *Zoologischer Anzeiger* **101:** 60–61.

Marcu, O. (1932c). Beitrag zur Kenntnis der Stridulationsorgane der Gattung *Ctenoscelis* Serv, *Zoologischer Anzeiger* **97:** 174–175.

Marcu, O. (1933). Zur Kenntnis der Stridulationsorgane der Curculioniden, *Zoologischer Anzeiger* **103:** 270–273.

Markl, H. (1968). Die Verstandigung durch Stridulationssignale bei Blattscheidenameisen. II. Erzeugung und Eigenschaften der Signale, *Zeitschrift fur Vergleichende Physiologie* **60:** 103–150.

Markl, H. (1973). Leistungen des Vibrationssinnes bei wirbellosen Tieren, *Fortschritte der Zoologie* **21:** 100–120.

Markl, H. (1983). Vibrational communication, In *Neuroethology and Behavioral Physiology*, Huber, F. and Markl, H., Eds., Springer, Berlin/Heidelberg, pp. 332–353.

Markl, H., Hölldobler, B., and Hölldobler, T. (1977). Mating behavior and sound production in harvester ants (*Pogonomirmex*, Formicidae), *Insectes Sociaux* **24:** 191–212.

Markl, H., Lang, H., and Wiese, K. (1973). Die Genauigkeit der Ortung eines Wellenzentrums durch den Rückenschwimmer *Notonecta glauca* L, *Journal of Comparative Physiology* **86:** 359–364.

Markl, H. and Wiese, K. (1969). Die Empfindlichkeit des Rückenschwimmers *Notonecta glauca* L. für Oberflächenwellen des Wassers, *Zeitschrift fur Vergleichende Physiologie* **62:** 413–420.

Marshall, T. (1833). Cause of sound emitted by *Cychrus rostratus*, *Entomologist's Monthly Magazine* **1:** 213–214.

Marshall, J. L., Arnold, M. L., and Howard, D. J. (2002). Reinforcement: the road not taken, *Trends in Ecology & Evolution* **17:** 558–563.

Marshall, D. C. and Cooley, J. R. (2000). Reproductive character displacement and speciation in periodical cicadas, with description of a new species, 13-year *Magicicada neotredecim*, *Evolution* **54:** 1313–1325.

Martens, J. and Nazarenko, A. A. (1993). Microevolution of eastern palaearctic grey tits as indicated by their vocalizations (*Parus poecile*, Paridae, Aves) I. *Parus montanus*, *Z. Zool. Syst. Evolutionsforsch* **31:** 127–143.

Martens, J. and Steil, B. (1997). Territorial songs and species differentiation in the Lesser Whitethroat superspecies *Sylvia* (*curruca*), *Journal fur Ornitholgie* **138:** 1–23.

Martin, S. D., Gray, D. A., and Cade, W. H. (2000). Fine-scale temperature effects on cricket calling song, *Canadian Journal of Zoology* **78:** 706–712.

Mason, C. A. (1991). Hearing in a primitive ensiferan: the auditoty system of *Cyphoderris monstrosa* (Orthoptera: Haglidae), *Journal of Comparative Physiology A* **168:** 351–163.

Masters, W. M. (1979). Insect disturbance stridulation: its defensive role, *Behavioral Ecology and Sociobiology* **5**(2), 187–200, illustr.

Masters, W. M. (1980). Insect disturbance stridulation: characterization of airborne and vibrational components of the sound, *Journal of Comparative Physiology* **135:** 259–268.

Masters, K. L. (1997). Behavioral and ecological aspects of inbreeding in natural animal populations: inferences from Umbonia treehoppers (Homoptera: Membracidae), Ph.D. Dissertation, Princeton University.

Masters, K. L., Masters, A. R., and Forsyth, A. (1994). Female-biased sex ratios in the Neotropical treehopper *Umbonia ataliba* (Homoptera: Membracidae), *Ethology* **96:** 353–366.

Masters, W.M. Tautz, J., Fletcher, N.H., and Markl, H. (1983). Body vibration and sound production in an insect (*Atta sexdens*) without specialized radiating structures. *Journal of Comparative Physiology A* **150:** 239–249.

Mayr, E. (1942). *Systematics and the Origin of Species*, Columbia University Press, New York.

McBrien, H. L., Čokl, A., and Millar, J. G. (2002). Comparison of substrate-borne vibrational signals of two congeneric stink bug species. *Thyanta pallidovirens* and *T. custator accerra* (Heteroptera: Pentatomidae), *Journal of Insect Behaviour* **15**(6): 715–738.

McBrien, H. L. and Millar, J. G. (1999). Phytophagous bugs, In *Pheromones of Non-lepidopteran Insects Associated with Agricultural Plants*, Hardie, J. and Minks, A. K., Eds., CABI, Walingford, UK, pp. 277–304.

McBrien, H. L. and Millar, J. G. (2003). Substrate-borne vibrational signals of the consperse stink bug (Hemiptera: Pentatomidae), *The Canadian Entomologist.* **135**: 555–567.

McEvoy, P. B. (1979). Advantages and disadvantages to group living in treehoppers (Homoptera: Membracidae), *Miscellaneous Publications of the Entomological Society of America* **11**: 1–13.

McGregor, P. K. (2000). Playback experiments: design and analysis, *Acta Ethologica* **3**: 3–8.

McIndoo, N. E. (1922). The auditory sense of the honey-bee, *Journal of Comparative Neurology* **34**: 173–199.

McKamey, S. H. and Deitz, L. L. (1996). Generic revision of the New World tribe Hoplophorionini (Hemiptera: Membracidae: Membracinae), *Systematic Entomology* **21**: 295–342.

Meier, T. and Reichert, H. (1990). Embryonic development and evolutionary origin of the orthopteran auditory organs, *Journal of Neurobiology* **21**(4): 592–610.

Meixner, A. J. and Shaw, K. J. (1979). Spacing and movement of singing *Neoconocephalus nebrascensis* males (Tettigoniidae: Copophorinae), *Annals of the Entomological Society of America* **72**: 602–606.

Mencinger, B. (1998). Prey recognition in larvae of the antlion *Euroleon nostras* (Neuroptera Myrmeleontidae), *Acta Zoologica Fennica* **209**: 157–161.

Menier, J. J. (1976). Occurrence of stridulatory organs in Platypodidae (Col.), *Annales de la Societe entomologique de France* **12**: 347–353.

Menzel, J. G. and Tautz, J. (1994). Functional morphology of the subgenual organ of the carpenter Ant, *Tissue Cell* **26**: 735–746.

Meyer, J. B. (1855). Aristoteles Thierkunde: Ein Beitrag zur Geschichte der Zoologie, *Physiologie und alten Philosophie*, Reimer, Berlin, p. 520.

Meyer, J. and Elsner, N. (1997). Can spectral cues contribute to species separation in closely related grasshoppers?, *Journal of Comparative Physiology Series A* **180**: 171–180.

Meyhöfer, R. and Casas, J. (1999). Vibratory stimuli in host location by parasitic wasps, *Journal of Insect Physiology* **45**: 967–971.

Meyhöfer, R., Casas, J., and Dorn, S. (1994). Host location by a parasitoid using leafminer vibrations: characterising the vibrational signals produced by the leafmining host, *Physiological Entomology* **19**: 349–359.

Meyhöfer, R., Casas, J., and Dorn, S. (1997a). Vibration-mediated interactions in a host-parasitoid system, *Proceedings of the Royal Society of London Series B* **264**: 261–266.

Meyhöfer, R., Casas, J. S., and Dorn, S. (1997b). Mechano- and chemoreceptors and their possible role in the host location behaviour of *Sympiesis sericeicornis* (Hymenoptera: Eulophidae), *Annals of the Entomological Society of America* **90**: 208–219.

Michel, K. (1975). Das Tympanalorgan *von Cicada orni* L.(Cicadina Homoptera). Eine licht- und elektronenmikroskopiche untersuchung, *Zoomorphologie* **82**: 63–78.

Michel, K., Amon, T., and Čokl, A. (1983). The morphology of the leg scolopidial organs in *Nezara viridula* (L.) (Heteroptera, Pentatomidae), *Revue Canadienne de Biologie Experimentale* **42**: 139–150.

Michelsen, A. (1966a). On the evolution of tactile stimulatory actions in longhorned beetles (Cerambycidae, Col.), *Zeitschrifte fur Tierpsychologie* **23**: 257–266.

Michelsen, A. (1966b). The sexual behaviour of some longhorned beetles (Col., Cerambycidae), *Entomologische Meddelelser* **34**: 329–355.

Michelsen, A. (1978). Sound reception in different environments, In *Sensory Ecology, Review and Perspectives*, Ali, M. A., Ed., Plenum Press, New York, pp. 345–373.

Michelsen, A. (1985). Environmental basis of sound communication in insects, In *Acoustic and Vibrational Communication in Insects*, Kalmring, K. and Elsner, N., Eds., Paul Parey, Berlin.

Michelsen, A. (1992). Hearing and sound communication in small animals: evolutionary adaptations to the laws of physics, *The Evolutionary Biology of Hearing*, Vol. 5: Webster, D. B., Fay, R. R., and Popper, A. N., Eds., Springer Verlag, Berlin, pp. 61–77.

Michelsen, A. (1993). The transfer of information in the dance language of honeybees: progress and problems, *Journal of Comparative Physiology A* **173**: 135–141.

Michelsen, A. (2003). Signals and flexibility in the dance communication of honeybees, *Journal of Comparative Physiology A* **189**: 165–174.

Michelsen, A., Fink, F., Gogala, M., and Traue, D. (1982). Plants as transmission channels for insect vibrational songs, *Behavioral Ecology and Sociobiology* **11**: 269–281.

Michelsen, A., Kirchner, W. H., Andersen, B. B., and Lindauer, M. (1986b). The tooting and quacking vibration signals of honeybee queens: a quantitative analysis, *Journal of Comparative Physiology A* **158**: 605–611.

Michelsen, A., Kirchner, W. H., and Lindauer, M. (1986a). Sound and vibrational signals in the dance language of the honeybee, *Apis mellifera*, *Behavioral Ecology and Sociobiology* **18**: 207–212.

Michelsen, A. and Larsen, O.-N. (1978). Biophysics of the ensiferan ear. I. Tympanal vibrations in bushcrickets (Tettigoniidae), studied with laser vibrometry, *Journal of Comparative Physiology* **123**: 193–203.

Michelsen, A. and Larsen, O.-N. (1983). Strategies for acoustic communication in complex environments, In *Neuroethology and Behavioral Physiology*, Huber, F. and Markl, H., Eds., Springer Verlag, Berlin, pp. 321–331.

Michelsen, A. and Nocke, H. (1974). Biophysical aspects of sound communication in insects, *Advances in Insect Physiology* **10**: 247–296.

Michelsen, A., Towne, W. F., Kirchner, W. H., and Kryger, P. (1987). The acoustic near field of a dancing honeybee, *Journal of Comparative Physiology A* **161**: 633–643.

Michener, C. D. (1974). *The Social Behavior of the Bees*, The Belknap Press of Harvard University Press, Massachusetts, p. 404.

Miklas, N., Čokl, A., Renou, M., and Virant-Doberlet, M. (2003a). Variability of vibratory signals and mate choice selectivity in the southern green stink bug, *Behavioural Processes* **61**: 131–142.

Miklas, N., Lasnier, T., and Renou, M. (2003b). Male bugs modulate pheromone emission in response to vibratory signals from conspecifics, *Journal of Chemical Ecology* **29**: 561–574.

Miles, R. M., Cocroft, R. B., Gibbons, C., and Batt, D. (2001). A bending wave simulator for investigating directional vibration sensing in insects, *Journal of the Acoustical Society of America* **110**: 579–587.

Miller, A. (1965). The internal anatomy and histology of the imago of Drosophila melanogaster, In *Biology of Drosophila*, Demerec, M., Ed., John Wiley & Sons, London, pp. 420–534.

Milne, L. J. and Milne, M. J. (1944). Notes on the behaviour of burying beetles (Nicrophorus spp.), *Journal of the New York Entomological Society* **52**: 311–327.

Mini, A. and Prabhu, V. K. K. (1990). Stridulation in the coconut rhinoceros beetle Oryctes rhinoceros (Coleoptera: Scarabaeidae), *Proceedings of the Indian Academy of Sciences, Animal Sciences* **99**: 447–455.

Mitchel, W. C. and Mau, R. F. L. (1969). Sexual activity and longevity of the southern green stink bug, *Nezara viridula*. *Annals of Entomological Society of America* **62**: 1246–1247.

Miyatake, T. and Kanmiya, K. (2004). Male courtship song in circadian rhythm mutants of *Bactrocera cucurbitae* (Tephritidae:Diptera), *Journal of Insect Physiology* **50**: 85–91.

Monro, J. (1953). Stridulation in the Quennsland fruit fly, Dacus (Strumeta) tryoni Frogg, *Australian Journal of Sciences* **16**: 60–62.

Monteleagre, Z. F. and Morris, G. K. (2004). The spiny devil katydids, *Panacanthus* Walker (Orthoptera: Tettigoniidae): an evolutionary study of acoustic behavior and morphological traits, *Systematic Entomology* **29**: 21–57.

Mook, J. H. and Bruggemann, C. G. (1968). Acoustical communication by *Lipara lucens* (Diptera Chloropidae), *Entomologia Experimentalis et Applicata* **11**: 397–402.

Moore, T. E. (1961). Audiospectrographic analysis of sounds of Hemiptera and Homoptera, *Annals of the Entomological Society of America* **54**: 273–291.

Moore, T. E. (1962). Acoustical behavior of the cicada Fidicina pronoe (Walker), *The Ohio Journal of Science* **62**(3): 113–119.

Moore, T. E. (1993). Acoustic signals and speciation in cicadas (Insecta: Homoptera: Cicadidae), In *Evolution Patterns and Processes*, Lees, D. R. and Edwards, D., Eds., Academic Press, London, pp. 269–284.

Moore, T. E., Huber, F., Weber, T., Klein, U., and Bock, C. (1993). Interaction between visual and phonotactic orientation during flight in *Magicicada cassini* (Homoptera: Cicadidae), *The Great Lakes Entomologist* **26**: 199–221.

Moore, B. P., Wodroffe, G. E., and Sanderson, A. R. (1956). Polymorphism and parthenogenesis in a ptinid beetle, *Nature* **177**: 847–848.

Morales, M. A. (2000). Mechanisms and density dependence of benefit in an ant-membracid mutualism, *Ecology* **81:** 482–489.

Moran, D. T. and Rowley, J. C. (1975). The fine structure of the cockroach subgenual organ, *Tissue Cell* **7:** 91–106.

Moron, M. A. (1995). Review of the Mexican species of *Golofa* hope (Col., Melolonthidae, Dynastinae), *Coleopterists Bulletin* **49:** 343–386.

Morris, G. K. (1970). Sound analysis of *Metrioptera sphagnorum* (Orthoptera: Tettigoniidae), *Canadian Entomologist* **102:** 363–368.

Morris, G. K. (1971). Aggression in male conocephaline grasshoppers (Tettigoniidae), *Animal Behaviour* **19:** 132–137.

Morris, G. K., Mason, A. C., and Wall, P. (1994). High ultrasonic and tremulation signals in neotropical katydids (Orthoptera: Tettigoniidae), *Journal of Zoology London* **233:** 129–163.

Morton, E. S. (1977). On the occurrence and significance of motivation-structural rules in some bird and mammal sounds, *American Naturalist* **111:** 855–869.

Moulds, M. S. (1990). *Australian cicadas*, New South Wales University Press, Sydney, p. 217.

Moulet, P. (1991). De la stridulation chez les Coreoidea, *Bull. Soc. Et. Sci. nat. Vaucluse* **1991:** 29–33.

Moulin, B., Rybak, F., Aubin, T., and Jallon, J.-M. (2001). Compared ontogenesis of courtship song components of males from sibling species, *D. melanogaster* and *D. simulans*, *Behavioral Genetics* **31:** 299–308.

Mound, L. A. and Halsey, S. H. (1978). *Whitefly of the world: A Systematic Catalogue of the Aleyrodidae (Homoptera) with Host Plant and Natural Enemy Data, British Museum (Natural History), London, UK*, John Wiley and Sons, Chichester.

Mousseau, T. A. and Howard, J.-H. (1998). Genetic variation in Cricket calling song across a hybrid zone between two sibling species, *Evolution* **52:** 1104–1110.

Mücke, A. (1991). Innervation pattern and sensory supply of the midleg of *Schistocerca gregaria* (Insecta, Orthopteroidea), *Zoomorphology* **110:** 175–187.

Muir, F. (1907). Notes on the stridulating organ and stink-glands of *Tessaratoma papillosa* Thunb, *Transactions of the Entomological Society of London* **1907:** 256–258.

Müller, H. J. (1942). Über Bau und Funktion des Legeapparates der Zikaden (Homoptera-Cicadina), *Zeitschrift fur Morphologie und Ökologie der Tiere* **38:** 534–629.

Müller, H. J. (1954). Der Saisondimorphismus bei Zikaden der Gattung *Euscelis* Brullé (Homoptera Auchenorrhyncha), *Beitrage zur Entomologie* **4:** 1–56.

Müller, H. J. (1957). Die Wirkung exogener Faktoren auf die zyklische Formenbildung der Insekten, insbesondere der Gattung *Euscelis* (Homoptera-Auchenorrhyncha), *Zoologische Jahrbucher, Abteilung Systematik* **85:** 317–430.

Murphey, R. K. (1971a). Motor control of orientation to prey by the waterstrider *Gerris remigis*, *Zeitschrift fur Vergleichende Physiologie* **72:** 150–167.

Murphey, R. K. (1971b). Sensory aspects of the control of orientation to prey by the waterstrider *Gerris remigis*, *Zeitschrift fur Vergleichende Physiologie* **72:** 168–185.

Murphey, R. K. (1973). Mutual inhibition and the organization of a non-visual orientation in *Notonecta*, *Journal of Comparative Physiology* **84:** 31–40.

Murphey, R. K. and Mendenhall, B. (1973). Localization of receptors controlling orientation to prey by the back-swimmer *Notonecta undulata*, *Journal of Comparative Physiology* **84:** 19–30.

Myers, J. G. (1926). New or little-known Australasian cicadas of the genus *Melampsalta*, with notes on songs by Iris Myers, *Psyche* **33:** 61–76.

Myers, J. G. (1928). The morphology of the Cicadidae (Homoptera), *Proceedings of the Zoological Society of London* **1928:** 365–472.

Myers, J. G. (1929). *Insect Singers. A Natural History of the Cicadas*, George Routledge and Sons, London, i–xix, pp. 1–304, pls 1–7.

Nachtigall, W. (2003). *Insektenflug*, Springer-Verlag, Berlin, Heidelberg, p. 510.

Nakao, S. (1952). On the diurnal rhythm of sound-producing activity in *Meimuna opalifera* Walker (Cicadidae Homoptera), *Physiol. Ecol.* **5:** 17–25.

Nakao, S. (1958). Dynamic structure of a cicada community in the level land of North Kyushu, *Insect Ecol.* **7:** 51–59, in Japanese.

Nartshuk, E. P. and Kanmiya, K. (1996). A new species of gall producing chloropid flies of the genus Lipara Meigen (Diptera Chloropidae) by the acoustic signals, *Entomol. Obozr.* **75:** 706–713, in Russian.

Nast, J. (1972). *Palaeartic Auchenorrhyncha (Homoptera), An Annotated Checklist*, Polish Scientific Publications, Warszawa, p. 550.

Nault, L. R., Wood, T. K., and Goff, A. M. (1974). Treehopper (Membracidae) alarm pheromones, *Nature* **249:** 387–388.

Nebeling, B. (1994). Darstellung der Morphologie von physiologisch charakterisierten auditorischen Interneuronen im Kopfbereich (Ober- und Unterschlundganglion), der Beißschreckenarten *Decticus albifrons* und *Decticus verrucivorus* (Ensifera, Orthoptera), Dissertation Marburg.

Neems, R. M. and Buttlin, R. K. (1993). Divergence in mate finding behavior between two subspecies of the meadow grasshopper *Chortippus parallelus* (Orthoptera: Acrididae), *Journal of Insect Behaviour* **6:** 421–430.

Neems, R. M., Dooher, K., Butlin, R. K., and Shorrocks, B. (1997). Differences in male courtship song among the species of the *quinaria* group of *Drosophila, Journal of Insect Behaviour* **10:** 237–262.

New, T. R. (1991). Neuroptera (Lacewings), In *The Insects of Australia (C.S.I.R.O.)*, 2nd ed.,, Waterhouse, D. F., Carne, P. B., and Naumann, I. D., Eds., Cornell University Press, Ithaca, NY, pp. 525–542.

New, T. R. and Theischinger, G. (1993). Megaloptera (Alderflies, Dobsonflies), *Handbuch der Zoologie*, Vol. 4 (Part 33): Walter de Gruyter, Berlin.

Newport, G. (1851). On the anatomy and affinities of *Pteronarcys regalis*, Newm.: with a postscript containing descriptions of some American Perlidae, together with notes on their habits, *Transactions of the Linnean Society, London* **20:** 425–451.

Nickle, D. A. and Walker, T. J. (1974). A morphological key to field crickets of southeastern United States (Orthoptera: Gryllidae: *Gryllus*), *Florida Entomologist* **57:** 8–12.

Nieh, J. C. (1993). The stop signal of honey bees: reconsidering its message, *Behavioral Ecology and Sociobiology* **33:** 51–56.

Nieh, J. C. (1996). A stingless bee, *Melipona panamica,* may use sound to communicate the distance and canopy height of a food source, Proceedings of the 10th International Meeting on Insect Sound and Vibration, Woods Hole, MA.

Nieh, J. C. (1998). The recruitment dance of the stingless bee, *Melipona panamica, Behavioral Ecology and Sociobiology* **43:** 133–145.

Nieh, J. C. (2000). The communication of 3-dimensional food location by a stingless bee, *Melipona panamica, Anais do IV Encontro sobre Abelhas*, Ribeirão Preto – SP, Brasil, pp. 142–143.

Nieh, J. C., Contrera, F. A. L., Rangel, J., and Imperatriz-Fonseca, V. L. (2003). Effect of food location and quality on recruitment sounds and success in two stingless bees, *Melipona mandacaia and Melipona bicolor, Behavioral Ecology and Sociobiology* **55:** 87–94.

Nieh, J. C. and Roubik, D. W. (1995). A stingless bee (*Melipona panamica*) indicates food location without using a scent trail, *Behavioral Ecology and Sociobiology* **37:** 63–70.

Nieh, J. C. and Roubik, D. W. (1998). Potential mechanisms for the communication of height and distance by a stingless bee, *Melipona panamica, Behavioral Ecology and Sociobiology* **43:** 387–399.

Nieh, J. C. and Tautz, J. (2000). Behaviour-locked signal analysis reveals weak 200–300 Hz comb vibrations during the honeybee waggle dance, *Journal of Experimental Biology* **203:** 1573–1579.

Niemits, C. (1972). Biochemische, verhaltensphysiologische und morphologische Untersuchungen an *Necrophorus vespillo, Forma et Functio* **5:** 209–230.

Nocke, H. (1971). Biophysik der Schallerzeugung durch die Vorderflügel der Grillen, *Zeitschrift fur Vergleichende Physiologie* **74:** 272–314.

Nocke, H. (1975). Physical and physiological properties of the tettigoniid (grasshopper), ear, *Journal of Comparative Physiology* **100:** 25–57.

Noor, M. A. F. and Aquadro, C. F. (1998). Courtship songs of *Drosophila pseudoobscura* and *D. persimilis*: analysis of variation, *Animal Behaviour* **56:** 115–125.

Noor, M. A. F., Williams, M. A., Alvarez, D., and Ruiz-Garcia, M. (2000). Lack of evolutionary divergence in courtship songs of *Drosophila pseudoobscura* subspecies, *Journal of Insect Behaviour* **13:** 255–262.

Numata, H., Kon, M., Fujii, H., and Hidaka, T. (1989). Sound production in the bean bug Riptortus clavatus Thunberg (Heteroptera: Alydidae), *Applied Entomology and Zoology* **24**(2): 169–173.

Oester, P. T., Rykar, L. C., and Rudinsky, J. A. (1978). Complex male premating stridulation of the bark beetle *Hylurgops rugipennis* (Mann), *Coleopterists Bulletin* **32**: 93–98.

Oeynhausen, A. and Kirchner, W. H. (2001). Vibrational signals of foraging bumblebees *(Bombus terrestris)* in the nest, In Proceedings of the Meeting of the European Sections of IUSSI, Berlin, p. 31.

Ogawa, K. (1992). Field trapping of male midges *Rheotanytarsus kyotoensis* (Diptera: Chironomidae) by sounds, *Japanese Journal of Sanitary Zoology* **43**: 77–80.

Ohaus, F. (1900). Bericht über eine entomologische Reise nach Centralbrasilien (Fortsetzung), *Stettiner Entomologische Zeitung* **61**: 164–191.

Ohtani, T. and Kamada, T. (1980). "Worker piping": the piping sounds produced by laying and guarding worker honeybees, *Journal of Apicultural Research* **19**: 154–163.

Ohya, E. (1996). External structure of the putative stridulatory apparatus of the fungus beetles, *Dacne japonica* and *D. picta* (Coleoptera: Erotylidae), *Aplied Entomolgy and Zoology* **31**: 321–325.

Ohya, E. and Kinuura, H. (2001). Close range sound communication of the oak platypodid beetle *Platypus quercivorus* (Murayama) (Col., Platypodidae), *Aplied Entomolgy and Zoology* **36**: 317–321.

Oldfield, B. P. (1982). Tonotopic organization of auditory receptors in Tettigoniidae (Orthoptera: Ensifera), *Journal of Comparative Physiology* **147**: 461–469.

Oldfield, B. P. (1983). Central projections of primary auditory fibres in Tettigoniidae (Orthoptera: Ensifera), *Journal of Comparative Physiology* **151**: 389–395.

Oldfield, B. P. (1984). Physiology of auditory receptors in two species of Tettigoniidae (Orthoptera: Ensifera): Alternative tonotopic organisations of the auditory organ, *Journal of Comparative Physiology* **155**: 689–696.

Oldfield, B. P. (1985). The role of the tympanal membrane in tuning of auditory receptors in Tettigoniidae (Orthoptera: Ensifera), *Journal of Experimental Biology* **116**: 493–497.

Oldfield, B. P. (1988). The effect of temperature on the tuning and physiology of insect auditory receptors, *Hearing Research* **35**: 151–158.

Olmstead, K. L. and Wood, T. K. (1990). Altitudinal patterns in species richness of neotropical treehoppers (Homoptera: Membracidae): the role of ants, *Proceedings of the Entomological Society of Washington* **92**: 552–560.

Olvido, A. E. and Mousseau, T. A. (1995). Effect of rearing environment on calling-song plasticity in the striped ground cricket, *Evolution* **49**: 1271–1277.

Oppenheim, A. V. and Schafer, R. W. (1975). *Digital Signal Processing*, Prentice-Hall, Englewood Cliffs, NJ, xiv, p. 585.

Örösi − Pál, Z. (1932). Wie tütet die Arbeitsbiene?, *Zoologischer Anzeiger* **98**: 147–148.

Ossiannilsson, F. (1946). On the sound-production and the sound-producing organ in Swedish Homoptera Auchenorhyncha, *Opuscula Entomologica* **11**: 82–84, a preliminary note.

Ossiannilsson, F. (1949). Insect Drummers, A study on the morphology and function of the sound-producing organ of Swedish Homoptera Auchenorrhyncha. *Opuscula Entomologica*, Suppl. X, Lund, p. 145.

Ossiannilsson, F. (1950). Sound-production in psyllids (Hem. Hom.), *Opuscula Entomologica* **15**(3), 202.

Ossiannilsson, F. (1953). On the music of some European leafhoppers (Homoptera-Auchenorrhyncha) and its relation to courtship. *Transactions IXth International Congress of Entomology* **2**: 139–141.

Ossiannilsson, F. (1992). The Psylloidea (Homoptera) of Fennoscandia and Denmark, *Fauna Entomologica Scandinavica*, Vol. 26: E.J. Brill, Leiden, New York, Köln, p. 347.

Oswald, J. D. (1993). Revision and cladistic analysis of the world genera of the family Hemerobiidae (Insect, Neuroptera), *Journal of the New York Entomological Society* **101**: 143–299.

Oswald, J. D. (2003). *Index to the Neuropterida Species of the World. College Station*, Texas A & M University, Texas.

Ota, D. and Čokl, A. (1991). Mate location in the southern green stink bug *Nezara viridula* (Heteroptera: Pentatomidae) mediated through substrate-borne signals on ivy, *Journal of Insect Behaviour* **4**: 441–447.

Otis, G. W., Patton, K., and Tingek, S. (1995). Piping by queens of *Apis cerana* Fabricius 1973 and *Apis koschevnikovi* v Buttel-Reepen 1906, *Apidologie* **26**: 61–65.

Otte, D. (1974). Effects and functions in the evolution of signaling systems, *Annual Review of Ecology and Systematics* **5**: 385–417.

Otte, D. (1977). *Communication in Orthoptera*, Indiana University Press, Bloomington, IN.

Otte, D. (1989). Speciation in Hawaian crickets, In *Speciation and its Consequences*, Otte, D. and Endler, J. A., Eds., Sinauer Associates Inc., Sunderland, MA, pp. 482–526.

Otte, D. (1994). *The Crickets of Hawaii: Origin, Systematics and Evolution*, The Orthopterists' Society, Academy of Natural Sciences of Philadelphia, Philadelphia, PA.

Otte, D., and Endler, J. Eds. (1989). *Speciation and its Consequences*, Sinauer Associates Inc., Sunderland, MA.

Otte, D., Nasckrecki, P., and Eades, D., Orthoptera Species File Online Version 2 (OSF), 2004, http://osf2.orthoptera.org/basic/hierarchy.asp.

Oudman, L., Landman, W., and Duijm, M. (1989). Genetic distance in the genus Ephippiger (Orthoptera, Tettigonioidea) — a reconnaissance, *Tijdschrift voor Entomologie* **132**: 177–181.

Päällysaho, S., Aspi, J., Liimatainen, J. O., and Hoikkala, A. (2003). Role of X chromosomal song genes in the evolution of species-specific courtship songs in *Drosophila virilis* group species, *Behavioral Genetics* **33**: 25–32.

Päällysaho, S., Huttunen, S., and Hoikkala, A. (2001). Identification of X chromosomal restriction fragment length polymorphism markers and their use in a gene localization study in *Drosophila virilis* and *D. littoralis*, *Genome* **44**: 242–248.

Padgham, M. (2004). Reverberation and frequency attenuation in forests-implications for acoustic communication in animals, *Annals of the Entomological Society of America* **115**: 402–410.

Palestrini, C., Pensati, F., Barbero, E., Reyes-Castillo, P., and Zunino, M. (2003). Differences in distress signals of adult passalid beetles (Coleoptera Passalidae), *Bollettino della Societa Entomologica Italiana* **135**: 45–53.

Panhuis, T. M., Butlin, R., Zuk, M., and Tregenza, T. (2001). Sexual selection and speciation, *Trends in Ecology & Evolution* **16**: 364–371.

Panizzi, A. R. (1997). Wild hosts of pentatomids: ecological significance and role in their pest status in crops, *Annual Review of Entomology* **42**: 99–122.

Panizzi, A. R., McPherson, J. E., James, D. G., Javahery, M., and McPherson, R. M. (2000). Stink bugs (Pentatomidae), In *Heteroptera of Economic Importance*, Schaefer, C. W. and Panizzi, A. R., Eds., CRC Press, Boca Raton, New York, chap. 13.

Parfin, S. I. (1952). The megaloptera and neuroptera of minnesota, *American Midland Naturalist* **47**: 421–434.

Paterson, H. E. (1980). A comment on "mate recognition systems", *Evolution* **34**: 330–331.

Paterson, H. E. H. (1985). The recognition concept of species, In *Species and Speciation*, Vrba, E. S., Ed., Transvaal Museum Monograph no. 4, Pretoria, pp. 21–29.

Paulo, O. S., Dias, C., Bruford, M. W., Jordan, W. C., and Nichols, R. A. (2001). The persistence of Pliocene populations through the Pleistocene climatic cycles: evidence from the phylogeography of an Iberian lizard, *Proceedings of the Royal Society of London B* **268**: 1625–1630.

Pavan, N. G. (1992). A portable PC-based DSP workstation for bioacoustical research, *Bioacoustics* **4**(1): 65–66.

Pavlovčič, P. and Čokl, A. (2001). Songs of *Holcostethus strictus* (Fabricius): a different repertoire among landbugs (Heteroptera: Pentatomidae), *Behavioural Processes* **53**: 65–73.

Pearson, G. A. and Allen, D. M. (1996). Vibrational communication in Eusattus convexus LeConte (Coleoptera: Tenebrionidae), *Coleopterists Bulletin* **50**: 391–394.

Peixoto, A. A. (2002). Evolutionary Behavioral genetics in Drosophila, *Advances in Genetics* **47**: 117–152.

Peixoto, A. A. and Hall, J. C. (1998). Analysis of temperature-sensitive mutants reveals new genes involved in the courtship song of *Drosophila*, *Genetics* **148**: 827–838.

Perdeck, A. C. (1957). The isolating value of specific song patterns in two sibling species of grasshoppers (*Chorthippus brunneus* Thunbg. and *C. biguttulus* L.), *Behaviour* **12**: 1–75.

Perdeck, A. C. (1958). The isolating value of specific song patterns in two sibling species of grasshoppers (*Chorthippus brunneus* Thunbg. and *C. biguttulus* L.), *Behaviour* **12**: 1–75.

Pereboom, J. J. M. and Sommeijer, M. J. (1993). Recruitment and flight activity of Melipona favosa, foraging on an artificial food source, *Proc. Exper. & Appl. Entomology* **4**: 73–78.

Péricart, J. (1998). *Hemiptères Lygaeidae euro-mèditerranèens*. Faune de France 84, Fèdèration Frangaise des Societes de sciences Naturelles, Paris.

Péricart, J. and Polhemus, J. T. (1990). Un appareil stridulatoire chez les Leptopodidae de l'ancien monde (Heteroptera), *Annales de la Societe entomologique de France (NS)* **26**(1): 9–17.

Perring, T. M., Cooper, A. D., Rodriguez, R. J., Farrar, C. A., and Bellows, T. S. (1993). Identification of a whitefly species by genomic and behavioral studies, *Science* **259:** 74–77.

Pfannenstiel, R. S., Hunt, R. E., and Yeargan, K. V. (1995). Orientation of a hemipteran predator to vibrations produced by feeding caterpillars, *Journal of Insect Behaviour* **8:** 1–9.

Pfau, H. K. (1988). Untersuchungen zur Stridulation und Phylogenie der Gattung Pycnogaster Graells, 1851 (Orthoptera, Tettigoniidae, Pycnogastrinae), *Mitt. Schweiz. Entomol. Ges.* **61:** 167–183.

Pfau, H. K. (1996). Untersuchungen zur Bioakustik und Evolution der Gattung Platystolus Bolivar (Ensifera Tettigoniidae), *Tijdschr. Entomol.* **139:** 33–72.

Pfau, H. K. and Pfau, B. (1995). Zur Bioakustik und Evolution der Pycnogastrinae (Orthoptera: Tettigoniidae): Pycnogaster valentini Pinedo and Llorente, 1986 und Pycnogaster cucullata (Charpentier, 1825), *Mitt. Schweiz. Entomol. Ges.* **68:** 465–478.

Pfau, H. K. and Pfau, B. (2002). Zur Bioakustik und Evolution der Gattung Calliphona (Orthoptera: Tettigoniidae), *Mitt. Schweiz. Entomol. Ges.* **75:** 253–271.

Pierce, G. W. (1948). *The Songs of Insects with related material on the Production, Propagation, Detection, and Measurement of Sonic and Supersonic Vibrations*, Harvard University Press, Cambridge, p. 329.

Pinto-Juma, G., Simões, P. C., Seabra, S. C., and Quartau, J. A. (2005). Calling song structure and geographic variation in *Cicada orni* Linnaeus (Hemiptera: Cicadidae), *Zoological Studies*, **44**(1): 81–94.

Pires, A. and Hoy, R. R. (1992a). Temperature coupling in cricket acoustic communication. I. Field and laboratory studies of temperature effects on calling song production and recognition in *Gryllus firmus*, *Journal of Comparative Physiology* **171A:** 69–78.

Pires, A. and Hoy, R. R. (1992b). Temperature coupling in cricket acoustic communication. II. Localization of temperature effects on song production and recognition networks in *Gryllus firmus*, *Journal of Comparative Physiology* **171A:** 79–92.

Pires, H. H. R., Lorenzo, M. G., Lazzari, C. R., Diotaiuti, L., and Manrique, G. (2004). The sexual behaviour of *Panstrongylus megistus* (Hemiptera: Reduviidae): an experimental study, *Memorias do Instituto Oswaldo Cruz* **99:** 295–300.

Pirisinu, Q., Spinelli, G., and Bicchierai, M. C. (1988). Stridulatory apparatus in the Italian species of the genus Laccobius Erichson (Coleoptera: Hydrophilidae), *International Journal of Insect Morphology & Embryology* **17:** 95–101.

Pittino, R. and Mariani, G. (1993). *Aphodius (Agrilinus) convexus* Erichson: a misinterpreted valid species from the western palaearctic fauna, *Bollettino della Societa Entomologica Italiana, Genova* **125:** 131–142.

Plant, C. W. (1991). An introduction to the British wax-flies (Neuroptera: Coniopterygidae) with a revised key to British species, *British Journal of Entomological Natural History* **4:** 157–162.

Poda, F. (1761). *Insecta Musei Graecensis*. Graz.

Pohlmann, K. C. (1989). *Principles of Digital Audio*, Howard W. Sams and Company, IN, USA, pp. 1–474.

Pollack, G. S. (1982). Sexual differences in cricket calling song recognition, *Journal of Comparative Physiology A* **146:** 217–221.

Pollack, G. S. (1986). Discrimination of calling song models by the cricket, *Teleogryllus oceanicus*: the influence of sound direction on neural encoding of the stimulus temporal pattern and on phonotactic behavior, *Journal of Comparative Physiology A* **158:** 549–561.

Pollack, G. S. (1998). Neural processing of acoustic signals, In *Comparative Hearing: Insects*, Hoy, R. R., Popper, A. N., and Fay, R. R., Eds., Springer, New York, chap. 5.

Pollack, G. S. and Hoy, R. (1981). Phonotaxis to individual rhythmic components of a complex cricket calling song, *Journal of Comparative Physiology A* **144:** 367–373.

Popov, A. V. (1975). The structure of the tymbals and characteristic of sound signals of singing cicadas (Homoptera Cicadidae) from the southern regions of the USSR, *Entomological Review of Washington* **54**(2): 7–35.

Popov, A. V. (1981). Sound production and hearing in the Cicada, *Cicadetta sinuatipennis* Osh., (Homoptera, Cicadidae), *Journal of Comparative Physiology A* **142:** 271–280.

Popov, A. V. (1997). Acoustic signals of the three morphologically similar species of singing cicadas (Homoptera, Cicadidae), *Entomological Review* **76**(1): 1–11.

Popov, A. V. (1998). Sibling species of the singing cicadas *Cicadetta prasina* (Pall.) and *C. pellosoma* (Uhler) (Homoptera: Cicadidae), *Entomological Review* **78:** 309–318.

Popov, A. V., Beganovic, A., and Gogala, M. (1997a). Bioacoustics of singing Cicadas of the western palaearctic: *Tettigetta brullei* (Fieber 1876), *Acta Entomologica Slovenica* **5**(2): 89–101.

Popov, A. V., Beganovic, A., and Gogala, M. (1997b). Bioacoustics of singing cicadas of the western Palaearctic: *Tettigetta brullei* (Fieber 1876) (Cicadoidea: Tibicinidae), *Acta Entomologica Slovenica* **5**: 89–101.

Popov, A. V. and Shuvalov, V. F. (1977). Phonotactic behavior of crickets, *Journal of Comparative Physiology* **119A:** 111–126.

Poramarcom, R. and Boake, C. R. (1991). Behavioural influences on male mating success in the Oriental fruit fly, *Dacus dorsalis Hendel, Animal Behaviour* **42:** 453–460.

Poulet, J. F. A. and Hedwig, B. (2003). A corollary discharge mechanism modulates central auditory processing in singing crickets, *Journal of Neurophysiology* **89:** 1528–1540.

Prager, J. and Larsen, O.-N. (1981). Asymmetrical hearing in the water bug *Corixa punctata* Ill. observed with laser vibrometry, *Naturwissenschaften* **68:** 579–580.

Prager, J. and Streng, R. (1982). The resonant properties of physical gill of *Corixa punctata* and their significance in sound reception, *Journal of Comparative Physiology* **148:** 323–335.

Pratt, S. C., Kühnholz, S., Seeley, T. D., and Weidenmüller, A. (1996). Worker piping associated with foraging in undisturbed queenright colonies of honey bees, *Apidologie* **27:** 13–20.

Prešern, J., Gogala, M., and Trilar, T. (2004). Comparison of *Dundubia vaginata* (Auchenorrhyncha: Cicadoidea) songs from Borneo and Peninsular Malaysia, *Acta Entomologica Slovenica* **12**(2), 239–248.

Prestwich, K., Brugger, K. E., and Topping, M. (1989). Energy and communication in three species of hylid frogs: power input, power output and efficiency, *Journal of Experimental Biology* **144:** 53–80.

Prestwich, K. N. and Walker, T. J. (1981). Energetics of singing in crickets: effect of temperature in three trilling species (Orthoptera: Gryllidae), *Journal of Comparative Physiology* **143B:** 199–212.

Price, T. (1996). Exploding species, *Trends in Ecology & Evolution* **11:** 314–315.

Price, J. J. and Lanyon, S. M. (2004). Patterns of song evolution and sexual selection in the oropendolas and caciques, *Behavioral Ecology* **15:** 485–497.

Priesner, H. (1949). *Curimosphena villosa* Haag, a sound producing beetle, (Col., Tenebrionidae), *Bull. Soc. Fouad 1er Ent. Cairo* **33:** 11–12.

Principi, M.M. (1949). Contributi allo studio dei Neurotteri Italiani, VIII, Morfologia, anatomia e funzionamento degli apparati genitali nel gen, *Chrysopa* Leach (*Chrysopa septempuntata* Wesm, e *Chrysopa formosa* Brauer), *Bollettino dell'Istituto di Entomologia della Università di Bologna*, Vol. 17, pp. 316–362.

Pringle, J. W. S. (1953). Physiology of song in cicadas, *Nature* **172:** 248–249, London.

Pringle, J. W. S. (1954). A physiological analysis of cicada song, *Journal of Experimental Biology* **31**(4), 525–560.

Pringle, J. W. S. (1955). The songs and habits of Ceylon cicadas, with a description of two new species, *Spolia Zeylanica* **27:** 229–238, pl. 1.

Pringle, J. W. S. (1957). *Insect Flight*, Cambridge University Press, Cambridge, p. 133.

Pugh, A. R. G. and Ritchie, M. G. (1996). Polygenic control of a mating signal in *Drosophila, Heredity* **77:** 378–382.

Puissant, S. (2001). Eco-éthologie *de Cicadetta montana* (Scopoli, 1772) (Auchénorhyncha, Cicadidae, Cicadettini), *EPHE, Biologie et Evolution des Insectes* **14:** 141–155.

Puissant, S. and Boulard, M. (2000). Cicadetta cerdaniensis, espèce jumelle de *Cicadetta montana* décryptée par l'acoustique (Auchenorhyncha, Cicadidae, Tibicininae), *EPHE, Biologie et Evolution des Insectes* **13:** 111–117, pl. 1.

Pumphrey, R. J. (1940). Hearing in insects, *Biological Reviews* **15:** 107–132.

Pupedis, R. J. (1980). Generic differences among New World spongilla-fly larvae and a description of the female of *Climacia areolaris* (Neuroptera: Sisyridae), *Psyche* **87:** 305–314.

Pupedis, R. J. (1985). The Bionomics and Morphology of New England Spongilla-flies (Neuroptera: Sisyridae), *Biological Sciences Group*, University of Connecticut, Connecticut, p. 308.

Puranik, P. G., Ahmed, A., and Siddiqui, M. A. (1981). The mechanism of sound production in the pentatomid bug Tessaratoma javanica Thunberg, *Proceedings of the Indian Academy of Sciences* **90**(2): 173–186.

Pureswaran, D. S. and Borden, J. H. (2003). Is bigger better? Size and pheromone production in the mountain pine beetle, Dendroctonus ponderosae Hopkins (Coleoptera: Scolytidae), *Journal of Insect Behaviour* **16**: 765–782.

Purvis, A. and Rambaut, A. (1995). Comparative analysis by independent contrasts (CAIC): an Apple Macintosh application for analyzing comparative data, *Computer Applications in the Biosciences* **11**: 247–251.

Pye, J. D. and Langbauer, W. R. Jr. (1998). Ultrasound and infrasound, In *Animal Acoustic Communication: Sound Analysis and Research Methods*, Hopp, S. L., Owren, M. J., and Evans, C. S., Eds., Springer-Verlag, Berlin, pp. 221–249.

Quartau, J. A. (1988). A numerical taxonomic analysis of interspecific morphological differences in two closely related species of *Cicada* (Homoptera, Cicadidae) in Portugal, *Great Basin Naturalist Memoirs* **12**: 171–181.

Quartau, J. A. (1995). Cigarras esses insectos quase desconhecidos, *Correio da Natureza* **38**: 19–33.

Quartau, J. A. and Fonseca, P. J. (1988). An annotated check-list of the species of cicadas known to occur in Portugal (Homoptera: Cicadoidea), In *Proceedings of the 6th Auchenorrhyncha Workshop*, Vidano, C. and Arzone, A., Eds., Torino, Italy, pp. 367–375.

Quartau, J. A., Rebelo, M. T., Simões, P. C., Fernandes, T. M., Claridge, M. F., Drosopoulos, S., and Morgan, J. C. (1999). Acoustic signals of populations of *Cicada orni* L. in Portugal and Greece (Hemiptera: Auchenorrhyncha: Cicadomorpha: Cicadidae), *Reichenbachia Staatliches Museum fuer Tierkunde Dresden* **33**(8): 71–80.

Quartau, J. A., Ribeiro, M., Simões, P. C., and Coelho, M. M. (2001). Genetic divergence among populations of two closely related species of *Cicada* Linnaeus (Hemiptera: Cicadoidea) in Portugal, *Insect Systematics and Evolution* **32**: 99–106.

Quartau, J. A., Ribeiro, M., Simões, P. C., and Crespo, A. (2000b). Taxonomic separation by isozyme electrophoresis of two closely related species of *Cicada* L. (Hemiptera: Cicadoidea) in Portugal, *Journal of Natural History* **34**: 1677–1684.

Quartau, J. A., Seabra, S., and Sanborn, A. (2000a). Effect of ambient air temperature on the calling song of *Cicada orni* Linnaeus, 1758 (Hemiptera: Cicadidae) in Portugal, *Acta Zoologica Cracoviensia* **43**(3–4): 193–198.

Quartau, J. A. and Simões, P. C. (2003). Bioacoustic and morphological differentiation in two allopatric species of the genus *Tibicina* Amyot (Hemiptera Cicadoidea) in Portugal, *Deut. Entomol. Z.* **50**(1): 113–119.

Quartau, J.A. and Simões, P.C., *Cicada cretensis* n. sp., a new cicada from southern Greece (Insecta: Hemiptera, Cicadidae), (submitted for publication).

Rabiner, L. R. and Gold, B. (1975). *Theory and Application of Digital Signal Processing*, Prentice-Hall, Englewood Cliffs, NJ, p. xv + 762.

Racovitza, E. G. (1907). Essai sure les problèmes biospéologiques, *Archives de Zoologie Expérimental Génerate* **4**(6): 371–488.

Ragge, D. R. (1965). *Grasshoppers, Crickets and Cockroaches of the British Isles*, Frederick Warne, London.

Ragge, D. R. (1987). Speciation and biogeography of some southern European Orthoptera, as revealed by their songs, In *Evolutionary Biology of Orthopteroid Insects*, Baccetti, B., Ed., Ellis Horwood, Chichester, pp. 418–426.

Ragge, D. R. and Reynolds, W. J. (1988). Songs and taxonomy of the grasshoppers of the *Chortippus biguttulus* group in the Iberian peninsular (Orthoptera: Acrididae), *Journal of Natural History* **22**: 897–929.

Ragge, D. R., and Reynolds, W. J. (1998). *The Songs of the Grasshoppers and Crickets of Western Europe*, Harley Books, Colchester, Essex, p. 591, in association with The Natural History Museum, London.

Rakitov, R. A. (1998). On differentiation of cicadellid leg chaetotaxy (Homoptera: Auchenorrhyncha: Membracoidea), *Russian Entomological Journal* **6**(3–4): 7–27.

Ray, J. (1710). *Historia Insectorum,* London.

Readio, P. A. (1927). Studies on the biology of the Reduviidae of America north of Mexico, *Kans. Univ. Sci. Bull.* **17**: 5–291.

Réaumur, R.A.F. (1734–1742). *Mémoires pour servir a l'histoire des insectes*, Six Tomes, De l'Imprimerie Royale, Paris.

Reeve, J. and Abouheif, E. (2003). *Phylogenetic Independence*, Version 2.0, Department of Biology, McGill University.

Reimarus, H. S. (1798). *Allgemeine Betrachtungen über die Triebe der Thiere, hauptsächlich über ihre Kunsttriebe*, 4th ed. (1st ed., 1760), K.E. Bohn, Hamburg.

Reinig, H.-J. and Uhlemann, H. (1973). Über das Ortungsvermögen des Taumelkäfers *Gyrinus substriatus* Steph, (Coleoptera, Gyrinidae), *Journal of Comparative Physiology* **84**: 281–298.

Remane, R. (1967). Zur Kenntnis der Gattung *Euscelis*, Brullé (Homoptera, Cicadina, Jassidae), *Entomologische Abhandlungen, Staatliches Museum fur Tierkunde, Dresden* **36**: 1–35.

Remane, R. and Hoch, H. (1988). Cave-dwelling Fulgoroidea (Homoptera Auchenorrhyncha) from the Canary islands, *Journal of Natural History* **22**: 403–412.

Remes Lenicov, A. M. de (1992). Fulgoroideos sudamericanos I. Un nuevo genero y especie de Cixidae cavernicola de la Patagonia (Insecta: Homoptera), *Neotropica* **38**(100): 155–160.

Remy, P. (1935). L'appareil stridulant du Coléoptère Ténebrionide *Clocrates abbreviatus* Ol., *Ann. Sci. Nat. Zool. 10th Ser.* **18**: 7–20.

Rence, B. G. and Loher, W. (1975). Arrhythmically singing crickets: thermoperiodic re-entrainment after bilobectomy, *Science* **190**: 385–387.

Rentz, D. C. F. (1975). Two new katydids of the genus *Melanonotus* from Costa Rica with comments on their life history strategies (Tettigoniidae: Pseudophyllinae), *Entomological News* **86**: 129–140.

Reyes-Castillo, P. and Jarman, M. (1983). Disturbance sounds of adult passalid beetles (Coleoptera: Passalidae): structure and functional aspects, *Annals of the Entomological Society of America* **76**: 6–22.

Rheinländer, J. (1975). Transmission of acoustic information at three neuronal levels in the auditory system of *Decticus verrucivorus* (Tettigoniidae, Orthopterea), *Journal of Comparative Physiology* **97**: 1–53.

Ribeiro-Costa, C. S. (1999). Description of two new species of *Amblycerus* Thunberg (Col.: Bruchidae) with a probable stridulatory mechanism, *Proceedings of the Entomological Society of Washington* **101**: 337–346.

Richard, G. (1950). L'innervation et les organes sensoriels de la patte du termite a cou jaune (*Calotermes flavicollis*), *Ann. Sci. Nat. Zool. Biol. Anim.* **11**: 65–83.

Richards, D. G. and Wiley, R. H. (1980). Reverberations and amplitude fluctuations in the propagation of sound in a forest: implication for animal communication, *American Naturalist* **115**: 381–399.

Riede, K. (1993). Monitoring biodiversity: analysis of Amazonian rainforest sounds. *Ambio* **22**: 546–548.

Riede, K. (1996). Diversity of sound-producing insects in a Bornean lowland rain forest, In *Tropical Rainforest Research*, Edwards, D. S. *et al.*, Eds., Kluwer Academic Publishers, Netherlands, pp. 77–84.

Riede, K. (1997). Bioacoustic monitoring of insect communities in a Bornean rainforest canopy, In *Canopy Arthropods*, Stork, N. E., Adis, J., and Didham, R. K., Eds., Chapman and Hall, London, pp. 442–452.

Riede, K. (1998). Acoustic monitoring of Orthoptera and its potential for conservation, *Journal of Insect Conservation* **2**: 217–223.

Riede, K. and Kroker, A. (1995). Bioacoustics and niche differentiation in two cicada species from Bornean lowland forest, *Zoologischer Anzeiger* **234**: 43–51.

Riede, K. and Stueben, P.E. (2000). The musical Acalles, Observations on the stridulations of Cryptorhynchinae of the Canaries (Col.: Curculionoidea). Cryptorhynchinae study 13. *Snudebiller — Studies on Taxonomy, Biology and Ecology of Curculionoidea (CURCULIO-Institut, Mönchengladbach)*, Vol. 1, pp. 307–317.

Riek, E. F. (1967). Structures of unknown, possibly stridulatory, function on the wings and body of Neuroptera; with an appendix on other endopterygote orders, *Australian Journal of Entomology* **15**: 337–348.

Ritchie, M. G. (1996). The shape of female preferences, *Proceedings of the National Academy of Science USA* **93**: 14628–14631.

Ritchie, M. G. and Gleason, J. M. (1995). Rapid evolution of courtship song pattern in *Drosophila willistoni* sibling species, *Journal of Evolutionary Biology* **8**: 463–480.

Ritchie, M. G. and Kyriacou, C. P. (1994). Genetic variability of courtship song in a population of *Drosophila melanogaster*, *Animal Behaviour* **48**: 425–434.

Ritchie, M. G., Saarikettu, M., Livingstone, S., and Hoikkala, A. (2001). Characterization of female preference functions for *Drosophila montana* courtship song and a test of the temperature coupling hypothesis, *Evolution* **55**: 721–727.

Ritchie, M. G., Sunter, D., and Hockman, L. R. (1998a). Behavioral components of sex role reversal in the tettigoniid bushcricket *Ephippiger ephippiger*, *Journal of Insect Behaviour* **11**: 481–491.

Ritchie, M. G., Townhill, R. M., and Hoikkala, A. (1998b). Female preference for fly song: playback experiments confirm the targets of sexual selection, *Animal Behaviour* **56:** 713–717.

Ritchie, M. G., Yate, V. H., and Kyriacou, C. P. (1994). Genetic variability of the interpulse interval of courtship song among some European populations of *Drosophila melanogaster*, *Heredity* **72:** 459–464.

Robert, D. (2001). Innovative biomechanics for directional hearing in small flies, *Biological Bulletin* **200:** 190–194.

Robert, D. and Göpfert, M. C. (2002). Acoustic sensitivity of fly antennae, *Journal of Insect Physiology* **48:** 189–196.

Robinson, D. (1990a). Acoustic communication between the sexes in bushcrickets, In *The Tettigoniidae: Biology, Systematics and Evolution*, Bailey, W. J. and Rentz, D. C. F., Eds., Springer, Berlin, pp. 71–97.

Robinson, D. (1990b). Acoustic communication between the sexes, In *The Tettigoniidae: Biology, Systematics and Evolution*, Bailey, W. J. and Rentz, D. C. F., Eds., Crawford House Press, Bathurst, pp. 112–129.

Roces, F. and Manrique, G. (1996). Different stridulatory vibrations during sexual behaviour and disturbance in the blood-sucking bug *Triatoma infestans* (Hemiptera: Reduviidae), *Journal of Insect Physiology* **42:** 231–238.

Roces, F., Tautz, J., and Hölldobler, B. (1993). Stridulation in leaf-cutting ants: short-range recruitment through plant-borne vibrations, *Naturwisseschaften* **80:** 521–524.

Rodríguez, R. L., Sullivan, L. E., and Cocroft, R. B. (2004). Vibrational communication and reproductive isolation in the *Enchenopa binotata* species complex of treehoppers (Hemiptera: Membracidae), *Evolution* **58:** 571–578.

Roeder, K. (1965). Moths and ultrasound, *Scientific American* **232:** 94–102.

Rohrseitz, K. (1998). Biophysikalische und ethologische Aspekte der Tanzkommunikation der Honigbienen (*Apis mellifera carnica* Pollm.), Ph.D. thesis, University of Würzburg, Germany.

Rohrseitz, K. and Kilpinen, O. (1997). Vibration transmission characteristics of the legs of freely standing honeybees, *Zoology* **100:** 80–84.

Römer, H. (1983). Tonotopic organization of the auditory neuropile in the bushcricket *Tettigonia viridissima*, *Nature* **305:** 29–30.

Römer, H. (1985). Anatomical representation of frequency and intensity in the auditory system of Orthoptera, In *Acoustic and Vibrational Communication in Insects*, Kalmring, K. and Elsner, N., Eds., Parey Verlag, Hamburg, pp. 25–32.

Römer, H. (1993). Environmental and biological constraints for the evolution of long-range signalling and hearing in acoustic insects, *Philosophical Transactions of the Royal Society of London Series B, biological Sciences* **340:** 179–185.

Römer, H. (1998). The sensory ecology of acoustic communication in insects, In *Comparative Hearing: Insects*, Hoy, R. R., Popper, A. N., and Fay, R. R., Eds., Springer-Verlag, Berlin, chap. 3.

Römer, H. and Bailey, W. J. (1986). Insect hearing in the field. II. Male spacing behaviour and corelated acoustic cues in the bushcricket *Mygalopsis marki*, *Journal of Comparative Physiology A* **159:** 627–638.

Römer, H., Bailey, W. J., and Dadour, I. (1989). Insect hearing in the field: III. Masking by noise, *Journal of Comparative Physiology A* **164:** 609–620.

Römer, H. and Lewald, J. (1992). High-frequency sound transmission in natural habitats: implications for the evolution of insect acoustic communication, *Behavioral Ecology and Sociobiology* **26:** 437–444.

Römer, H., Marquart, V., and Hardt, M. (1988). Organization of a sensory neuropile in the auditory pathway of two groups of Orthoptera, *Journal of Comparative Neurology* **275:** 201–215.

Ronacher, B. and Stumpner, A. (1988). Filtering of behaviourally relevant temporal parameters of a grasshopper's song by an auditory interneuron, *Journal of Comparative Physiology A* **163:** 517–523.

Rössler, W. (1992a). Postembryonic development of the complex tibial organ in the foreleg of the bushcricket *Ephippiger ephippiger* (Orthoptera, Tettigoniidae), *Cell and Tissue Research* **269:** 505–514.

Rössler, W. (1992b). Functional morphology and development of the complex tibial organs in the legs I, II and III of the bushcricket *Ephippiger ephippiger* (Insecta, Ensifera), *Zoomorphology* **112:** 181–188.

Rössler, W., Hübschen, A., Schul, J., and Kalmring, K. (1994). Functional morphology of bushcricket ears: comparison between two species belonging to the Phaneropterinae and Decticinae (Insecta, Ensifera), *Zoomorphology* **114:** 39–46.

Rössler, W. and Kalmring, K. (1994). Similar structural dimensions in bushcricket auditory organs in spite of different foreleg size: consequences for auditory tuning, *Hearing Research* **80:** 191–196.

Rotenberry, J. T., Zuk, M. T., Simmons, L. W., and Hayes, C. (1996). Phonotactic parasitoids and cricket song structure: an evaluation of alternative hypotheses, *Evolutionary Ecology* **10:** 233–243.

Roth, L. M. (1948). A study of mosquito behavior: an experimental laboratory study of the sexual behavior of *Aedes aegypti* (Linnaeus), *American Midland Naturalist* **40:** 265–352.

Rovner, J. S. and Barth, F. G. (1981). Vibratory communication through living plants by a tropical wandering spider, *Science* **214:** 464–466.

Rudinsky, J. A. and Michael, R. R. (1973). Sound production in Scolytidae — Stridulation by female Dendroctonus beetles, *Journal of Insect Physiology* **19:** 689–705.

Rudinsky, J. A. and Ryker, L. C. (1976). Sound production in Scolytidae — rivalry and pre-mating stridulation of male douglas-fir beetle, *Journal of Insect Physiology* **22:** 997–1008.

Rudolph, P. (1967). Zum Ortungsverfahren von *Gyrinus substriatus* Steph, *Zeitschrift fur Vergleischende Physiologie* **56:** 341–375.

Rupprecht, R. (1968). Das Trommeln der Plecopteren, *Zeitschrift fur Vergleischende Physiologie* **59:** 38–71.

Rupprecht, R. (1969). Zur Artsperzificität der trommelsignale der Plecopteren (Insects), *Oikos* **20:** 26–33.

Rupprecht, R. (1975). Die Kommunikation von *Sialis* (Megaloptera) durch Vibrationssignale, *Journal of Insect Physiology* **21:** 305–320.

Rupprecht, R. (1976). Struktur und funcktion der bauchblase und des hammers von Plecopteren, *Zool. Jb. Anat.* **96:** 9–80.

Rupprecht, R. (1981). A new system of communication within Plecoptera and a signal with a new significance, *Biology of Inland Waters* **2:** 30–35.

Rupprecht, R. (1995). Anmerkungen zurn Paarungsverhalten von *Sisyra*. 3). Treffen deutschsprachiger Neuropterologen, *Galathea* **2:** 15–17.

Ryan, M. J. and Brenowitz, E. A. (1985). The role of body size, phylogeny and ambient noise in the evolution of bird song, *American Naturalist* **126:** 87–100.

Ryan, M. A., Čokl, A., and Walter, G. H. (1995). Differences in vibratory communication between a Slovenian and Australian population of *Nezara viridula* (L.) (Heteroptera: Pentatomidae), *Behavioural Processes* **36:** 183–193.

Ryan, M. J. and Rand, S. (1993). Species recognition and sexual selection as a unitary problem in animal communication, *Evolution* **47:** 647–657.

Ryan, M. J. and Wilczynski, W. (1991). Evolution of intraspecific variation in the advertisement call of a cricket frog (*Acris crepitans*, Hylidae), *Biological Journal of the Linnean Society* **44:** 249–271.

Ryker, L. C. (1975). Calling chirps in *Tropisternus natator* (D'Orchymont) and *T. lateralis* limbatus (Say) (Col., Hydrophilidae), *Entomological News* **86:** 179–186.

Ryker, L. C. (1976). Acoustic behavior of *Tropisternus ellipticus, T. columbianus* and *T. lateralis limbalis* in Western Oregon, *Coleopterists Bulletin* **30:** 147–156.

Ryker, L. C. and Rudinsky, J. A. (1976). Sound production in Scolytidae — aggressive and mating behavior of mountain pine beetle, *Annals of the Entomological Society of America* **69:** 677–680.

Sakagami, S. F. (1960). Preliminary report on the specific difference on behaviour and other ecological characters between European and Japanese honeybees, *Acta Hymenopterol.* **1:** 171–198.

Sales, G. and Pye, D. (1974). *Ultrasonic Communication by Animals*, Chapman and Hall, London.

Samways, M. J. (1976a). Song modification in the orthoptera. I. Proclamation songs of *Platycleis* spp., (Tettigoniidae), *Physiological Entomology* **1:** 131–149.

Samways, M. J. (1976b). Song modification in the Orthoptera. Part 3, micro syllabic schemes in Platycleis spp (Tettigoniidae), *Physiological Entomology* **1:** 299–303.

Samways, M. J. and Harz, K. (1982). Biogeography of intraspecific morphological variation in the bush crickets *Decticus verrucivorus* (L.) and D. *albifrons* (F.) (Orthoptera: Tettigoniidae), *Journal of Biogeography* **9:** 243–254.

Sanborn, A. F. (1997). Body temperature and the acoustic behavior of the cicada *Tibicen winnemanna* (Homoptera: Cicadidae), *Journal of Insect Behaviour* **10:** 257–264.

Sanborn, A. F. (1998). Thermal biology of cicadas (Homoptera: Cicadoidea), *Trends in Entomology* **1:** 89–104.

Sanborn, A. F. (2000). Comparative thermoregulation of sympatric endothermic and ectothermic cicadas (Homoptera: Cicadidae: *Tibicen winnemanna* and *Tibicen chloromerus*), *Journal of Comparative Physiology* **186A:** 551–556.

Sanborn, A. F. (2001). Timbal muscle physiology in the endothermic cicada *Tibicen winnemanna* (Homoptera: Cicadidae), *Comparative Biochemistry and Physiology* **130(4):** 9–19.

Sanborn, A. F. (2002). Cicada thermoregulation (Hemiptera, Cicadoidea), *Denisia* **4:** 455–470.

Sanborn, A. F. (2004). Thermoregulation and endothermy in the large western cicada *Tibicen cultriformis* (Hemiptera: Cicadidae), *Journal of Thermal Biology* **29:** 97–101.

Sanborn, A. F., Breitbarth, J. H., Heath, J. E., and Heath, M. S. (2002a). Temperature responses and habitat sharing in two sympatric species of *Okanagana* (Homoptera: Cicadoidea), *Western North American Naturalist* **62:** 437–450.

Sanborn, A. F., Heath, J. E., and Heath, M. S. (1992). Thermoregulation and evaporative cooling in the cicada *Okanagodes gracilis* (Homoptera: Cicadidae), *Comparative Biochemistry and Physiology* **102A:** 751–757.

Sanborn, A. F., Heath, M. S., Heath, J. E., and Noriega, F. G. (1995a). Diurnal activity, temperature responses, and endothermy in three South American cicadas (Homoptera: Cicadidae: *Dorisiana bonaerensis*, *Quesada gigas* and *Fidicina mannifera*), *Journal of Thermal Biology* **20:** 451–460.

Sanborn, A. F., Heath, J. E., Noriega, F. G., and Heath, M. S. (1995b). Thermoregulation by endogenous heat production in two South American grass dwelling cicadas (Homoptera: Cicadidae: *Proarna*), *Florida Entomologist* **78:** 319–328.

Sanborn, A. F. and Maté, S. (2000). Thermoregulation and the effect of body temperature on call temporal parameters in the cicada *Diceroprocta olympusa* (Homoptera: Cicadidae), *Comparative Biochemistry and Physiology* **125A:** 141–148.

Sanborn, A. F., Noriega, F. G., and Phillips, P. K. (2002b). Thermoregulation in the cicada *Platypedia putnami* var. *lutea* with a test of a crepitation hypothesis, *Journal of Thermal Biology* **27:** 365–369.

Sanborn, A. F. and Phillips, P. K. (1992). Observations on the effect of a partial solar eclipse on calling in some desert cicadas (Homoptera: Cicadidae), *Florida Entomologist* **75:** 285–287.

Sanborn, A. F. and Phillips, P. K. (1999). Analysis of the acoustic signals produced by the cicada *Platypedia putnami* variety *lutea* (Homoptera: Tibicinidae), *Annals of the Entomological Society of America* **92:** 451–455.

Sanborn, A. F., Phillips, P. K., and Villet, M. H. (2003a). Analysis of the calling songs of *Platypleura hirtipennis* (Germar, 1834) and *P. plumosa* (Germar, 1834) (Homoptera: Cicadidae), *African Entomology* **11:** 291–296.

Sanborn, A. F., Villet, M. H., and Phillips, P. K. (2003b). Hot-blooded singers: endothermy facilitates crepuscular signaling in African platypleurine cicadas (Homoptera: Cicadidae: *Platypleura* spp.), *Naturwissenschaften* **90:** 305–308.

Sanborn, A. F., Villet, M. H., and Phillips, P. K. (2004). Endothermy in African platypleurine cicadas: influence of body size and habitat (Hemiptera: Cicadidae), *Physiological and Biochemical Zoology* **77:** 816–823.

Sandberg, J. B. and Stewart, K. W. (2003). Continued studies of drumming in North American Plecoptera; evolutionary implications, In *Research Update on Ephemeroptera & Plecoptera*, Gaino, E., Ed., University of Perugia, Perugia, Italy, pp. 73–81.

Sandeman, D. C., Tautz, J., and Lindauer, M. (1996). Transmission of vibration across honeycombs and its detection by bee leg receptors, *Journal of Experimental Biology* **199:** 2585–2594.

Sanford, G. M., Lutterschmidt, W. I., and Hutchinson, V. H. (2002). The comparative method revisited, *BioScience* **52:** 830–836.

Satokangas, P., Liimatainen, J. O., and Hoikkala, A. (1994). Songs produced by the females of the *Drosophila virilis* group of species, *Behavioral Genetics* **24:** 263–272.

Sattman, D. A. and Cocroft, R. B. (2003). Phenotypic plasticity and repeatability in signals of *Enchenopa* treehoppers, with implications for gene flow among host-shifted populations, *Ethology* **109:** 981–994.

Schaefer, C. (1980). The sound-producing structures of some primitive Pentatomoidea (Hemiptera: Heteroptera), *Journal of the New York Entomological Society* **88:** 230–235.

Schaefer, C. and Pupedis, R. J. (1981). A stridulatory device in certain Alydinae (Hemiptera: Heteroptera: Alydidae), *Journal of the Kansas Entomological Society* **54**(1)**:** 143–152.

Schatral, A. (1990). Interspecific acoustic interactions in bushcrickets, In *The Tettigoniidae: Biology, Systematics and Evolution*, Bailey, W. J. and Rentz, D. C. F., Eds., Springer, Berlin, pp. 152–165.

Schatral, A. and Kalmring, K. (1985). The role of the song for spatial dispersion and agonistic contacts in male bushcrickets, In *Acoustical and Vibrational Communication in Insects*, Kalmring, K. and Elsner, N., Eds., Paul Parey, Berlin, pp. 111–116.

Schatral, A. and Latimer, W. (1988). A field study on the acoustic behaviour of mobile singers in the bushcricket *Psorodonotus illyricus*, *Journal of Natural History* **22:** 297–310.

Schatral, A., Latimer, W., and Broughton, B. (1984). Spatial dispersion and agonistic contacts of male bushcrickets in the biotope, *Zeitschrifte fur Tierpsychologie* **65:** 201–214.

Schatral, A. and Yeoh, P. B. (1990). Spatial distribution, calling and interspecific acoustic interactions in two species of the Australian tettigoniid genus *Tymphanophora* (Orthoptera: Tettigoniidae), *Journal of Zoology* **221:** 375–390.

Schilman, P. E., Manrique, G., and Lazzari, C. R. (2001). Comparison of disturbance stridulations in five species of triatominae bugs, *Acta Tropica* **79:** 171–178.

Schiödte, G. C. (1874). Note sur les organes de stridulation chez les larves des Coléoptères Lamellicornes, *Annales de la Societe entomologique de France (NS), 5th ser.* **4:** 39–41.

Schluter, D. (2000). *The Ecology of Adaptive Radiation*, Oxford University Press, Oxford.

Schmitt, M. (1991). Stridulatory devices of leaf beetles (Chrysomelidae) and other Coleoptera, In *Advances in Coleopterology*, Zunino, M., Belles, X., and Blas, M., Eds., Asociación Europea de Coleopterología, Barcelona, pp. 263–280.

Schmitt, M. and Traue, D. (1990). Bioacoustic and morphological aspects of the stridulation in Criocerinae (Col., Chrysomelidae), *Zoologischer Anzeiger* **225:** 225–240.

Schneider, W. (1950). Über den Erschütterungssinn von Käfern und Fliegen, *Zeitschrift fur Vergleischende Physiologie* **32:** 287–302.

Schneider, P. (1972). Akustische Signale bei Hummeln, *Naturwissenschaften* **59:** 168–169.

Schneider, P. (1975). Versuche zur Erzeugung des Verteidigungstones bei Hummeln, *Zoologische Jahrbuecher Physiologie* **79:** 111–127.

Schneider, P. and Kloft, W. (1971). Beobachtungen zum Gruppenverteidigungsverhalten der Östlichen Honigbiene *Apis cerana Fabr*, *Zeitschrifte fur Tierpsychologie* **29:** 337–342.

Schnorbus, H. (1971). Die subgenualen Sinnesorgane von *Periplaneta americana* und Vibrationsschwellen, *Zeitschrift fur Vergleichende Physiologie* **71:** 14–48.

Schofield, C. J. (1977). Sound production in some triatomine bugs, *Physiological Entomology* **2:** 43–52.

Schofield, C. J. (1979). The behaviour of Triatominae (Hemiptera: Reduviidae): A review, *Bulletin of Entomological Research* **69:** 363–379.

Schön, A. (1911). Bau und Entwicklung des tibialen Chordotonalorgans bei der Honigbiene und bei Ameisen, *Zool. Jb., (Anat)* **31:** 439–472.

Schuh, R. T. (1974). The Orthotylinae and Phylinae (Hemiptera: Miridae) of South Africa with a phylogenetic analysis of the ant-mimetic tribes of the two subfamilies for the world, *Entomologica Americana* **47:** 1–332.

Schuh, R. T. and Slater, J. A. (1974). *True Bugs of the World* (*Hemiptera: Heteroptera*). Cornell University Press, Ithaca and London.

Schul, J. (1994). Untersuchungen zur akustischen Kommunikation bei drei Arten der Gattung Tettigonia (Orthoptera, Tettigoniidae), Dissertation im Fachbereich Biologie der Philipps-Universität Marburg.

Schul, J. (1998). Song recognition by temporal cues in a group of closely related bushcricket species (genus *Tettigonia*), *Journal of Comparative Physiology A — Sensory Neural and Behavioral Physiology* **183:** 401–410.

Schultz, R. J. (1969). Hybridization, unisexuallity and polyploidy in the teleost *Poeciliopsis* (Poeciliidae) and other vertebrates, *Animal Behaviour* **103:** 605–619.

Schultz, R. J. (1973). Unisexual fish: laboratory synthesis of a "species", *Science* **179:** 180–181.

Schultz, R. J. and Kallman, K. D. (1968). Triploid hybrids between the all-female teleost *Poecilia formosa* and *Poecilia sphenops*, *Nature* **219:** 280–282.

Schumacher, R. (1973a). Beitrag zur Kenntnis der Stridulationsapparate einheimischer Necrophorus-Arten (*Necrophorus humator* Ol, *Necrophorus investigator* Zetterst, *Necrophorus vespilloides* Herbst) (Insecta, Col.), *Z. Morphol. Tiere* **75:** 65–75.

Schumacher, R. (1973b). Morphologische Untersuchungen der tibialen Tympanalorgane von neun einheimischen Laubheuschrecken-Arten (Orthoptera, Tettigonioidae), *Z. Morph. Tiere* **75**: 267–282.

Schumacher, R. (1975). Scanning-electron-microscope description of the tympanal organ of the Tettigoniidae (Orthoptera Ensifera), *Z. Morph. Tiere* **81**: 209–219.

Schumacher, R. (1979). Zur funktionellen Morphologie des auditiven Systems der Laubheuschrecken (Orthoptera, Tettigonioidae), *Entomologia Generalis* **5**: 321–356.

Schuster, J. (1975). *Comparative Behavior, Acoustical Signals, and Ecology of New World Passalidae (Col.)*, Ph.D. dissertation, University of Florida.

Schuster, J. (1983). Acoustical signals of passalid beetles: complex repertoires, *Florida Entomologist* **66**: 486–496.

Schuster, S. M. and Wade, M. J. (2004). *Mating Systems and Strategies*, Princeton University Press, Princeton, NJ.

Scudder, S. H. (1886). Notes on the stridulation of some New England Orthoptera, *Proceedings of the Boston Society of Natural History* **11**(1866–68): 306–313.

Seabra, S. G., Simões, P. C., Drosopoulos, S., and Quartau, J. A. (2000). Genetic variability and differentiation in two allopatric species of the genus *Cicada* L. (Hemiptera, Cicadoidea) in Greece, *Deutsche Entomologische Zeitschrift* **47**(2): 143–145.

Seabra, S. G., Wilcock, H. R., Quartau, J. A., and Bruford, M. W. (2002). Microsatellite loci isolated from the Mediterranean species *Cicada barbara* (Stal) and *C. orni* L. (Hemiptera, Coicadoidea), *Molecular Ecology Notes* **2**: 173–175.

Searcy, W. A. and Andersson, M. (1986). Sexual selection and the evolution of song, *Annual Review of Ecology and Systematics* **17**: 507–533.

Seeley, T. D. (1992). The tremble dance of the honey bee: message and meanings, *Behavioral Ecology and Sociobiology* **28**: 277–290.

Seeley, T. D. (1995). *The Wisdom of the Hive*, Harvard University Press, Cambridge, London, p. 309.

Seeley, T. D. (1998). Thoughts on information and integration in honey bee colonies, *Apidologie* **29**: 67–80.

Seeley, T. D., Morse, R. A., and Visscher, P. K. (1979). The natural history of the flight of honey bee swarms, *Psyche* **86**: 103–113.

Seeley, T. D. and Tautz, J. (2001). Worker piping in honey bee swarms and its role in preparing for liftoff, *Journal of Comparative Physiology A* **187**: 667–676.

Seeley, T. D., Seeley, R., and Akranatakul, P. (1982). Colony defense strategies of honeybees in Thailand, *Ecological Monographs* **52**: 43–63.

Selander, J. and Jansson, A. (1977). Sound production associated with mating behaviour of the large pine weevil, *Hylobius abietis* (Coleoptera, Curculionidae), *Annales Entomologicae Fennicae* **43**: 66–75.

Sen Sarma, M., Fuchs, S., Werber, C., and Tautz, J. (2002). Worker piping triggers hissing for coordinated colony defence in the dwarf honeybee *Apis florea*, *Zoology* **105**: 215–223.

Seymour, C., Lewis, B., Larsen, O. N., and Michelsen, A. (1978). Biophysics of the ensiferan ear. II. The steady-state gain of the hearing trumpet in bushcrickets, *Journal of Comparative Physiology* **123**: 205–216.

Shapiro, L. H. (1998). Hybridisation and geographic variation in two meadow katydid contact zones, *Evolution* **52**: 784–796.

Shaw, K. C. (1968). An analysis of the phonoresponse of males of the true katysis, *Pterophylla camellifolia* (Fabricius) (Orthoptera: Tettigoniidae), *Behaviour* **31**: 203–260.

Shaw, S. R. (1994). Detection of airborne sound by a cockroach "vibration detector": a possible missing link in insect auditory evolution, *Journal of Experimental Biology* **193**: 13–47.

Shaw, K. L. (1996a). Sequential radiations and patterns of speciation in the Hawaiian cricket genus *Laupala* inferred from DNA sequences, *Evolution* **50**: 237–255.

Shaw, K. L. (1996b). Polygenic inheritance of a behavioral phenotype: interspecific genetics of song in the Hawaiian cricket genus *Laupala*, *Evolution* **50**: 256–266.

Shaw, K. L. (2000). Further acoustic diversity in Hawaiian forests: two new species of Hawaiian cricket (Orthoptera: Gryllidae: Trigoniinae: *Laupala*), *Zoological Journal of the Linnean Society* **129**: 73–91.

Shaw, K. C. and Carlson, O. V. (1969). The true kaytdid *Pterophylla camellifolia* (Orthoptera Tettigoniidae) in Iowa: 2 populations which differ in behavior and morphology, *Iowa State Journal of Science* **44**: 193–200.

Shaw, K. L. and Herlihy, D. P. (2000). Acoustic preference functions and song variability in the Hawaiian cricket *Laupala cerasina*, *Proceedings of the Royal Society of London Series B* **267**: 577–584.

Shaw, K. C., North, R. C., and Meixner, J. A. (1981). Movement and spacing of singing *Amblyocorypha parvipennis* males, *Annals of the Entomological Society of America* **74**: 436–444.

Shaw, K.-C., Vargo, A., and Carlson, O.V. (1974). Sounds and associated behaviour of some species of *Empoasca* (Homoptera: Cicadellidae), *Journal of the Kansas Entomological Society*, **47**: 284–307.

Shimozawa, T., Mruakami, J., and Kumagai, T. (2003). Cricket wind receptors: thermal noise for the highest sensitivity known, In *Sensors and Sensing in Biology and Engineering.*, Barth, F., Humphrey, J., and Secomb, T., Eds., Springer Verlag, Wien, pp. 145–157.

Shorey, H. H. (1962). The nature of the sound produced by *Drosophila melanogaster* during courtship, *Science* **137**: 677–678.

Sickmann, T. (1996). Vergleichende funktionelle und anatomische Untersuchungen zum Aufbau der Hör- und Vibrationsbahn im thorakalen Bauchmark von Laubheuschrecken, Dissertation, Marburg, 1996.

Sickmann, T., Kalmring, K., and Müller, A. (1997). The auditory-vibratory system of the bushcricket *Polysarcus denticauda* (Phaneropterinae Tettigoniidae). I. Morphology of the complex tibial organs, *Hearing Research* **104**: 155–166.

Silver, S., Kalmring, K., and Kühne, R. (1980). The responses of central acoustic and vibratory interneurones in bushcrickets and locusts to ultrasonic stimulation, *Physiological Entomology* **5**: 427–435.

Simmons, L. W. (2001). *Sperm Competition and its Evolutionary Consequences in the Insects*, Princeton University Press, Princeton and Oxford.

Simmons, L. W., Beesley, L., Lindhjem, P., Newbound, D., Norris, J., and Wayne, A. (1999). Nuptial feeding by male bushcrickets: an indicator of male quality?, *Behavioral Ecology* **10**: 263–269.

Simmons, L. W. and Gwynne, D. T. (1991). The refractory period of female katydids (Orthoptera: Tettigoniidae): sexual conflict over the re-mating interval?, *Behavioral Ecology* **2**(4): 276–282.

Simmons, L. W. and Ritchie, M. G. (1996). Symmetry in the songs of crickets, *Proceedings of the Royal Society of London Series B* **263**: 305–311.

Simmons, L. W., Teale, R. J., Maier, M., Standish, R. J., Bailey, W. J., and Withers, P. C. (1992). Some costs of reproduction for male bushcrickets, Requena verticalis (Orthoptera: Tettigoniidae): allocating resources to mate attraction and nuptial feeding, *Behavioral Ecology and Sociobiology* **31**: 57–62.

Simões, P. C., Boulard, M., Rebelo, M. T., Drosopoulos, S., Claridge, M. F., Morgan, J. C., and Quartau, J. A. (2000). Differences in the male calling songs of two sibling species of Cicada (Hemiptera: Cicadoidea) in Greece, *European Journal of Entomology* **97**: 437–440.

Simpson, J. (1964). The mechanism of honey-bee queen piping, *Zeitschrift fur Vergleischende Physiologie* **48**: 277–282.

Simpson, J. and Cherry, S. M. (1969). Queen confinement, queen piping and swarming in *Apis mellifera* colonies, *Animal Behaviour* **17**: 271–278.

Sirot, L. K. (2003). The evolution of insect mating structures through sexual selection, *Florida Entomologist* **86**: 124–132.

Sismondo, E. (1979). Stridulation and tegminal resonance in the tree cricket *Oecanthus nigricornis* (Orthoptera: Gryllidae: Oecanthinae), *Journal of Comparative Physiology* **129A**: 269–279.

Sivinski, J. (1988). What do fruit fly songs mean?, *Florida Entomologist* **71**: 462–466.

Sivinsky, J., Burk, T., and Webb, J. C. (1984). Acoustic courtship signals in the Caribbean fruit fly, *Anastrepha suspensa* (Loew), *Animal Behaviour* **32**: 1011–1016.

Sivinsky, J. and Webb, J. C. (1985). The form and function of acoustic courtship signals of the Papaya fruit fly, *Toxotrypana curvicauda* (Tephritidae), *Florida Entomologist* **68**: 634–641.

Sivinsky, J. and Webb, J. C. (1986). Changes in a Caribbean fruit fly acoustic signal with social situation (Diptera: Tephritidae), *Annals of the Entomological Society of America* **79**: 146–149.

Skals, N. and Surlykke, A.M. (1999). Song production by abdominal tymbal organ in two moth species: the green silver-line and the scarce silver-line (Noctuoidea: Nolidae: Chloeophorinae), *Journal of Experimental biology* **202**: 2937–2949.

Skovmand, D. and Pedersen, S. B. (1983). Song recognition and song pattern in a shorthorned grasshopper: a parameter carrying specific behavioral information, *Journal of Comparative Physiology* **153A**: 393–401.

Slobodchikoff, C. N. and Spangler, H. G. (1979). Two types of sound production in Eupsophulus castaneus (Coleoptera: Tenebrionidae), *Coleopterists Bulletin* **33**: 239–244.

Smith, R. C. (1922). The biology of the Chrysopidae, *Memoirs of the Cornell University Agricultural Experiment Station* **58:** 1287–1372.

Smith, R. L. (1973). Aspects of the biology of three species of the genus *Rhantus*, with special reference to the acoustical behavior of two, *The Canadian Entomologist* **105:** 909–919.

Smith, W. J. (1979). The study of ultrasonic communication, *American Zoologist* **19:** 531–538.

Snodgrass, R. E. (1925). Insect Musicians, their music, and their instruments, *Annual Report of The Smithsonian Institution* **2738:** 405–452.

Snodgrass, R. E. (1956). *Anatomy of the Honey Bee*, Cornell University Press, Ithaca, New York, p. 334.

Snyder, M. R. (1998). A functionally equivalent artificial neuronal network model of the prey orientation behavior of waterstriders (Gerridae), *Ethology* **104:** 285–297.

Sokal, R. R. and Rohlf, F. J. (1995). *Biometry*, Freeman and Company, New York.

Sokolowski, M. B. and Turlings, T. C. J. (1987). *Drosophila* parasitoid-host interactions: vibrotaxis and ovipositor searching from the host's perspective, *Canadian Journal of Zoology* **65:** 461–464.

Soltavalta, O. (1947). The flight-tone (wing-stroke frequency) of insects, *Acta Entomologica Fennica* **4:** 1–117.

Soltavalta, O. (1952). Wing-stroke frequency of insects, *Ann. Zool. Botan. Fenn.* **15:** 1–47.

Souroukis, K., Cade, W. H., and Rowell, G. (1992). Factors that possibly influence variation in the calling song of field crickets: temperature, time, and male size, age, and wing morphology, *Canadian Journal of Zoology* **70:** 950–955.

Souza, N. A. de, Ward, R. D., Hamilton, J. G. C., Kyriacou, C. P., and Peixoto, A. A. (2002). Copulation songs in three siblings of *Lutzomyia longipalpis* (Diptera: Psychodidae), *Transactions of the Royal Society of Tropical Medicine and Hygiene* **96:** 102–103.

Spangler, H. G. (1984). Silence as a dense against predatory bats in two species of calling insects, *Soutwestern Naturalist* **29:** 481–488.

Spangler, H. G. (1986). High-frequency sound production by honeybees, *Journal of Apicultural Research* **25:** 213–214.

Spangler, H. G. (1988). Hearing in tiger beetles (Cicindelidae), *Physiological Entomology* **13:** 447–452.

Spieth, H. T. (1974). Courtship behavior in *Drosophila*, *Annual Review of Entomology* **19:** 385–405.

Spooner, J. D. (1968). Pair-forming acoustic systems of phaneropterine katydids (Orthoptera, Tettigoniidae), *Animal Behaviour* **16:** 197–212.

Sprecher-Uebersax, E. and Durrer, H. (1998). Stridulation of stag beetle larvae (*Lucanus cervus* L), *Mitt. Schweiz. Entomol. Ges.* **71:** 471–479.

Stein, W. and Sauer, A. E. (1999). Physiology of vibration-sensitive afferents in the femoral chordotonal organ of the stick insect, *Journal of Comparative Physiology* **184:** 253–263.

Stephen, R. O. and Hartley, J. C. (1991). The transmission of bush-cricket calls in natural environments, *Journal of Experimental Biology* **155:** 227–244.

Stephen, R. O. and Hartley, J. C. (1995). Sound production in crickets, *Journal of Experimental Biology* **198:** 2139–2152.

Stewart, K. W. (1994). Theoretical considerations of mate finding and other adult behaviors of Plecoptera, *Aquatic Insects* **16:** 95–104.

Stewart, K. W. (1997). Vibrational communication in insects: epitome in the language of stoneflies? *American Entomologist* **43:** 81–91.

Stewart, K. W. (2001). Vibrational communication (drumming) and mate-searching behavior of stoneflies (Plecoptera); evolutionary considerations, In *Trends in Research in Ephemeroptera and Plecoptera*, Dominguez, E., Ed., Kluwer Academic/Plenum Publishers, Dordrecht/New York, pp. 217–225.

Stewart, K. W., Abbott, J. C., and Bottorff, R. L. (1995). The drumming signals of two stonefly species *Cosumnoperla bypocrena* (Perlodidae) and *Paraperla wilsoni* (Chloroperlidae); a newly discovered duet pattern in Plecoptera, *Entomological News* **106:** 13–18.

Stewart, K. W. and Maketon, M. (1990). Intraspecific variation and information content of drumming in three Plecoptera species, In *Mayflies and Stoneflies*, Campbell, I. C., Ed., Kluwer Academic Publishers, Dordrecht, pp. 259–268.

Stewart, K. W. and Maketon, M. (1991). Structures used by Nearctic stoneflies (Plecoptera) for drumming, and their relationship to behavioral pattern diversity, *Aquatic Insects* **13:** 33–53.

Stewart, K. W. and Zeigler, D. D. (1984). The use of larval morphology and drumming in Plecoptera systematics, and further studies of drumming behavior, *Annales de Limnologie* **20:** 105–114.

Stiedl, O. (1991). Akusto-vibratorische Verhaltensuntersuchungen an Ephippigerinen im Labor und Biotop, Dissertation im Fachbereich Biologie der Philipps-Universität Marburg.

Stiedl, O. and Bickmeyer, U. (1991). Acoustic behaviour of Ephippiger ephippiger Fiebig (Orthoptera, Tettigoniidae). within a habitat of Southern France, *Behavioural Processes* **23**: 125–135.

Stiedl, O. and Kalmring, K. (1989). The importance of song and vibratory signals in the behaviour of the bushcricket *Ephippiger ephippiger* Fiebig (Orthoptera Tettigoniidae): taxis by females, *Oecologia* **80**: 142–144.

Stoddard, P. K. (1990). Audio computers, theory of operation and guidelines for selection of systems and components, *Bioacoustics* **2**(3): 217–239.

Stölting, H., Moore, T. E., and Lakes-Harlan, R. (2002). Substrate vibrations during acoustic signalling in the cicada *Okanagana rimosa*, *Journal of Insect Behaviour* **2**: 2.

Stölting, H. and Stumpner, A. (1998). Tonotopic organization of auditory receptors of the bushcricket *Pholidoptera griseoaptera* (Tettigoniidae, Decticinae), *Cell and Tissue Research* **294**: 377–386.

Stone, L. D. (1989). *Theory of Optimal Search*, Operation Research Society of America, Arlington, VA, p. 277.

Storm, J. (1998). The dynamics and flow field of the wagging dancing honeybee, PhD Thesis, Odense University, Denmark.

Storm, J. and Kilpinen, O. (1998). Modelling the subgenual organ of the honeybee *Apis mellifera*, *Biological Cybernetics* **78**: 175–182.

Strausfeld, N. J. and Bacon, J. P. (1983). Multimodal convergence in the central nervous system of dipterous insects, Fortschritte der Zoologie, Bd. 28, Horn (Hrsg), *Multimodal Convergency in Sensory Systems*, Gustav Fischer Verlag, Stuttgart, New York.

Stritih, N., Stumpner, A., and Čokl, A. (2003). Vibration sensitive interneurones of the primitive ensiferan (*Troglophilus neglectus*, Rhaphidophoridae) and their homology to acoustic interneurones of Ensifera, In *Proceedings of the 29th Göttingen Neurobiology Conference*, Elsner, N. and Zimmermann, H., Eds., Thieme, Stuttgart, pp. 415–416.

Stritih, N., Virant-Doberlet, M., and Čokl, A. (2000). Green stink bug *Nezara viridula* detects differences in amplitude between courtship song vibrations at stem and petiolus, *European Journal of Physiology* **439**(Suppl.): R190–R192.

Strong, W. J. and Palmer, E. P. (1975). Computer-based sound spectrograph system, *Journal of the Acoustical Society of America* **58**(4): 899–904.

Strong, F. E., Sheldahl, J. A., Hughes, P. R., and Hussein, E. M. K. (1970). Reproductive biology of Lygus hesperus Knight, *Hilgardia* **40**: 105–143.

Strübing, H. (1958). Lautäußerung — der entscheidende Faktor für das Zusammenfinden der Geschlechter bei Kleinzikaden (Homoptera-Auchenorrhyncha), *Zoologische Beiträge (N.F.)* **4**: 15–21.

Strübing, H. (1959). Lautgebung und Paarungsverhalten von Kleinzikaden, Verhandlung der Deutschen Zoologischen Gesellschaft in Münster/Westfalen.

Strübing, H. (1960). Paarungsverhalten und Lautäußerung von Kleinzikaden, demonstriert an Beispielen aus der Familie der Delphacidae (Homoptera: Auchenorrhyncha) *Verhandlungen, XI International Congress of Entomology* **3**: 12–14.

Strübing, H. (1963). Lautäußerung von Euscelis-Bastarden (Homoptera-Auchenorrhyncha), *Verh. Deutsch. Zool. Ges. München* 268–281.

Strübing, H. (1967). Zur Untersuchungsmethodik der Lautäußerungen von Kleinzikaden (Homoptera-Cicadina), *Zoologische Beiträge (N.F.)* **13**: 265–284.

Strübing, H. (1970). Zur Artberechtigung *von Euscelis alsius* Ribaut gegenüber *Euscelis plebejus* Fall. (Homoptera-Cicadina) — Ein Beitrag zur Neuern Systematik, *Zoologische Beiträge (N.F.)* **16**: 441–478.

Strübing, H. (1977a). Lauterzeugung oder Substratvibration als Kommunikationsmittel bei Kleinzikaden? (diskutiert am Beispiel *von Dictyophara europaea* — Homoptera-Cicadina: Fulgoroidea), *Zoologische Beiträge (N.F.)* **23**: 323–332.

Strübing, H. (1977b). *Neue Ergebnisse zur Kommunikation bei Kleinzikaden, Verhandlung der Deutschen Zoologischen Gesellschaft*, Vol. 336: Gustav Fischer Verlag, Stuttgart.

Strübing, H. (1978). *Euscelis lineolatus* Brullé 1832 und Euscelis ononidis Remane 1967. 1. Ein ökologischer, morphologischer und bioakustischer Vergleich, *Zoologische Beiträge (N.F.)* **24**: 123–154.

Strübing, H. (1980). *Euscelis remanei*, eine neue *Euscelis*-Art aus Südspanien im Vergleich zu anderen *Euscelis*-Arten (Homoptera-Cicadina), *Zoologische Beiträge (N.F.)* **26**: 383–404.

Strübing, H. (1983). Die Bedeutung des Kommunikationssignals für die Diagnose von *Euscelis*-Arten (Homoptera-Cicadina), *Zoologische Jahrbuecher Physiologie* **87**: 343–351.

Strübing, H. (1992). Vibrationskommunikation im Paarungsverhalten der Büffelzikade *Stictocephala bisonia* Kopp and Yonke 1977 (Homoptera-Auchenorrhyncha, Membracidae), *Mitteilungen der Deutschen Gesellschaft für allgemeine und angewandte Entomologie* **8**: 60–62.

Strübing, H. (1995). Artentstehung und — verdrängung am beispiel dreier Kleinzikadenarten (Homoptera — Auchenorrhyncha), *Mitteilungen der Deutschen Gesellschaft für allgemeine und angewandte Entomologie* **10**: 619–624.

Strübing, H. and Hasse, A. (1974). Zur Artberechtigung von Euscelis alsius Ribaut (Homoptera-Cicadina) II. Die Aedeagus-Variabilität, *Zoologische Beiträge (N.F.)* **20**: 527–542.

Strübing, H. and Rollenhagen, T. (1988). Ein neues Aufnehmersystem für Vibrationssignale und seine Anwendung auf Beispiele aus der Familie Delphacidae (Homoptera-Cicadina) — A new recording system for vibratory signals and its application to different species of the family Delphacidae (Homoptera-Cicadina), *Zoologische Jahrbücher, Abteilung Allgemeine Zoologie und Physiologie der Tiere*, Gustav Fischer Verlag Jena, Vol. 92, pp. 245–268, 1988.

Strübing, H. and Rollenbach, T. (1992). Vibratory communication of *Stictocephala bisonia* Kopp and Yonke (1977) and *Dictyphara europaea L.* (Dictyophalidae) — Homoptera — Auchenorryhnca, *Abstracts of the 8th International Meeting on Insect Sound*, Pommersfelden, pp. 69–70.

Strübing, H. and Schwartz-Mittelstaedt, G. (1986). The vibratory membrane in the genus *Euscelis*, *Proceeding of the second Congress on Rhynchota Balkan*, Mikrolimni, Greece, pp. 49–52.

Stumpner, A. (2002). A species-specific frequency filter through specific inhibition, not specific excitation, *Journal of Comparative Physiology A — Sensory Neural and Behavioral Physiology* **188**: 239–248.

Stumpner, A. and Meyer, S. (2001). Songs and the function of song elements in four duetting bushcricket species (Ensifera, Phaneropteridae, *Barbitistes*), *Journal of Insect Behaviour* **14**: 511–534.

Stumpner, A., Ronacher, B., and von Helversen, O. (1991). Auditory interneurons in the metathoracic ganglion of the grasshopper *Chorthippus biguttulus*: II Processing of temporal patterns of the song of the male, *Journal of Experimental Biology* **158**: 391–410.

Stumpner, A. and von Helversen, O. (1992). Recognition of a two-element song in the grasshopper *Chorthippus dorsatus* (Orthoptera: Gomphocerinae), *Journal of Comparative Physiology A* **171**: 405–412.

Stumpner, A. and von Helversen, O. (1994). Song production and song recognition in a group of sibling grasshopper species (*Chorthippus dorsatus*, *Ch. dichrous* and *Ch. loratus*: Orthoptera, Acrididae), *Bioacoustics* **6**: 1–23.

Stürzl, W., Kempter, R., and van Hemmen, J. L. (2000). Theory of arachnid prey localization, *Physical Review Letters* **84**: 5668–5671.

Štys, P. (1961). The stridulatory mechanism in Centrocoris spiniger (F.) and some other Coreidae (Heteroptera), *Acta ent. Mus. Nat. Prag.* **34**: 427–431.

Štys, P. (1964). Thaumastellidae — a new family of pentatomoid Heteroptera, *Acta Soc. Entomol. Cechoslov.* **61**(3): 238–253.

Štys, P. (1966). Revision of the genus Dayakiella Horv. and notes on its systematic position (Heteroptera, Colobathristidae), *Acta Bohemoslov.* **63**: 27–39.

Sueur, J. (2000). Description d'une espèce nouvelle de Cigale de la région de los Tuxtlas (Veracruz Mexique) et étude de son émission sonore, *Bulletin de la Société entomologique de France* **105**: 217–222.

Sueur, J. (2002a). Cicada acoustic communication: potential sound partitioning in a multispecies community from Mexico (Hemiptera: Cicadomorpha: Cicadidae), *Biological Journal of the Linnean Society* **75**: 379–394.

Sueur, J. (2002b). Étho-écologie de la communication sonore des Cigales: le Modèle *Tibicina* Amyot, 1847 (Cicadidae, Tibicininae), Thèse, École Pratique des Hautes Études, Paris, MNHN, p. 345.

Sueur, J. (2003). Indirect and direct acoustic aggression in cicadas: first observations in the Palaeartic genus *Tibicina* Amyot (Hemiptera: Cicadomorpha: Cicadidae), *Journal of Natural History* **37**: 2931–2948.

Sueur, J. and Aubin, T. (2002). Acoustic communication in the Palaearctic red cicada *Tibicina haematodes*: chorus organisation, calling song structure, and signal recognition, *Canadian Journal of Zoology* **80**: 126–136.

Sueur, J. and Aubin, T. (2003). Specificity of cicada calling songs in the genus *Tibicina* (Hemiptera, Cicadidae), *Systematic Entomology* **28**: 481–492.

Sueur, J. and Aubin, T. (2004). Acoustic signals in cicada courtship behaviour (order Hemiptera, genus Tibicina), *Journal of Zoology London* **262**: 217–224.

Sueur, J., Aubin, T., and Bourgoin, T. (2002). Bioacoustique et systématique des Insectes, *Mém. Soc. Ent. Fr.* **6**: 45–62.

Sueur, J. and Puissant, S. (2002). Spatial and ecological isolation in cicadas: first data from *Tibicina* (Hemiptera: Cicadoidea) in France, *European Journal of Entomology* **99**: 477–484.

Sueur, J. and Puissant, S. (2003). Analysis of sound behaviour leads to new synonymy in Mediterranean Cicadas (Cicadidae, *Tibicina*), *Deutsche Entomologische Zeitschrift* **50**: 121–127.

Sueur, J., Puissant, S., and Pillet, J. M. (2003). An Eastern Mediterranea cicada in the West: first record of *Tibicina steveni* (Krynicki, 1837) in Switzerland and France (Hemiptera, Cicadidae, Tibicinae), Revue française d'Entomologie **25**: 105–111.

Sueur, J., Puissant, S., Simões, P. C., Seabra, S., Boulard, M., and Quartau, J. A. (2004). Cicadas from Portugal: revised list of species with eco-ethological data (Hemiptera: Cicadidae), *Insect Systematics and Evolution* **35**(2), 177–187.

Sueur, J. and Sanborn, A. F. (2003). Ambient temperature and sound power of cicada calling song (Hemiptera Cicadidae *Tibicina*), *Physiological Entomology* **28**: 340–343.

Suga, N. (1966). Ultrasonic production and its reception in some neotropical Tettigoniidae, *Journal of Insect Physiology* **12**: 1039–1050.

Suvanto, L., Hoikkala, A., and Liimatainen, J. (1994). Secondary courtships songs and inhibitory songs of *Drosophila virilis* group males, *Behavioral Genetics* **24**: 85–94.

Suvanto, L., Liimatainen, J. O., Tregenza, T., and Hoikkala, A. (2000). Courtship signals and mate choice of the flies of inbred *Drosophila montana* strains, *Journal of Evolutionary Biology* **13**: 583–592.

Swammerdam, J. (1669). *Historia Insectorum Generalis*, Utrecht, p. 216.

Sweet, M. H. (1967). The tribal classification of the Rhyparochrominae (Heteroptera: Lygaeidae), Annals of the Entomological Society of America **60**: 208–226.

Sweet, M. H. and Schaefer, C. W. (2002). Parastrachiinae (Hemiptera: Cydnidae) raised to family level, *Annals of the Entomological Society of America* **95**(4): 441–448.

Synave, H. (1953). Une cixiide troglobie découvert dans les galéries souterraines de Namoroka (Hemiptera — Homoptera), *Le Naturaliste Malgache* **5**(2): 175–179.

Szczytko, S. W. and Stewart, K. W. (1979). Drumming behavior of four western Nearctic *Isoperla* (Plecoptera) species, *Annals of the Entomological Society of America* **72**: 781–786.

Tauber, C. A. (1969). Taxonomy and biology of the lacewing genus *Meleoma* (Neuroptera: Chrysopidae), *University of California Publications in Entomology* **58**: 1–86.

Tauber, E., Cohen, D., Greenfield, M. D., and Pener, M. P. (2001). Duet singing and female choice in the bushcricket *Phaneroptera nana*, *Behaviour* **138**: 411–430.

Tauber, E. and Eberl, D. F. (2001). Song production in auditory mutants of *Drosophila*: the role of sensory feedback, *Journal of Comparative Physiology A* **187**: 341–348.

Tauber, E. and Eberl, D. F. (2003). Acoustic communication in *Drosophila*, *Behavioural Processes* **64**: 197–210.

Tauber, M. J., Tauber, C. A., Hoy, R. R., and Tauber, P. J. (1991). Life history, mating behavior and courtship songs of the endemic Hawaiian *Anomalochrysa maclachlani* (Neuroptera: Chrysopidae), *Canadian Journal of Zoology* **68**: 1020–1026.

Tautz, J. (1979). Reception of particle oscillation in a medium — an unorthodox sensory capacity, *Naturwissenschaften* **66**: 452–461.

Tautz, J. (1996). Honey bee waggle dance: recruitment success depends on the dance floor, *Journal of Experimental Biology* **199**: 1375–1381.

Tautz, J., Casas, J., and Sandeman, D. C. (2001). Phase reversal of vibratory signals in honeycomb may assist dancing honeybees to attract their audience, *Journal of Experimental Biology* **204**: 3737–3746.

Tautz, J. and Rohrseitz, K. (1998). What attracts honeybees to a waggle dancer?, *Journal of Comparative Physiology A* **183**: 661–667.

Tautz, J., Rohrseitz, K., and Sandeman, D. C. (1996). One-strided waggle dance in bees, *Nature* **382**: 32.

Taylor, K. R. (1962). The Australian genera *Cardiaspina* Crawford and *Hyalinaspis* Taylor (Homoptera: Psyllidae), *Australian Journal of Zoology* **10**: 307–348.

Taylor, K. R. (1985). A possible stridulatory organ in some Psylloidea (Homoptera), *Journal of Australian Entomological Society* **24:** 77–80.

Tembrock, G. (1960). Stridulation und Tagesperiodik *bei Cerambyx cerdo* L, *Zoologische Beiträge (N.F.)* **5:** 419–441.

Templeton, A. R. (1981). Mechanisms of speciation — a population genetic approach, *Annual Review of Ecology and Systematics* **12:** 23–48.

Terradas, I. (1999). Les Cigales et le rythme des jours: la symbolique cicadéenne des îles Andaman (Inde), *EPHE, Biologie et Evolution des Insectes* **11/12:** 19–54.

Theiss, J. (1982). Generation and radiation of sound by stridulating water insects as exemplified by the corixids, *Behavioral Ecology and Sociobiology* **10:** 225–235.

Theiss, J., Prager, J., and Streng, R. (1983). Underwater stridulation by corixids: stridulatory signals and sound production mechanism in *Corixa dentipes* and *Corixa punctata*, *Journal of Insect Physiology* **29:** 761–771.

Thiele, D. R. and Bailey, W. J. (1980). The function of sound in male spacing behaviour in bushcrickets (Tettigoniidae, Orthoptera), *Australian Journal of Ecology* **5:** 275–286.

Thom, C., Gilley, D. C., and Tautz, J. (2003). Worker piping in honey bees (*Apis mellifera*): the behavior of piping nectar foragers, *Behavioral Ecology and Sociobiology* **53:** 199–205.

Thomas, E. S. and Alexander, R. D. (1957). *Nemobius melodius*, a new species of cricket from Ohio (Orthoptera, Gryllidae), *Ohio Journal of Science* **57:** 148–152.

Thornhill, R. (1983). Cryptic female choice and its implications in the scorpionfly *Haprobittacus nigriceps*, *American Naturalist* **122:** 765–788.

Thornhill, R. and Gwynne, D. T. (1986). The evolution of sexual differences in insects, *American Scientist* **74:** 382–389.

Thorpe, K. V. and Harrington, B. J. (1981). Sound production and courtship behavior in the seed bug Ligyrocoris diffusus, *Annals of the Entomological Society of America* **74**(4): 369–373.

Tieu, T. D. (1996). Mechanical modeling of the vibration of the treehopper *Umbonia crassicornis*, Master's thesis, State University of New York at Binghamton.

Tishechkin, D. Yu. (1989). Acoustic communication in the psyllids (Homoptera, Psyllinea) from Moscow district, *Moscow University Bulletin, Series 16, Biology* **4:** 20–24, in Russian with English summary.

Tishechkin, D. Yu. (2000). Vibrational communication in Aphrodinae leafhoppers (Deltocephalinae auct., Homoptera: Cicadellidae) and related groups with notes on classification of higher taxa, *Russian Entomological Journal* **9**(1): 1–66.

Tishechkin, D. Yu. (2001). Vibrational communication in Cicadellinae sensu lato and Typhlocybinae leafhoppers (Homoptera: Cicadellidae) with notes on classification of higher taxa, *Russian Entomological Journal* **9**(4): 283–314.

Tishechkin, D. Yu. (2002). Review of the species of the genus *Macropsis* Lewis, 1834 (Homoptera: Cicadellidae: Macropsinae) from European Russia and adjacent territories, *Russian Entomological Journal* **11:** 123–184.

Tishechkin, D. Yu. (2003). Vibrational communication in leafhoppers from Ulopides subfamilies group (Homoptera: Cicadellidae) and Membracidae with notes on classification of higher taxa, *Russian Entomological Journal* **12**(1): 11–58.

Tishechkin, D. Yu. (2004). Vibrational communication in Cercopoidea and Fulgoroidea (Homoptera: Cicadina) with notes on classification of higher taxa, *Russian Entomological Journal* **12**(2): 129–181.

Tomaru, M., Doi, M., Higuchi, H., and Oguma, Y. (2000). Courtship song recognition in the *Drosophila melanogaster* complex: heterospecific songs make females receptive in *D. melanogaster*, but not in *D. sechellia*, *Evolution* **54:** 1286–1294.

Tomaru, M. and Oguma, Y. (1994a). Differences in courtship song in the species of the *Drosophila auraria* complex, *Animal Behaviour* **47:** 133–140.

Tomaru, M. and Oguma, Y. (1994b). Genetic basis and evolution of species-specific courtship song in the *Drosophila auraria* complex, *Genetical Research, Cambridge* **63:** 11–17.

Tomaru, M., Matsubayashi, H., and Oguma, Y. (1995). Heterospecific inter-pulse intervals of courtship song elicit female rejection in *Drosophila biauraria*, *Animal Behaviour* **50:** 905–914.

Toms, R. B. (1992). Effects of temperature on chirp rates of tree crickets (Orthoptera, Oecanthidae), *South African Journal of Zoology* **27:** 70–73.

Toms, R. B., Ferguson, J. W. H., and Becker, S. (1993). Relationship between body temperature in stridulating male crickets (Orthoptera, Gryllidae), *South African Journal of Zoology* **28**: 71–73.

Toschi, C. A. (1965). The taxonomy, life histories, and mating behavior of the green lacewings of Strawberry Canyon (Neuroptera: Chrysopidae), *Hilgardia* **36**: 391–433.

Townw, W. F. and Kirchner, W. H. (1989). Hearing in honeybees: detection of air-particle oscillations, *Science* **244**: 686–688.

Traue, D. (1978a). Vibrationskommunikation bei *Euides speciosa* Boh., (Homoptera-Cicadina: Delphacidae) — Vibrational communication of *Euides speciosa* Boh., (Homoptera-Cicadina: Delphacidae), *Verhandlung der Deutschen Zoologischen Gesellschaft*, Gustav Fischer Verlag, Stuttgart, p. 167.

Traue, D. (1978b). Zur Biophysik der Schallabstrahlung bei Kleinzikaden am Beispiel von *Euscelis incisus* Kb. (Homoptera-Cicadina: Jassidae), *Zoologische Beiträge. (N.F.)* **24**: 1–20.

Traue, D. (1980). Neue Ergebnisse zur Substratvibration bei Kleinzikaden (Homoptera-Cicadina) — New results of structure-borne sounds of small cicadas (Homoptera-Cicadina), *Verhandlung der Deutschen Zoologischen Gesellschaft*, Gustav Fischer Verlag, Stuttgart, p. 371.

Travassos, M. A. and Pierce, N. E. (2000). Acoustics, context and function of vibrational signalling in a lycaenid butterfly-ant mutualism, *Animal Behaviour* **60**: 13–26.

Trilar, T. and Holzinger, W. E. (2004). Bioakustische Nachweise von drei Arten des Cicadetta montana-Komplexes aus Österreich (Insecta: Hemiptera: Cicadoidea), *Linzer Biol. Beitr.* **36**(2): 1383–1386.

Tschirnhaus, M. (1971). Unbekannte Stridulationsorgane bei Dipteren und ihr Bedeutung fur Taxonomie und Phylogenetik der Agromyziden, *Beitr. Entomol.* **21**: 551–579.

Tucker, V. A. (1969). Wave-making by whirligig beetles (Gyrinidae), *Science* **166**: 897–899.

Tuckerman, J. F., Gwynne, D. T., and Morris, G. K. (1993). Reliable acoustic cues for female mate preference in a katydid (*Scudderia curvicauda*, Orthoptera: Tettigoniidae), *Behav. Ecol.* **4**(2): 106–113.

Turelli, M., Barton, N. H., and Coyne, J. A. (2001). Theory and speciation, *Trends in Ecology & Evolution* **16**: 330–343.

Tuxen, S. L. (1964). *Insekt-Stemmer*, (Dyrenes Liv, Bind II) Rhodos, Copenhagen: p. 163.

Tyrer, M. N. and Gregory, G. E. (1982). A guide to the neuroanatomy of locust suboesophagal and thoracic ganglia, *Transactions of the Royal Society, London* **297**: 91–123.

Unrast, C. (1996). Untersuchung des Sinnessystems der tympanalen und atympanalen Tibialisorgane von Larven und Adulten verschiedener Laubheuschreckenarten; Rezeptorphysiologie, Projektion und zentrale Verarbeitung, Dissertation, Marburg.

Usinger, R. L. (1954). A new genus of Aradidae from the Belgian Congo, with notes on stridulatory mechanisms in the family, *Ann. Mus. Congo Tervuren (Zool.)* **1**: 540–543.

Usinger, R. L. and Matsuda, R. (1959). *Classification of the Aradidae (Hemiptera, Heteroptera)*, British Museum (N.H.), London.

Van Dijken, M. J. and Van Alpen, J. J. M. (1998). The ecological significance of differences in host detection behaviour in coexisting parasitoid species, *Ecological Entomology* **23**: 265–270.

Van Duzee, E. P. (1915). A preliminary review of the West coast Cicadidae, *Journal of the New York Entomological Society* **23**: 21–44.

van Staaden, M. J. and Römer, H. (1997). Sexual signaling in bladder grasshoppers: tactical design for maximizing calling range, *Journal of Experimental Biology* **200**: 2597–2608.

Van Tassell, E. R. (1965). An audiospectrographic study of stridulation as an isolating mechanism in the genus *Berosus* (Col., Hydrophilidae), *Annals of the Entomological Society of America* **58**: 407–413.

Van Wyk, J. W. and Ferguson, J. W. H. (1995). Communicatory constraints on field crickets *Gryllus bimaculatus* calling at low ambient temperatures, *Journal of Insect Physiology* **41**: 837–841.

Varley, G. C. (1939). Unusual methods of stridulation in a cicada (*Clidophleps distanti* (Van D.)) and a grasshopper (*Oedaleonotus fuscipes* Scud.) in California, *Proceedings of the Royal Entomological Society of London, A* **14**(7–8): 97–100.

Vedenina, V. Yu. and Bukhvalova, M. A. (2001). Contributions to the study of acoustic signals of grasshoppers (Orthoptera, Acrididae, Gomphocerinae) from Russia and adjacent countries. 2. Calling songs of widespread species recorded in different localities, *Russian Entomological Journal* **10**(2): 93–123.

Vedenina, V. Y. and von Helversen, O. (2003). Complex courtship in a bimodal grasshopper hybrid zone, *Behavioral Ecology and Sociobiology* **54**: 44–54.

Vedenina, V. Yu. and Zhantiev, R. D. (1990). Recognition of acoustic signals in sympatric species of locusts, *Zoologicheskiy zhurnal* **69**(2): 36–45, in Russian with English summary.

Verhoeff, K. W. (1902). Ueber die Zusammengesetzte Zirpvorrichtung von *Geotrupes*, *Sber. Ges. Naturforsch. Freunde Berlin* 149–155.

Vieillard, J. (1993). Recording wildlife in tropical rainforest, *Bioacoustics* **4**: 305–311.

Villatoro, K. (2002). Revision of the Neotropical genus *Trizogeniates* Ohaus (Col., Scarabaeidae: Rutelinae: Geniatini), *Entomotropica* **17**: 225–294.

Villella, A., Gailey, D. A., Berwald, B., Ohshima, S., Barnes, P. T., and Hall, J. C. (1997). Extended reproductive roles of the *fruitless* gene in *Drosophila melanogaster* revealed by behavioral analysis of new *fru* mutants, *Genetics* **147**: 1107–1130.

Villet, M. (1988). Calling songs of some African cicadas (Homoptera: Cicadoidea), *South African Journal of Zoology* **23**(2): 71–77.

Villet, M. (1992). Responses of free-living cicadas (Homoptera: Cicadidae) to broadcasts of cicada songs, *Journal of Entomological Society of South Africa* **55**(1): 93–97.

Villet, M. (1995). Intraspecific variability in SMRS signals: some causes and implications in acoustic signaling systems, In *Speciation and the Recognition Concept: Theory and Application*, Lambert, D. and Spencer, H., Eds., John Hopkins U.P., Baltimore, pp. 422–439.

Villet, M. H., Sanborn, A. F., and Phillips, P. K. (2003). Endothermy and chorusing behaviour in the African platypleurine cicada *Pycna semiclara* (Germar 1834) (Hemiptera: Cicadidae), *Canadian Journal of Zoology* **81**: 1437–1444.

Virant-Doberlet, M. (1989). Analysis of vibratory stimuli in the central nervous system of a cricket (*Gryllus campestris*), M.Sc. Thesis, University of Ljubljana, Ljubljana.

Virant-Doberlet, M. and Čokl, A. (2004). Vibrational communication in insects, *Neotropical Entomology* **33**: 121–134.

Visscher, P. K. (1993). A theoretical analysis of individual interests and intercolony conflict during swarming of honey bee colonies, *Journal of Theoretical Biology* **165**: 191–212.

Vogel, R. (1923). Uber ein tympanales Sinnesorgan, das mutmassliche Hörorgan der Singzikaden, *Zeitschrift für die gesammte Anatomie* **67**: 190–231.

Voisin, J.-F. (coord.). (2003). *Atlas des Orthoptères et des Mantides de France*, Collection Patrimoines Naturels, Vol. 60, Muséum national d'Histoire naturelle, Paris.

von Buddenbrock, W. (1928). *Grundriß der vergleichenden Physiologie*, Gebrüder Bornträger, Berlin, p. 830.

von Frisch, K. (1921). Über den Sitz des Geruchsinnes bei Insekten, *Zoologische Jarbucher – Abteilung fur Allgemeine Zoologie und Pysiologie der Tiere* **38**: 449–516.

von Frisch, K. (1965). *Tanzsprache und Orientierung der Bienen*, Springer, Berlin Heidelberg, p. 578.

von Helversen, D. (1984). Parallel processing in auditory pattern recognition and directional analysis by the grasshopper *Chorthippus biguttulus* L (Acrididae), *Journal of Comparative Physiology, A* **154**: 837–846.

von Helversen, O. (1986). Gesang und Balz bei Feldheuschrecken der *Chorthippus albomarginatus*-Gruppe (Orthoptera: Acrididae), *Zoologische Jahrbucher, Abteilung Systematik* **113**: 319–342.

von Helversen, O. and Elsner, N. (1977). The stridulatory movements of acridid grasshoppers recorded with an opto-electronic device, *Journal of Comparative Physiology* **122**: 53–64.

von Helversen, D. and von Helversen, O. (1981). Korrespondenz zwischen Gesang und auslösendem Schema bei Feldheuschrecken, *Nova Acta Leopold. N.F.* **54**(245): 449–462.

von Helversen, D. and von Helversen, O. (1983a). Species recognition and acoustic localization in acridid grasshoppers: a behavioral approach, In *Neuroethology and Behavioral Physiology*, Huber, F. and Markl, H., Eds., Springer, Heidelberg, pp. 95–105.

von Helversen, D. and von Helversen, O. (1983b). Species recognition and acoustic localization in Acridid grasshoppers: a behavioral approach, In *Neuroethology and Behavioral Physiology*, Huber, F. and Markl, H., Eds., Springer-Verlag, Berlin, chap. II.3.

von Helversen, O. and von Helversen, D. (1994). Forces driving coevolution of song and song recognition in grasshoppers, In *Neural Basis of Behavioural Adapatations*, *Progress in Zoology*, Vol. 39: Schildberger, K. and Elsner, N., Eds., Gustav Fischer Verlag, Stuttgart, pp. 253–284.

von Helversen, D. and von Helversen, O. (1998). Acoustic pattern recognition in a grasshopper: processing in the time or frequency domain?, *Biological Cybernetics* **79**: 467–476.

von Helversen, D. and Wendler, G. (2000). Coupling of visual to auditory cues during phonotactic approach in the phaneropterine bushcricket *Poecilimon affinis*, *Journal of Comparative Physiology, A* **186**: 729–736.

von Schilcher, F. (1976). The function of pulse song and sine song in the courtship of *Drosophila melanogaster*, *Animal Behaviour* **24:** 622–625.

Wäckers, F. L., Mitter, E., and Dorn, S. (1998). Vibrational sounding by the pupal parasitoid *Pimpla (Coccygomimus) turionellae*: an additional solution to the reliability–detectability problem, *Biological Control* **11:** 141–146.

Waddington, K. D. and Kirchner, W. H. (1992). Acoustical and behavioral correlates of profitability of food sources in honey bee round dances, *Ethology* **92:** 1–6.

Wagner, W. (1907). Psychobiologische Untersuchungen an Hummeln, *Zoologica* **19:** 1–239.

Wagner, W. (1939). Die Zikaden des Mainzer Beckens, *Jb. Nass. Ver. Naturk.* **86:** 77–212.

Wagner, W. (1962). Dynamische Taxonomie, angewandt auf die Delphaciden Mitteleuropas, *Mitt. Hamburg, Zool. Mus. Inst.* **60:** 111–180.

Wagner, W. E. and Hoback, W. W. (1999). Nutritional effects on male calling behaviour in the variable field cricket, *Animal Behaviour* **57:** 89–95.

Wakabayashi, T. and Hagiwara, S. (1953). Mechanical and electrical events in the main sound muscle of cicada, *Japanese Journal of Physiology* **3:** 249–253.

Wakabayashi, T. and Ikeda, K. (1961). Interrelation between action potential and miniture electrical oscillation in the tymbal muscle of the cicada, *Japanese Journal of Physiology* **11:** 585–595.

Waldron, I. (1964). Courtship sound production in two sympatric sibling *Drosophila* species, *Science* **144:** 191–193.

Wales, W., Clarac, F., and Laverack, M. S. (1971). Stress detection at the autotomy plane in decapod crustacea. I. Comparative anatomy of the receptors of the basi-ischiopodite region, *Zeitschrrift fur Vergleich. Physiologie.* **73:** 357–382.

Walker, T. J. (1957). Specificity in the response of female tree crickets (Orthoptera, Gryllidae, *Oecanthinae*) to calling songs of the males, *Annals of the Entomological Society of America* **50:** 626–636.

Walker, T. J. (1962a). Factors responsible for intraspecific variation in the calling songs of crickets, *Evolution* **16:** 407–428.

Walker, T. J. (1962b). The taxonomy and calling songs of United States tree crickets (Orthoptera: Gryllidae: Oecanthinae), I, the Genus *Neoxabea* and the *niveus* and *varicornis* groups of the genus *Oecanthus*, *Annals of the Entomological Society of America* **55:** 303–322.

Walker, T. J. (1963). The taxonomy and calling songs of United States tree crickets (Orthoptera: Gryllidae: Oecanthinae). II. The *nigricornis* group of the genus *Oecanthus*, *Annals of the Entomological Society of America* **56:** 772–789.

Walker, T. J. (1964). Cryptic species among sound-producing Ensiferan Orthoptera (Gryllidae and Tettigoniidae), *Quarterly Review of Biology* **39:** 345–355.

Walker, T. J. (1969a). Acoustic synchrony: two mechanisms in the snowy tree cricket, *Science* **166:** 891–894.

Walker, T. J. (1969b). Systematics and acoustic behavior of United States crickets of the genus *Cyrtoxipha* (Orthoptera: Gryllidae), *Annals of the Entomological Society of America* **62:** 945–952.

Walker, T. J. (1969c). Systematics and acoustic behavior of United States crickets of the genus *Orocharis* (Orthoptera: Gryllidae), *Annals of the Entomological Society of America* **62:** 752–762.

Walker, T. J. (1974). Character displacement and acoustic insects, *Am. Zool.* **14:** 1137–1150.

Walker, T. J. (1975a). Effects of temperature, humidity, and age on stridulatory rates in *Atlanticus* spp. (Orthoptera: Tettigoniidae: Decticinae), *Annals of the Entomological Society of America* **68:** 607–611.

Walker, T. J. (1975b). Effects of temperature on rate in poikilotherm nervous systems: Evidence from calling songs of meadow katydids (Orthoptera: Tettigoniidae: *Orchelium*) and re-analysis of published data, *Journal of Comparative Physiology* **101:** 57–69.

Walker, T. J. (1983). Diel patterns of calling in nocturnal Orthoptera, In *Orthopteran Mating Systems: Sexual Selection in a Diverse Group of Insects*, Gwynne, D. T. and Morris, G. K., Eds., Westview Press, Boulder, Colorado, chap. 3.

Walker, T. J. (1998). Trilling field crickets in a zone of overlap (Orthoptera: Gryllidae: *Gryllus*), *Annals of the Entomological Society of America* **91:** 175–184.

Walker, T. J. (2000). Pulse rates in the songs of trilling field crickets (Orthoptera: Gryllidae: *Gryllus*), *Annals of the Entomological Society of America* **93:** 565–572.

Walker, T. J. (2004). The *uhleri* group of the genus *Ambylocorypha* (Orthoptera: Tettigoniidae): extraordinary complex songs and new species, *Journal of Orthoptera Research* **13**(2): 169–183.

Walker, T. J., Forrest, T. G., and Spooner, J. D. (2003). The *rotundifolia* complex of the genus *Amblycorypha* (Orthoptera: Tettigoniidae: Phaneropterinae): songs reveal new species, *Annals of the Entomological Society of America* **96**: 443–457.

Walker, T. J., Moore, T.E. (2004). Singing Insects of North America, 2004, http://buzz.ifas.ufl.edu, 22/06/2004.

Walker, T. J., Whitesell, J. J., and Alexander, R. D. (1973). The robust conhead: two widespread sibling species (Orthoptera: Tettigoniidae: *Neoconocephalus* «*robustus*»), *Ohio Journal of Science* **73**: 321–330.

Wallace, M. S. and Deitz, L. L (2004). Phylogeny and systematics of the treehopper subfamily Centrotinae (Hemiptera: Membracidae), *Memoirs on Entomology*, International Associated Publishers, FL.

Ward, J. P. (1981). A comparison of the behavioural responses of the haematophagous bug, *Triatoma infestans* to synthetic homologues of two naturally occurring chemicals (*n*- and isobutyric acid), *Physiological Entomology* **6**: 325–329.

Ward, R. D., Phillips, A., Burnet, B., and Marcondes, C. B. (1988). The *Lutzomyia longipalpis* complex: reproduction and distribution. In Service, M. W., Ed., *Biosystematics of Haematophagus Insects*, Systematics Association Special Volume 37, Clarendon Press, Oxford.

Waser, P. M. and Waser, M. S. (1977). Experimental studies of primate vocalization: specializations for long-distance propagation, *Zeitschrifte für Tierpsychologie* **43**: 239–263.

Watanabe, N. (2000). The fine structure of the stridulatory apparatus of the water scavenger beetle *Regimbartia attenuata* (Fabricius) (Coleoptera Hydrophilidae), *Elytra* **28**: 275–284.

Webb, J. C., Calkins, C. O., Chambers, D. L., Schwienbacher, W., and Russ, K. (1983). Acoustical aspects of behavior of Mediterranean fruit fly, *Ceratitis capitata*: analysis and identification of courtship sounds, *Entomologia Experimentalis et Applicata* **33**: 1–8.

Webb, J. C., Sharp, J. L., Chambers, D. L., McDow, J. J., and Benner, J. C. (1976). Analysis and identification of sounds produced by the male Caribbean fruit fly, *Anastrepha suspensa*, *Annals of the Entomological Society of America* **69**: 415–420.

Webb, J. C., Sivinski, J., and Litzkow, C. (1984). Acoustic behavior and sexual success in the Caribbean fruit fly, *Anastrepha suspensa* (Loew) (Diptera: Tephritidae), *Environmental Entomology* **13**: 650–656.

Weber, H. (1931). Levensweise und Umweltbeziehungen von *Trialeurodes vaporariorum* (Westwood) (Homoptera-Aleurodina), *Zeitschrift fur Morphologie und Okologie de Tiere* **23**: 575–753.

Weidemann, S. and Keuper, A. (1987). Influence of vibratory signals on the phonotaxis of the gryllid *Gryllus bimaculatus* De Geer (Ensifera: Gryllidae), *Oecologia* **74**: 316–318.

Wells, M. M. (1993). Laboratory hybridization in green lacewings (Neuroptera, Chrysopidae, Chrysoperla): evidence for genetic incompatibility, *Canadian Journal of Zoology* **71**: 233–237.

Wells, M. M. (1994). Small genetic distances among populations of green lacewings of the genus *Chrysoperla* (Neuroptera: Chrysopidae), *Annals of the Entomological Society of America* **87**: 737–744.

Wells, M. M. and Henry, C. S. (1992). The role of courtship songs in reproductive isolation among populations of green lacewings of the genus *Chrysoperla* (Neuroptera: Chrysopidae), *Evolution* **46**: 31–42.

Wells, M. M. and Henry, C. S. (1994). Behavioral responses of hybrid lacewings (Neuroptera: Chrysopidae) to courtship songs, *Journal of Insect Behavior* **7**: 649–662.

Wells, M. M. and Henry, C. S. (1998). Songs, reproductive isolation and speciation in cryptic species of insects: a case study using green lacewings, In *Endless Forms: Species and Speciation*, Howard, D. and Berlocher, S., Eds., Oxford University Press, New York, pp. 217–233.

Wenner, A. M. (1962a). Communication with queen honey bees, *Science* **138**: 446–447.

Wenner, A. M. (1962b). Sound production during the waggle dance of the honeybee, *Animal Behaviour* **10**: 79–95.

Wessel, A. and Hoch, H. (1999). Remane's statistic species criterion applied to Hawaiian cave planthoppers (Hemiptera: Auchenorrhyncha: Fulgoromorpha: Cixiidae), *Reichenbachia* **33**(3): 27–35.

West-Eberhard, M. J. (1983). Sexual selection, social competition, and speciation, *Quarterly Review of Biology* **58**: 155–183.

West-Eberhard, M. J. (1984). Sexual selection, social communication, and species specific signals in insects, In *Insect Communication*, Lewis, T., Ed., Academic Press, London, pp. 283–324.

Wheeler, D. A., Kyriacou, C. P., Greenacre, M. L., Yu, Q., Rutila, J. E., Rosbash, M., and Hall, J. C. (1991). Molecular transfer of a species-specific behavior from *Drosophila simulans* to *Drosophila melanogaster*, *Science* **251**: 1082–1085.

Whitesell, J. J. and Walker, T. J. (1978). Photoperiodically determined dimorphic calling songs in a katydid, *Nature* **274**: 887–888.

Whiting, M. (2002). Phylogeny of the holometabolous insect orders: molecular evidence, *Zool. Scr.* **31**: 3–15.

Wiese, K. (1972). Das mechanorezeptorische Beuteortungssystem *von Notonecta*. I. Die Funktion des tarsalen Scolopidialorgans, *Journal of Comparative Physiology* **78**: 83–102.

Wiese, K. (1974). The mechano-receptive system of prey localization in *Notonecta*. II. The principle of prey localization, *Journal of Comparative Physiology* **92**: 317–325.

Wiese, K. and Schmidt, K. (1974). Mechanorezeptoren im Insektentarsus. Die Konstruktion des tarsalen Scolopidialsorgan bei *Notonecta* (Hemiptera, Heteroptera), *Zeitschrifte fur Morphologie der Tiere* **79**: 47–63.

Wilcox, R. S. (1972). Communication by surface waves. Mating behavior of a water strider (Gerridae), *Journal of Comparative Physiology* **80**: 255–266.

Wilcox, R. S. and Spence, J. R. (1986). The mating system of two hybridizing species of water-striders (Gerridae). I. Ripple signal functions, *Behavioral Ecology and Sociobiology* **19**: 79–85.

Wiley, R. H. (1993). The evolution of communication: information and manipulation, In *Animal Behaviour, Communication*, Vol. 2: Halliday, T. J. and Slater, P. J. B., Eds., Blackwells, Oxford, pp. 156–189.

Wiley, R. H. and Richards, D. G. (1978). Physical constraints on acoustic communication in the atmosphere: implications for the evolution of animal vocalization, *Behavioral Ecology and Sociobiology* **3**: 69–94.

Willemse, F. (1980). Three new species and some additional notes on *Parnassiana Zeuner* from Greece (Orthoptera, Ensifera, Decticinae), *Entomol. Ber. Amsterdam* **40**: 103–112.

Willemse, F. (1992). *Catalogue of the Orthoptrera of Greece, Fauna Graeciae*, Hellenic Zoological Society, Athens, p. 1, i-xii, 1-275, 1984.

Willemse, F. (1985) *A Key to the Orthoptera Species of Greece*, Fauna Graeciae 2, Helenic Zoological Society, Athens.

Willemse, F. and Heller, K.-G. (1992). Notes on systematics of Greek species of *Poecilimon* Fischer, 1853 (Orthoptera: Phaneropterinae), *Tijdschrift voor Entomologie* **135**: 299–315.

Willemse, F. and Heller, K.-G. (2001). Two new species of *Eupholidoptera* Maran (Orthoptera Tettigoniidae) from Crete with a checklist and key to the species, *Tijdschrift voor Entomologie* **144**: 329–343.

Willemse, L. and Willemse, F. (1987). *Platycleis (Parnassiana) nigromarginata* sp. nov. from Greece (Orthoptera: Decticinae), *Entomol. Ber. Amsterdam* **47**: 105–107.

Williams, M. A., Amanda, G. B., and Noor, M. A. F. (2001). Courtship songs of *Drosophila pseudoobscura* and *D. persimilis*. II. Genetics of species differences, *Heredity* **86**: 68–77.

Willmann, R. (1990). The phylogenetic position of the Rhachiberothidae and the basal sister-group relationships within the Mantispidae (Neuroptera), *Systematic Entomology* **15**: 253–265.

Wilson, E. O (1971). *The Insect Societies*, Harvard University Press, Cambridge, p. 558.

Wilson, L. M., Henry, C. S., Johnson, J. B., and McCaffrey, J. P. (1993). Sound production in Phrydiuchus tau (Col., Curculionidae), *Annals of the Entomological Society of America* **86**: 621–630.

Winking-Nikolay, A. (1975). Untersuchungen zur Bio-Akustik des Waldmistkäfers *Geotrupes stercorosus*, *Z. Tierpsychol.* **37**: 515–542.

Winston, M. L. (1987). *The Biology of the Honey Bee*, Harvard University Press, Cambridge, London, p. 224.

Withers, P. C. (1992). *Comparative Animal Physiology*, Saunders College Publishing, New York, p. 123.

Withycombe, C. L. (1922). *Parasemidalis annae* Enderlein, a coniopterygid new to Britain, with notes on some other British Coniopterygidae, *Entomologist* **55**: 169–172.

Withycombe, C. L. (1925). Some aspects of the biology and morphology of the Neuroptera, with special reference to the immature stages and their possible phylogenetic significance, *Transactions of the Royal Entomological Society, London* **72**: 303–411.

Woglum, R. S. and McGregor, E. A. (1958). Observations on the life history and morphology of *Agulla bractea* Carpenter (Neuroptera: Raphidiodea: Raphidiidae), *Annals of the Entomological Society of America* **51**: 129–141.

Wolda, H. (1993). Diel and seasonal patterns of mating calls in some Neotropical Cicadas. Acoustic interference?, *Proceedings of the Koninklijke Nederlandse Akademie van Wetenschappen* **96**(3): 369–381.

Wolf, H. and Burrows, M. (1995). Proprioreceptive sensory neurons of a locust leg receive rhythmic presynaptic inhibition during walking, *Journal of Neuroscience* **15**: 5623–5636.

Wood, T. K. (1974). Aggregating behavior of *Umbonia crassicornis* (Homoptera: Membracidae), *The Canadian Entomologist* **106**: 169–173.

Wood, T. K. (1979). Sociality in the Membracidae (Homoptera), *Miscellaneous Publications of the Entomological Society of America* **11**: 15–22.

Wood, T. K. (1980). Divergence in the *Enchenopa binotata* Say complex (Homoptera: Membracidae) effected by host plant adaptation, *Evolution* **34**: 147–160.

Wood, T. K. (1983). Brooding and aggregating behavior of the treehopper, *Umbonia crassicornis*, *National Geographic Society Research Report* **15**: 753–758.

Wood, T. K. (1984). Life history patterns of tropical membracids (Homoptera: Membracidae), *Sociobiology* **8**: 299–344.

Wood, T. K. (1993a). Speciation of the *Enchenopa binotata* complex (Insecta: Homoptera: Membracidae), In *Evolutionary Patterns and Processes*, Lee, D. R., Ed., Academic Press, London, pp. 299–317.

Wood, T. K. (1993b). Diversity in the new world membracidae, *Annual Review of Entomology* **38**: 409–435.

Wood, T. K. and Dowell, R. (1985). Reproductive behavior and dispersal in *Umbonia crassicornis* (Homoptera: Membracidae), *Florida Entomologist* **68**: 151–158.

Wood, T. K. and Guttman, S. I. (1983). *Enchenopa binotata* complex: sympatric speciation?, *Science* **220**: 310–312.

Wood, T. K., Guttman, S. I., and Taylopr, M. C. (1984). Mating behavior of *Platycotis vittata* (Fabricius) (Homoptera: Membracidae), *American Midland Naturalist* **112**: 305–313.

Wood, T. K., Olmstead, K. L., and Guttman, S. I. (1990). Insect phenology mediated by host-plant water relations, *Evolution* **44**: 629–636.

Wood, T. K. and Tilmon, K. J. (1993). The plant phenology hypothesis of speciation through host plant shifts: a test, In *Proceedings of the 8th Auchenorrhyncha Congress*, Drosopoulos, S., Petrakis, P. V., Claridge, M. F., and de Vrijer, P. W. F., Eds., Delphi, Greece, pp. 108–109, 9–13 August.

Wood, T. K., Tilmon, K. J., Shantz, A. B., Harris, C. K., and Pesek, J. (1999). The role of host-plant fidelity in initiating insect race formation, *Evolutionary Ecology Research* **1**: 317–332.

Woods, E. F. (1956). Queen piping, *Bee World* **10**: 185–194, 216–219.

World Health Organization, (2002). *TDR Strategic Directions: Chagas disease*, WHO, Geneve, p. 6.

Yager, D. D. (1999). Structure, development, and evolution of insect auditory systems, *Microscopy Research and Technology* **47**: 380–400.

Yager, D. D. and Spangler, H. G. (1995). Characterization of auditory afferents in the tiger beetle, *Cicindela marutha Dow*, *Journal of Comparative Physiology* **176**: 587–599.

Yamada, H., Sakai, T., Tomaru, M., Doi, M., Matsuda, M., and Oguma, Y. (2002). Search for species-specific mating signal in courtship songs of sympatric sibling species, *Drosophila ananassae* and *D. pallidosa*, *Genes & Genetic Systems* **77**: 97–106.

Yang, M.-M., Yang, Ch.-T., and Chao, J.-T. (1986). Reproductive isolation and taxonomy of two taiwanese *Paurocephala* species (Homoptera: Psylloidea), *Monograph of Taiwan Museum* **6**: 176–203.

Yersin, A. (1854). Mémoire sur quelques faits relatifs à la stridulation des Orthoptères et à leur distribution géographique en Europe, *Bull. Soc. Vaud. Sci. Nat.* **4** (1853–55), 108–128.

Yost, W. A. and Nielsen, D. W. (1985). *Fundamentals of Hearing: An Introduction*, 2nd ed., Holt, Rinehart and Winston, New York.

Young, D. (1972). Analysis of songs of some Australian cicadas (Homoptera: Cicadidae), *Journal of Australian Entomological Society* **11**: 237–243.

Young, D. (1975). Chordotonal organs associated with the sound apparatus of cicadas (Insecta Homoptera), *Zeitschrift für Morphologie und Ökologie der Tiere* **81**(2): 111–136.

Young, A. M. (1981). Temporal selection for communicatory optimization: The dawn-chorus as an adaptation in tropical cicadas, *American Naturalist* **117**: 826–829.

Young, D. and Bennet-Clark, H. C. (1995). The role of the tymbal in cicada sound production, *Journal of Experimental Biology* **198**: 1001–1019.

Young, D. and Josephson, R. K. (1983a). Mechanisms of sound production and muscle-contraction kinetics in cicadas, *Journal of Comparative Physiology* **152A:** 183–195.

Young, D. and Josephson, R. K. (1983b). Pure-tone songs in cicadas with special reference to the genus *Magicicada, Journal of Comparative Physiology* **152A:** 197–207.

Zabel, J. (1941). Die Kamelhals-Fliege, *Natur und Volk* **71:** 187–195.

Zeigler, D. D. and Stewart, K. W. (1977). Drumming behvior of eleven Nearctic stonefly (Plecoptera) species, *Annals of the Entomological Society of America* **70:** 495–505.

Zeigler, D. D. and Stewart, K. W. (1985). Age effects on drumming behavior of *Pteronarcella badia* (Plecoptera) males, *Entomological News* **96:** 157–160.

Zeigler, D. D. and Stewart, K. W. (1986). Female response thresholds of two stonefly (Plecoptera) species to computer-simulated and modified male drumming calls, *Animal Behaviour* **34:** 929–931.

Zhantiev, R. D. (1981). *Bioacoustics of Insects*, Moscow State University Press, Moscow, chap. 6, in Russian.

Zhantiev, R. D. and Korsunovskaya, O. S. (1978). Morpho-functional organization of tympanal organs in *Tettigonia cantans* (Orthoptera Tettigoniidae), *Zoologichersky Zhurnal* **57:** 1012–1016.

Zimmermann, U., Rheinlaender, J., and Robinson, D. (1989). Cues for male phonotaxis in the duetting bushcricket *Leptophyes punctatissima, Journal of Comparative Physiology, A* **164:** 621–628.

Zorović, M., Čokl, A., and Virant-Doberlet, M. (2004). The interneural network underlying vibrational communication in the green stinkbug *Nezara viridula* (L.): processing of the female calling song with applications for orientation and species recognition, *The 7th Congress of the International Society for Neuroethology, Program and Abstracts*, University of Southern Denmark, Odense, PO201.

Zorović, M., Virant-Doberlet, M., and Čokl, A. (2003). The vibratory interneurons in the central ganglion of the southern green stink bug *Nezara viridula* (L.) (Heteroptera: Pentatomidae), In *Proceedings of the 29th Göttingen Neurobiology Conference*, Elsner, N. and Zimmermann, H., Eds., Thieme, Stuttgart, pp. 416–417.

Zuk, M. L., Rotenberry, J. T., and Simmons, L. W. (1998). Calling songs of field crickets (*Teleogryllus oceanicus*) with and without phonotactic parasitoid infections, *Evolution* **52:** 166–171.

Zuk, M., Rotenberry, J. T., and Simmons, L. W. (2001). Geographical variation in calling song of the field cricket *Teleogryllus oceanicus*: the importance of spatial scale, *Journal of Evolutionary Biology* **14:** 731–741.

Zuk, M., Simmons, L. W., and Cupp, L. (1993a). Calling characteristics of parasitized and unparasitized populations of the field cricket *Teleogryllus* oceanicus, *Behavioral Ecology and Sociobiology* **33:** 339–343.

Zuk, M., Simmons, L. W., and Cupp, L. (1993b). Calling songs of field crickets (*Teleogryllus oceanicus*) with and without phonotactic parasitoid infection, *Evolution* **52:** 166–171.

Zunino, M. and Ferrero, S. (1989). El dimorfismo sexual de los aparatos estriduladores en el genero Megatrupes (Coleoptera: Scarabaeoidea Geotrupidae), *Folia Entomologica Mexicana* **76:** 65–72.

Zwick, P. (1973). *Insects: Plecoptera. Phylogenetisches system und katalog*, Das Tierreich 94, Walter de Gruyter, Berlin.

Zwick, P. (1980). *Plecoptera (Steinfliegen), Handbuch der Zoologie 26*, Walter de Gruyter, Berlin, pp. 1–115.

Index

A

A. albogeniculata, 203
A. aucubae, 368
A. cerana, 432
A. contaminatus, 410
A. dorsata, 432
A. fasciatus Crickets, 215
A. fimetarius, 410
A. florea, 432
A. impicticorne, 284
A. lapponum, 410
A. polygoni, 359
A. polygoni, 360
A. prodromus, 410
A. scrutator, 410
A. sorinii, 371
A. strepitans, 343
A/D converter, 23
Abdominal tremulation, 155, 158, 160
Abdominal vibrations, 157, 166, 383
Abdominal-substrate stridulation, 181
Abroma inaudibilis, 344
Absence of females, 309
Acalyptrate flies, 386
Acanthobemisia distylii, 370
Acanthosomatidae, 290
Acceleration amplitude, 100
Acceleration sensitivity, 79
Accelerometer (ACC), 14, 21, 308, 366
Accessory gland products, 132
Accidental invasions, 190, 365
Accordion-like movements, 338
Acer pseudoplatanus, 87
Acherontia atropos, Death's Head Hawk, 6
Aconophora mexicana, 313
Aconophora, 314
Aconurella diplachnis, 324
Acoustic analysis, methodology used for, 11
Acoustic appearance, 199, 319
Acoustic behaviour of *Picromerus bidens*, 288
Acoustic communication in bees, 422
Acoustic cues during courtship, 168
Acoustic development of the male vibrational signals, 392
Acoustic efficiency, 14
Acoustic ethograms, 417
Acoustic identity cards, 339
Acoustic impedance, 44
Acoustic memory, 133
Acoustic mimicry in some predators, 402
Acoustic neurones, 68
Acoustic niche in species diagnosis, 205, 209, 362

Acoustic properties are likely to be under genetic control, 365
Acoustic repertoire, 6
Acoustic shortcircuit, 100
Acoustic sibling species in several Orders of insects, 216
Acoustic signals (piping) of honeybee queens (*Apis mellifera*), 426
Acoustic signals in *Aphodius*, 415–416
Acoustic signals in bees, 426
Acoustic signals of bugs, 277
Acoustic system of a male cicada, 334
Acoustic trachea in *G. gratiosa*, 51
Acoustically conspicuous species of genus *Cicada*, 227
Acoustics, definition of, 11
Acquisition devices, 16
Acridid grasshoppers typically sing even when males and females are in close contact, 149
Acrididae, 51, 80, 93, 95, 319
Acridids of the genus *Chorthippus*, 138
Acrometopa macropoda, 143
Acroneuria abnormis, 182
Acroneuria evoluta, 184, 185
Acrosternum hilare, 104 , 284
Activation pattern, 121
Active defense, 312
Acutalis tartarea, 102
Acymbalic cicadas, 333
Adaptation for communicatory optimisation, 124
Adaptive process of innate larval feeding activities, 396
Adaptive shift hypothesis, 190
Adaptive species strategies, 392
Adaptive value in the underground environment, 188
Adult parasitoids, 9
Advertisement signals, 307, 309, 312
Advertising signal, 132
Aedeagus line, 260
Aedeagus structure, 255
Aegean Archipelago, 225–226
Aegean Sea, distribution of *Lyristes* spp. in, 220
Aegiali blanchardi, 410
Aegialia, 407–409, 412
Aegialinae, 406, 407–409
Aelia acuminata, 285
Aerial insect singers, effects of temperature on, 8
Aerial maneuvering, 383
Aetalion reticulatum treehopper, 101, 102, 315
Aetalion, 101, 315
affinisdisjuncta, 173
African Hyaliodinae, 292
African Mirinae, 292
African species, 343
African Sphingid moths, 6
African *Ugada giovanninae*, 346

QL 496.5 .I57 2006

Insect sounds and
communication

DATE DUE
